T0321809

ROCKET PROPULSION

A modern pedagogical treatment of the latest industry trends in rocket propulsion, developed from the authors' extensive experience in both industry and academia.

Students are guided along a step-by-step journey through modern rocket propulsion, beginning with the historical context and an introduction to top-level performance measures, and progressing to in-depth discussions of the chemical aspects of fluid flow combustion thermochemistry and chemical equilibrium, solid, liquid, and hybrid rocket propellants, mission requirements, and finally to an overview of electric propulsion.

Key features include:

- A wealth of homework problems, with a solutions manual for instructors online.
- Numerous real-life case studies and examples.
- An appendix detailing key numerical methods, and links to additional online resources.
- Design approaches for all types of chemical rockets (including solid, liquid, and hybrid propellants) and electric propulsion.
- Techniques for analysis of engine and motor operation, and system and vehicle level performance.

This is a must-have book for senior and first-year graduate students looking to gain a thorough understanding of the topic along with practical tools that can be applied in industry.

Stephen D. Heister is the Raisbeck Distinguished Professor of Aeronautics and Astronautics at Purdue University and has previously worked in industry. He has served as an Associate Editor of *The AIAA Journal of Propulsion & Power* and is an elected Fellow of the AIAA.

William E. Anderson is a Professor of Aeronautics and Astronautics at Purdue University and is former Associate Head of the Undergraduate Program. He is an Associate Fellow of the AIAA.

Timothée Pourpoint is an Associate Professor of Aeronautics and Astronautics at Purdue University. He is an Associate Fellow of the AIAA.

R. Joseph Cassady is the Executive Director for Space in Aerojet Rocketdyne Washington Operations. He is an Associate Fellow of the AIAA and former Chair of the Electric Propulsion Technical Committee. He serves as Vice President of the Electric Rocket Propulsion Society, and Executive Vice President of ExploreMars Inc.

ROCKET PROPULSION

STEPHEN D. HEISTER
Purdue University, Indiana

WILLIAM E. ANDERSON
Purdue University, Indiana

TIMOTHÉE POURPOINT
Purdue University, Indiana

R. JOSEPH CASSADY
Aerojet Rocketdyne

Shaftesbury Road, Cambridge CB2 8EA, United Kingdom

One Liberty Plaza, 20th Floor, New York, NY 10006, USA

477 Williamstown Road, Port Melbourne, VIC 3207, Australia

314–321, 3rd Floor, Plot 3, Splendor Forum, Jasola District Centre, New Delhi – 110025, India

103 Penang Road, #05–06/07, Visioncrest Commercial, Singapore 238467

Cambridge University Press is part of Cambridge University Press & Assessment,
a department of the University of Cambridge.

We share the University's mission to contribute to society through the pursuit of
education, learning and research at the highest international levels of excellence.

www.cambridge.org
Information on this title: www.cambridge.org/9781108422277

DOI: 10.1017/9781108381376

First published 2019

A catalogue record for this publication is available from the British Library

Library of Congress Cataloging-in-Publication data
Names: Heister, Stephen D., author. | Anderson, William E. (William Edward), author. | Pourpoint, Timothee
(Timothee Louis), author. | Cassady, R. Joseph, (Joe), author.
Title: Rocket propulsion / Stephen D. Heister (Purdue University, Indiana), William E. Anderson (Purdue
University, Indiana), Timothee Pourpoint (Purdue University, Indiana), Joseph Cassady (Aerojet Rocketdyne).
Description: First edition. | New York : Cambridge University Press, [2018] | Includes bibliographical
references.
Identifiers: LCCN 2018036914 | ISBN 9781108422277
Subjects: LCSH: Rockets (Aeronautics) | Jet propulsion. | Rocket engines.
Classification: LCC TL782 .H45 2018 | DDC 621.43/56–dc23
LC record available at https://lccn.loc.gov/2018036914

ISBN   978-1-108-42227-7   Hardback

Additional resources for this publication at www.cambridge.org/heister.

CONTENTS

PREFACE

BACKGROUND

While its roots extend back centuries, the modern foundations of chemical propulsion were developed in the World War II era and in a bit more than two short decades later these systems pushed American astronauts to the moon. The engineering challenges for the development of these systems are numerous because the volumetric energy release is perhaps the highest of any device constructed by humans. This factor, combined with the need to minimize weight for devices that are being flown, leads to a demanding set of requirements for engineers. As a result, the knowledge set required to contribute in this field demands a broad set of considerations including combustion, single and multiphase fluid flow, material properties and compatibility, structural analysis, and design as important considerations.

This textbook grew out of course notes I developed over a 20-plus year period after joining Purdue in 1990. Purdue already had had two chemical rocket propulsion courses on its books at that time: a senior-level course (AAE439) that parallels those taught in other aerospace engineering schools around the globe, and a more unique graduate-level treatment (AAE539). The AAE439 notes have been published locally on an informal basis and have provided an opportunity to correct the many errors that inevitably creep in in a document of this size. The graduate course was initially developed by my predecessor, Professor Robert Osborn. This course delved heavily into solid rocket propellant combustion – Osborn's main area of research over his storied career. Given the broader needs of the community, I set out to include liquid and hybrid propulsion as well. At present, about half of the graduate-level material from AAE539 is included in the book. Perhaps my colleagues and I will add more of this material in future editions if market interest and energy of the authors so warrant. Other than this limitation, I believe that the text is up to date at this point with the latest information on new vehicles under development and the role that additive manufacturing is playing in our industry.

MAIN FEATURES OF THIS BOOK

This text was developed for propulsion *students* and the tone of the book emphasizes the tutorial nature of this task. I strongly believe that engineering students learn material by working problems; nearly 300 are included in the book for this purpose. Many of these problems are analogous to a cursory analysis that might be done by a practicing engineer in industry and many are motivated by the authors' collective experience of issues and studies performed in industry. The abundance of practice material distinguishes this text from those that are currently available or have been published over the years. Chapters 9, 10 and 12 are exceptions in this regard. It is difficult to assign homework problems on liquid propellants and the combustion instability topic is more advanced and not typically something we would assign students in coursework. The turbomachinery chapter came together over the last year or so and we are currently building up the problem database in that area.

My co-authors and I made another conscious decision to emphasize analytic approaches to problems in order to arm students with tools that they might apply in an industrial career. Even in this modern computer age with numerous multidimensional computer codes, it is often useful to check results against a simpler zero or one-dimensional result and in doing so gain insight into the complexity of the problem at hand.

COURSE PREREQUISITES AND THEIR RATIONALE

A student in a course that uses this book should already have taken thermodynamics and compressible fluid flow courses that are typically taught in sophomore and junior years in aerospace engineering curricula. Knowledge of sophomore-level calculus is also assumed. While many students in mechanical engineering or combined ME/AE programs gain background in heat transfer, this subject is absent from many AE schools and for that reason a focused attempt is made to provide top-level principles in Chapter 6 of the book. This material is not intended to replace a full semester course in heat transfer, but to motivate for more formal coursework in the area and to provide some background on rocket-related analyses that are commonplace in our industry. Because heat transfer is an integral performance element in a staged combustion rocket engine, it is really necessary to provide a sufficient level of detail such that students understand the fundamentals of this broad area. The broad topic of combustion is addressed similarly and focuses on specific approaches that are used in the rocket industry. Here at Purdue, we encourage all our propulsion students to try to get additional coursework in the combustion and heat transfer fields due to their importance in our profession.

STRUCTURE OF THE BOOK

As I formulated a plan toward manuscript development, it became apparent that my colleagues here could strengthen the work substantially. I engaged Bill Anderson and Tim Pourpoint early on, to improve the book and specifically to add material in combustion instability and liquid rocket propellants per their extensive expertise in these areas. Bill helped in many other ways and made suggestions/improvements to numerous chapters. We also engaged long-time friend and Purdue alumnus Joe Cassady to provide valuable input on electric propulsion. With a 30-plus year career at Aerojet Rocketdyne, and substantial experience developing educational materials, he was a natural choice.

We open with a brief history of rocketry and introduce the types of rocket propulsion systems commonly in use today. Chapters 2 and 3 contain material on mission requirements and trajectory analyses that are not strictly germane to the propulsion student, but we like to have the student have an understanding of how the propulsion analyst works with mission/trajectory designers in a real-world application. This approach also allows one to conduct simple sizing exercises without running to mission design experts, thereby providing a comprehensive capability for the student.

Chapter 4 provides fundamentals of compressible flow through nozzles with a review of nozzle design alternatives and treatment of flow separation, two-phase flow, and a brief discussion of method of characteristics. Chapter 5 provides classical introduction to combustion thermochemistry with a large number of examples to work out chemical equilibrium by hand in order to gain perspective for use of our standard industry computer codes. Chapter 6 provides introductory background on heat transfer and a focused discussion on some of the more important problems facing the rocket propulsion analyst.

Chapters 7–11 provide in-depth discussion of solid, liquid, and hybrid rocket propulsion systems. An entire chapter (Chapter 9) is dedicated to liquid propellants for liquid rocket or hybrid propulsion systems. Turbomachinery discussions are also broken out into a dedicated chapter (Chapter 10). Once again, the emphasis in these chapters is to provide students with analytic approaches to computing behaviors of a given design and propellant selection. Particular emphasis is placed on problems that we still struggle with as a community in order to provide students with a glimpse of the issues they will face as they enter the field.

Chapter 12 is Bill's extended discussion on one of our most vexing topics: combustion instability. In addition to providing overviews of the problem, design guidelines for achieving stability, and methods for measurement and rating, an introduction to linear analysis necessary to understand the most common techniques in use today is covered in some detail. Admittedly, this is higher-level material and would certainly be omitted in a first course in most instances.

Finally, we close out the book with Joe's perspectives on electric propulsion in Chapter 13. As with the chemical propulsion discussion, he provides perspective on the "hard problems" faced by the EP community.

A single appendix is included in the text. We defer to our modern age by including a list of website links rather than reams of paper that regurgitate the compressible flow and thrust coefficients typically seen in a text of this nature. We believe this is a superior way to access these things in today's highly connected environment. Numerous other links are also included and the entire list of links is provided on our Purdue Propulsion Web Page (https://engineering.purdue.edu/~propulsi/propulsion/). In addition, substantial background is provided on numerical methods that are required to solve some of the homework problems in the text. We wanted the book to stand on its own in this regard as many of our students have little or no experience in these methods when they enter our classroom.

ACKNOWLEDGMENTS

We have had the help of countless Purdue students over the years in "debugging" the problems we have constructed, and we are most grateful to this esteemed group. In particular, our teaching and research assistants have served a pivotal role in this venture. In fact, some of the problems have been formulated as a result of interactions with these individuals who now hold responsible positions within the US rocket propulsion industry. A few specific contributors who are current students or recent graduates include Nate Byerly, Dr. Jim Hilbing, Dr. Brandon Kan, Jenna Humble, Dr. Enrique Portillo, Dr. David Stechmann, Dr. David Reese, Gowtham Tamananpudi, Stephen Kubicki, and Chris Zascek.

In addition, we are indebted to colleagues at NASA and in industry who have aided immensely in the compilation of the text. Specifically, Jim Goss, Dean Misterek, Chad Schepel, Ken Miller, Matt Smith, and Tom Feldman of Blue Origin provided invaluable guidance. Dr. John Tsohas, Sam Rodkey, and Jon Edwards at SpaceX have also provided help with images and editing. James Cannon at NASA MSFC also provided useful inputs on turbomachinery as did Bill Murray at Ursa Major Technologies. We thank Gerald Hagemann of Ariane Group for his guidance on nozzles. Also thanks to Andy Hoskins and Roger Myers of Aerojet Rocketdyne for helpful discussions on some of the electric propulsion materials. And to Frank Curran for sharing some charts from his AIAA short course.

We thank three very important people who had a large role in typing, re-typing, formatting, and re-formatting the numerous versions of original course notes that eventually became this text. Sharon Wise typed much of the early notes in the antiquated TRoff software. More recently, Leslie Linderman and Audrey Sherwood helped with the current version that was eventually sent to our capable publishers at Cambridge Press. We thank these women for their efforts.

Finally, as we approach the 50th anniversary of the historic lunar landing, we leave you with an image of undergraduate student Neil Armstrong as a source of inspiration for all who wish to help mankind travel to space.

CHAPTER 1

CLASSIFICATION OF ROCKET PROPULSION SYSTEMS AND HISTORICAL PERSPECTIVE

1.1 Introduction

This text is intended to provide undergraduate and first-year graduate students with an introduction to the principles governing the design and performance of rocket propulsion systems. Readers of the text are expected to have a good working knowledge of thermodynamics and the principles of fluid flow in addition to a background in chemistry at least to the college freshman level. Fundamentally, the text has been developed as a compilation of resources from two courses offered at Purdue University: AAE439, *Rocket Propulsion* and AAE539, *Advanced Rocket Propulsion*. The former course is taught at the Senior level, while the latter is targeted to Seniors and first-year graduate students. The primary emphasis of the text is placed in the area of chemical rocket propulsion systems, the realm that includes solid rocket motors, liquid rocket propulsion devices, and hybrid systems that utilize one liquid propellant and one solid propellant. Electric and nuclear propulsion devices are not discussed in detail, but are highlighted briefly within this chapter to place them within the broad context of existing or future rocket propulsion devices. Specifically, this book places emphasis on the propellant systems, combustion processes, and thermodynamic and fluid flow processes occurring within the rocket combustion chamber and nozzle. In addition, overall system sizing is presented as well as the interaction of propulsion system performance with vehicle trajectory calculations. Control and structural aspects of a rocket are not considered in detail in this course, but structural issues and sizing considerations are discussed at a mainly conceptual level.

In this introductory chapter, we will briefly discuss the history of rocketry and will introduce the various types of rocket propulsion systems in use today.

1.2 A Brief History of Rocketry

It would be difficult to envision a student who has absolutely no exposure to the world of rockets since we are clearly living in the "space age" with literally hundreds of space flights and missile firings occurring during the year. While we all have some sense of the current state of affairs, it is useful to take a brief look into the past to understand how we managed to end up in the state we see today. The field of rocketry has a rich and storied history, and there are a number of good resources

Figure 1.1 Photo/sketch of Chinese "fire arrow" battery dating to the thirteenth century. We can only imagine the hazards faced by those who were asked to ignite such a system (Source: Wikipedia).

for those interested in gaining additional background in this subject area. Many of these resources are listed in the further reading section of this chapter, so in the interest of brevity only a brief compendium will be provided here.

The earliest rockets date back as far as the tenth century Chinese "fire arrows," which were similar to the fireworks we observe today in the USA on July 4 (or other suitable celebration) with the exception that they were mounted on an arrow. By the thirteenth century, the Chinese had developed sizable rocket batteries for use against enemies. These rockets utilized a solid propellant, which was essentially gunpowder, and were launched as unguided projectiles against the enemy (see Figure 1.1).

Rockets similar to these were used in various civilizations through the ages; in fact every major nation in Europe had rocket brigades by the nineteenth century. The British contribution here was designed by Sir William Congreve and hence dubbed the Congreve rocket. Congreve rockets were developed in a variety of sizes ranging from 6 to 300 lb and were employed in the Napoleonic Wars and the War of 1812. Unguided rockets used extensively in the early 1800s. During the Civil War, both the Union and Confederate Armies worked to develop rockets to use against the opposing forces. There are accounts that the Confederate army built a 12-foot tall rocket and launched the device towards Washington DC. The missile quickly ascended out of sight and the impact point was never recorded.

In the context of today's systems, all these devices would be characterized as military weapons and unguided missiles. Beginning in the late 1800s, another group of individuals began to gain some attention from their developments. These pioneers, whose inspiration was drawn from science fiction works such as those of H. G. Wells and Jules Verne (Wells wrote *The War of the Worlds* in 1898 and Verne wrote *From the Earth to the Moon* in 1863), dreamt of sending humans into space. Today, we regard three of these individuals: Konstantin Tsiolkovsky, Herman Obert, and Dr. Robert Goddard as "Fathers of Modern Rocketry."

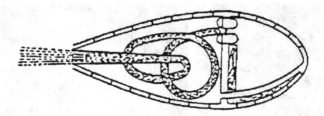

Figure 1.2 Early rocket/spaceship design (1903) of Tsiolkovsky using a circuitous delivery system to the combustion chamber. The overall shape of the rocket is quite similar to those shown in Jules Verne's books (Source: nasaimages.org).

Tsiolkovsky (also spelled Tsiolkovskii) was a Russian schoolteacher and mathematician who developed much of the theory for the requirements for orbiting the Earth and pursuing spaceflight. He is credited with derivation of what we call "The Rocket Equation" or "The Tsiolkovsky Equation," which relates a rocket's speed to its mass properties and fuel efficiency (see Chapter 3) and was one of the first to suggest the development of liquid propellant systems and propellants such as liquid oxygen and liquid hydrogen for spaceflight. He also is credited with discovering the advantages of using multiple stages for increasing rocket performance. Figure 1.2 shows the design of one of his liquid rocket propulsion devices – a far cry from today's engines with the circuitous path to the combustion chamber and the absence of a throated nozzle. In 1935, he published *Na Lune* (*On the Moon*); a copy of the cover from this book is shown in Figure 1.3. (For more about the history of rocket propulsion, see Smith, 2014.)

Ask a German or Romanian about the father of modern rocketry, and you will likely hear the name "Herman Oberth" in response. Herman Oberth was a Romanian-born scientist who studied medicine and physics, spent much of his life in Munich and was inspired by Verne's works even as a teenager. His doctoral dissertation, *Wege zur Raumschiffahrt* ("Ways to Spaceflight") was initially rejected as being too far-flung but this setback did not deter Oberth from publishing his works.[1] Oberth's 1923 book, *The Rocket into Planetary Space*, contained detailed accounts of many aspects required for such a mission (see also Oberth, 1972). Figure 1.4 provides a photo of the author. This work served as a particular inspiration for an energetic and passionate teenager by the name of Werner von Braun. Von Braun eventually came to work with Oberth for a period of time at the Technical University of Berlin. While mainly a theoretician, Oberth worked on numerous development programs during his long career and eventually came to work with his former student, Dr. Werner von Braun, in Huntsville, Alabama after World War II.

In America, Dr. Robert Goddard is acknowledged as the father of modern rocketry due to his extensive testing of solid and liquid propellant rockets between 1915 and 1941. While Tsiolkovsky and Oberth are generally known for their theoretical contributions, Dr. Goddard

---

[1]   He eventually was granted his Doctoral degree for this work by the Romanian Babes-Bolyai University.

Figure 1.3 (a) The cover of Tsiolkovsky's book *Na Lune* (1935). Note that he hadn't considered the lack of a lunar atmosphere in his research. (b) Konstantin Tsiolkovsky (Source: en.wikipedia.org).

Figure 1.4 Dr. Herman Oberth (Source: NASA).

had a strong experimental component to his studies. A graduate of Clark University, he performed much of his research as a physics professor within this institution and conducted extensive testing of solid and liquid propellant rockets between. Like Tsiolkovsky, he recognized the potential for liquid propellants and worked to develop and fly the first liquid propellant rocket in 1926. He is also credited with the first patent on multi-stage rockets, for the use of jet vanes and gimballing systems for guidance and control, and for proving that rockets can provide thrust in a vacuum. This latter proof resulted from a famous exchange that Dr. Goddard had with the *New York Times*. In a reaction to Goddard's now famous paper, "A Method of Reaching Extreme Altitudes," the *Times* published an editorial berating the work and the notion that rockets could successfully provide thrust in a vacuum (see Goddard, 1919). Goddard devised an experiment using a .22 caliber revolver that was mounted on a turntable and placed in a vacuum bell jar. A blank cartridge was loaded into the gun and once a vacuum was established, he fired the gun. The impulse generated caused the table/gun pair to rotate, thereby proving his point. The *New York Times* eventually published a retraction on July 17, 1969, three days before Neil Armstrong set foot on the surface of the moon for the first time.

Figure 1.5 provides a scale drawing of the 1926 Model A rocket that utilized liquid oxygen (LOX) and gasoline propellants. The interesting arrangement, with tankage placed below the thrust chamber, was undoubtedly employed by Goddard to enhance vehicle stability. The arrangement has generated substantial debate on modern websites. If the rocket was rigid with a thrust that did not gimbal, the fact that the thrust center lies above the center of mass is insufficient to guarantee stability. However, the flexibility of the long structure that supports the engine and feed plumbing plays a role in providing a pseudo-gimbal for the thruster, thereby aiding overall stability of the system.

While Oberth, Goddard, and Tsiolkovsky proposed peaceful uses for rocket-propelled vehicles, as World War II approached it became clear that the military establishment also had a great deal of interest in these devices. While we mentioned that virtually all the world's powers possessed rocket-propelled weapons well before this time, it was the advent of inertial guidance systems that really opened a new realm of rocketry.

The Germans were the first to incorporate this innovation into a long-range rocket, the V-2. The V-2 was clearly the most advanced rocket in the world at the time it was built and was an amazing achievement considering the technologies available at the time. It was over 46 feet (14 m) tall, weighed 27,000 pounds (12,250 kg), and had a range of 200 miles (320 km) with a payload of 2000 pounds (900 kg) (see Figure 1.6). The famous German scientist, Dr. Werner Von Braun, was instrumental in the development of the V-2. The missile incorporated two gyroscopes and an integrating gyro-accelerometer in its revolutionary guidance system, and utilized LOX and alcohol as propellants.

The V-2 rocket served as a basis for both American and Russian missile designs after the war. The German scientists involved in the rocket program at the famed Peenemunde, Germany site on the Baltic Sea aided in Russian, American, and British developments. Most notably, Dr. Von

(a)

(b)

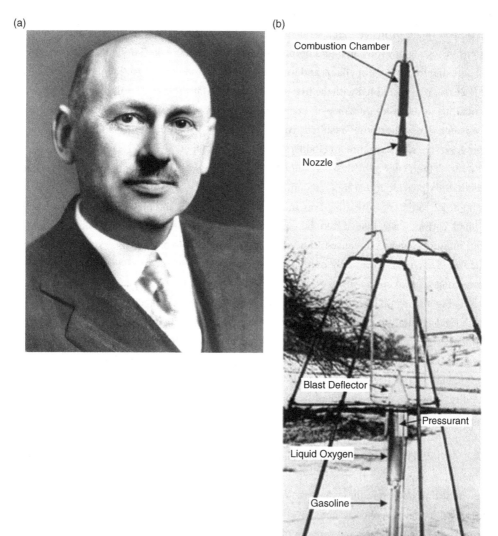

Figure 1.5 (a) Dr. Robert Goddard and (b) his Model A Rocket. Note that he placed the combustion chamber above the propellant tanks in order to maintain stability with the rocket's center of gravity below its center of thrust.

(a)

(b)

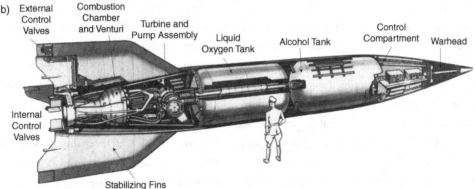

Figure 1.6 (a) Dr. Werner Von Braun. (b) The German V-2 rocket (Source: www.nationalmuseum.af.mil/).

Braun came to the United States and eventually became the director of NASA's Marshall Space Flight Center. Dr. Von Braun was an instrumental part of the Apollo program, which eventually put men on the moon.

In spite of the end of World War II, military applications still ushered in most of the rocket developments of the 1940s and 1950s. In the mid 1950s the cold war was in full swing and both Russia and America raced to develop intercontinental ballistic missiles (ICBMs) with nuclear capabilities. Out of these efforts came the US ballistic missile fleet summarized in Table 1.1.

Table 1.1   Early US ballistic missiles

Missile, program start date	Propulsion system	Range (miles)
Atlas-A, 1954	LOX-kerosene	2,500
Titan I, 1954	LOX-kerosene	6,000
Titan II, 1955	N_2O_4-aerozine 50[*]	7,000
Polaris, 1957	Solid propellant	1,400
Thor/Delta, 1957	LOX-kerosene	1,600
Minuteman I, 1957	Solid [ropellant	6,500

[*] Aerozine 50 is a blend of monomethyl hydrazine and unsymmetrical dimethyl hydrazine.

To say that rocket developments were occurring at a frenzied state in the late 1950s is an understatement. As Table 1.1 demonstrates, six separate ICBM/SLBM programs were initiated over a three-year period with missiles going through design and development in a matter of months. The Atlas, Titan, and Thor/Delta (now known as Delta) rockets have evolved to the backbone of the US unmanned launch vehicle force we have today. This fact highlights the importance of the military in developing viable rocket-propelled vehicles.

It is also worth noting that parallel development of a number of rocket-assisted takeoff, or more commonly called jet-assisted takeoff (JATO), devices were in development to provide augmented thrust for aircraft sporting the new gas turbine/jet engines in development at the time. While the jet engines provided the aircraft the ability to attain higher speeds, early models lacked sufficient takeoff thrust to lift a fully laden aircraft, and numerous rocket systems were built to provide thrust augmentation during this time. While some of these systems used cryogenic liquid oxygen, non-cryogenic oxidizers such as nitric acid and variants red-fuming and white-fuming nitric acid were developed for these applications.

The Soviets were certainly not standing still while the US was feverishly working on several types of rockets. In fact, much of the vigor within the US industry was aimed at "catching up" with the Russians in large rocket technology. The true evidence of the success of the Soviet space program occurred on October 4, 1957 when a modified SS-6 ICBM launched a small satellite called "Sputnik" into orbit around the Earth. On this day, the space age was born. The US response wasn't far behind; on January 31, 1958 Explorer 1 was placed into orbit using a Juno launcher.

In April 1958, President Eisenhower signed a declaration creating the National Aeronautics and Space Administration (NASA), and the civilian space program was born. Since derivatives of ICBMs did not promise enough capability for manned missions beyond low Earth orbit, NASA embarked on the development of the Saturn family of vehicles which culminated in the creation of the gigantic Saturn V used for the Apollo lunar missions. As most of you probably know, the Saturn V enabled Neil Armstrong to take the first steps on the surface of the moon on the evening of July 20, 1969.

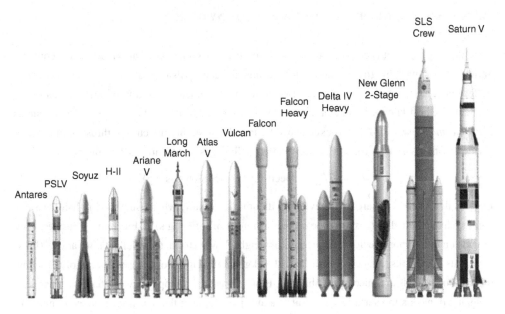

Figure 1.7 Comparison of launch vehicles from around the world (Source: Blue Origin).

Figure 1.7 describes much of the current landscape in terms of vehicles that are actively flying or in major development. The famed Saturn V lunar launch vehicle, which stood 363 feet (111 m) tall is also shown as a basis of comparison. The payload of the vehicle is defined as the useful mass that is delivered to low Earth orbit (LEO) and this ranges from roughly 1 metric ton for the smallest vehicles shown to over 140 metric tons for the Saturn V. The gross liftoff weight (GLOW) of this compilation ranges from 300 to 3000 metric tons. Dividing these two numbers (payload mass/GLOW) gives the payload fraction for the vehicle and we can see that it is a small number typically less than 2%. It is an unfortunate consequence that the energy level of propellants we have available (I_{sp} as defined in Eq. 1.1 below) and the tremendous gravity well represented by the Earth creates a situation where we must use very large devices to place relatively small devices in orbit. In Chapters 2 and 3 we explain the physics leading to this conclusion.

Many of the vehicles employ *strap-on boosters* that utilize either solid rocket motor or liquid rocket engines to augment first stage thrust. The strap-on systems typically burn for about 1–2 minutes during the ascent, then are jettisoned from the main vehicle as it climbs to higher altitudes. All of the rockets utilize multiple *stages* in order to deliver payloads into low earth orbit and beyond. By jettisoning some of the structural mass part way through the mission, we can reduce the overall size of the vehicle since the dropped mass need not be accelerated to the highest velocities required for orbit. In Chapters 2 and 3 we investigate the issue of staging and its benefits in reducing overall vehicle size. The compilation in Figure 1.7 is by no means exhaustive and interested readers are referred to a wealth of online information as well as to *International Reference Guide to Launch Systems* by Isakowitz *et al.* (2004).

1.3 CLASSIFICATION OF ROCKET PROPULSION SYSTEMS

As with any product, it is desirable to develop measures of "goodness" that can be used to compare various alternatives. In order to classify the various rocket propulsion systems in terms of a general performance parameter, we first need to define this parameter in terms of known quantities. Probably the best-known parameter used to describe the performance of rocket propulsion systems is its *specific impulse* or I_{sp}. A rocket's specific impulse is simply the current thrust of the rocket, divided by the mass flow (or weight-flow) of propellant gases currently exiting the nozzle:

$$I_{sp} = F/\dot{m} = \text{specific impulse (s or N-s/kg)} \qquad (1.1)$$

In general, rockets types are quite cavalier with units, especially when using the English measurement system. Formally, I_{sp} should be reported in units of lbf/(slug/s) or lbf/(lbm/s), but generally propulsion folks report the flow rate in terms of weight-flow such that the resultant units are lbf/(lbf/s) = seconds. In SI units, the proper measure is N-s/kg, but if weight-flow is used instead of mass flow the units of seconds are once again obtained. From a practical standpoint, we determine masses by weighing things on the surface of the Earth, and it is this notion that has driven the community to adopt this convention for reporting I_{sp}.

The specific impulse of a rocket is a measure of its propellant usage efficiency since it measures the flow rate required to attain a given thrust level. The system I_{sp} is a measure of the energy of the propellant chosen as well as the specific nozzle selected. We can see where the term specific impulse comes from if we define the total impulse as the thrust integrated over the rocket's burn time (t_b):

$$I = \text{total impulse} = \int_0^{t_b} F \, dt = Ft_b \ \text{ for } F = \text{constant} \qquad (1.2)$$

Substituting Eq. 1.2 into 1.1 assuming that I_{sp} is constant gives:

$$I = I_{sp} \int_0^{t_b} \dot{m} \, dt = I_{sp} \, m_p \qquad (1.3)$$

where m_p is the total propellant mass (weight) consumed during the firing. If we solve Eq. 1.3 for I_{sp} we see that this quantity is the amount of impulse obtained from a unit mass of propellant, hence the term "specific impulse."

For aerospace systems, or anything that we have to lift, the weight is always a critical parameter. Having a high I_{sp} is great, but if the system that provides such performance is very heavy, it will not be a good performer when compared against other alternatives. Typically, the weight or structural efficiency of a rocket propulsion system is prescribed in terms of a parameter called the *propellant mass fraction, λ*. This quantity measures the structural efficiency of the system by determining the proportion of the propulsion system made up of propellant. The remainder (non-propellant portion) of the system is effectively inert mass which we can denote m_i. Using this notation, the propellant mass fraction is:

$$\lambda = \text{propellant mass fraction} = m_p/m_o = \frac{m_p}{m_p + m_i} \tag{1.4}$$

where m_o is the overall mass of the loaded propulsion system. Later, we will see that high λ values are desirable to enhance the performance of rocket-propelled vehicles. This fact is obvious if you realize that the inert weight does not contribute to payload so that substantial penalties exist for carrying large amounts of inert components.

By these two basic measures, the ideal system will exhibit both high I_{sp} and λ values. However, these are in some sense mutually exclusive requirements. As we will see later, obtaining high I_{sp} typically requires high flame temperature and high combustion pressures which will require thicker and better insulated chamber and nozzles thereby lowering mass fraction. For this reason, there are *always* trades between performance and weight in these systems. Some of these trades are highlighted as we introduce some of the more common rocket propulsion systems in the coming subsections.

1.3.1 The Solid Rocket Motor

Figure 1.8 highlights one of the common types of rocket propulsion systems in use today: the solid rocket motor or SRM. An SRM is conceptually a very simple device with few moving parts. The entire propellant charge, known as a *propellant grain*, is contained within the pressure vessel generally called the *motor case*. The case is protected from hot combustion gases by internal insulation. The nozzle serves to convert the thermal energy of the combustion gases to thrust. The nozzle may in some cases be vectorable in order to provide vehicle control. As we shall see in Chapter 6, the thrust history of an SRM can be controlled by carefully assigning the initial grain

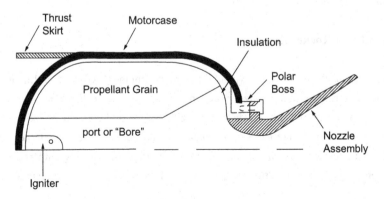

Figure 1.8 Solid rocket motor (SRM).

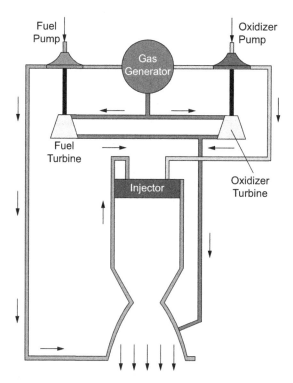

Figure 1.9 Liquid rocket engine (LRE) utilizing a gas-generator power cycle.

geometry. While this approach can give thrust-time variations, thrust levels of SRMs are in general not controllable during the course of a firing.

While the performance of today's solid propellant/motor combinations leads to I_{sp} values that do not in general compare favorably with liquid systems, the high mass fraction of SRM systems due to simplicity and high fuel density does make these devices attractive for many applications. For example, SRMs are used in many launch vehicles (Space Shuttle, Ariane, Atlas/Delta) to augment first stage thrust.

1.3.2 The Liquid Rocket Engine

Figure 1.9 highlights the features of a typical liquid rocket engine (LRE) system. Liquid propellant combinations requiring fuel and oxidizer stored in separate tanks are called *bipropellant systems* as in Figure 1.9. Some propellants (such as hydrazine, N_2H_4 or hydrogen peroxide, H_2O_2) can decompose to high-temperature gases upon contact with a suitable catalyst material. These types of propellants are termed *monopropellants*. Bipropellant systems typically outperform monopropellant systems significantly in terms of I_{sp}. As we will see in later chapters, the relatively simple monopropellant systems have applications in small LREs used as satellite propulsion, while the larger LREs are generally bipropellant systems.

The engine in Figure 1.9 utilizes combustion products from a *gas generator* in order to drive a turbine, which provides the power required by both pumps. The gas generator obtains a small amount of fuel and oxidizer from the main propellant lines as is shown in the figure. The method used to obtain energy to drive the turbine (or turbines) is referred to as the *engine power cycle*. The gas generator power cycle is one of several alternatives we will address in more detail in Chapter 7.

Since propellants do not initially reside in the combustion chamber for these liquid engines, we require a feed system to force the liquid propellants into the chamber. In Figure 1.9, a high-pressure gas source, such as an inert gas stored in a separate bottle, is used to force propellants into the chamber. This *pressure-fed* arrangement is contrasted with a *pump-fed* arrangement in which turbopumps are installed between tank outlets and the chamber. Most large systems are pump-fed due to the enormous amount of pressurant gas which would be required to displace the large quantity of propellants.

Finally, we note that while the performance of bipropellant liquid systems is significantly higher than solid rocket motors, the propellant mass fraction is usually markedly lower for these complex devices, i.e. tankage, pumps, and plumbing all act as inert weights in Eq. 1.4.

1.3.3 Hybrid Rocket Engines

Figure 1.10 highlights the features of a rocket, which combines the SRM and LRE ideas into a single system, the hybrid rocket engine (HRE). The fuel is cast in solid form into a chamber, which will serve as the combustion chamber. The oxidizer is then injected at either the head or aft end of the chamber and reacted with the fuel. Since the engine thrust is proportional to the oxidizer flow rate, thrust can be modulated using a valve on the oxidizer. The hybrid engine is inherently safer than an SRM since the oxidizer and fuel are stored separately until engine operation. The device is simpler than bipropellant liquid systems since all the fuel is stored in the chamber. Like the LRE, the hybrid engine has several alternatives (other than the gas generator shown) for providing the energy required to drive the oxidizer into the high-pressure combustion chamber.

Performance and propellant mass fractions for hybrid systems typically lie somewhere between the SRM and LRE extremes. Typical specific impulse values lie between those of solid and liquid bipropellant systems, while mass fractions are higher than most liquid systems and lower than most solid systems. The most notable issue relative to combustion of hybrid propellants lies in the fact that the relative amounts of fuel and oxidizer are not controlled in this system. The rate of consumption of the fuel depends on the local conditions in the fuel port and therefore, the oxidizer-to-fuel ratio can change with time. In contrast, we can control the mixture ratio in solid propellants very accurately by carefully weighing all ingredients. In liquid bipropellant systems we have individual control of both fluids and can therefore set their relative mixture with reasonable accuracy. We'll provide more insight into these distinctions in coming chapters.

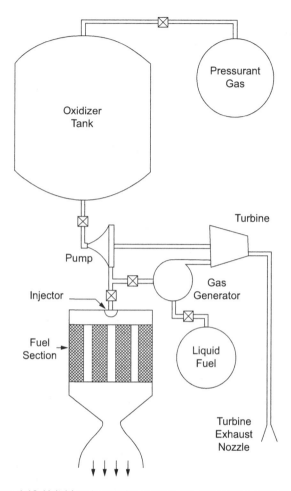

Figure 1.10 Hybrid rocket engine using a gas-generator power cycle.

1.3.4 Nuclear Rocket Engines

Nuclear rockets can be separated into two general categories. In a *nuclear thermal rocket*, energy from a nuclear process is used to heat a working fluid (usually hydrogen) to high temperatures and the hot gases are then expanded through a typical de Laval nozzle. A schematic of this type of rocket is shown in Figure 1.11. In this concept, the turbine is driven by fuel, which has been heated as a result of passing through a cooling jacket surrounding the engine. Fuel passing through the turbine is then fed into the chamber for energy addition and expansion through the main nozzle (expander cycle). The *nuclear electric rocket* utilizes electricity generated from a nuclear powerplant to generate thrust through electric rocket principles as discussed in the following subsection.

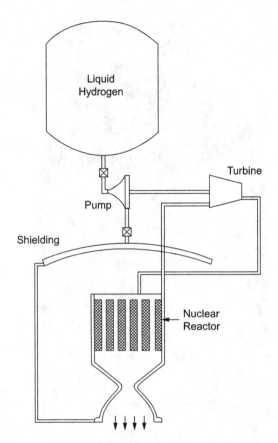

Figure 1.11 Nuclear thermal rocket using an expander cycle.

Nuclear thermal rockets have demonstrated I_{sp} values in excess of 800 seconds, which is much higher than either solid or liquid systems. However, the mass fraction of the nuclear devices can be quite low since heavy shielding is required to avoid high amounts of radiation from the reactor core. In spite of this fact, nuclear thermal rockets are attractive alternatives for interplanetary missions where high I_{sp} is a necessity and presently are receiving increased attention for human missions to Mars. While nuclear rockets have been ground-tested, and experimental flights have been conducted by the Soviet Union, no production system has yet been developed.

Nuclear thermal rockets have been the subject of extensive work in the US and USSR/ Russia mainly in the 1950s–1970s. In the US the NERVA (Nuclear Engine for Rocket Vehicle Application) program ran from 1955 to 1973 and a total of 20 reactors were designed, built, and tested at a Nevada site known as Jackass Flats. The largest of the reactors was dubbed Phoebus and put out over 4000 MW of thermal power with an engine thrust level of 210,000 lbf. Figure 1.12 shows Professor Heister next to a NERVA XE engine that is on display at NASA Marshall Spaceflight Center.

Figure 1.12 Professor Heister adjacent to NERVA XE engine on display at NASA Marshall Spaceflight Center.

1.3.5 Electric Rocket Engines

Figure 1.13 highlights the two most common electric propulsion concepts. In an *arcjet* or *electro-thermal thruster*, an electric potential is "arced" across a gap between two electrodes, much like the operation of a spark plug in your car. Fluid passing through the arc is superheated to extremely high temperatures and expanded through a nozzle. Typical I_{sp} values for the electrothermal devices lie

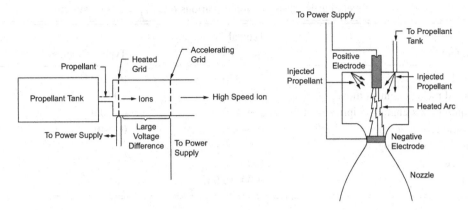

Figure 1.13   Electric rockets.

between 400 and 1500 seconds. Hydrogen, ammonia, and helium have all been used successfully as working fluids for these devices. Currently, arcjet thrusters are being flown aboard communication satellites for the US and other nations.

In *electrostatic*, or *ion thrusters*, heavy atoms (usually xenon) are ionized via electric heating coils, and the resulting positively charged ions are accelerated using an electric potential to very high velocities. The ions are "neutralized" by recombination with an electron beam at the exit of the thruster. Specific impulse values for these engines are measured in the thousands of seconds, with values of 10,000 seconds attainable under certain conditions.

While the specific impulse of the electric devices is certainly attractive, these devices utilize large amounts of electric power to generate very small thrust values. For example, thrust levels of these devices are usually measured in millipounds with power requirements in the kilowatt to megawatt range. In spite of the low thrust, these devices are attractive for satellite maneuvering and possibly interplanetary travel applications. Use of a nuclear generator has been proposed to supply the large amounts of power to these types of thrusters.

1.3.6 Summary of Today's Rocket Propulsion Systems

It is difficult to make an overall quantitative assessment of the various propulsion options discussed above, although it is obviously useful to the reader. Both the specific impulse and mass fraction of a given system are strongly dependent on its size, the propellant combinations employed, and the materials and structural technologies folded into the fabrication of the device. Having said that, Table 1.2 provides a rough comparison compiled from a database of existing systems mainly in the space propulsion area. There are certainly examples that lie outside the quoted ranges, but the information can be used by students to provide a rough assessment of their computed results and to give a perspective on the relative performance of various propulsion options. As mentioned

Table 1.2   Overall gross classification of various rocket propulsion systems

System	Typical I_{sp} range (seconds)	Typical λ range
Solid rocket motor	150–350	0.8–0.95
Liquid rocket engine – monopropellant	200–250	<0.5
Liquid rocket engine – bipropellant	300–460	0.7–0.9
Hybrid rocket engine	300–350	0.85–0.9
Nuclear thermal engine	800–1000	?
Arcjet	400–1500	<0.5
Ion	1000–10,000	<0.5

previously, both performance (I_{sp}) and weight (λ) are important to any system designs and the overlapping characteristics of various design options make for interesting trades in many instances.

1.3.7 Advanced Concepts

The subsections above discussed the most common rocket propulsion concepts available today. There are many other advanced concepts, which have been theorized or demonstrated on a laboratory basis. We will discuss only a few of these concepts below.

Laser Propulsion
Laser propulsion systems utilize energy from a laser to heat a working fluid and exhaust the high-temperature fluid through a nozzle. These concepts have large power requirements similar to electric thrusters.

Magnetoplasmadynamic (MPD) Thrusters
These thrusters utilize a magnetic field to accelerate a plasma formed by subjecting a working fluid to large amounts of energy. Power requirements are large for these devices as well.

Solar Propulsion Systems
These systems utilize solar energy generally focused with a mirror, to heat working fluid to high temperatures for exhausting through a nozzle. While the performance of such systems is quite good, the large mirror mass generally prohibits economical use of such a system in space. The solar sail, which uses the momentum of reflected photons (solar pressure) as a thrust source, is another low-thrust alternative for space propulsion missions.

Antimatter Propulsion
This concept relies on the incredible amount of energy produced when recombining hydrogen with anti-hydrogen inside a "combustion" chamber. The theoretical performance of these devices can be

in the thousands of seconds in specific impulse. The main problem is the containment and manufacture of antimatter in substantial quantities.

FURTHER READING

Alway, P. (1996) *Retro Rockets: Experimental Rockets 1926–1941*. Saturn Press.

Alway, P. (1993) *Rockets of the World: A Modeler's Guide*. Peter Alway.

Barrere, M., Jaumotte, A., De Veubeke, B. F., and Vandenkerckhove, J. (1960) *Rocket Propulsion*. Elsevier Publishing Company.

Doyle, S. E. (ed.) (1992) *History of Liquid Rocket Engine Development in the United States, 1955–1980, AAS History Series, Vol. 13*. AAS Rocket Propulsion History Colloquia, Vol. 1. AAS Publications.

Feodosiev, V. I., and Siniarev, G. B. (1959) *Introduction to Rocket Technology*, Academic Press.

Goddard, R. H. (1919) "A Method of Reaching Extreme Altitudes," Smithsonian Miscellaneous Collections, Vol. 71, no. 2. Smithsonian Institution, City of Washington.

Goddard, R. H. (2002) *Rockets*. AIAA Reprint of 1946 American Rocket Society Publication.

Humble, R. W., Henry, G. N. and Larson, W. J. (1995) *Space Propulsion Analysis and Design*. McGraw Hill Publishers.

Isakowitz, S. J., Hopkins, J. B., and Hopkins, J. P. Jr. (2004) *International Reference Guide to Space Launch Systems, 4th edn*. AIAA.

Kuentz, Craig (1964) *Understanding Rockets and Their Propulsion*. John F. Ryder Publishers.

Neufield, M. J. (1995) *The Rocket and the Reich*. Harvard University Press.

Oberth, H. (1972) *Ways to Spaceflight*. No. 622. National Aeronautics and Space Administration.

Ordway, F. I., III (ed.) (1989) *History of Rocketry and Astronautics*. American Astronautical Society History Series, Vol. 9.

Smith, M. G. (2014) *Rockets and Revolution: A Cultural History of Early Spaceflight*. University of Nebraska Press.

Sutton, G. P., and Biblarz, O. (2016) *Rocket propulsion elements*. John Wiley & Sons.

Von Braun, W., and Ordway, F. I. (1985) *Space Travel: A History*. Harper and Row.

Wertz, J. R. and Larson, W. J. (eds.) (1991) *Space Mission Analysis and Design*. Klewer Academic Publishers.

www.clarku.edu/research/archives/goddard/ (a special collection of Goddard's papers).

www.nasa.gov/centers/goddard/about/history/dr_goddard.html (a collection of Goddard's works and history on the Clark University website).

CHAPTER 2   MISSION ANALYSIS FUNDAMENTALS

As you know, the propulsion system is only a portion of a rocket-propelled vehicle. We must have a good working knowledge of the mission of the particular vehicle when designing the propulsion system as it affects the demands put upon it. If tankage is counted as part of the propulsion system, you will find that propulsion components make up the bulk of the size and weight of most rockets. Clearly it is appropriate for us to discuss the overall vehicle requirements, since for many missions propulsion experts will be responsible for the preliminary sizing of practically the entire vehicle.

In this chapter, we will restrict ourselves to pure rockets that carry both fuel and oxidizer on-board and make no use of atmospheric oxygen. For this reason, we will not address *combined cycle* propulsion systems that may contain an airbreathing component, although extension to these systems is rather straightforward as will be pointed out in Chapter 3. In Section 2.1, we will discuss the four major categories of rocket vehicles; while in Sections 2.2–2.5 we discuss the various types of rocket-propelled vehicles in use today and preliminary techniques to determine the mission requirements for them. Finally, we summarize this chapter in Section 2.6.

2.1 CLASSIFICATION OF ROCKET-PROPELLED VEHICLES

Most rocket propulsion missions can be broadly categorized within four distinct groups with unique attributes:

- launch vehicles;
- upper-stage or orbital transfer vehicles (OTV);
- interceptors; and
- ballistic missiles.

Note that here we have ignored small thrusters used in maneuvering satellites (or vehicles) since they do not typically have a large impact on the overall vehicle design. We shall discuss these systems in more detail in Chapter 9.

The components of these basic vehicle types can be placed into one of three major functional groups:

- *Inert components* – These are non-propellant components within the propulsion system which are required for structural integrity and proper system performance. Examples of inert components include: engines, tanks, pumps, interstage structure, motor/engine insulation, propellant lines, pressurization systems, starting systems, etc. In other words, inert components are essentially all non-propellant components of the propulsion system.

- *Propellant* – The fuel and oxidizer expended to fulfill the primary acceleration of the vehicle. While necessary for vehicle functionality, propellants utilized for attitude control effectively act as inert components that are jettisoned at regular intervals during the mission; their impulse does not lead to the primary acceleration of the vehicle in the desired direction. In the case of a nuclear thermal rocket, the propellant here would be the working fluid being accelerated through the primary nozzle.

- *Payload components* – The object or system we are trying to accelerate with our propulsion system. For a multi-stage vehicle, the "payload" of the first stage would be the mass of all upper stages including the actual useful payload of the rocket. There are many distinctions here and we must be careful to distinguish between the terms *payload* and *useful payload*. For example, a three-stage launch vehicle may boost a satellite that is in fact its useful payload, but for the propulsion system designer, the payload fairing, any accessories/batteries/cabling, satellite cradle structure, and any other attached hardware would all represent payload that the third-stage motor/engine must push.

You will see in Chapter 3 why we choose to classify components in these three categories, but basically it is because of the fact that we have some preliminary background involving the ratio of propulsion system inert weights and propellant weights. Recall from Chapter 1 that the propellant mass fraction, λ, is written in terms of propellant and inert masses associated with the propulsion system:

$$\lambda = \frac{m_\mathrm{p}}{m_\mathrm{p} + m_\mathrm{i}} \qquad (2.1)$$

where m_i is the sum of all inert component weights and m_p is the propellant mass. Many preliminary design exercises make use of historical propellant mass fractions in order to estimate the amount of inert weight for a given propellant weight. As you might expect, the propellant mass fraction gets closer to one as the system grows in size. The reason for this is due to the fact that many components are limited due to "minimum gauge" thickness constraints as system sizes get smaller.

Let's assume that we want to design a rocket from the four categories noted above. We would employ the following general process:

1. *Select a propulsion system* – There may be a set of top-level system requirements that are prescribed by the customer that will impact the type of propulsion system that may be selected. For example, there may be handling issues that would necessitate a solid propulsion system rather than a liquid, or environmental concerns that would effectively eliminate some choices. Inherent in this selection is vehicle topology, i.e. how many

stages are to be employed and what is their general arrangement and operational scheme?

2. *Derive the mission requirements* – This is the subject of the present chapter. A top-level system requirement might specify a vehicle capable of launching a given useful payload mass to a certain orbit; we need to break this down into an engineering-level requirement for the vehicle topology (i.e. number and type of stages) selected.

3. *Preliminary design and trade studies* – This step is where the basic system design parameters are decided. Such items as chamber pressures, mixture ratios, propellant types, engine types, nozzle types and expansion ratios, and materials of fabrication are decided at this stage of the process.

4. *Detailed design and fabrication* – Preliminary analyses are refined and supplemented with component test data in order to provide information for detailed design and fabrication.

5. *Demonstration and qualification testing* – Depending on the maturity of the system relative to historical cousins, the program may require one or more demonstration firings. The purpose of these firings is to validate the overall concept and operational environment for the design. As the demonstration tests are full-scale tests, there is a strong motivation to remove this aspect in many cases on the basis of cost. In order to take this route, more proven technologies are sometimes preferred over less mature (but potentially higher performing) approaches. Qualification testing follows successful demonstration testing or is initiated immediately depending on the program. In general, qualification tests are conducted to verify successful operation at the required ambient temperature extremes. NASA refers to this step as certification testing in that the system/ vehicle is being certified for flight.

6. *Iteration as required* – If test results are satisfactory, design is complete; if not return to Step 4.

In this book, we will focus primarily on the first three steps in this process.

We select a propulsion system based on the requirements for the vehicle, from a cost standpoint and many other factors. For instance, an application requiring engine throttling would lead us to first consider a liquid engine since SRMs do not generally have this capability. If design simplicity and a high degree of launch readiness is desired (such as in a ballistic missile), we may favor SRMs since we don't have to worry about taking time to fill fuel tanks for a LRE.

Of course, the specific impulse of the particular propulsion system chosen is also an important parameter. Recall that in Chapter 1 we defined this parameter:

$$I_{sp} = F/(\dot{m}g) \qquad (2.2)$$

By integrating the thrust of the engine over the burning duration, t_b, we may write the *total impulse*:

$$I = \text{ total impulse } = \int_0^{t_b} F dt = F t_b \text{ for } F = \text{constant} \tag{2.3}$$

Substituting Eq. 2.2 into 2.3, we can gain insight into importance of I_{sp}

$$I = \int_0^{t_b} I_{sp} \dot{m} \; g dt = I_{sp} \int_0^{t_b} \dot{m} g dt = I_{sp} m_p g \tag{2.4}$$

where m_p is the total mass of propellant ($m_p g$ being the *weight* of propellant) expended during the burning duration, t_b.

Therefore $I_{sp} = I/(m_p g)$ is the impulse provided per lb (kgf or N) of propellant. Hence the term, *specific* impulse. Since the mission requirements will amount to a specification of the total impulse, we can see that this knowledge, combined with knowledge of the specific impulse of the propulsion system selected, will enable us to determine the amount of propellant required to accomplish the mission.

The *mission requirements* can generally be expressed in terms of a velocity increment to be imparted on the payload. This velocity increment, often called Δv or "delta-vee," is a function of the type of mission selected. Now, in the previous paragraph, we said that the mission requirement essentially amounted to a specification of I, not Δv. We can see that the two quantities are related by thinking of the case where we push with constant force F on an object of mass m for a duration t_b seconds. In this case, we can integrate $F = ma$ to obtain:

$$\Delta v = \frac{F}{m} t_b = \frac{I}{m} \tag{2.5}$$

which shows that the velocity increment is directly related to the total impulse. Now, in a rocket, the mass is constantly changing so Eq. 2.5 is not exactly correct. In Chapter 3 we will derive the proper form of Eq. 2.5 for a rocket-propelled vehicle.

In any case, we have shown that velocity increment and total impulse are related, so that if we specify the mission requirement in terms of a velocity change, we should be able to relate this to the total impulse later. To help illuminate this issue, let us now consider each of the four rocket types individually and try to calculate in a preliminary sense the velocity increment mission requirement for each of these systems.

2.2 MISSION REQUIREMENTS FOR LAUNCH VEHICLES

Launch vehicles are rockets designed to carry a payload to low Earth orbit (LEO) or to other bodies in the solar system (or galaxy). We already discussed many of the world's vehicles in Chapter 1 of this text. In general, a launch vehicle can be ground, air, or sea launched and contains two or three stages for LEO missions. Candidate propulsion systems for these types of vehicles include SRMs, LREs, and hybrid engines due to their high thrust capabilities. Nuclear thermal rockets are not normally considered in these types of systems due to the possibility of nuclear contamination in the event of a failure.

Launch vehicles can be totally or partly reusable, or expendable. Reusable systems are more complex due to added recovery features and the fact that components must be designed for many uses. However, these systems do enjoy cost advantages in terms of the amount of labor (amortized over many missions) as compared to fully expendable vehicles. As you may be aware, the tradeoff between reusable and expendable systems is still a subject of debate and it is still not clear which system is preferable.

Some of the critical design parameters governing the design of launch vehicles include:

- *The burning time of each stage.* Since the vehicle must work against Earth's gravity, minimizing the flight time is desirable. This desire must be balanced against drag and possible weight increases associated with short duration designs.
- *Burnout acceleration.* As the vehicle nears the end of burn it is very light since most of the propellant has been burned. We must limit the acceleration in this region to prevent structural failures and/or damage to the payload.
- *Maximum dynamic pressure, or "Max. q."* All vehicle aerodynamic loads are propor-tional to this parameter so we need to design the ascent to keep dynamic pressure below acceptable limits.
- *Series versus parallel burn.* In a series burn only one stage burns at a time, while in a parallel burn an upper stage can be burning at the same time as the lower stage. The space shuttle was an example of a parallel burn since the orbiter engines are thrusting during operation of the first-stage SRMs.

2.2.1 Mission Requirements for LEO Missions

First we will consider a LEO mission. When the rocket is in its final state, the payload will orbit the Earth in a stable trajectory. Let us assume a circular orbit for simplicity. For this orbit, the centripetal force (F_c) must balance the gravitational force (F_g) exactly as shown in Figure 2.1.

Therefore, we can express this requirement in terms of the orbiting mass, m, and the circular velocity, v_c, and the altitude, z:

$$F_c = \frac{mv_c^2}{r_e + z} = F_g = \frac{Gm_e m}{(r_e + z)^2} \tag{2.6}$$

where r_e is the Earth's radius (about 20 million feet), m_e is the mass of the Earth, and G is *Newton's universal gravitation constant*:

$$G = 6.674 \times 10^{-11} \ \frac{\text{N m}^2}{\text{kg}^2} \tag{2.7}$$

The quantity $Gm_e \equiv \mu$ is referred to as the *gravitational constant* for Earth. Gravitational constants and radii of several bodies within our solar system are given in Table 2.1.

Table 2.1   Characteristics of bodies in the solar system

| Body | Gravitational constant, μ | | Radius at equator | | Rotational period (rev/day) |
	10^5 km^3/s^2	10^{16} ft^3/s^2	km	nautical miles	
Sun	1,327,000	468,600	696,000	376,000	0.0394
Mercury	0.2203	0.07800	2,440	1,317	0.01705
Venus	3.249	1.147	6,052	3,268	0.004115[*]
Earth	3.986	1.407	6,378	3,444	1.00274
Mars	0.4283	0.1513	3,397	1,834	0.9747
Jupiter	1,266.9	447.4	71,490	38,600	2.4182
Saturn	379.2	133.9	60,270	32,540	2.2522
Uranus	58.02	20.49	25,560	13,800	1.3921
Neptune	68.50	24.19	24,760	13,370	1.3202
Pluto	0.009831	0.003472	1,151	621.5	0.1566
Moon	0.04908	0.01733	1,737	937.9	0.0366

[*]   Venus rotates in a westerly direction – all other bodies rotate in an easterly direction.

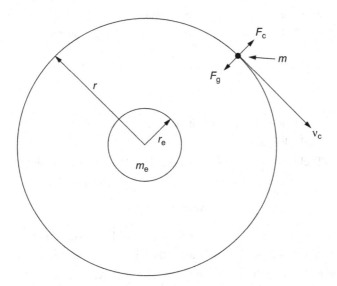

Figure 2.1 Forces acting on a body orbiting Earth.

Solving Eq. 2.6 in terms of μ gives:

$$v_c = \sqrt{\frac{\mu}{r_e + z}} \tag{2.8}$$

which is the required initial velocity for launching our payload to an altitude z above the Earth's surface. Note that this velocity is independent of the size (mass) of the payload. One can determine

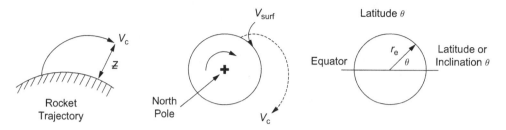

Figure 2.2 Surface velocity for an easterly launch.

the period of the circular orbit, τ, by noting that this parameter corresponds to the time required to traverse a distance $2\pi(r_e + z)$ at speed v_c:

$$\tau = \frac{2\pi(r_e + z)}{v_c} = \frac{2\pi(r_e + z)^{3/2}}{\sqrt{\mu}} \tag{2.9}$$

Since we are assuming launch from the surface of the Earth, the launch vehicle has an initial velocity equivalent to the rotational velocity of the Earth's surface at a latitude of θ degrees. Since the Earth rotates from east to west, the situation for an easterly launch is depicted in Figure 2.2. From this figure, one can see that the surface velocity, v_{surf}, can be expressed:

$$v_{\text{surf}} = \frac{2\pi r_e \cos(\theta)}{1 \text{ day}} \tag{2.10}$$

At Kennedy Space Center, which has a latitude of 28.5°, the surface velocity is about 1320 ft/s (400 m/s) in an easterly direction.

The overall velocity increment required to attain a low earth orbit is simply the difference of v_c and v_{surf}:

$$\Delta v = v_c - v_{\text{surf}} \tag{2.11}$$

The actual Δv value will be 20–25% higher than this ideal result due to losses from drag, vehicle steering, and gravitational deceleration during ascent. You can see that we benefit by launching to the east due to the Earth's rotation. If we launch in other directions, the Earth's surface velocity can work against us and drive up the Δv requirement. To be perfectly honest, this estimate is very crude; it is included here to elucidate the magnitude of velocities involved for orbital missions. To get a more detailed prediction of Δv for launch vehicles, we will need to perform a trajectory simulation as described in Chapter 3. A good estimate for an easterly launch from Cape Kennedy to a low earth orbit lies in the 29,000–29,500 ft/s (8.8–9.0 km/s) range. These are large velocities and these estimates do a lot to explain the size of launch vehicles.

2.2.2 Mission Requirements for Earth Escape Missions

We begin our discussion with Newton's law of gravitation. At the surface of the Earth, we may write the gravitational force:

$$F_g = \frac{Gm_e m}{r_e^2} = m\, g_e \tag{2.12}$$

where g_e is the acceleration at the Earth's surface ($g_e = 32.2 \text{ f/s}^2 = 9.81 \text{ m/s}^2$). From Eq. 2.12 we can write the gravitational constant, μ:

$$\mu = Gm_e = g_e r_e^2 \tag{2.13}$$

Using this relation, we can write the gravitational force at any distance $r > r_e$ away from the center of the Earth:

$$F_g = mg_e(r_e/r)^2 \quad \text{or} \quad g \equiv g_e(r_e/r)^2 \tag{2.14}$$

Equation 2.14 is essentially a restatement of Newton's law in that it says that the gravitational force decreases as the inverse square of the distance from its origin.

Now we can determine the potential energy required to do work against the gravitational force of Eq. 2.14. Let's say we want to raise an object of mass m to an altitude z where we would then have $r = r_e + z$. In this case, the potential energy would be the integral of the gravitational force over the interval z:

$$\text{PE} = -\int \frac{mg_e r_e^2}{(r_e + z)^2}\, dz = \frac{mg_e r_e z}{r_e + z} \tag{2.15}$$

where we have set PE = 0 at the Earth's surface. We could exchange this potential energy with some initial kinetic energy of our mass. For instance, the kinetic energy can be written:

$$\text{KE} = \frac{m}{2} v_z^2 \tag{2.16}$$

Now if we wish to use this energy to raise the height of the mass we note that the total energy KE + PE is constant. Assuming that the mass has zero velocity at its highest point (apogee), the kinetic energy there is zero. Similarly, if we let the potential energy be zero at the Earth's surface (as we assumed in our integration), we can simply equate Eqs. 2.15 and 2.16 to give:

$$v_z = \sqrt{2g_e r_e z/(r_e + z)} \tag{2.17}$$

Here v_z is the initial velocity of our mass at the Earth's surface, which will attain an altitude z. This solution is analogous to finding the height of a rock that you throw straight up (with initial velocity u_z). If we let $z \to \infty$, Eq. 2.17 gives:

$$v_\infty = \sqrt{2g_e r_e} = 36,700 \text{ ft/s} \ (= 11.2 \text{ km/s}) \tag{2.18}$$

Therefore, if you could throw the rock at a speed of 36,700 ft/s (what an arm!), the rock would totally escape Earth's gravitational field and come to a complete stop at infinite altitude.

Of course, since we have neglected drag and the fact that we cannot impulsively launch our rocket at such high speeds, this prediction represents a lower bound on Earth's *escape velocity*. Since Eq. 2.17 holds for arbitrary altitudes, we could find the velocity required to reach any point above the surface of the Earth (or any other body). Finally, note that once again our results are independent of the mass of our object and that atmospheric drag is not included in the simple estimates.

2.3 Mission Requirements for Upper-Stage or Orbital Transfer Vehicles

This type of rocket-propelled vehicle is used to transfer payloads (usually satellites) from LEO to higher orbits. Upper-stage propulsion systems are also used to accelerate planetary missions, such as Galileo, Voyager, Magellan, and Ulysses, out of the Earth's gravitational field. These systems typically contain one or two stages and can utilize solid, liquid, electric, or nuclear propulsion systems.

Since the entire mission of an upper-stage rocket takes place in the vacuum of space, we need not worry about the aerodynamics of these systems. The systems must be simple and highly reliable since they are remotely controlled. In addition, since there may be long "coast periods" the systems must contain adequate shielding and insulation from the Sun's radiation. Some of the main parameters driving the designs of upper-stage systems include:

- *The length and size of the system*. Package-ability is crucial for these systems since they must be carried within the launch vehicle cargo bay. Accommodations must be made for a proper interface with the launch vehicle, which may include electrical, mechanical, and venting considerations.
- *Propulsion system weight*. Since the entire system must be lifted by the launch vehicle, overall weight is a crucial consideration. Therefore, high-energy propellants are often utilized and the high-I_{sp} electric propulsion systems appear as viable candidates (although none have been used to this date).
- *Guidance considerations*. Most upper stages utilize advanced guidance systems to ensure accurate *payload insertion* into the desired orbit. As you may have noted from early Space Shuttle missions, some of the upper stages utilizing SRMs are *spin stabilized*, which means that the entire system is "spun up" to a rate usually between 40 and 120 RPM prior to motor ignition. Use of the spinning mechanism effectively reduces any small thrust misalignments that may be present due to slight inaccuracies in motor manufacture. Most modern systems employ three-axis control such that spinning/despinning is not required.

Figure 2.3 shows an OTV propulsion system consisting of two solid rocket motors. Note the efficient packaging of the system whereby the first-stage motor fits within the interstage structure to

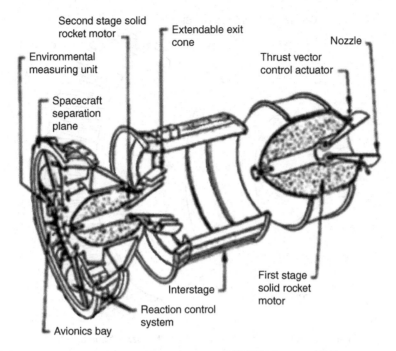

Figure 2.3 The inertial upper stage (IUS) OTV (Source: AIAA).

provide a minimal gap with the second-stage nozzle. To further reduce length, the nozzles employ *extendible exit cones*, or EECs as shown in Figure 2.4. In the past, designers have also used a gas-deployed skirt to enhance the expansion ratio of the nozzle without increasing nozzle length. Figure 2.4 provides a schematic representation of the skirt in its deployed position. Refractory metals such as columbium/niobium are generally used in these devices. This inertial upper stage (IUS) system utilizes a three-axis guidance system with nozzle *gimballing* in order to achieve high-accuracy insertion of the payload. Solid rocket systems such as these have seen reduced usage in modern vehicles as many launch vehicles place payloads into highly elliptic transfer orbits that necessitate a single propulsive firing to circularize the orbit. However, solid rocket space motors such as these have still seen usage in interplanetary space exploration missions in recent history.

A primary liquid system in use in US missions is the Centaur upper stage. Boosted by variants of the RL 10 engine using liquid oxygen and liquid hydrogen propellants, this system can provide for multiple starts/burns to conduct orbital transfer missions. Figure 2.5 provides a photo of the Centaur system on its way to the launch complex.

One of the more common missions for OTVs involves boosting a satellite to an orbit that has a period the same as the length of a complete day on Earth. This orbit, called a *geosynchronous Earth orbit*, or GEO, is very useful since the satellite will effectively remain fixed over a given position at the surface of the Earth. If we let $\tau = 1$ day, we can use

Columbium Gas Deployed Skirt

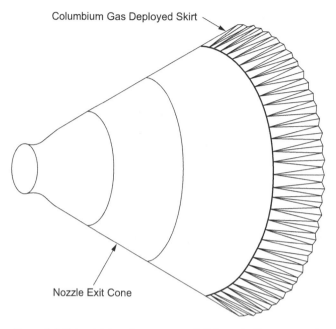

Nozzle Exit Cone

Figure 2.4 Extendable exit cone on an IPSM nozzle (Source: AIAA).

Figure 2.5 Photo of the Centaur upper stage using the RL 10 engine for propulsion being offloaded by NASA and ULA crew for NASA's Mars Science Laboratory (MSL) mission (Source: https://mars.nasa.gov /mars2020/multimedia/images/?ImageID=3638&s=6).

Figure 2.6 R-4D liquid apogee engine using storable bipropellant combination for orbit circularization and in-flight maneuvers (Source: Aerojet Rocketdyn).

Eq. 2.9 to note that the required altitude for such an orbit is about 22,200 miles (35,800 km) above the Earth's surface. Many communication and Earth observation satellites are currently in orbits at this altitude. For spacecraft being transported to this destination, the current launch vehicle suppliers are initially placing the payload in a *geosynchronous transfer orbit* (GTO). The evolution of launch vehicle capabilities, in terms of throttleable and restartable liquid propulsion stages, has permitted access to the GTO orbit in modern systems. The GTO is a highly elliptic orbit with apogee at GEO and perigee in LEO.

Spacecraft manufacturers may integrate a propulsion system into the payload itself in order to provide the final apogee impulse required to circularize this highly elliptic orbit. Oftentimes, liquid apogee engines (LAEs) are used for this purpose. Both the US and European space propulsion communities offer LAEs that utilize storable liquid bipropellants, as discussed in detail in Chapter 8. Figure 2.6 shows a 445 N (100 lbf) thrust class LAE manufactured by Aerojet

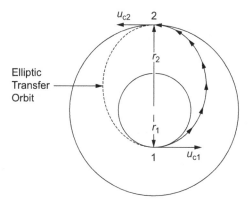

Figure 2.7 Geometry for a Hohmann transfer orbit.

Rocketdyne. The lower thrust LAE systems might require several apogee burns to eventually raise the perigee and circularize the orbit.

2.3.1 Mission Requirements for OTVs

Consider the typical mission outlined in Figure 2.5. Using Eq. 2.8, we can calculate the stable circular velocity for the lower orbit (call it v_{c1}) and upper orbit, v_{c2}. If we consider a two-stage system (such as the IUS) and can assume that the burning time of each stage is much less than the period of the orbit ($t_b \ll \tau$), we can determine the velocity increment using a maneuver called a *Hohmann transfer*. In a Hohmann transfer, the first-stage burn creates an elliptic orbit with an apogee equivalent to the desired altitude for the final orbit. The second-stage motor is ignited when the vehicle reaches the apogee and *circularizes* this *elliptic transfer orbit*, which is shown as the dashed line in Figure 2.7. Solid rocket motors used in Hohmann transfers are sometimes referred to as *perigee kick motors* (PKM) and *apogee kick motors* (AKM) for the first- and second-stage burns respectively.

Since this is not a text in astrodynamics, we will not derive the Hohmann transfer equations directly, but the resulting velocity gains are determined by assessing potential (gravitational) and kinetic energy contributions at the various states. Those interested in further reading on this subject might consult Bate *et al.*'s book, *Fundamentals of Astrodynamics* (1971) or other suitable texts. The result is valid when the burning times are small compared to the orbital periods:

$$\Delta v = \Delta v_1 + \Delta v_2 \tag{2.19}$$

$$\Delta v_1 = v_{c1} \left[\sqrt{\frac{2r_2}{r_1 + r_2}} - 1 \right] \; ; \quad \Delta v_2 = v_{c1} \left[\sqrt{\frac{r_1}{r_2}} - \sqrt{\frac{2r_1}{r_2(1 + r_2/r_1)}} \right] \tag{2.20}$$

As mentioned above, electric propulsion devices also appear as candidate propulsion systems for the OTV mission. By using solar panels to supply the electricity, an electric propulsion stage could be an efficient means of raising payloads to higher orbits. In this case, the low thrust level of the electric engines implies that the transfer process could take several months to perform. Since $t_b \gg \tau$ for this transfer, we cannot use a Hohmann technique. The trajectory in this case is a spiral; Edelbaum (1961) has estimated the velocity increment for such a transfer. The result from this analysis is quoted below:

$$\Delta v = \sqrt{v_{c1}^2 - 2v_{c1}v_{c2} + v_{c2}^2} \qquad (2.21)$$

As a result of the increased transfer time, Eq. 2.20 has a higher velocity requirement than Eq. 2.19. Due to this fact, some of the benefit associated with the high-I_{sp} electric OTV is lost since this vehicle must supply more velocity increment than a high-thrust chemical propulsion stage.

Finally, we should note that both Eq. 2.19 and 2.20 hold in the case where no plane change maneuver is performed. This means that the higher orbit is assumed to lie at the same *inclination* (latitude) as the lower orbit.

Plane Changes

For a spiral transfer involving a plane change of α degrees, Edelbaum (1961) derived the velocity requirement:

$$\Delta v = \sqrt{v_{c1}^2 - 2v_{c1}v_{c2} \cos\left(\frac{\pi\alpha}{2}\right) + v_{c2}^2} \qquad (2.22)$$

where α must be in radians.

For a Hohmann transfer, no simple relation of this form exists. For our purposes, we can approximate the Δv for the plane change maneuver separately by assuming the Hohmann transfer has already been performed. In this case, Bate *et al.* (1971) show:

$$\Delta v = 2v_c \sin(\alpha/2) \qquad (2.23)$$

which gives the velocity requirement for the plane change maneuver (at a fixed altitude implied by v_c) alone. Since this velocity increment depends on v_c, it is best to perform plane changes at the upper orbit where v_c is lower in magnitude.

Finally, we note that Eq. 2.23 is a general result, which can be utilized to determine velocity requirements for plane change maneuvers for launch vehicles or any other space-borne propulsion system.

As mentioned above, the LAEs used on GEO missions may execute several firings to circularize the orbit. Fundamentally, this approach is required to minimize the overall propellant usage and maintain the $t_b \ll \tau$ assumption consistent with the Hohmann transfer. Longer burns necessitate additional propellant due to the smearing effect of not thrusting exactly at the apogee of the orbit. In addition, the advanced guidance systems employed in modern systems permit simultaneous plane change and perigee raising maneuvers such that the LAE system can precisely

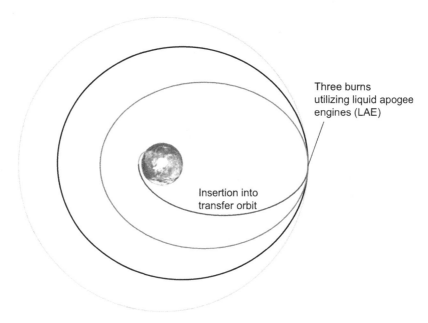

Three burns
utilizing liquid apogee
engines (LAE)

Insertion into
transfer orbit

Figure 2.8 Circularization of a GEO orbit using multiple LAE burns.

place the payload in the desired final orbit. Figure 2.8 provides a schematic of the orbital changes
that can occur during this process.

2.4 MISSION REQUIREMENTS FOR BALLISTIC MISSILES

As you are probably aware, ballistic missiles are currently utilized to deliver a conventional or
nuclear warhead to a fixed location a given range from the launch site. They are referred to as
ballistic missiles because the bulk of their flight is spent in a *ballistic trajectory* in which only
gravity acts on the projectile. As pointed out in Chapter 1, modern ballistic missiles began with the
V-2 missile developed by Germany in World War II. Figure 2.8 provides a schematic representation
of an MX Peacekeeper ballistic missile. This design is typical of long-range ballistic missiles in that
it incorporates three propulsive stages followed by a maneuvering stage called a *post-boost vehicle*,
or PBV. Shorter-range systems, such as the Pershing missile, incorporate one or two stages.

The primary parameters influencing the design of ballistic missiles include:

- *The burning time of each stage*. Since gravity is acting against the missile as the
  propellant is burned, it is advantageous to expend the propellant as quickly as possible.
  This requirement must be balanced against the fact that nozzle and case weights tend to
  increase with decreasing burning time.
- *Responsiveness of the system*. Generally, ballistic missiles must respond to a threat in
  a short period of time. For this reason, the missile must be capable of operation on short

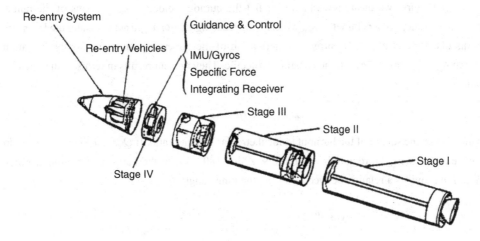

Figure 2.9 Schematic of Peacekeeper ballistic missile.

notice, which tends to be the reason that many of the systems employ SRMs rather than liquid or hybrid propulsion systems. Storable liquid propellants (those that are liquid at ambient conditions) also are viable candidates and are used in many of the Russian systems.

- *Inert weights.* High-velocity requirements for these systems make weight reduction a necessity.

- *Guidance systems.* Advanced guidance systems enhance the accuracy of these devices. Ballistic missiles incorporate some of the most sophisticated guidance equipment available.

- *Reentry heating.* One of the big design drivers for the payloads involves the extreme thermal environment encountered during reentry. Advanced composite materials are employed to withstand the high temperatures near the leading edge of reentry bodies.

Figure 2.9 highlights many of the parameters defining the trajectory of a ballistic missile. Ballistic flight is defined as a "zero-lift" trajectory in which the missile always flies at zero angle of attack. For our simple analysis, we shall neglect the rotation of the Earth and any trajectory contributions within the Earth's atmosphere. Under these assumptions, the trajectory of the missile is elliptical with one of the focal points of the ellipse lying at the center of the Earth. The range of the missile can be approximated by the *free-flight range angle*, Ψ, which is the angle subtended by the ends of the trajectory not effected by atmospheric drag. Of course the range of the missile is simply:

$$\text{Range} = r_e \Psi \tag{2.24}$$

if Ψ is measured in radians. The burnout point in Figure 2.8 corresponds to the point where the last stage of the missile expires – beyond this point only gravity acts on the missile as it coasts through the vacuum of space. Typical burnout altitudes lie 250–500 miles above the Earth's surface.

Of course we desire a means to predict the burnout velocity, v_{bo}, in terms of the range, burnout altitude, and a parameter ϕ_{bo}, which is the local *flight path angle* at the instant the last stage burns out. Bate *et al.* (1971) provide a derivation of the governing equations for this situation assuming only gravity acts on the coasting projectile. They present results in terms of the velocity ratio:

$$Q_{bo} = v_{bo}^2 r_{bo}/\mu = (v_{bo}/v_c)^2 \tag{2.25}$$

which is just the square of the burnout to local circular velocity ratio. If $Q_{bo} = 1$ the missile would have enough energy to remain in orbit halfway $\Psi = 180°$ around the Earth. The *free-flight range equation* provides a relation between Q_{bo} and the range angle:

$$\cos(\Psi/2) = \frac{1 - Q_{bo}\cos^2\phi_{bo}}{\sqrt{1 + Q_{bo}(Q_{bo} - 2)\cos^2\phi_{bo}}} \tag{2.26}$$

To solve for the flight path angle, we may use:

$$\sin(2\phi_{bo} + \Psi/2) = \frac{2 - Q_{bo}}{Q_{bo}}\sin(\Psi/2) \tag{2.27}$$

Now it turns out that the solution of Eq. 2.27 for given Q_{bo} and Ψ gives two possible flight path angles satisfying this condition. The smaller angle is called the *low-trajectory solution*, while the larger angle gives the *high-trajectory solution*. Long-range missiles will typically use the high trajectory to minimize drag losses in flight. If one is interested in the fast (shortest flight time) solution, the low trajectory is utilized.

Physically, one can visualize this situation by considering an ordinary garden hose. For a given point (within the range of the hose as determined by its "exhaust" velocity), we can choose a low or high trajectory for the water. As we move the target point away from the hose, we will have to increase our low trajectory angle and decrease our high trajectory angle until we reach the point (maximum range) where both solutions converge to a single angle. If we move the point beyond this location, we will be out of the range of the hose. For a ballistic missile, the maximum range condition is given by:

$$Q_{bo} = \frac{2\sin(\Psi/2)}{1 + \sin(\Psi/2)} \quad \text{maximum range} \tag{2.28}$$

We can use either the combination of Eqs. 2.26 and 2.27 or 2.26 and 2.28 to solve for Q_{bo} and ϕ_{bo} for a given range requirement. The actual burnout velocity, corresponding to the total velocity increment (including losses) input by the propulsion system, can be determined from Eq. 2.25 provided that the burnout altitude is given. The maximum altitude attained during the flight is then given by:

$$\text{Alt}_{max} = \frac{r_{bo}}{2 - Q_{bo}}[1 + \sqrt{(1 + Q_{bo}(Q_{bo} - 2)\cos^2\phi_{bo})}] - r_e \tag{2.29}$$

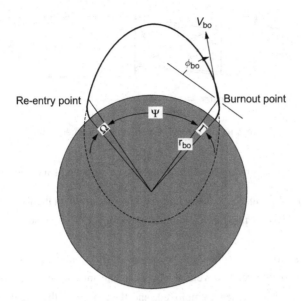

Figure 2.10 Geometry of the ballistic missile trajectory.

Those interested in a more in-depth treatment of this subject are referred to Bates *et al.* (1971), or other suitable astrodynamics texts. For preliminary design purposes, the above procedure will give adequate results.

2.5 MISSION REQUIREMENTS FOR INTERCEPTORS

An interceptor is a unique type of rocket in that it is designed to close in on a moving target. Air-to-air and surface-to-air missiles are typical examples of interceptors. In addition, space-based missiles have been considered for an interceptor function due to the ability to address large regions of Earth from LEO. A schematic of this type of missile that employs a kinetic kill vehicle (KKV) that is boosted by two axial propulsion SRMs is shown in Figure 2.10. The KKV provides highly responsive lateral thrusters to coordinate a direct hit of the device with the target such that the kinetic energy of the collision is the destructive force. Note that since drag is not an issue in this design, streamlined shapes need not be employed.

Interceptors are typically one- to three-stage devices with a high degree of maneuverability (at least on the terminal stage). Propulsion systems for interceptors must be able to withstand high acceleration loads during maneuvers, and must be ready to respond at a moment's notice. For these reasons, solid rocket motors and liquid rocket systems using storable propellant systems are optimal. In addition, liquid systems using "gel propellants" have been considered due to the inherent safety advantages of a "semi-liquid" system. The gels are made by adding chemical agents to current formulations to effectively increase the propellant viscosity by large amounts. Chapter 8 provides additional discussion on gelled propellants.

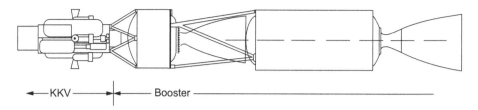

Figure 2.11 Schematic of a space-based interceptor.

Some of the major considerations in designing an interceptor include:

- *Efficient packaging.* Since many interceptors are carried on the wings or in payload bays of aircraft, designs must utilize a minimum amount of space so as to minimize aircraft drag.
- *Target maneuverability and guidance systems.* The agility required for the interceptor is a function of the relative velocity between the missile and its target, the particular intercept guidance scheme employed, and the agility of the target. Some interceptors are capable of lateral accelerations of the order of 100 g's. These high acceleration levels have a definite impact on the design of the propulsion system.
- *Stage burning time.* Since missile velocity (and hence maneuverability) is decreased after propulsion system burnout, engagements must be carried out within a short duration of this event. For space-based interceptors which cannot use aerodynamics to aid maneuvering, additional "divert engines" must be utilized when the main propulsion system shuts down.
- *Safety.* Since many interceptors are carried aboard aircraft and ground vehicles, they must be insensitive to inadvertent shocks and should not explode if punctured by other projectiles. Safety requirements have a large impact on the type of propellant utilized and have motivated designers to consider the gel propellants discussed above.

A mission requirement for an interceptor can be thought of as the traversing of a given range in a given duration of time. Determining a Δv requirement for interceptors is quite difficult because there are infinitely many Δv values that can give a required range over a certain duration of time. Figure 2.11 illustrates this point considering a number of alternatives and neglecting aerodynamic drag. The range is the integral of the missile velocity over the mission time, t_m, so while all the curves have the same average velocity, the missile burnout velocity (Δv requirement) increases as the coast time decreases.

For this reason, it is desirable to accelerate to high velocities in a short period of time with the first stage of a multi-stage interceptor. This approach will tend to minimize the Δv provided by the propulsion system and maximize the range for a given mission time. Including aerodynamic drag further complicates the situation and in general full-trajectory simulations are required even for the

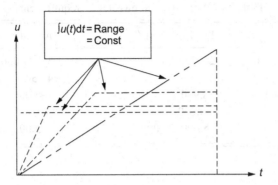

Figure 2.12 Mission requirements for interceptors.

early stages of a missile that is functioning in the atmosphere. A series of engagements must be simulated for a variety of initial launch platform and target locations and velocity states and the resulting performance of a certain design can be assessed in this fashion. For the terminal intercept stage, the desire to burn out early must be balanced against maneuverability requirements near the intercept point.

To illustrate some of the tradeoffs, let us consider a simple case where the missile range, z, and the mission time are specified and atmospheric drag is neglected. Under these circumstances, the horizontal range of the missile can be expressed:

$$z = \int_0^{t_b} v(t) \ dt + \Delta v(t_m - t_b) \tag{2.30}$$

since $\Delta v = v(t_b)$ by definition. As an example, let us assume that the velocity during the firing is simply $v(t) = kt$ where k is some constant. In this case, the range of the missile is:

$$z = \frac{k}{2}t_b^2 + kt_b(t_m - t_b) \tag{2.31}$$

If k is known, then we could solve for the burning time of the propulsion system:

$$t_b = t_m \pm \sqrt{t_m^2 - 2z/k} \tag{2.32}$$

where we must choose the negative sign to have $t_b < t_m$. If t_b is known (or varied parametrically), the expression for k becomes:

$$k = \frac{z}{t_b(t_m - t_b/2)} \tag{2.33}$$

which corresponds to the curves shown in Figure 2.10. As noted previously, the Δv requirements for atmospheric interceptors really necessitate complete trajectory simulations. In Chapter 3, we summarize the approach used in this case.

Table 2.2   Some typical mission Δv requirements

Mission	Ideal Δv (ft/s)	Actual Δv (ft/s)
Satellite orbit about Earth	26–35,000	30–41,000
Escape from Earth	36,700	42,500
Earth to Moon and back	88,900	105,000
Space-based interceptor	10,000	10,000
Launch to LEO	25,000–27,000	29,000–30,000

2.6 SUMMARY

The techniques described in the previous sections enable the designer to get a quick preliminary estimate of the velocity increment required by the propulsion system. This preliminary estimate ignores drag and gravity losses present during the firing, and therefore is called an *ideal* Δv. Detailed trajectory calculations (discussed in Chapter 3) would be required to further refine the design. A summary of ideal and actual Δv requirements for several missions is provided in Table 2.2. Note that for space-based interceptors, drag and gravity losses are not present, and the ideal and actual Δv values are the same.

FURTHER READING

Barrere, M., Jaumotte, A., Frais de Veubeke, B., and Vandenkerckhove, J. (1960) *Rocket Propulsion*. Elsevier Publishing Company.

Bate, R. R., Mueller, D. D., and White, J. E. (1971) *Fundamentals of Astrodynamics*. Dover Publishing Company.

Edelbaum, T. S. (1961) "Propulsion Requirements for Controllable Satellites," *ARS Journal*, 31(8), August.

Hammond, W. E. (1999) *Space Transportation: A Systems Approach to Analysis and Design*. AIAA.

Larson, W. J. and Wertz, J. R. (1992) *Space Mission Analysis and Design*. No. DOE/NE/32145–T1. Microcosm, Inc., Torrance, CA (US).

Larson, W. J., Henry, G. N. and Humble, R. W., eds. (1995) *Space Propulsion Analysis and Design*. McGraw-Hill.

Sellers, J. J. *et al.* (2000) *Understanding Space: An Introduction to Astronautics*. Primis.

Wood, K. D. (1964) *Spacecraft Design*. Johnson Publishing Co.

HOMEWORK PROBLEMS

2.1   Assume that you are a deer hunter sitting around a mountain campfire recalling the day's events. Your companion starts in with one of those "big fish stories" and claims that she once shot a deer on a ridge over 4000 feet above her location in the valley. Knowing the muzzle velocity of her gun is 500 ft/s, either confirm or refute this story.

2.2 Some interceptors make use of a two-pulse design (with a coast phase in between) in order to optimize performance. Let's assume our interceptor is launched horizontally from an aircraft flying 200 m/s and that each pulse provides 500 m/s Δv. Furthermore, assume that both pulses are 20 seconds in duration and that the velocity history during the pulse is given by:

$$v = v_0\left(1 + a\left(t - t_{ig}\right)^2\right)$$

where a is a constant, v_0 is the velocity at the start of the pulse, and t_{ig} is the time the pulse is initiated. During the coast phase, drag acts to reduce the missile velocity according to:

$$v = \frac{v_0}{1 + 0.01\left(t - t_c\right)^2}$$

where t_c is the time of initiation of the coasting phase. Finally, our mission requires that we cover a range of 30 km in the one minute flight time of the device.

(a) Assuming the first pulse is fired as the missile is launched from the aircraft, determine the coast time required to accomplish the mission.

(b) Sketch the velocity history for this flight, noting velocity values at the start and end of each flight segment.

(c) Sketch range as a function of time for the flight, noting velocity values at the end and start of each flight segment.

2.3 In 1997, the Mars Global Surveyor was captured into a highly elliptic orbit about the planet. Over the next few months, the spacecraft used the Martian atmosphere as an aerobrake to eventually arrive at a low-altitude circular orbit about the planet. Assuming that the initial elliptic orbit had an apogee of 30,000 km and a perigee of 50 km, estimate the Δv saved by using the aerobrake maneuver, i.e. what Δv would have been required if Mars didn't have an atmosphere?

2.4 Suppose we wish to resupply the new space station using Boeing's Sea Launch Vehicle. Using a Sea Launch, we don't have any inclination changes, and we can launch from the equator to minimize Δv.

(a) Assuming the space station is in a 200 km circular orbit, determine the total Δv required using a conventional launch vehicle approach.

(b) An engineer has an idea to resupply the station using a ballistic trajectory in which the payload is deposited on the station as the missile reaches apogee. Assuming a burnout flight path angle of 30° and a burnout altitude of 10 km, determine the required burnout velocity of the missile in this case.

(c) How much Δv is saved by using option (b)? Is this a good idea? You may wish to calculate the horizontal velocity at apogee:

$$v = \sqrt{2\left(\frac{\mu}{r_a} - \frac{\mu}{r_{bo}}\right) + v_{bo}^2}$$

Is there an optimal flight path angle for the ballistic missile option?

2.5 Typical ballistic missiles have a burnout height of 100 miles with a maximum range of 8000 miles. How much extra Δv capability would be required to use the missile as a launch vehicle to orbit payloads in polar orbits (90° inclination) at the 100 mile altitude?

2.6 A small interceptor is launched horizontally from an aircraft flying at $M = 0.8$ at an altitude of 40,000 ft. The rocket motor operates over a one second duration after release from the aircraft. The missile velocity history during this time is given by

$$v = v_0\left(1 + 2\sin\left(\frac{\pi t}{2}\right)\right) \qquad 0 \le t \le 1$$

where v_0 is the aircraft velocity at the time of release. Guidance experts indicate that the missile will have adequate agility to intercept its moving target as long as its velocity is at least 1000 f/s. Assume that we can neglect drag during the brief boost phase. During the coast phase, assume the following:

Missile mass = 300 lb.
Average drag coefficient = 0.2 (i.e. drag *not* negligible during coast)
Reference area = 50 in^2

Determine:

(a) the range at the end of the boost phase; and
(b) the total range of the missile.

2.7 In this problem, consider an orbital transfer vehicle raising an orbit of radius r_1 to a higher orbit of radius r_2 with no plane change. We would like to determine the Δv penalty (or benefit) in using a spiral trajectory versus a standard Hohmann transfer for this mission. Let Δv_S and Δv_H represent the *total* Δv required for the spiral and Hohmann transfers, respectively.

(a) Show that the ratio $\Delta v_S/\Delta v_H$ depends only on the radius ratio r_1/r_2 and determine the form of that dependence.
(b) If $r_1/r_2 = 0.2$, how much Δv penalty (benefit) do we incur by employing the spiral transfer (as compared to Hohmann transfer)

2.8 In the Apollo lunar missions, the lunar lander had to rendezvous with the command module which was orbiting the Moon at an altitude of 80 nmi (1 nautical mile (nmi) = 6076 ft). Estimate the Δv required to accomplish this mission assuming that no plane change is required and that gravity losses amount to 20% of the ideal velocity increment.

2.9 A ballistic missile has a maximum range of 3000 km when launched on the Earth. If we use an identical missile on Jupiter, what will its maximum range be? You may assume that the burning time of the missile is so short that it effectively burns out at the surface of the planet (i.e. zero altitude).

2.10 A ballistic missile is launched so as to maintain a 70° angle with respect to the local horizon during the burn. The acceleration history for the vehicle is given by

$$a = 10 + 0.8t \quad 0 \le t \le 50 \text{ seconds}$$

where a is in m/s^2. Determine the range of this missile.

2.11 The US launches the bulk of its satellites from Cape Kennedy (inclination 28.5°) while the French launch the Ariane from French Guiana (inclination 3°).

  (a) Determine the advantage the French have in terms of lower ideal velocity requirements if an 85 nmi equatorial orbit is desired.

  (b) How do the two launch sites compare for a 90 nmi orbit which passes over the center of the continental US (inclination 35°).

2.12 Compare the ideal Δv between a Hohmann and spiral transfer from LEO to GEO. Assume the LEO orbit is circular at 150 km altitude for your calculation.

2.13 An interceptor has a velocity history:

$$v(t) = at - bt^2 \quad a, b \text{ constants} \quad t \le t_b$$

  (a) If the total mission time is t_m, determine the range of the interceptor, z, in terms of a, b, t_b, and t_m.

  (b) If $a < 2bt_m$, determine the burning time that maximizes the range.

  (c) What is the maximized range, z_{max}, corresponding to the optimized burning time derived in part (b) above?

2.14 As many of you know, airbreathing engines (scramjets, etc.) have been proposed as a means to develop a single-stage-to-orbit (SSTO) vehicle. The airbreather operates over a portion of the trajectory at which point the vehicle transitions to a rocket-propelled mode. The other option is a pure rocket vehicle. Assume a Scramjet engine can accelerate the vehicle to Mach 15 at 150,000 feet, and that a 90 nmi circular orbit is desired (no plane change). What fraction of the ideal energy is constant? Suppose the mass varies with time (as it really does). How would this fraction be effected?

2.15 What Δv is required to travel from Earth's orbit (around the sun) to Mars' orbit? The radii from the Sun to Earth and Mars are 1.5×10^8 km and 2.28×10^8 km, respectively.

2.16 Determine the burnout flight path angle and burnout velocity for a ballistic missile with a maximum range of 5000 miles. Plot your results as a function of the height at burnout (h_{bo}) for 10 miles < h_{bo} < 100 miles.

2.17 Estimate the Δv required to reach a 100 nmi polar orbit assuming southerly launch from Vanderberg AFB.

2.18 A very important orbit utilized by many satellites has a period of 24 hours. This orbit, called a geostationary earth orbit (GEO), is such that a satellite remains fixed with respect to a point on the ground (obviously desirable for communications). Determine the altitude of this orbit in kilometers and statute miles.

2.19 Prove that the ideal Δv required to travel from Earth to the Moon and return to Earth is 88,900 ft/s. What is the ideal value assuming we can use the Earth's atmosphere as an "aerobrake" during the descent phase in the return to Earth?

2.20 The Titan Launch vehicle (replaced by the Delta IV and/or Atlas V in the current US fleet) boosts an IUS and payload to a 85×90 nmi equatorial parking orbit. Assuming the first stage of the IUS fires from the apogee of this orbit, determine the following:

   (a) The Δv required by the IUS first- and second-stage burns required to place the satellite in GEO orbit.

   (b) The ideal Δv input by Titan to attain the parking orbit including contributions for the plane change to equatorial orbit.

2.21 Suppose we wish to replace the IUS with an electric OTV. What Δv would be required by this vehicle assuming the same initial and final orbits?

2.22 What ideal Δv is required to travel from a LEO of 200 km to Mars and back? Include Δv budget to actually land on the Martian surface.

2.23 An interceptor is required to travel a distance z in a mission time, t_m. Because the rocket motor on the interceptor has a thrust which decreases with time, the velocity of the missile during rocket operation can be expressed:

$$v(t) = K_1 t^{0.5} + K_2 \quad 0 < t < t_b$$

where K_1 and K_2 are constants and t_b is the motor burning time.

   (a) Derive an expression for z in terms of K_1, K_2, t_m, and t_b.

   (b) If the missile is initially at rest, determine values for K_1 and K_2 in terms of z, t_m, and t_b

   (c) Suppose $K_1 = 30$ ft/s$^{1.5}$, $K_2 = 100$ ft/s, $t_m = 30$ s, and $t_b = 9$ s. Determine the range z at $t = t_m$ and the portion of this range attained during the coasting period.

2.24 An ambitious student wishes to launch a rocket into low Earth orbit from his/her backyard. Nearby power lines inhibit an easterly launch, so the student elects to launch in a westerly direction. The I_{sp} of the vehicle is 300 s. Assuming the launch site is at a latitude of 35 degrees, and a 200 km circular orbit is desired, what ideal velocity increment would be required to complete this mission?

2.25 We would like to modify our analysis of mission requirements for Earth escapes to take into account dissipation due to atmospheric drag. Assume we can write the average drag during the ascent as

$$\overline{D} = c_D \frac{\overline{\rho}}{2} \left(\frac{u_z}{2}\right)^2 A$$

where

$$\overline{D} = \text{average drag force}$$
$$c_D = \text{drag coefficient}$$
$$\overline{\rho} = \text{average atmospheric density during ascent}$$
$$u_z/2 = \text{average velocity during ascent}$$
$$A = \text{vehicle cross-sectional area}$$

(a) Using this expression, derive a relation (analogous to Eq. 2.17) for the impulsive velocity to attain a height z (u_z) in terms of the parameters above, the body mass (m) and known characteristics of planet Earth.

(b) How much additional impulsive velocity (above the case with no drag) will be required if $C_D\overline{\rho}A/m = 1 \times 10^{-5}\,\mathrm{m}^{-1}$ and $z = 100$ km?

2.26 What Δv is required to escape Jupiter's gravitational field assuming one starts on the surface of the planet and drag is neglected?

2.27 The initial portion of the return trajectory for a future Mars probe would involve a launch from the surface of Mars to a temporary circular orbit 150 miles above the planet's surface. Assuming the probe is departing from the equator, estimate the Δv required to accomplish this portion of the mission. [*Hint*: The length of a Martian day is 97.5% that of a day on Earth.]

2.28 The "*g-t*" loss for a typical launch from Earth is roughly 5000 ft/s. Can you estimate the loss for a launch from the surface of Venus? Carefully state all assumptions required to make this estimate.

2.29 What altitude would be attained by a projectile launched vertically at half of Earth's escape velocity?

2.30 We have had considerable discussion on the Δv estimates for launch vehicles. The highly simplified result that $\Delta v = v_c - v_{\mathrm{surf}}$ (with v_c evaluated at Earth's surface) works okay for orbits very near the surface of the Earth (neglecting gravity and drag losses, of course). For Kennedy Space Center, this technique gives Δv of 7.5 km s^{-1}.

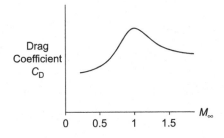

Figure 2.13 Diagram for Problem 2.32.

(a) How does this estimate compare with a Hohmann transfer from Kennedy Space Center to a 150 km orbit inclined at 28.5°?

(b) Which approach do you prefer and why? Do you have a better suggestion?

2.31 Typical ballistic missiles have a burnout height of 100 miles with a maximum range of 8000 miles. How much extra Δv capability would be required to use the missile as a launch vehicle to orbit payloads in polar orbits (90° inclination) at the 100 mile altitude?

2.32 A solid rocket motor is to be utilized for propulsion on an air-launched missile whose drag characteristics are shown in Figure 2.13. Assuming the missile is launched at $M_\infty = 0.9$, discuss the implications of the SRM grain design on the missile's performance (for a description of grain design, see Chapter 7). Which grain design should be used (regressive, neutral, or progressive), assuming that each one provides the same total impulse?

CHAPTER 3 TRAJECTORY ANALYSIS AND ROCKET DESIGN

In Chapter 4, we will discuss how to determine propulsion system performance (I_{sp}) based on chamber, nozzle, and propellant characteristics. In Chapter 2, we discussed simple techniques to determine the mission requirements (Δv) for various rocket-propelled vehicles. The one piece of the puzzle still missing is the physical size and mass of a vehicle which has known payload mass, Δv, and I_{sp} values. This information is normally obtained through analysis of the *rocket trajectory*. In the simplest sense, the overall forces acting on a rocket can be integrated into an expression that relates vehicle mass to its performance and mission requirement – this famous expression is called "*The Rocket Equation*," or alternatively "*The Tsiolkovsky Equation*" after Konstantin Tsiolkovsky, the Russian theoretician who is often credited with the original derivation of the relationship. Since the rocket equation serves as a mechanism to compute rocket mass, it also serves as the first step in a preliminary design since mass is related to size and sizing is what preliminary design is all about.

While we admit the above procedure is a bit of a simplification, much can be learned by simple trajectory analysis of a vehicle with a known mission requirement. Detailed designs must be developed by considering interactions of the trajectory and the mission requirements. For example, the Δv for a launch vehicle can vary drastically as the trajectory changes even if we consider the final orbit to be fixed. Therefore, the detailed design must certainly consider these variations, through a process called *trajectory optimization.*

In Section 3.1 we will consider simple vertical trajectories starting at the surface of the Earth and will derive the *rocket* or *Tsiolkovsky* equation. In Section 3.2 we assess the effects of burning time and acceleration levels on vertically launched rockets. In Section 3.3 we will show why staging is desirable for rockets with large Δv requirements. Finally, in Section 3.4 we will discuss the general form of the trajectory equations suitable for more detailed analysis, and we provide a few words of caution for users of the rocket equation.

3.1 VERTICAL TRAJECTORIES – THE ROCKET EQUATION

Consider the rocket shown in Figure 3.1 at some arbitrary point within a vertical trajectory. The vehicle trajectory equation is fundamentally derived from Newton's second law:

$$\Sigma \text{ forces} = m \, \frac{dv}{dt} \tag{3.1}$$

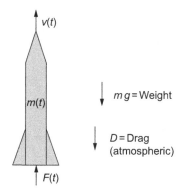

Figure 3.1 Forces on a rocket in vertical flight.

which states that the sum of forces on a body must be equal to the instantaneous mass times the change in velocity of the body. Referencing Figure 3.1, we can see that drag (D), the rocket's weight (mg), and the thrust ($F = \dot{m}gI_{\mathrm{sp}}$) are the only three forces acting on the body. Since we are assuming a vertical trajectory here, all forces are assumed to be aligned with the axis of the vehicle that is coincident with the gravitational force vector. Side forces stemming from such items as wind gusts will be considered in a generalized trajectory discussion in Section 3.4. Using the simple force balance in Figure 3.1, Eq. 3.1 becomes:

$$-mg - D + \dot{m}gI_{\mathrm{sp}} = m\frac{\mathrm{d}v}{\mathrm{d}t} \tag{3.2}$$

Note that since the rocket may attain high altitudes, in general the gravitational acceleration g must be considered a variable (i.e. a function of altitude or time).

Now if the rocket is expelling only propellant gases, then $\dot{m} = -\mathrm{d}m/\mathrm{d}t$ since the vehicle mass must be decreasing at the rate of propellant expulsion. Therefore, Eq. 3.2 becomes:

$$\mathrm{d}v = -gI_{\mathrm{sp}}\frac{\mathrm{d}m}{m} - \frac{D}{m}\mathrm{d}t - g\,\mathrm{d}t \tag{3.3}$$

which is the governing equation for a vertical trajectory. We integrate Eq. 3.3 between initial ($()_{\mathrm{o}}$) and final ($()_{\mathrm{f}}$) states to give:

$$\Delta v = g_{\mathrm{e}}\,I_{\mathrm{sp}}\,\ln(m_{\mathrm{o}}/m_{\mathrm{f}}) - \int_0^{t_{\mathrm{b}}} g\,\mathrm{d}t - \int_0^{t_{\mathrm{b}}} (D/m)\,\mathrm{d}t \tag{3.4}$$

Equation 3.4 is the basic equation relating rocket performance, (I_{sp}), mission requirements (Δv), and system masses (m_{o}, m_{f}). Propulsion scientists refer to this result as *the rocket equation* or *the Tsiolkovsky equation.* This relationship serves as the basis for rocket design as for a given mission (Δv) and rocket performance (I_{sp}) the relationship is telling us something about the *mass ratio*, $m_{\mathrm{o}}/m_{\mathrm{f}}$, required to attain the prescribed amount of velocity gain. Note that the result in

Eq. 3.4 holds for constant I_{sp}. If I_{sp} is varying in time, the more fundamental relations (Eq. 3.2) would need to be integrated numerically (see Section 3.4).

The first term on the left-hand side of Eq. 3.4 represents the velocity gain, which is sensed by the payload. The first term on the right-hand side of Eq. 3.4 is the *ideal velocity gain*, Δv_{id}, imparted by the propulsion system:

$$\Delta v_{id} = g_e\, I_{sp} \ln(m_o/m_f) \tag{3.5}$$

If we were in the vacuum of interstellar space, the propulsion system would provide this ideal level of velocity gain; however, for most missions we have losses due to gravity, drag, and other factors. The second term on the right-hand side of Eq. 3.4 is the *gravity* or *"g-t" loss*. This loss stems from the fact that during the rocket burn, the vehicle is working to overcome the acceleration due to gravity. It is sometimes difficult for students to understand how gravity can represent a loss in the system, but think about the case where the rocket is hovering. In this case, there is no velocity gain imparted to the payload and there is no drag, so any propellant expended in countering the rocket's current weight is due to the countering of the local gravitational field. We can refer to the gravity term as Δv_g, i.e.:

$$\Delta v_g = \int_0^{t_b} g\, dt \tag{3.6}$$

The last term on the right-hand side of Eq. 3.4 is called the *drag loss*, and it stems from the fact that the vehicle acceleration is retarded by the action of drag when in a planet's atmosphere. As in the case of aircraft, we write the vehicle drag:

$$D = -C_D A_f \frac{\rho v^2}{2} \tag{3.7}$$

where A_f is the reference area (generally the frontal area for rockets), and C_D is the drag coefficient. Note that since both ρ and v are functions of time, the drag will vary as the rocket ascends through the atmosphere. Typically, the drag (and all other aerodynamic forces or torques) is a maximum at "max. q" where the dynamic pressure $\rho v^2/2$ is maximized. The rocket drag coefficient is a function of geometry, local Mach and Reynolds numbers, and angle of attack. For large rockets, turbulent boundary layers exist over the bulk of the body, so Reynolds number influence is rather minor. For this reason, drag data is typically presented as a function of Mach number for various angles of attack as shown in Figure 3.2.

Equation 3.7 indicates that the drag of the rocket is proportional to an area, or a characteristic length squared (L^2). However, we see from Eq. 3.3 that the important factor is D/m. Since the mass of the rocket is proportional to L^3, we can write:

$$D/m \propto L^2/L^3 = 1/L \tag{3.8}$$

which indicates that drag losses become less important for larger rockets. As with the gravity loss, we can define a velocity change due to drag:

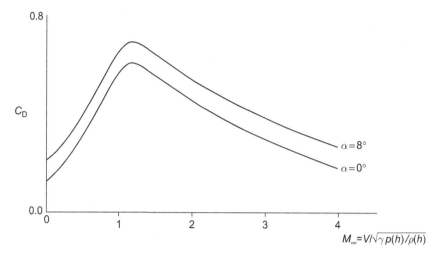

Figure 3.2 Typical drag curves as a function of Mach number.

$$\Delta v_{\mathrm{D}} = \int_0^{t_b} (D/m)\, \mathrm{d}t \tag{3.9}$$

Because we considered a vertical trajectory with thrust aligned with the vehicle axis, we missed out on one other loss mechanism in Eq. 3.4. In general, there are *steering losses* that result from the fact that the thrust vector may not be aligned with the vehicle axis when the vehicle is maneuvering. Think of the case where the nozzle of the rocket is gimbaling to provide control of the vehicle – some component of the thrust vector will then act perpendicular to the vehicle axis and therefore will not contribute to vehicle acceleration along its main axis. In general, steering losses are much lower than either drag or gravity losses – at least for launch vehicles. Steering losses can be much higher for missiles as they may need to vector thrust substantially to intercept a target. As with the other loss mechanisms, we can think about a velocity loss, Δv_s to account for steering losses.

If we combine Eqs. 3.4–3.6 and 3.8 an alternate form of the rocket equation can be written:

$$\Delta v_{\mathrm{id}} = g_e\, I_{\mathrm{sp}} \ln(m_0/m_f) = \Delta v + \Delta v_{\mathrm{g}} + \Delta v_{\mathrm{D}} + \Delta v_{\mathrm{S}} \tag{3.10}$$

where it becomes obvious that the ideal velocity gain imparted by the propulsion system must account for the actual velocity gain to be imparted to the payload plus all losses from gravity, drag, and steering. In other words, we must size the propulsion system to accomplish the desired velocity gain *and* to overcome any losses obtained in achieving that velocity gain. For a given I_{sp}, the lightest system results from the one that achieves the mission objective with minimum losses and the field of trajectory shaping/design attempts to achieve just that (see Section 3.3).

Now we have said that the rocket equation (either Eq. 3.4 or 3.10 version) serves as the basis for rocket design, yet it is certainly not apparent from this simple result. The key factor that gives information about the size of the rocket (or its parts) is hidden inside the mass ratio m_0/m_f that

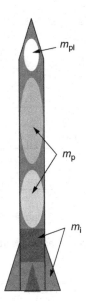

Figure 3.3 Functional decomposition of masses in rocket into payload (top), propellant (middle), and inert (bottom) masses.

represents the ratio of initial to final masses of the entire rocket. Recall in Chapter 2 we said that one can think of the rocket as being comprised of payload, inert, and propellant masses as illustrated in Figure 3.3. The propulsion system is comprised of propellant (m_p) and inert (m_i) masses and its structural efficiency is measured by the *propellant mass fraction, λ*:

$$\lambda = \frac{m_p}{m_p + m_i} \tag{3.11}$$

In general, we tend to know something about λ; at a minimum we can assess what was achieved historically in prior rocket programs. We know that as systems get bigger, their structural efficiency increases due to the "cubed-squared law." This law simply states that since the volume of a system goes as the cube of some linear dimension and the surface area of the system goes as the square of some linear dimension, then its packaging efficiency tends to rise as the system gets bigger. These trends are borne out by actual data on rocket systems as shown in Figures 3.4–3.6. Figures 3.4 and 3.5 show λ trends for solid rockets, while Figure 3.6 has λ values for several current launch vehicles that include both solid and liquid propulsion stages.

As Figure 3.4 shows, very small systems tend to have low propellant mass fractions. Here the cubed–squared law comes into play as well as the fact that we may be reaching minimum gauges in the thicknesses of some of the parts of the system, i.e. we can't make it as thin as it needs to be to retain the design loads. Things like batteries and electronics do not typically scale directly with propellant mass and this is another reason that smaller systems tend to suffer in mass fraction. Larger solids, shown in Figure 3.5, can be very efficient devices with mass fractions well above 0.9.

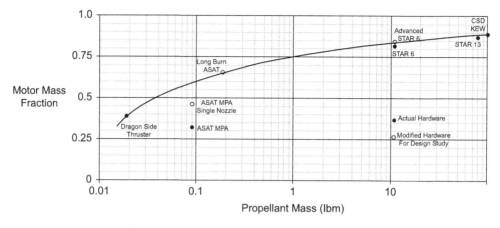

Figure 3.4 Solid rocket propellant mass fractions – small motors.

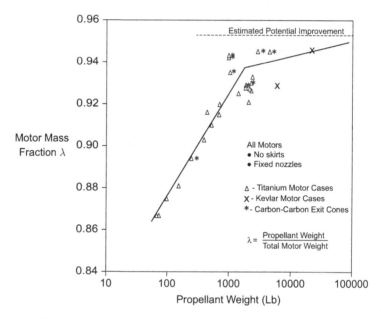

Figure 3.5 Solid rocket propellant mass fractions – larger motors.

As a word of caution, one must always be mindful of the fact that mass fractions are reflective of the manufacturing and materials technologies used within the device in question, and that they may or may not include all of the system inert components. For example, Figure 3.5 was developed for solid motors, but does not include the weight of any thrust vector control system that may or may not be required for steering of the rocket. One must be judicious in the use of these data, but the huge advantage they provide in allowing rapid system sizing still makes the approach attractive for conceptual and preliminary designs.

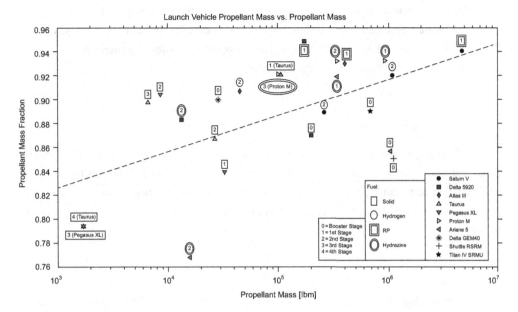

Figure 3.6 Mass fractions of several launch vehicle stages.

Figure 3.6 shows results for a number of launch vehicle stages and the large diversity of systems tends to obscure trends with increased propellant mass and comparisons between the various technologies. In general, storable propellants (propellants that are liquids at ambient pressure and temperature) tend to give higher λ systems than cryogenic fluids, and solid stages generally provide higher λ values than liquids. However, the exact loads on the structure, the type of trajectory flown, and the overall mission requirements can lead to large differences and obscure these presumed trends. In spite of these facts, historical λ data are often used in rough system sizing – let's investigate how to do this.

Oftentimes, we wish to express the mass ratio in terms of the inert, propellant, and payload (m_{pl}) masses. We can do this by incorporating the definition of propellant mass fraction from Eq. 3.11 into the relationship for mass ratio:

$$\text{MR} = \frac{m_o}{m_f} = \frac{m_{pl} + m_p + m_i}{m_{pl} + m_i} = \frac{\lambda m_{pl} + m_p}{\lambda m_{pl} + m_p(1 - \lambda)} \tag{3.12}$$

Now if we think of a situation where the payload mass and λ values are known, we can rearrange Eq. 3.12 to immediately solve for the amount of propellant required to achieve the mission objectives:

$$m_p = m_{pl} \frac{\text{MR} - 1}{\text{MR} - (\text{MR} - 1)/\lambda} \tag{3.13}$$

Equation 3.12 is hugely useful in the rapid sizing of a system for which mission requirements, payload mass, and propulsion system performance are known. The general process is as follows:

1. Pick a propulsion system and estimate λ and I_{sp} for the technology you select.
2. Pick a mission and estimate any losses or be given a Δv_{id} from mission designers.
3. Pick a payload mass or be given a requirement from mission designers.
4. Solve Eq. 3.10 for the required mass ratio:

$$\mathrm{MR} = e^{\Delta v_{id}/(g_e I_{sp})} \tag{3.14}$$

5. Use Eq. 3.13 to solve for the required propellant mass.

This simple process allows one to begin to size a rocket vehicle. From the required propellant mass and its density, once can determine the volume/size of propellant tanks. Given a burning time and chamber pressure, one can then start to size combustion chambers, valves, and plumbing systems. For example, since the average mass flow is m_p/t_b, we may determine the nozzle throat area:

$$A_t = \frac{m_p c^*}{g_e t_b \, p_c} \tag{3.15}$$

where c^* is the characteristic velocity, which is a propellant combustion characteristic defined in Chapter 4, and p_c is the chamber pressure the designer chooses for the rocket system. Choosing a nozzle expansion ratio that is suitable for the mission then allows one to provide a preliminary nozzle design. Using this information and other selected system pressures, the nozzle, combustion chamber, and other individual components can be designed subject to the structural environments they must survive. Individual inert weights can then be assigned and the resultant total can then be compared with the λ value that was assumed in starting the process. At this point, a new iteration on the design can commence.

An important thing to note about Eqs. 3.13 and 3.14 is that the masses in the system increase exponentially with the required Δv. Figure 3.7 illustrates this result for a variety of different I_{sp} values. The sensitivity of system masses at high Δv levels makes these missions very difficult as a small change in performance can lead to large changes in delivered payload for a system that is well into its design and development. Remember from Chapter 2 that the rough requirement for an orbital mission is a Δv of 29,000 ft/s (17.7 km/s). The classic example of this problem is the single-stage-to-orbit design. With the propellants we have available (i.e. the values of I_{sp} we have available with today's propellants), a single-stage-to-orbit mission lies on a very steep part of these curves. Any growth in structural weight as the system is developed can lead to huge losses in payload. A solution to this problem is to employ multiple stages for ambitious missions like space access – see the Section 3.2.

One last item worth mentioning relative to our derivation of the rocket equation is something that some of you may have noticed relative to our "incorrect" application of Newton's second law in Eq. 3.1. As you may recall from your introductory physics course, Newton's law relates the sum of forces to the change in *momentum* (*mv*). Since both vehicle mass and vehicle velocity are changing with time, the rate of change of momentum must reflect this fact, and Eq. 3.1 should then

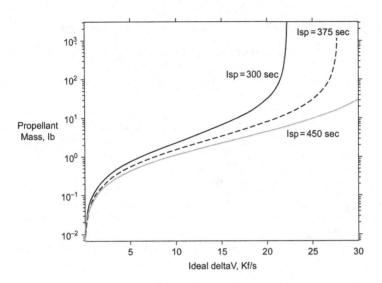

Figure 3.7 Propellant mass required to accelerate a 1 lbm payload, λ = 0.9.

be written Σ forces $= m \frac{dv}{dt} + v \frac{dm}{dt}$. Use of this form of the second law requires that we think about an inertial coordinate system relative to a stationary observer, and in this system the jet thrust must be reflective of the fact that the effective velocity exiting the nozzle from the observer standpoint is $v_e - v$. Using this factor in correcting the thrust as viewed in the inertial frame, it can be shown that one recovers the same result in Eq. 3.2 such that either avenue provides the same result.

To illustrate the power of the use of the rocket equation, we will work a simple example below. The example draws from performance analysis techniques developed in Chapter 4, so students will need to read this chapter before looking through the example. While it is a bit awkward, we thought it best to present this analysis in this chapter as it speaks to the ability of the designer to rapidly determine key sizes and weights.

Example: Preliminary Design of Interceptor Vehicle
Given:

$$m_{pl} = 100 \, lb$$

$$\Delta v_{id} = 3000 \, ft/s, \, \lambda = 0.8$$

$$c^* = 4900 \, ft/s$$

$$\gamma = 1.2$$

$$t_b = 3 \, s, p_c = 1000 \, psi \quad \text{(sea-level operation)}$$

Find: $A_t, A_e \, m_p, m_{in}$

To maximize performance at S/L, we want $p_e = p_a$ at $p_a = 14.7$ psi:

$$p_e/p_c = 0.0147 \Rightarrow \epsilon = 9 \; c_f = c_{f,\,\text{opt}} = 1.6$$

$$I_{\text{sp}} = \frac{c_{f,\,\text{opt}} \; c^*}{g} = 243 \text{ s}$$

$$\Delta v_{\text{id}} = g I_{\text{sp}} \ln MR = g I_{\text{sp}} \ln \frac{m_p + m_{\text{in}} + m_{\text{pl}}}{m_{\text{in}} + m_{\text{pl}}}$$

$$\lambda = \frac{m_p}{m_p + m_{\text{in}}} \; m_{\text{in}} = m_p \left(\frac{1}{\lambda} - 1 \right)$$

Using g =3 2.2ft/s^2:

$$e^{\frac{\Delta v}{g I_{\text{sp}}}} = \frac{\frac{m_p}{\lambda} + m_{\text{pl}}}{m_p \left(\frac{1}{\lambda} - 1 \right) + m_{\text{pl}}} \; \overset{MR = \; e^{\Delta v / g I_{\text{sp}}} \; = 1.466}{\Longleftrightarrow} \; \left(m_p (1/\lambda - 1) + m_{\text{pl}} \right) MR = m_p / \lambda + m_{\text{pl}}$$

$$m_{\text{pl}} (MR - 1) = m_p \left(\frac{1}{\lambda} - \frac{m_p}{\lambda} + MR \right) m_p$$

$$= m_{\text{pl}} \frac{MR - 1}{(1 - MR)/\lambda + MR} = 52.8 \text{ lb}$$

$$m_{\text{in}} = m_p \left(\frac{1}{\lambda} - 1 \right) = 13.2 \text{ lb}$$

$$\dot{m} \cong \frac{m_p}{t_{\text{b}}} = 17.6 \text{ lb/s}$$

$$A_t = \frac{\dot{m} \; c^*}{g p_c} = 2.7 \text{ in}^2 D_t = 1.85 \text{ in}$$

$$A_e = \epsilon A_t = 24 \text{ in}^2 D_e = 5.50 \text{ in}$$

We can see here that, in just a few minutes, we can determine critical weights and sizes that allow the design process to proceed. Knowing the mass of propellant and its density, the propellant volume can then be determined, and the tanks or chambers required to house this volume begin to take shape.

3.2 BURNING TIME AND ACCELERATION EFFECTS

Clearly, in order the vehicle to liftoff at $t = 0$, we require:

$$\eta_0 = F/(m_o g_e) > 1 \tag{3.16}$$

where $\eta = a/g_e$ is the thrust-to-weight ratio measured in "g." Launch vehicles typically have initial acceleration values, η_0, between 1.1 and 1.5g.

The burning time of the rocket also is related to the rocket acceleration since reducing this parameter forces increases in propellant mass flow (and hence thrust) for a fixed propellant mass.

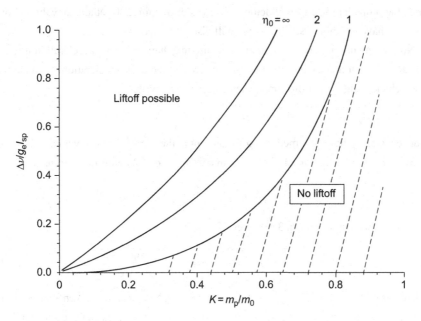

Figure 3.8 Performance of a vertically launched rocket.

Since the structure making up the vehicle and payload may have limits as to the acceleration loadings, we cannot reduce t_b indiscriminately even though reducing this parameter does minimize $g-t$ losses. For a constant flow rate rocket:

$$t_b = m_p/\dot{m} = m_p I_{sp}/F \tag{3.17}$$

We can get some analytic results that show trends in burning time/vehicle acceleration if we neglect drag. In addition, for trajectories relatively near the Earth's surface, we can use $g = g_e = $ constant with good accuracy. In this case, Eq. 3.4 becomes:

$$\Delta v = g_e I_{sp} \ln(m_o/m_f) - g_e t_b \tag{3.18}$$

Let's define the quantity $m_p/m_o = K$ and rewrite Eq. 3.15 using Eqs. 3.14 and 3.17. In this case, Eq. 3.18 can then be written:

$$\Delta v = -g_e I_{sp} \ln(1-K) - \frac{K \cdot g_e I_{sp}}{\eta_o} \tag{3.19}$$

which is valid only for a constant mass flow and no drag vertical trajectory. Figure 3.8 depicts solutions to Eq. 3.19 for various η_o values. $K = m_p/m_o$ has a maximum value of unity for a rocket made entirely of propellant.

Figure 3.8 demonstrates that for given system mass characteristics (K and I_{sp}), there is a limit to the amount of Δv even at infinite vehicle acceleration. Basically, the system is unable to accelerate the propellant itself to velocities in excess of this limit. To attain Δv values above this limit, we will have to employ staging as we will discuss in Section 3.3.

Now we mentioned that the vehicle structure may have an upper acceleration limit, which we will call η_{max}. For a constant thrust (flow rate) rocket, the peak acceleration occurs at burnout since the vehicle is lightest at this time. Therefore, we may write:

$$m_f \eta_{max} g_e = F \tag{3.20}$$

where once again we have assumed that g is constant at the Earth's surface value.

Since $m_f = m_o - m_p = m_o(1 - K)$, we can combine Eq. 3.20 with Eq. 3.16 to write:

$$\eta_{max} = \frac{\eta_o}{1 - K} \tag{3.21}$$

Substitution of this result into Eq. 3.19 gives:

$$\Delta v = -g_e \, I_{sp} \ln(1 - K) - \frac{K g_e I_{sp}}{(1 - K)\eta_{max}} \tag{3.22}$$

This relation is plotted for various η_{max} values in Figure 3.9. Note that for a given acceleration limit, Δv is maximized at some unique mass ratio combination. This maximum corresponds to the optimal payload/propellant combination such that $g - t$ losses are minimized. If more propellant is added beyond this limit, the vehicle becomes too heavy to attain this maximum Δv, if less propellant is loaded there is insufficient impulse to accelerate the lighter vehicle to the maximum attainable Δv.

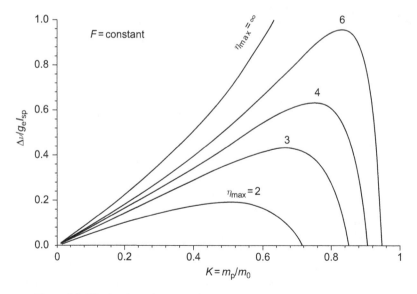

Figure 3.9 Effects of maximum vehicle acceleration on rocket performance.

3.2.1 Vertical Height Calculations

To estimate the height of the rocket at burnout, we will assume that we are launching from the Earth's surface so that $v_0 = h_0 = 0$ (where h = height). If we return to Eq. 3.4 and assume the integration has been performed from $t = 0$ to some arbitrary time $t = t$, this gives the instantaneous velocity as a function of the current rocket mass:

$$v(t) = -g_e I_{sp} \ln(m(t)/m_0) - g_e t \tag{3.23}$$

Integrating this velocity over $0 < t < t_b$ and noting that $m(t) = m_0 - \dot{m}$ for constant mass flow gives the height at burnout, h_{bo}:

$$h_{bo} = g_e I_{sp} t_b \left(\frac{m_f}{m_0 - m_f} \ln(m_f/m_0) + 1 \right) - \frac{g_e}{2} t_b^2 \tag{3.24}$$

which holds only in our special case of a vertical launch from the Earth's surface, neglecting drag and changes in g. Note that Eq. 3.23 can be rewritten as:

$$h_{bo} = g_e I_{sp} t_b \left(\frac{1-K}{K} \ln(1-K) + 1 \right) - \frac{g_e}{2} t_b^2 \tag{3.25}$$

To maximize the height of our rocket, there are various constraints we would employ. For instance, if we hold K fixed, we can differentiate Eq. 3.25 with respect to t_b and set the result equal to zero to give an optimum t_b and maximum burnout height:

$$h_{bom} = \frac{1}{2} g_e t_b^2, \quad K = \text{constant} \tag{3.26}$$

Unfortunately, to fix K and vary t_b, we are forced to vary payload mass and η_{max}.

A more realistic case would consider a fixed payload and inert mass but have $K = K(t_b)$ to account for changes in propellant mass with burn time. Note that this is more realistic since:

$$K(t_b) = \frac{m_p}{m_p + m_i + m_{pl}} = \frac{\dot{m} t_b}{\dot{m} t_b + m_i + m_{pl}} \tag{3.27}$$

We can differentiate h_{bo} with respect to t_b again, but this time we must include dK/dt_b terms since K is no longer constant. Setting the result equal to zero will give the optimum burning time for this situation. Plugging this result into Eq. 3.28 gives:

$$h_{bom} = g_e I_{sp}^2 [(1-K)\ln(1-K) + K - K^2/2], \quad K = K(t_b) \tag{3.28}$$

Finally, we can find the increase in height due to the coasting process after burnout. If we assume that we are launching from the Earth's surface ($v_0 = 0$), the burnout velocity is simply equal to Δv:

$$v_{bo} = g_e I_{sp} [\ln(1/(1-K)) - 1/\eta_0] \tag{3.29}$$

To determine the additional height attained in the coasting phase, we equate the kinetic energy at burnout to the potential energy at the apogee to give:

$$h_c = \frac{1}{2} v_{bo}^2 / g_e = \frac{1}{2} g_e \, I_{sp}^2 [\ln(1 - K) + K/\eta_o]^2 \tag{3.30}$$

where we have assumed that $g = g_e$ in the coast phase. The total height (to the apogee of the vertical trajectory) would simply be the sum of h_{bo} and h_c.

3.3 MULTISTAGE ROCKETS

Up to now, we have concentrated on the trajectory and performance of a single-stage rocket. As you know from our previous discussions, many vehicles incorporate more than one stage to achieve mission objectives. As we pointed out in Figures 3.7 and 3.8, we may have mission requirements, performance, and weight characteristics such that a single-stage device cannot accelerate the payload to the desired Δv. Under these conditions, we have no alternative but to select a multistage vehicle for our mission. The advantage of staging comes through the fact that we can discard structural (inert) elements of lower stages at some point in the trajectory, thereby reducing vehicle weight for the remainder of the mission.

To analyze the effects of staging, consider the N-stage rocket shown in Figure 3.10. To define the mass properties of this rocket, we will employ the following definitions for an arbitrary stage $n < N$:

- m_{fn} = burnout mass of stage n
- m_{on} = initial mass of stage n
- ε_n = stage structural factor = m_{fn}/m_{on}
- β_n = payload factor = $\sum_{n+1}^{N} m_{on} / \sum_{n}^{N} m_{on}$

Note that for the top stage ($n = N$), the structural and payload factors are:

$$\varepsilon_N = \frac{m_{fN} - m_{pl}}{m_{oN} - m_{pl}} \; ; \quad \beta_N = m_{pl}/m_{oN} \tag{3.31}$$

As an example, for a two-stage rocket we would have:

$$\varepsilon_1 = \frac{m_{f1}}{m_{o1}} \; ; \; \varepsilon_2 = \frac{m_{f2} - m_{pl}}{m_{o2} - m_{pl}} \; ; \; \beta_1 = \frac{m_{o2}}{m_{o1} + m_{o2}} \; ; \; \beta_2 = m_{pl}/m_{o2} \tag{3.32}$$

In general, ε and β would vary for each stage. We desire small ε values but we are limited by the strength of materials making up the rocket. Smaller β values imply higher Δu values or reduced payload capabilities. Since the burnout velocity of the rocket (v_{boN}) is simply the sum of the velocity increments for the N stages:

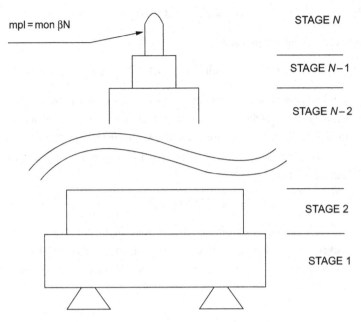

Figure 3.10 General rocket containing N stages.

$$v_{boN} = \sum_{j=1}^{N} \Delta v_j \qquad (3.33)$$

we need to determine the velocity increment for each stage of the vehicle.

Recall that for a vertical trajectory with no drag we had:

$$\Delta v = g_e I_{sp} \ln(m_o/m_f) - \int_0^{t_b} g\,dt \qquad (3.34)$$

so that if we knew the I_{sp}, burn time, and stage mass ratio, we could determine Δv for each stage and sum the contributions to get the overall change in velocity. In other words, for the jth stage:

$$\Delta v_j = g_e I_{spj} \ln(MR_j) - \overline{g}_j t_{bj} \qquad (3.35)$$

where $\overline{g} = \int_o^{t_b} g\,dt/t_b$ and MR_j is defined as the *stage mass ratio*. Using our definition of the mass ratio, we may write:

$$MR_j = \frac{\displaystyle\sum_{i=j}^{N} m_{oi}}{\displaystyle\sum_{i=j+1}^{N} m_{oi} + m_{fi}} \qquad (3.36)$$

By doing a little algebra, we can write this results as

$$MR_j = [\varepsilon_j(1 - \beta_j) + \beta_j]^{-1} \tag{3.37}$$

so that the stage velocity increment becomes:

$$\Delta v_j = -g_e I_{spj} \ln(\varepsilon_j(1 - \beta_j) + \beta_j) - \overline{g}_j t_{bj} \tag{3.38}$$

For simplicity, we shall now invoke the *similar stages* assumption where we will assume that ε, β, I_{sp}, and t_b are the same for each of the N stages in the rocket. In addition, we will assume that gravity variations are negligible so that $\overline{g} = g_e$ for each stage. Under these assumptions, inspection of Eq. 3.38 indicates that each stage will have an identical velocity increment. For this special case, we may write the final burnout velocity:

$$v_{boN} = -N g_e I_{sp} \ln(\varepsilon (1 - \beta) + \beta) - g_e \tau_b \tag{3.39}$$

where $\tau_b = N t_b$ is the total power-on time for the N-stage flight. In other words, each stage in a similar stage rocket provides the same velocity increment ($\Delta v_j = v_{boN}/N$, $j = 1, N$).

To investigate the behavior of Eq. 3.39 for various N values, let us define the following parameters:

$$G = \frac{\text{liftoff mass}}{\text{payload mass}} = \frac{\displaystyle\sum_{i=1}^{N} m_{oi}}{m_{pl}} \tag{3.40}$$

$$S = \frac{v_{boN} + g_e \tau_b}{g_e I_{sp}} = -N \cdot \ln(\varepsilon(1 - \beta) + \beta) \tag{3.41}$$

The parameter G is a measure of the overall performance of the rocket in terms of the total liftoff mass required to transport a unit mass of payload. The parameter S represents a dimensionless mission requirement for a vehicle with a given I_{sp} propulsion system. Note that the mass fraction in Eq. 3.40 can be written:

$$\sum_{1}^{N} m_{oi}/m_{pl} = \frac{\displaystyle\sum_{1}^{N} m_{oi}}{\displaystyle\sum_{2}^{N} m_{oi}} \frac{\displaystyle\sum_{2}^{N} m_{oi}}{\displaystyle\sum_{3}^{N} m_{oi}} \cdots \frac{m_{oN}}{m_{pl}} \tag{3.42}$$

Noting that the fractions on the right-hand side of 3.42 represent β values for the various stages, we may write the performance parameter:

$$G = \prod_{j=1}^{N} 1/\beta_j = \beta^{-N} \tag{3.43}$$

We wish to evaluate changes in masses (G) with a variable number of stages (N) for a fixed mission requirement (S) and reasonable structural factor (ε). We can eliminate β by combining Eqs. 3.41 and 3.43:

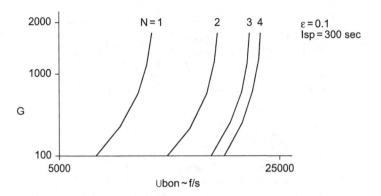

Figure 3.11   Effects of staging on rocket performance.

$$G = [(1 - \varepsilon)/(e^{-S/N} - \varepsilon)]^N \tag{3.44}$$

To assess the effects of increasing the number of stages, we should find the limit of G as $N \to \infty$. It turns out that it is easier to determine the behavior of the logarithm of G:

$$\ln G = N \ln((1 - \varepsilon)/(e^{-S/N} - \varepsilon)) \tag{3.45}$$

Using L'Hopital's rule on Eq. 3.45 for fixed S, ε, we obtain:

$$\lim_{N \to \infty} \ln G = e^{S/(1-\varepsilon)} \tag{3.46}$$

The importance of the result shown in Eq. 3.46 is that even as the number of stages approaches infinity, our performance parameter G approaches a finite limit. The implication of this result is that at some point the addition of more stages has a negligibly small effect on reducing overall vehicle mass. This point is displayed graphically in Figure 3.11, which presents G vs. v_{boN} for a given I_{sp} and structural factor. As is evidenced in this figure, the addition of a second stage permits a large increase in velocity for a given G while the third and fourth stages have smaller and smaller effect in the overall velocity attainable. This effect is sometimes called a *law of diminishing returns* for multistage vehicles.

Clearly, at some point, the advantage of adding an additional stage does not offset the additional cost and complexity associated with such a design. In addition, in real vehicles we will not necessarily have the same λ or I_{sp} value for each stage. This is the fundamental tradeoff conducted by engineers during the preliminary design of a vehicle. Figure 3.12 provides a qualitative description of results of a real trade study. For missions with low Δv requirements, a single-stage vehicle is optimal, as multistage vehicles would have very small second and third stages with very low (i.e. inefficient) λ values. As the required Δv increases, there will be some range of values where a two-stage system provides the lightest alternative, i.e. in this range the third stage is so small and inefficient it is not beneficial. Finally at even higher values of Δv, a three-stage system may provide the lightest solution. As it is expensive to develop an entirely new stage, one

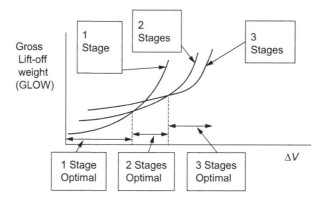

Figure 3.12 Schematic representation of a trade study on a multistage vehicle.

may accept a slightly higher weight system using a smaller number of stages as a "minimum cost" approach.

The trade space shifts with changes in I_{sp} and mass properties. Generally, higher I_{sp} propulsion systems will "optimize out" with fewer stages than a low I_{sp} system since the incremental Δv will be larger for the higher performing system.

3.4 GENERALIZED TRAJECTORIES

Using the simplified vertical trajectories discussed in the previous sections, we were able to indicate the effect of vehicle acceleration, burn time, coasting, and the inclusion of multiple stages. Conclusions we had drawn about these effects hold (at least qualitatively) for launch vehicles with arbitrary trajectories. In the general case however, we may wish to incorporate effects such as variable thrust (and I_{sp}), drag losses, and non-vertical flight. In this case, the equations of motion cannot be integrated directly, but must be integrated numerically to give vehicle velocity and position as a function of time. Advanced trajectory simulations also consider the effect of Earth's rotation and the oblateness of the Earth (it's not a perfect sphere).

Generalized trajectory equations (in two dimensions) can be derived from a summation of forces parallel and perpendicular to the flight direction. The flight direction corresponds to the direction parallel to the vehicles current velocity vector. A summary of the forces and angles required for generalized trajectories is given in Figure 3.13 for 2-D flight. In Figure 3.13 the horizon is a plane that is parallel to the local surface of the Earth; the angle θ is called the horizon reference angle; the angle α is the angle of attack of the wing (if present), and the angle $\psi - \theta$ is the flight path angle.

Aerodynamic lift and drag forces (L, D in Figure 3.13) are assumed to act normal and parallel to the flight direction, respectively. Note that in general both lift and drag are present even if the rocket does not have a wing. The weight of the rocket acts along a line normal to the horizon.

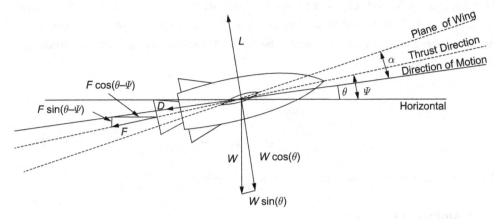

Figure 3.13 Forces on a rocket in generalized 2-D flight.

Along the flight direction, the acceleration is simply the time rate of change of the velocity, so that writing $\sum F = ma$ in this direction gives:

$$m\frac{dv}{dt} = F\cos(\psi - \theta) - C_D\,\rho\,v^2A/2 - mg\sin\theta \qquad (3.47)$$

where $D = C_D\rho v^2A/2$ and C_D is the drag coefficient. Note that a portion of the vehicle weight acts as an equivalent drag for $\theta \neq 0$.

In the direction perpendicular to the flight path, the acceleration comes from centrifugal effects:

$$\text{Acceleration} = \frac{v^2}{R} = v\frac{d\theta}{dt}\quad \text{normal to flight path} \qquad (3.48)$$

Writing $\sum F = ma$ in this plane gives:

$$mv\frac{d\theta}{dt} = F\sin(\psi - \theta) + C_L\,\rho\,v^2A/2 - mg\cos\theta \qquad (3.49)$$

In launch vehicle applications, lift and drag coefficients are typically tabulated (or parameterized) as a function of angle of attack and Mach number as shown in Figure 3.2. Reynolds number dependence is not normally included in these force coefficients since they are typically well into the turbulent range for the bulk of the flight. (Recall that, at high Reynolds number, C_D, C_L are relatively insensitive to changes in this parameter.)

Equations 3.47 and 3.49 can be integrated numerically to give $v = v(t)$ and $\theta = \theta(t)$ provided that suitable initial conditions are provided, for example $v(0) = 0$, $\theta(0) = 90°$, for a vertical launch from Earth. Since trajectory integrations can extend for long time periods, numerical accuracy can be an important consideration, i.e. a small error in acceleration or velocity vector at early parts in the trajectory can add up to large errors much later in the trajectory.

See Appendix A for a discussion of numerical methods that can be applied to this problem. In addition, vehicle thrust and mass histories, as well as information about aerodynamic forces and atmospheric conditions, are assumed to be known. For a non-rotating Earth, the altitude and range can be calculated:

$$h = \int_0^t v \cdot \sin \theta \, dt = \text{ altitude} \tag{3.50}$$

$$x = \int_0^t v \cdot \cos \theta \, dt = \text{ range} \tag{3.51}$$

Note that h must be calculated since $\rho = \rho(h)$ and $g = g(h)$.

3.4.1 Additional Factors

Equations 3.47 and 3.49 are valid for a simple 2-D trajectory analysis. Detailed simulations may consider many other factors, such as the following:

- *3-D effects* – Additional aerodynamic yaw forces can be handled by defining a sideslip angle and an appropriate aerodynamic coefficient.
- *Actual atmospheric conditions* – Preliminary analyses often utilize a standard atmosphere model for simulations, but actual flight predictions may make use of actual (or predicted) day of launch conditions. For example, crosswinds and actual atmospheric pressure and density profiles could be incorporated into the analysis if such data were available. For example, SpaceX derives an optimal trajectory for each launch of the Falcon 9 vehicle based on atmospheric wind measurements a day or two prior to that launch.
- *Thrust vectoring or gimbaling* – Most vehicles utilize the propulsion system for steering by changing the thrust vector a small amount. This effect would be incorporated by including the gimbal angle in the analysis.
- *Rotating and oblate Earth* – The effects of oblateness and rotation are included in advanced analysis codes. The range (Eq. 3.50) is affected since the launch point is not fixed for a rotating Earth simulation.
- *Staging* – Advanced codes can handle weight jettison and propulsion performance changes associated with staging. These codes can also handle payload fairing jettison, which typically occurs after the vehicle leaves the bulk of Earth's atmosphere.

3.4.2 Trajectory Optimization

Developing a trajectory to launch a satellite or spacecraft into low Earth orbit (LEO) is one of the more challenging exercises in rocketry. Figure 3.14 highlights some of the main milestones in

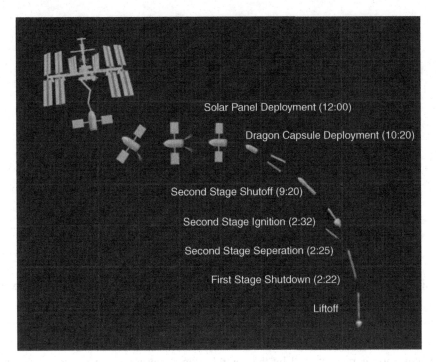

Solar Panel Deployment (12:00)

Dragon Capsule Deployment (10:20)

Second Stage Shutoff (9:20)

Second Stage Ignition (2:32)

Second Stage Seperation (2:25)

First Stage Shutdown (2:22)

Liftoff

Figure 3.14 Trajectory of a Falcon 9 rocket (Source: SpaceX).

a launch of a Global Positioning System (GPS) satellite on the Titan IV launch vehicle used up into the late 1990s. As the GPS constellation is at mid-altitude, a Hohmann transfer is required to boost the satellite into the desired location once the launch vehicle has accessed a LEO "parking orbit." In general, most launch vehicles access LEO in about 8–9 minutes. This burning time turns out to be optimal in terms of minimizing gravity and drag losses over the entire trajectory.

To illustrate the issues in trajectory optimization, consider the three candidate trajectories shown in Figure 3.15 for a vehicle launched into an orbit about the Earth (or the celestial body of your choice). To limit drag losses, we may wish to fly straight up so as to exit the atmosphere in a minimal amount of time (trajectory A in Figure 3.15). Unfortunately, this approach would lead to high $g - t$ losses since the gravity vector is directly opposed to thrust in this case. The other extremes (trajectory C) would attempt to minimize $g - t$ losses by flying horizontal to the Earth's surface. Using this approach, drag losses would be high due to the large portion of the trajectory spent in the atmosphere.

Clearly, the optimal trajectory lies between these two extremes. The early part of the flight will be nearly vertical to reduce drag effects while the end of the trajectory is near horizontal to limit gravity losses and approach the desired circular velocity for the orbit. Trajectory designers are tasked with defining this path by using advanced simulation codes. Typically, the optimization involves finding the trajectory which maximizes payload to a given orbit for given vehicle

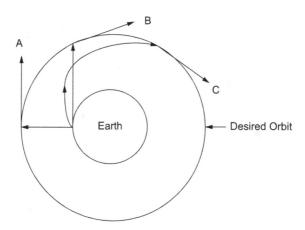

Figure 3.15 Trajectory optimization for a launch vehicle.

propulsion weight and drag characteristics. Constraints in such an analysis include maximum allowable dynamic pressure and aerodynamic heating limits, angle of attack limits during staging and possibly payload related constraints (acceleration limits, vibration limits induced by atmospheric forces). Optimization variables include the time at which the pitch maneuver is executed and various angular rates applied by the control systems during ascent. Typical optimization computer simulations can take many, many hours on the fastest machines currently available, as the long time integrations demand higher-order accuracy numerical methods and there are many variables to optimize.

3.4.3 A Few Words of Caution

Using propellant mass fractions to size vehicles provides a very rapid and powerful ability to generate preliminary designs, but with its simplicity also come a few faults. In viewing the mass fraction data on Figures 3.4–3.6, one can see large dispersions about the trend line – this implies that there is substantial uncertainty introduced when one uses data like this. Do not be naïve in applying this methodology and making strong conclusions about the relative merits of various types of systems (solids vs. liquids, differing propellant types, etc.) as the methodology is in general not precise enough to make fine judgments about systems with similar performance levels. These judgments become especially difficult when we are on the steep part of the curves in Figure 3.7 – small changes in velocity requirement, I_{sp}, or structural weights can lead to very substantial changes in overall system mass for high-velocity gain stages. When structural efficiency is high, small increases in λ can be had only by very significant reductions in structural mass. For example, to increase λ from 0.9 to 0.91 requires a full 10% reduction in structural weight! This behavior is sometimes referred to as the tyranny of the rocket equation; the fact that masses change exponentially with velocity gain makes our job particularly difficult. Mass fraction based sizing is best used for "getting you in the ballpark." If you want to know what base you are on in that ballpark you had better sharpen your pencil and begin to size the actual components that comprise the system inert mass, and you had better have an excellent understanding of your propulsion system performance and the Δv required.

HOMEWORK PROBLEMS

3.1 Most of you have probably seen the movie, *Rocketeer*, in which a rocket propulsion system is used to propel a man through the air. In real life, these systems use hydrogen peroxide (H_2O_2) as a propellant and decompose it to steam $+O_2$ in a catalytic chamber. Given the following information:

$$I_{sp} = 135 \text{ s (delivered)}, c* = 1100 \text{ m/s}$$
$$\rho_{H_2O_2} = 1.4\rho_{H_2O}$$
$$P_c = 2 \text{ MPa}$$
$$\lambda = 0.5$$
Mass of man $= 70\text{kg}$
Twin Propellant tanks 0.7m long by 0.4m diameter

we wish to determine the hover time capability assuming an initial boost to an altitude of 10 m. You may neglect any gravity losses during the short boost phase, but you must retain enough propellant to land after the hover phase. What throat and exit diameters are required for the nozzle?

3.2 Let's say you're not satisfied with the hover time of the "jet pack" design in Problem 3.1. One way to improve performance is to increase the chamber pressure (to 3 MPa average). Let's assume you have done this and you use the same nozzle as in Problem 3.1. What happens? Suppose you redesign the nozzle for optimal performance at the new pressure. How much does the hover time increase?

3.3 A two-stage rocket is to be designed to provide an ideal velocity increment 10,000 ft/s to a payload weighing 500 lb. The propulsion systems for first and second stages have:

$$\begin{array}{llll}\text{1st stage}: & I_{sp} = & 270 \text{ s} & \lambda = 0.9 \\ \text{2nd stage}: & I_{sp} = & 350 \text{ s} & \lambda = 0.8\end{array}$$

and the interstage weight is 200 lb. Assuming the velocity gain is equally distributed between the stages determine:

(a) the propellant and inert masses (weights) for each stage; and
(b) the rocket gross liftoff weight (GLOW).

3.4 A rocket with a liftoff mass of 100 kg containing 50 kg of propellant is to be launched on a vertical trajectory. If the delivered I_{sp} of the rocket is 150 seconds, determine the burning time which provides a 20,000 m rocket height at apogee. You may neglect drag in your analysis and may assume the rocket has constant thrust during the powered portion of the flight.

3.5 Write a computer code to predict an arbitrary 2-D ballistic trajectory on a flat non-rotating Earth. Consider a rocket with the following characteristics:

$$C_D = 0.3, \qquad C_L = 0, \qquad\qquad D_{ref} = 3.1 \text{ in}$$
$$F = 18 - 9t, \quad 0 \le t \le 2.0 \text{ s} \quad (F \sim \text{lb})$$
$$m_o = 1.1 \text{ lb}, \quad I_{sp} = 165 \text{ s}$$

(a) Derive values for m_f and $\dot{m}(t)$ based on the information given above.

(b) Write your code assuming constant atmospheric conditions, arbitrary wind velocity, and negligible change in gravity. Use Huen's method for the integration and attach a listing of the code.

(c) Verify the algorithm by comparing the altitude obtained with the vertical trajectory solution discussed in class by setting $C_D = 0$ and $F = 9$ constant, with $t_b = 2$ s. Tabulate your burnout and total heights using both techniques.

(d) Having verified the code, predict and plot the vehicle trajectory (altitude vs. range) for initial launch angles of 70, 75, and 80° from the horizon directed into a 10 mph crosswind. Include drag in the simulation and provide a tabulated listing of vehicle velocity, acceleration, range, and altitude as a function of time for the $\theta = 75°$ case. Assume a five foot launch rod for your calculations.

3.6 Consider a 1 lb object with a drag coefficient (based on an 8 inch radius circle) of 1.5.

(a) Calculate the terminal velocity for such an object. (Note: the terminal velocity is the maximum velocity attained if the object is dropped and is influenced only by gravity and drag.)

(b) Consider the object suspended in a wind tunnel with a wind speed equal to the object's terminal velocity. At $t = 0$ the object is released. How long will it take for the object to attain 90% of the wind tunnel velocity? You may neglect the influence of gravity, which will tend to pull the object toward the bottom wall of the tunnel.

(c) How far will the object have travelled at the time it reaches the 90% terminal velocity point?

(d) How fast will the object be going at a location 50 feet downstream from the release point?

3.7 Due to the large differences in burning/mission time, there are interesting tradeoffs in choosing an electric vs. chemical propulsion OTV. One of our historical chemical systems, the inertial upper stage (IUS) employs two solid rocket motors with the following characteristics:

$$\text{Stage 1}: \quad \lambda = 0.91 \qquad I_{sp} = 295 \text{ s}$$
$$\text{Stage 2}: \quad \lambda = 0.89 \qquad I_{sp} = 305 \text{ s}$$

In the near term, arcjects provide the only viable electric propulsion options. Consider two candidates:

$$
\begin{array}{lccl}
10 \text{ kW ammonia arcjet}: & F = 2 \text{ N} & \lambda \cong 0.7 & I_{\text{sp}} = 500 \text{ s} \\
10 \text{ kW } H_2 \text{ thruster}: & F = 1 \text{ N} & \lambda \cong 0.6 & I_{\text{sp}} = 1000 \text{ s}
\end{array}
$$

While the hydrogen thruster has a higher performance, storage of H_2 is much more difficult than NH_3 – a factor leading to the reduced mass fraction for the H_2 case. Both systems must carry solar panels to provide power – a major factor influencing λ values for these systems. Consider a baseline mission to boost a payload from 100 nm LEO to GEO.

(a) Determine the Δv requirements for each system (chemical and elective).

(b) Plot vehicle gross weight as a function of payload weighted, for payloads between 500 and 5000 lb for each of the three options.

(c) For the electric OTVs, plot transfer time as a function of payload weight over the same range as in (b).

(d) Discuss your results. Which option is preferred for which range of payloads?

3.8 An air-launched missile is designed to operate at an altitude of 20,000 feet using a solid rocket propellant which delivers a c^* of 5000 ft/s at $\gamma = 1.2$. The missile design is such that the nozzle expansion ratio is limited to 30 so as to have the nozzle exit fit within the desired envelope. The missile is loaded with 250 lb of propellant and has a 2.0 inch diameter throat.

(a) Since the missile will primarily be used to "shoot up" at targets at higher altitudes, we desire for the nozzle to be on the verge of separation at ignition. Assuming a separation pressure of one-third the local ambient pressure, determine the chamber pressure required to force this condition.

(b) Assuming constant chamber pressure operation at 20,000 feet using the value for P_c obtained in (a), find the motor thrust and burning time.

(c) Determine the sea-level thrust and burning time for this missile using the P_c value from (a).

3.9 Many satellite thruster firings use so little fuel that the overall change in satellite mass is very small. If we assume a constant thrust firing for a fixed duration, show that in this case the rocket equation $\Delta v = g I_{\text{sp}} \ln(\text{MR})$ reduces to Newton's second law, $F = m_s a$, where F is the engine thrust, m_s is the satellite mass, and a is the acceleration imparted to the satellite. (Hint: you can use the fact that $\ln(1 + x) \cong x$ if $|x| \ll \ll 1$.)

3.10 Consider a constant thrust rocket launched vertically in the absence of drag. Assuming constant gravitational acceleration $(g = g_e)$ over the entire flight, show that the height of the rocket at burnout is:

$$
h_{\text{bo}} = g_e I_{\text{sp}} t_b \left[\frac{M_f \ln\left(M_f / M_o \right)}{M_o - M_f} + 1 \right] - \frac{g_e}{2} t_b^2
$$

where M_f and M_o are the final and initial masses, and t_b is the burn time of the engine.

3.11 (a) In class we derived an expression for the burnout (h_{bo}) and total (h_{tot}) height attained by a rocket launched vertically under the assumption of no drag and constant mass flow. How does this expression change assuming a mass flow history of the form $\dot{m} = a - bt$, where $a, b = $ const, for $0 \leq t \leq t_b$? That is, derive expressions for h_{bo} and h_{tot} in terms of this mass flow, I_{sp}, m_o, $\dot{m}_f$, and t_b.

(b) Suppose we have a model rocket with a liftoff mass and burnout mass of 1.1 and 0.9 lbm, respectively and the engine burns for two seconds with an I_{sp} of 170 s. Determine h_{bo} and h_{tot} for this case.

3.12 On an airplane flight, I was able to watch the movie, *Armageddon*. In this film, Bruce Willis and cohorts endure an 11 minute burn at $9g$ acceleration in order to catch up to an Earth-killer asteroid. Such a long-duration, high-acceleration burn implies massive amounts of propellant (Hollywood chose not to show any tankage on the shuttle-type vehicle used for this firing). Let's be really forgiving and assume a burnout mass of 5000 lb (which seems a bit light considering all the men and equipment on board). Furthermore, let's assume the propulsion system sports an electric-engine-type I_{sp} of 1000 seconds. Assuming constant acceleration over the duration of the burn, determine the propellant load required. (Note: don't be too surprised here if you find lots of zeros spitting out of your calculator.)

3.13 A sounding rocket has a liftoff mass of 50 kg and a burnout mass of 20 kg using a propulsion system with an I_{sp} of 250 s. By varying the chamber pressure, the designer has the latitude to control thrust level and burn time. Neglecting drag and assuming a vertical constant thrust trajectory, determine the burn time which maximizes the burnout height of the rocket. What burnout height corresponds to this condition?

3.14 At Rocket World, a new amusement park for rocket scientists, we have abandoned the conventional chain drive used to hoist a roller coaster to the top of a hill. In this case, we fire a rocket that accelerates the vehicle to the required speed to crest the hill just as we encounter the incline (see Figure 3.16). Presume that we know the empty mass of the vehicle, M_e, which includes everything except the propellant load. The engine I_{sp} and hill geometry (h, θ) as shown below are also known. In addition, we must account for friction on the rails with a known friction coefficient, μ.

(a) Using a force balance on the roller coaster, derive an expression for the velocity V_o in terms of h, θ, and μ.

(b) Derive an expression for the propellant mass (M_p) required to accomplish this mission in terms of M_e, h, θ, and μ.

3.15 You have been directed to decide on propulsion system alternatives for a vertically launched sounding rocket with a payload of 150 kg and an empty weight of 1500 kg (excluding the payload). The rocket employs twin liquid engines with a storable propellant combination that provides an average I_{sp} of 280 s. The payload contains sensitive instruments, which cannot withstand acceleration levels in excess of $4g$.

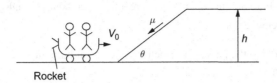

Rocket

Figure 3.16 Diagram for Problem 3.14.

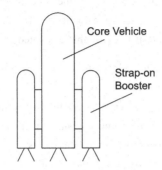

Core Vehicle

Strap-on
Booster

Figure 3.17 Diagram for Problem 3.16.

(a) Assuming negligible drag and that the engines are not throttleable, determine the propellant load, engine burn time, and engine thrust level that will maximize the performance
(Δv) of this vehicle. What is the Δv achieved under these conditions?

(b) Being the bright young engineer that you are (must be a Purdue graduate), you suggest
that the rocket may have additional performance if we are able to shut one engine down at
some appropriate point in the flight. Determine the burning times, propellant consumption, thrust level, and Δv achieved during both phases of the flight.

3.16 Many launch vehicles employ a "parallel burn" during initial operation through the use of
strap-on boosters as illustrated in Figure 3.17. In this case, both boosters and the core vehicle
are thrusting. Suppose the vehicle has an initial mass, m_o of 500,000 kg and operates for $t_b = $
100 s prior to burnout of the boosters. Furthermore, the performance of both strap-ons and
core are known:

$$\text{Strap-ons}: \quad F_s = 2.5 \text{ MN} \quad \text{(each booster)}, \quad I_{sps} = 240 \text{ s}$$
$$\text{Core}: \quad F_c = 1.0 \text{ MN}, \quad I_{spc} = 280 \text{ sec}$$

(a) Assuming a vertical trajectory with no drag, derive the applicable form of the rocket
equation for this application. Let the velocity gain be Δv and final vehicle mass be m_f in
your equation.

(b) Using your equation, compute the Δv attained during operation of the strap-on boosters.

3.17 A suborbital science mission using a 5000 lb instrumentation package (the payload) to study how meteorites burn up in the mesosphere (about 200,000 ft high) has the following trajectory requirements:

Actual (delivered to the payload) velocity increment at burnout = 15,000 ft/s
Burn time = 40 s

An existing one-stage rocket is being used for the mission. The performance and mass characteristics of the stage are: $I_{sp} = 280$ s; $\lambda = 0.9$. The characteristic velocity, c^*, is 5500 ft/s, the thrust coefficient, C_f, is 1.64, the specific heat ratio γ is 1.2, the chamber pressure is 1500 psia, and the nozzle expansion ratio is 12. The aerodynamicist who analyzed the results of the wind tunnel tests told you that estimated drag losses are five percent of Δv.

(a) How much propellant should be loaded?
(b) What is the thrust-to-weight ratio at takeoff?
(c) More wind tunnel data become available late into the project, which indicate that the drag loss is actually 7.0% (whoops). Concisely state the impact and some potential solutions that can be used to meet the trajectory requirements.
(d) Pick one of your solutions from (c), and explain how it would be done, and do the calculations that prove it will work.

Show all work and state assumptions.

3.18 In the 1970s, Bell Aerospace manufactured a "Rocket Belt" which utilized decomposed hydrogen peroxide to allow a person to hover and maneuver for short durations (it was a big hit at halftime during football games). Assume the I_{sp} of the steam/oxygen exhaust is about 180 s and that the propulsion system has a mass fraction of only 0.6 in this application due to the fact that the frame supporting the tank/engine is also factored into this variable. Furthermore, assume that the pilot has a mass of 70 kg and that he/she can carry a payload (the total loaded propulsion system) of 35 kg. Using this data, determine:

(a) the total mass of hydrogen peroxide at fully loaded conditions
(b) the maximum amount of time the person can hover starting with a fully loaded propulsion system.

3.19 A solid rocket motor is being used to accelerate a payload on a horizontal rocket sled as shown in Figure 3.18. A wagon-wheel grain design has been selected to provide the following regressive mass flow history:

$$\dot{m}(t) = \frac{k}{(t+1)^2} \;\; ; \quad k = \text{const.} \; ; \quad 0 \leq t \leq t_b$$

For a fixed initial vehicle mass, m_o, increasing k corresponds to increasing the grain length (and hence burn surface, $\dot{m}$, and thrust). Suppose that we know the I_{sp} of the rocket and wish to maintain an acceleration level below η_{max} (in g) for our payload on the sled.

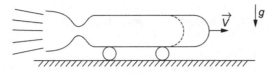

Figure 3.18 Diagram for Problem 3.19.

(a) Derive an expression for the k value that meets these constraints.

(b) Suppose we want to estimate the Δv imparted to this vehicle. Assume we can neglect drag but must consider the friction of the sled against the track (given a friction coefficient, μ) under these assumptions, and derive an expression for the Δv of the vehicle in terms of g, Isp, k, μ, and t_b.

3.20 Determine the burnout and apogee heights the sounding rocket in Problem 3.16 attains for both conditions stated in that problem.

3.21 You are designing a three-stage launch vehicle to place a 10,000 lb payload into a 100 nmi orbit. The propulsion systems available for the three stages include:

$$\text{System A}: \quad I_{sp} = 430 \text{ s}, \quad \lambda = 0.8$$
$$\text{System B}: \quad I_{sp} = 260 \text{ s}, \quad \lambda = 0.88$$
$$\text{System C}: \quad I_{sp} = 330 \text{ s}, \quad \lambda = 0.80$$

(a) Calculate the total Δv required for the mission assuming $g - t$ and drag losses amounting to 4000 ft/s.

(b) Determine the vehicle gross liftoff weight (GLOW) of the optimal design as well as a tabulation of propellant and inert weights for all three stages.

(c) Determine the thrust levels of all stages assuming constant mass flow.

(d) Can you explain why the upper stage of most launch vehicles is cryogenic?

3.22 A rocket propulsion system designed for an interceptor has a thrust history of

$$F = at - bt^2; \quad a, b = \text{constants}; \quad 0 \leq t \leq 20 \text{ s}$$

over its 20 second firing duration. Also, the specific impulse of the rocket is 250 seconds. The propulsion system mass fraction is predicted to be 0.8 if we account for all payload attach provisions. If the desired payload and ideal Δv are 100 kg and 500 m/s respectively, determine:

(a) the mass of propellant and propulsion system inert components;

(b) the constants a and b in the thrust expression;

(c) the acceleration (in g) at the time of peak thrust; and

(d) is the acceleration in (c) the maximum acceleration level in the flight? Explain your answer.

3.23 Two rockets (A and B) with identical I_{sp} and liftoff mass have differing propellant mass flows:

$$\dot{m}_A = 150 - 5t \qquad 0 < t < 30$$
$$\dot{m}_B = 12.5 + 10t \qquad 0 < t < 20$$

where $\dot{m}_A$, $\dot{m}_B$ are the mass flows for rockets A and B respectively.

(a) Neglecting gravity and drag – which of these two rockets gives the higher ΔV?
(b) Suppose both rockets were launched vertically in the absence of drag. Which rocket would have the highest burnout velocity?

3.24 The "g–t" loss for a typical launch from Earth is roughly 5000 ft/s. Can you estimate the loss for a launch from the surface of Venus? Carefully state all assumptions required to make this estimate.

3.25 In class, we were forced to neglect drag in order to integrate the equation of motion for a vertical trajectory to get the rocket equation. If we assume $\dot{m}$ = constant and that $D = \overline{D}$ is the average drag during the burn, we can actually integrate the equation of motion and determine the drag loss. Define the average drag, $\overline{D}$, as a function of the drag coefficient, C_D, the reference area, A, the average atmospheric density (during the burn), $\overline{\rho}$, and the average vehicle velocity. Assume the rocket starts at rest and that $g = g_e$ = constant. Under these assumptions, derive a new form of the rocket equation to give $\Delta V = \Delta V(C_D, A, Isp, g, M_o, M_f, \overline{\rho}, t_b)$.

3.26 The United Nations has learned that another nation is currently developing a nuclear missile utilizing storable liquid propellants: N_2O_4 as oxidizer and monomethyl hydrazine as fuel. In addition to the geometry shown below, the following information (or estimates) have been obtained:

$$t_b = 60 \text{ s}, \qquad \varepsilon = 40, \qquad M_{PL} = 1500 \text{ kg (includes payload fairing)}$$
$$I_{sp} = 270 \text{ s}, \quad \lambda = 0.8, \quad \text{burnout altitude} = 100 \text{ km}$$

Using this information (and Figure 3.19), determine:

(a) The ideal ΔV capability of the missile.
(b) The delivered (actual) Δv capability assuming negligible drag and essentially vertical trajectory during powered flight.
(c) The maximum range capability of the missile.

3.27 Consider a two-stage rocket designed to provide an ideal Δv of 3000 m/s. The following information is known:

$$\text{Payload mass} = 100\text{kg}$$
$$\text{1st stage}: \quad \lambda = 0.9, \qquad I_{sp} = 250 \text{ s}, \qquad 40\% \text{ of total } \Delta v$$
$$\text{2nd stage}: \lambda = 0.85, \qquad I_{sp} = 300 \text{ s}, \qquad 60\% \text{ of total } \Delta v$$

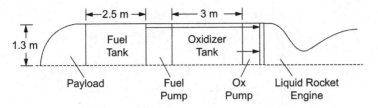

Figure 3.19 Diagram for Problem 3.26.

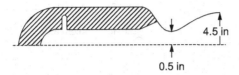

4.5 in

0.5 in

Figure 3.20 Diagram for Problem 3.28.

While the λ values quoted above account for the propulsion systems, the interstage structure is generally treated separately. Assume that the weight of this component is equivalent to 5% of the mass of the 2nd stage/payload combination it supports. Determine:

(a) the 2nd stage propellant and inert masses;
(b) the interstage mass; and
(c) the total vehicle mass.

3.28 The solid rocket motor shown in Figure 3.20 has a chamber pressure history given by:

$$P_c = 300 + 30t - 0.75t^2; \quad 0 \leq t \leq 45 \text{ s}; \quad P_c \sim \text{ psi}$$

In addition, the following propellant characteristics are known:

$$c* = 4800 \text{ ft/s}, \, \rho_p = 0.06 \text{ lb/in}^3, \, \gamma = 1.2$$

The motor is to be used to launch a space-based missile. Telemetry data reveal a velocity increment of 5000 ft/s during a horizontal flight test of the device. Finally, the motor mass fraction (λ) is known to be 0.8. Using this information, determine:

(a) the mass (weight) of propellant in the SRM;
(b) the maximum thrust of the SRM; and
(c) the payload of the missile.

3.29 The maximum acceleration of a rocket with constant thrust (and hence constant mass flow) occurs at the end of the burn when the vehicle is lightest. If $\dot{m} \neq$ constant, this conclusion is not necessarily true. Consider a rocket with constant I_{sp} and a mass flow given by:

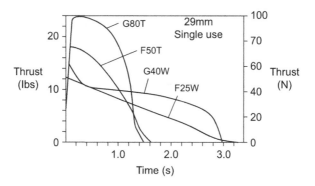

Figure 3.21 Diagram for Problem 3.30.

$$\dot{m} = \frac{2m_\mathrm{p}}{3t_\mathrm{b}} \left(2 - \frac{t}{t_\mathrm{b}} \right) \qquad 0 \le t \le t_\mathrm{b}$$

where m_p is the total propellant mass and t_b is the burning duration.

(a) Derive an expression for vehicle acceleration, η, in terms of the parameters above and the initial vehicle mass, m_o.

(b) Find the dimensionless burning time t/t_b at which the acceleration of this rocket is a maximum for the case where $m_o/m_\mathrm{p} = 2$ and $I_\mathrm{sp} = 300$ s.

(c) Does a maximum acceleration (within the interval $0 \le t/t_\mathrm{b} \le 1$) exist for all m_o/m_p?

3.30 In Purdue's annual Rocket Launch Competition, Team B had the misfortune of having their F-50 engine fail on ignition, and were forced to fly an F-25 engine instead. This engine has a lower thrust, but longer burntime as illustrated in Figure 3.21. The engine is designed to have the same total impulse and parachute ejection delay as the F-50. Given that Team B had calibrated their launch code using the F-50 engine, discuss the implications of the change of engine on the trajectory. Would you expect the new engine to require a higher launch angle (more toward vertical) or a lower launch angle (more toward horizontal)? Consider the repercussions of this change in terms of burnout height, gravity, and drag losses in assessing the differences attributable to the F-25 design. Write a few sentences discussing and justifying your choice.

3.31 Current estimates on the cost of launching one lb of payload into low Earth orbit are as high as $10,000 (I think Ariane is supposed to be ~ $5K/lb, but it is difficult to get an accurate number of the "real" cost to launch). Suppose you think you know how to increase the I_sp of the Space Shuttle main engines by one percent, but it would cost $100,000,000 to implement it. Provide an analysis that shows how many flights it would take to reclaim your investment. Table 2.10 in Larson *et al.*'s *Space Propulsion Analysis and Design* (McGraw-Hill, 1995) says that the Δv for the Shuttle is 9086 m/s. Assume all of the Δv comes from the solid rocket motors (SRMs) and the main engines. Each of the Shuttle's two SRM's burns a total of 1.1 million pounds in 120 seconds and produces an average thrust of 2.7×10^6 lbf. There are three main engines with a vacuum specific impulse of 450 s. The nominal payload of the Shuttle was roughly 50 klbf.

3.32 Consider a vehicle containing three stages with the following characteristics:

$$Stage\ 1 : I_{sp} = 270\ s, \quad \lambda = 0.92$$
$$Stage\ 2 : I_{sp} = 320\ s, \quad \lambda = 0.89$$
$$Stage\ 3 : I_{sp} = 420\ s, \quad \lambda = 0.82$$

Assume that each imparts an equal fraction of the total Δv.

(a) Suppose the vehicle has a payload of 5000 kg. Plot the vehicle gross liftoff weight (GLOW) vs. Δv for Δv between 0 and 10 km/s.

(b) Repeat (a) for a two-stage vehicle (the second stage above is removed) and a one-stage vehicle (stages 1 and 3 above are removed), and plot on the same curve as in (a). Over what range of Δv are one- or two-stage vehicles desirable?

3.33 Consider a solid rocket motor to be optimized for a space propulsion mission to impart a Δv to a 1000 lb payload. The motor contains 5000 lb of propellant, but its inert mass will vary depending on the burn time we choose. Suppose the inert mass can be written:

$$M_{in} = 100 + M_c + M_{ins} + M_{noz}\ (in\ lbs)$$

The case (M_c), insulation (M_{in}), and nozzle (M_{noz}) masses are all functions of burn time:

$$M_c = 100 + 500/t_b$$
$$M_{ins} = 50 + 0.2t_b$$
$$M_{noz} = 200 + 0.01t_b$$

where all masses are in lb, and t_b is in seconds.

(a) Assuming the nozzle contour (i.e. A_t, ε values) and I_{sp} are fixed, find the t_b value which maximizes the performance (Δv) of this motor.

(b) Explain why the case, insulation, and nozzle masses have the functional dependence on t_b as noted in the equations above, i.e. what physical mechanisms lead to this behavior with t_b?

3.34 The rocket equation states that for a given I_{sp} and propulsion system mass fraction, λ, propellant mass increased exponentially with increasing Δv If the payload mass is fixed, derive an expression for the maximum Δv attainable for a given rocket with fixed I_{sp} and λ values. You may neglect gravity and drag losses for your analysis.

3.35 An OTV will be utilized to boost a satellite to an equatorial GEO orbit from the initial circular LEO orbit of 150 km altitude. Two options are being considered:

(a) Two-stage solid rocket

$$\text{Stage 1}: \quad I_{sp} = 300 \text{ s}, \quad \lambda = 0.94$$
$$\text{Interstage mass} = 100 \text{ kg}$$
$$\text{Stage 2}: \quad I_{sp} = 310 \text{ s}, \quad \lambda = 0.92$$

(b) Electric OTV

$$I_{sp} = 500 \text{ s}, \quad \lambda = 0.92$$

Determine the payload capabilities of both of these systems assuming a gross initial mass of 10,000 kg.

3.36 A sounding rocket has a liftoff mass of 50 kg and a burnout mass of 20 kg using a propulsion system with an I_{sp} of 250 s. By varying the chamber pressure, the designer has the latitude to control thrust level and burn time. Neglecting drag and assuming a vertical constant thrust trajectory, determine the burn time that maximizes the burnout height of the rocket. What burnout height corresponds to this condition?

3.37 A two-stage vehicle is to incorporate identical (same I_{sp} and inert masses) solid rockets for each stage. To enhance performance, the steel motor case is replaced with a lightweight graphite-epoxy case on one of the two motors. If you have the option to stack the motors in either order to maximize Δv, which one would you use as the first stage of the vehicle (steel or graphite case)? Explain the rationale for your decision.

4   ROCKET NOZZLE PERFORMANCE

The performance of rocket engines is governed by the thermal energy contained within the combustion gases, as well as the efficiency of the nozzle in converting this thermal energy into kinetic energy and thrust. Therefore, performance analyses emphasize the effect of the nozzle design and propellant ingredients on thrust and specific impulse. In this chapter, we will concentrate on the nozzle contributions, while in Chapter 5 we will discuss the combustion process in more detail. While the derivations in the first part of this chapter generally hold for idealized flow situations (which we will discuss in a moment), they serve as an upper bound on the engine/nozzle performance and therefore represent a good starting point for the calculation of real engine behavior. The relations between nozzle shape and local flow conditions are very important and form the cornerstone of rocket performance analysis.

In Section 4.1 we will briefly review the fundamentals of compressible flow for perfect gases. Governing parameters for use in description of rocket performance will be derived in Section 4.2, while we will address the design of nozzle aerodynamic contours in Section 4.3. In Section 4.4 some of the other unconventional rocket nozzle designs are described.

Factors leading to deviations from ideal performance are considered in Sections 4.5–4.8. Two-dimensional effects, shocks and separation losses, two-phase flow losses, and boundary layer losses are discussed in detail in these sections.

4.1 Review of Compressible Flow of a Perfect Gas

It is presumed that students using this text have already studied gas dynamics or compressible fluid flow. Students should be familiar with one-dimensional (1-D) compressible flow concepts as a basis for the developments in this chapter. The assumptions normally employed in preliminary propulsion system analysis are summarized below:

- 1-D steady flow of homogeneous perfect gas;
- no friction or heat transfer (isentropic processes); and
- no chemical reactions in nozzle.

We will talk about the last assumption in Chapter 5, but for the time being, be aware that the large temperature changes that occur as the gas exits the nozzle can cause large changes in the composition of the fluid due to recombinations of various radical species that exist in the high-temperature combustion chamber. For now, let us ignore this effect and concentrate on a gas with constant composition, and therefore constant molecular weight, $\mathfrak{M}$.

The first assumption above permits us to utilize the perfect gas law wherever it may be convenient. Assumption of a perfect gas is quite reasonable for most chemical rocket combustors, but substantial deviations from perfect gas behavior are noted in the emptying of high-pressure bottles sometimes used to force propellants into the chamber. In general, a gas will display perfect gas behavior if the vast majority of molecular collisions are with the walls of the container and not with other molecules of gas. While rocket combustors can be high pressure, they are also typically high temperature such that gas densities lie in the range where perfect gas behavior can be assumed. The perfect gas law relates gas pressure, p, temperature, T, and gas density, ρ:

$$p = \rho R T = \rho (R_u / \mathfrak{M}) T \tag{4.1}$$

where R_u is the universal gas constant:

$$R_u = 1545.3 \frac{\text{ft lb}}{\text{lb-mol}^\circ\text{R}} = 8317 \frac{\text{Nm}}{\text{kg-mol K}} \tag{4.2}$$

Specific heats for perfect gases (or mixtures of perfect gases) may be written in terms of the gas constant and the ratio of specific heats, $\gamma = C_p / C_v$:

$$C_p = \frac{\gamma R}{\gamma - 1}, \qquad C_v = \frac{R}{\gamma - 1} \tag{4.3}$$

The second assumption above implies that we may assume isentropic flow and that stagnation conditions may be related to local static conditions through the Mach number:

$$M = v/a = \frac{v}{\sqrt{\gamma R T}} \tag{4.4}$$

where v is the local fluid velocity and a is the local speed of sound in the gas. In most rocket engines, flow velocities in the chamber are quite low so that the chamber can be assumed to be at a stagnation condition. While this is a convenient assumption that is frequently employed to ease manipulation of the equations, we can relax the assumption and arrive with similar results. If there is a finite Mach number in the chamber, then the dynamic pressure is a non-negligible part of the total pressure and the resultant kinetic energy of gases flowing in the chamber should be taken into account.

Assigning the subscript "c" to denote stagnation or "chamber" conditions, the isentropic flow relations may be written as follows:

$$p_c / p = (1 + \frac{\gamma - 1}{2} M^2)^{\gamma/(\gamma-1)} \tag{4.5}$$

$$T_c / T = (1 + \frac{\gamma - 1}{2} M^2) \tag{4.6}$$

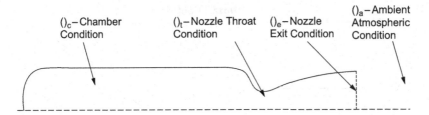

Figure 4.1 Schematic of a rocket combustion chamber with de Laval nozzle.

$$\rho_{\mathrm{c}}/\rho = (1 + \frac{\gamma - 1}{2}M^2)^{1/(\gamma-1)} \tag{4.7}$$

In Eqs. 4.5–4.7, M is the Mach number and γ is the ratio of specific heats. Since we are assuming a steady flow, the mass flow at any point in the nozzle must be the same. Using this condition and the isentropic relations above yields the Mach number/area ratio relation:

$$A/A_x = \frac{M_x}{M} \left\{ \frac{2 + (\gamma - 1)M^2}{2 + (\gamma - 1)M_x^2} \right\}^{\frac{\gamma+1}{2(\gamma-1)}} \tag{4.8}$$

where A and M represent the cross-sectional area and Mach number at one location in the nozzle and A_x and M_x represent analogous quantities at another nozzle location. Since the pressure in the chamber of practically all rocket engines is quite high compared to the local ambient conditions, rocket nozzle throats are practically always choked. In general, the nozzle will be unchoked only during early portions of the start transient and late portions of the shutdown transient where the steady-state assumption would not be terribly accurate anyway. Figure 4.1 provides a schematic denoting the locations of chamber, throat, and nozzle exit planes. Recall that the choking condition for a nozzle implies that the Mach number at the throat is identically equal to unity (i.e. a sonic condition). Therefore, if we substitute $A_x = A_t$ as the throat area, then $M_x = 1$ (sonic flow at throat) the Mach number/Area Ratio relation can be expressed as:

$$\frac{A}{A_t} = \frac{1}{M} \left\{ \frac{2 + (\gamma - 1)M^2}{(\gamma + 1)} \right\}^{\frac{\gamma+1}{2(\gamma-1)}} \tag{4.9}$$

which is equivalent to an expression for A/A^* in the isentropic fluid flow tables that appear in most compressible flow texts. Since $A \propto M^2$, there are two Mach number solutions for a given area ratio in Eq. 4.9. Note that A/A_t grows very large if the Mach number approaches zero or the Mach number tends toward a large value, i.e. at a given A/A_t, a low Mach number and a high Mach number solution exist. These two solutions correspond to the regions upstream and downstream of the throat in a converging/diverging de Laval nozzle. The upstream solution gives subsonic results, while flow in the downstream portion is supersonic. Of course at $A = A_t$ only a single solution exists since this corresponds to the sonic point in the nozzle.

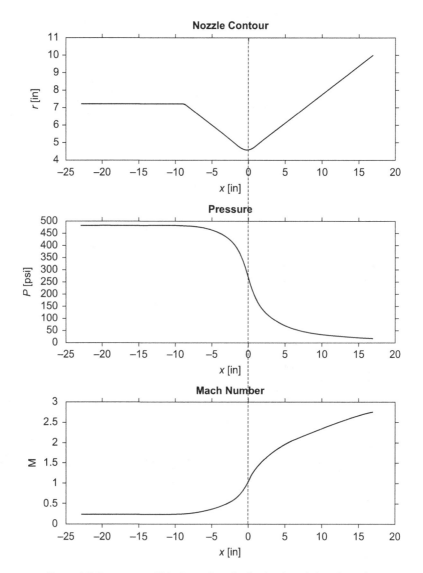

Figure 4.2 Pressure and Mach number distribution in a de Laval nozzle.

Figure 4.2 presents the pressure and Mach number distribution along the length of a de Laval nozzle. Let us first think about the case where the chamber pressure (or upstream stagnation pressure) is only slightly higher than the ambient pressure. In this case, the flow will accelerate through the throat region, but will remain subsonic. The subsonic flow downstream of the throat will sense the exit pressure and equilibrate accordingly at the nozzle exit (condition "a" in Figure 4.1).

As we increase the chamber pressure, acceleration in the throat will increase until the choking condition $M_t = 1$ is reached. Under this condition the throat pressure, p_t, is:

$$p_c/p_t = [1 + \frac{\gamma - 1}{2} 1^2]^{\gamma/(\gamma-1)} = [\frac{\gamma + 1}{2}]^{\gamma/(\gamma-1)} = 1.89 \text{ (for } \gamma = 1.4) \tag{4.10}$$

Therefore, if the pressure ratio is minutely above this limit, the flow will remain subsonic and will reach condition "a" (ambient) at the nozzle exit. If the chamber pressure is increased ever so slightly above this condition, the nozzle will choke and lead to supersonic flow downstream of the throat as indicated by the bottom plot in Figure 4.2. Virtually all rocket nozzles will operate in this manner.

If the nozzle is poorly designed, the possibility of shocks forming in the supersonic portion exists for a range of upstream pressures. Assuming a normal shock is present, the shock will position itself within the nozzle so that the subsonic flow downstream of the shock will expand to the local ambient pressure at the nozzle exit. Increasing the chamber pressure further will cause the shock to move toward the nozzle exit plane. Increasing chamber pressure beyond this point will cause the shock system to occur outside the nozzle, as we will discuss later in this chapter. It is in this regime that functional rocket propulsion systems operate – the chamber pressures are high enough to maintain a supersonic flow within the entire nozzle exit cone. From a practical perspective, this is a requirement for an efficient nozzle design, as flows with shocks in the exit cone will have dramatically less thrust. While the normal shock solutions are possible under the assumption of 1-D flow, multidimensional effects generally will cause oblique shocks to form within the nozzle under the pressure ratio conditions where such solutions are preferred. Section 4.3 provides additional insight into this area.

4.2 ROCKET PERFORMANCE FUNDAMENTALS

To the designer, performance implies the amount of thrust generated by the system for a given mass flow and ambient pressure/flight condition. In Chapter 1, we introduced specific impulse (thrust per unit propellant flow) as an important efficiency measure that also describes system performance and efficiency. In this section, we derive relationships for these parameters as a function of the nozzle design, propellant combustion characteristics, and operating altitude. The chamber pressure (assumed to be the stagnation pressure in the combustion chamber) is the most important parameter set by the designer and also one of the most important measurements one can make to assess engine operation. For this reason, we are motivated to express our performance results in terms of this parameter.

4.2.1 Thrust of a Rocket Engine

There are a number of different ways to derive the thrust generated by rocket propulsion devices. Figure 4.3 provides a control volume with which we can determine the thrust of a rocket engine that

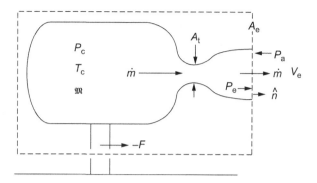

Figure 4.3 Control volume for determination of rocket thrust.

is mounted to a strut as one might see in a ground test of the device. Note that for this simple situation, the continuity equation for a steady flow is simply:

$$\dot{m} = \rho v A = \text{constant} \tag{4.11}$$

where v is the velocity of gases parallel to the centerline of our rocket chamber/nozzle. If we are thinking of a liquid rocket engine in our example, $\dot{m}$ will then correspond to the combined flow rate of fuel and oxidizer being fed into the engine. Remember, for a steady flow, the mass flow at all points along the flow-path must be constant.

Now we turn to the momentum equation. Recall from fluid mechanics that the sum of forces on a control volume is simply equal to the integral of the momentum fluxes through the surfaces of the control volume assuming that the flow is steady. For the situation in Figure 4.3 the sum of forces corresponds to the inverse of the force on the test stand $\left(-(-F)\right)$, and the pressure force exerted at the nozzle exit plane. Therefore, for the configuration in Figure 4.3, the momentum equation becomes:

$$F + \int_{cs} p \cdot \mathbf{n} dA = \int_{cs} \rho \mathbf{v}(\mathbf{v} \cdot \mathbf{n}) dA \tag{4.12}$$

where $\mathbf{n}$ is the unit normal vector pointed outward from the control surface (in this case outward from the nozzle exit plane).

Now, the only net momentum flux and pressure force in the axial direction occur over the exit plane of the nozzle. If the flowpath is truly axisymmetric, then radial components of the pressure integral in Eq. 4.12 will cancel (i.e. we have an equal contribution of radial force at any azimuthal plane) such that there are no side forces generated by the nozzle. This is the situation of interest to us in general, except in situations where we wish to generate side forces for thrust vector control, i.e. maneuvering of the vehicle. We'll discuss this aspect in Section 7.6.

Assuming that the exit pressure is p_e and that the exhaust velocity, v_e is in the axial direction, the thrust becomes:

$$F = \dot{m}v_e + (p_e - p_a)A_e \qquad (4.13)$$

which is our final expression for the ideal thrust of a rocket propulsion device. The first term on the right-hand side of this equation is the *jet thrust*, while the second term is the *pressure thrust*. Typically, the jet thrust contributes the lion's share of the overall force; this is the force we tend to associate with rockets, as the high-speed plume ejected from the nozzle can create large amounts of noise and vibration giving us the sense of the power of the device. The pressure thrust is the lesser known term, but is non-negligible in most instances. Depending on the relative levels of the nozzle exit pressure, p_e, and the ambient pressure, p_a, this term can either be positive, negative, or zero. For a rocket ascending through the atmosphere, the ambient pressure will decrease with altitude and in general all three regimes (positive, zero, negative pressure thrust) will be realized within a given flight. Note that the thrust is a maximum in a vacuum, where $p_a = 0$.

For those of us who have experience with airbreathing propulsion systems, the expression will look quite similar to the thrust equation of a jet engine. For rocket applications, there is no ram drag present, as we are not ingesting air into the system. The overall effect of the air on a rocket system is the resultant drag force created by forcing this air around the sides of the vehicle. Note that since a rocket engine does not take in any air, it has no ram drag and hence its thrust is independent of the flight speed. Thrust is dependent on altitude due to the presence of p_a in the relation.

If we divide Eq. 4.13 by the propellant weight-flow $\dot{m}g$, we can write the I_{sp} as:

$$I_{sp} = \frac{F}{\dot{m}g} = v_e/g + \frac{A_e}{\dot{m}g}(p_e - p_a) \qquad (4.14)$$

so we can see that I_{sp} is related to the exhaust velocity of the engine and that high exhaust velocities imply high performance. However, as with the thrust itself, I_{sp} depends on the pressure thrust and this term must always be considered for accurate evaluation of performance. Many texts define an *effective exhaust velocity*, c, as the product gI_{sp}:

$$c = gI_{sp} = v_e + \frac{A_e}{\dot{m}}(p_e - p_a) \qquad (4.15)$$

This relation indicates that the effective exhaust velocity increases above the exit velocity if the pressure thrust is positive. Since c differs from I_{sp} by only a constant, we will not bother to make much use of it in this text. The reader should be aware that the two variables are essentially interchangeable and that some astrodynamics texts prefer the use of effective exhaust velocity.

Vacuum Conditions

If we define I_{spv} as the vacuum I_{sp} (corresponding to the case where $p_a = 0$), we may write:

$$I_{sp} = I_{spv} - p_a A_e/(\dot{m}g) \qquad (4.16)$$

Equation 4.16 shows that it is simple to correct the specific impulse to any altitude from the vacuum result, provided that the mass flow and nozzle exit area are known. A similar correction exists for thrust itself:

$$F = F_\mathrm{v} - p_\mathrm{a} A_\mathrm{e} \tag{4.17}$$

Oftentimes, the propulsion engineer expresses data in terms of these vacuum conditions to alleviate problems with changing atmospheric conditions. For instance, test data measured at sea level can be corrected to vacuum using Eq. 4.17. Using the vacuum thrust, the trajectory designers can make the simple correction within their analysis since the altitude–time history is one result of a trajectory simulation.

At this point, we have examined two fairly simple expressions for thrust and specific impulse for a general ambient pressure (Eqs. 4.13 and 4.14) or more specifically for vacuum conditions (Eqs. 4.16 and 4.17). It is now worth discussing the rationale for the coming derivations. When we look at the four equations noted and think about evaluating the terms in those equations, we may have trouble evaluating nozzle exit velocity and mass flow. We could certainly think about measuring the nozzle exit velocity, but this is not a trivial measurement by any means, as a probe placed in the exhaust would possibly burn up. While we may have alternate means to compute the mass flow pumped into a liquid rocket engine, how do we evaluate this result for a solid rocket motor?

For these reasons, we are motivated to express nozzle mass flow and exit velocity in terms of chamber conditions. It is quite reasonable for us to measure the chamber pressure in any engine (solid or liquid propellant), and you'll see in Chapter 5 that we can accurately compute the combustion chamber temperature by knowing about the specific propellant formulation being burned and the pressure at which it is being burned. If we can express the nozzle mass flow and exit velocity in terms of these parameters, we can tie our ideal rocket performance to measurable/ calculable quantities, thereby providing a mechanism to measure the efficiency of the system relative to these ideal estimates. It is with this motivation that we now work to relate nozzle mass flow and exit velocity to chamber conditions.

Nozzle Exit Velocity

If we assume isentropic flow throughout the device, we can express I_sp and other performance parameters in terms of nozzle geometry and the conditions "c" in Figure 4.3. Conditions "c" are often called chamber conditions, with p_c representing the *chamber pressure* and T_c representing the *chamber or "flame" temperature*. Actually, T_c is nearly always assumed to be the adiabatic flame temperature of the propellant combination being considered, as we will see in Chapter 5.

To express the nozzle exit velocity in terms of these chamber conditions, let us start with the energy equation (first law of thermodynamics) written in terms of specific enthalpy h (in Btu/lbm or cal/gm):

$$\mathrm{d}h + v\,\mathrm{d}v = \mathrm{d}q - \mathrm{d}w_\mathrm{sh} \tag{4.18}$$

where dq is the amount of heat added during the process (per unit mass) and dw_{sh} is the amount of shaft work performed *by the fluid* during the process. The left-hand side of Eq. 4.18 represents the total change in energy as measured in terms of thermal energy (dh) and kinetic energy (v dv). In rocket propulsion devices, nuclear sources of energy are negligible (we aren't splitting or fusing atoms) and electrical sources of energy are also neglected (we aren't generating plasmas). In many electric propulsion devices, the latter effect must be considered as charged particles within the chamber are then subject to forces imposed by electric or magnetic fields.

The right-hand side of Eq. 4.18 represents the amount of heat added and work done on gases lying within the control volume. Since we stated above that we would assume that our process is adiabatic (d$q = 0$), and there is no shaft work interaction, we have the simple result:

$$dh + v\,dv = 0 \qquad (4.19)$$

The adiabatic assumption deserves some discussion, as it can be a point of confusion. One might ask, how can I neglect heat transfer in a collection of gases that are combusting at several thousand degrees? The answer to this question lies in the definition of dq. Formally, dq represents heat transferred from the gases in the control volume into the walls of the container. For an adiabatic process, we are assuming that this heat transfer is negligible and that all the heat present in the chamber gases remains within those gases. The chemical processes within the combustion event itself are neglected here as h is measuring the thermal energy (and the p–dV work) of chamber gases that have already been combusted.

It is interesting to think about the relevance of the adiabatic assumption as intuition might tell us that this could surely be suspect when we have such high-temperature gases flowing at high speeds within our container. In fact, large rocket propulsion devices sport some of the most impressive volumetric energy release rates of any manmade device. For example, according to NASA, the Space Shuttle main engine develops the power of 40 locomotives within a combustor that is roughly the size of a trash can. Even a small portion of the energy release from the combustion process represents a huge amount of energy as far as chamber walls are concerned. As a practical consequence of this, a successful rocket combustor must achieve very low heat transfer to the walls, or it will burn up.

Equation 4.19 can be integrated between states "c" and "e" in Figure 4.2 to give:

$$h_e - h_c = \frac{1}{2}(v_c^2 - v_e^2) \qquad (4.20)$$

As we have mentioned before, the flow within the chamber is at very low Mach numbers so that p_c and T_c can be assumed to be stagnation conditions, and $v_c \approx 0$. Under these conditions, we may write:

$$v_e = \sqrt{2(h_c - h_e)} \qquad (4.21)$$

From Eq. 4.21 we can see that the nozzle exit velocity is related to the enthalpy drop from the chamber to the nozzle exit plane. Physically, the nozzle is converting the high thermal energy in the chamber gases to kinetic energy within the exhaust. Now if we assume perfect gases, we can write $dh = c_p \, dT = 2\gamma R dT/(\gamma - 1)$, so that:

$$v_e^2 = \frac{2\gamma R_u T_c}{\mathfrak{M}(\gamma - 1)}(1 - T_e/T_c) \tag{4.22}$$

where R_u is the universal gas constant and $\mathfrak{M}$ is the molecular weight of the mixture of gases within the chamber as noted previously. Finally, if we write the temperature ratio in terms of the pressure ratio using isentropic flow relations (Eqs. 4.5 and 4.6), we have:

$$v_e^2 = \frac{2\gamma R_u T_c}{\mathfrak{M}(\gamma - 1)}[1 - (p_e/p_c)^{(\gamma-1)/\gamma}] \tag{4.23}$$

which gives us the ideal exit velocity for a rocket nozzle. The quantity p_e/p_c in Eq. 4.23 can be determined immediately for a given nozzle geometry. Let's say we specify the *nozzle expansion ratio*, $\varepsilon = A_e/A_t$, where A_t is the area at the nozzle throat. Assuming choked flow at the throat, we could then use Eq. 4.9 to solve (albeit implicitly) for the Mach number at the exit plane, M_e (we would take the supersonic solution from the two solutions possible). Since for isentropic flow, the ratio p_c/p_e is a function of M_e alone (per Eq. 4.5), we can say that the nozzle exit pressure is a function of the nozzle expansion ratio as we have claimed above. Therefore, for a given nozzle design and chamber pressure, Eq. 4.23 provides the important result $v_e \propto \sqrt{T_c/\mathfrak{M}}$ so that we will desire propellants with high flame temperatures and low molecular weights for high performance. Physical reasoning tends to tell us that hotter propellant combustion temperatures increase performance, but the low molecular weight requirement should not be forgotten. Physically, minimizing $\mathcal{M}$ corresponds to maximizing C_p of the combustion gases. Higher heat capacity in the chamber gases means that more heat can be stored at a given chamber temperature and we would then have more heat to convert to kinetic energy in the nozzle.

Nozzle Mass flow

As we mentioned above, we are motivated to write nozzle mass flow in terms of chamber conditions in order to make comparisons with readily measurable/calculable parameters. From Eq. 4.11, using the perfect gas law and definition of Mach number, we may write:

$$\dot{m} = \rho v A = \frac{\gamma p M A}{a} \tag{4.24}$$

where a is the local speed of sound at the point of interest in the nozzle. Let's concentrate on the mass flow through the throat. Since the mass flow is constant at all points in the nozzle and since $M_t = 1$ for choked flow we may write:

$$\dot{m} = \frac{\gamma \, p_t \, A_t}{a_t} \tag{4.25}$$

Now we can write the throat pressure using Eq. 4.5 and recalling that sonic velocity can be expressed $a = \sqrt{\gamma R T}$, we may then write:

$$p_c/p_t = \left(1 + \frac{\gamma-1}{2}1^2\right)^{\gamma/(\gamma-1)} = \left[\frac{\gamma+1}{2}\right]^{\gamma/(\gamma-1)} \tag{4.26}$$

$$(a_c/a_t)^2 = T_c/T_t = \frac{\gamma+1}{2} \tag{4.27}$$

Now we can substitute Eqs. 4.26 and 4.27 into 4.28 to write the mass flow in terms of chamber conditions and nozzle throat area:

$$\dot{m} = p_c A_t \sqrt{\frac{\gamma\mathfrak{M}}{R_u T_c}} [2/(\gamma+1)]^{\frac{\gamma+1}{2(\gamma-1)}} \tag{4.28}$$

The quantity involving $\mathfrak{M}$ and variables under the radical is like the inverse of a velocity. Define this quantity as c^* and we may write:

$$c^* = \frac{p_c A_t}{\dot{m}} = \sqrt{\frac{R_u T_c}{\gamma\mathfrak{M}}} [2/(\gamma+1)]^{\frac{-(\gamma+1)}{2(\gamma-1)}} \tag{4.29}$$

This quantity is known as the *characteristic velocity* for a rocket. Note that $c^* = c^*(\gamma, T_c, \mathfrak{M})$ so that the characteristic velocity is a function of the propellant combination only. Therefore, c^* is a measure of the energy available in the propellant combination. From the definition of c^* we can write the mass flow:

$$\dot{m} = p_c A_t/c^* = g p_c A_t/c^* \quad \text{(English/Imperial units)} \tag{4.30}$$

where the latter form suitable for computations in English/Imperial units produces a weight-flow (lbf/s). From Eq. 4.30, we can see that a high c^* value reduces the mass flow required to sustain a pressure p_c with the throat area A_t. For this reason, propellants with high characteristic velocity are desirable.

Finally, note from Eq. 4.29 that c^* is independent of pressure. In many analyses we will assume this is true to permit integration of the governing equations. In real life there is a weak pressure dependence since the chamber composition (and hence $\mathfrak{M}$, γ) and temperature do vary a small amount with pressure due to recombination of radical species at higher pressures. Therefore, c^* will tend to increase slightly with pressure. We will discuss this issue in more detail in Chapter 5.

We once again see the important result that c^* is maximized when $\sqrt{T_c/\mathfrak{M}}$ is maximized. This fact shows why hydrogen is such an excellent rocket fuel – its low molecular weight increases performance even though its combustion temperature is not drastically different than other fuels. We are not simply working to maximize combustion temperature to get high performing rockets and must always be cognizant of the molecular weight of combustion gases produced by the propellants we use.

Thrust Coefficient

Using our definition of thrust and characteristic velocity we may write:

$$F = \dot{m}v_e + (p_e - p_a)A_e = p_c A_t v_e/c^* + A_e(p_e - p_a) \tag{4.31}$$

Now if we substitute for our relations we have derived for v_e and c^* (Eqs. 4.23 and 4.29) we may write:

$$F = p_c A_t \left[\frac{2\gamma^2}{\gamma - 1} \left(\frac{2}{\gamma + 1} \right)^{\frac{\gamma+1}{\gamma-1}} (1 - (p_e/p_c)^{(\gamma-1)/\gamma}) \right]^{1/2} + (p_e - p_a)A_e \qquad (4.32)$$

which indicates that $F \propto p_c A_t$. We can introduce a dimensionless parameter called *thrust coefficient* by dividing Eq. 4.32 by $p_c A_t$:

$$c_f = \frac{F}{p_c A_t} = \left[\frac{2\gamma^2}{\gamma - 1} \left(\frac{2}{\gamma + 1} \right)^{\frac{\gamma+1}{\gamma-1}} (1 - (p_e/p_c)^{(\gamma-1)/\gamma}) \right]^{1/2} + (p_e/p_c - p_a/p_c)\varepsilon \qquad (4.33)$$

where c_f is the thrust coefficient, and $\varepsilon = A_e/A_t$ is the *nozzle expansion ratio* as discussed previously. We can see from Eq. 4.33 that $c_f = c_f(\gamma,\varepsilon,p_e/p_c,p_a/p_1)$ and since we had said that p_e/p_c was a function of ε, we can see that c_f is really a measure of nozzle performance for a given ambient condition. Physically, the first term on the right-hand side represents the dimensionless jet thrust, while the second term represents the dimensionless pressure thrust. As the nozzle grows larger, gases expand further in the exit cone and p_e gets smaller and smaller. For this reason, the jet thrust term continues to grow as ε increases. However, we can see the opposite effect for the pressure thrust term, as nozzle size grows, pressure thrust drops. Because of this competing behavior, an optimal nozzle size (ε) exists for given chamber and ambient pressures. We can prove that c_f is maximized under the perfectly expanded condition when $p_e = p_a$.

There are three general cases of interest in describing the behavior of the thrust coefficient:

1. $p_e > p_a$. This is the case for an *underexpanded* nozzle, where we obtain positive pressure thrust.

2. $p_e < p_a$. This is the case for an *overexpanded* nozzle, where we obtain negative pressure thrust.

3. $p_e = p_a$. This is the case for a *perfectly expanded* nozzle where we obtain no pressure thrust.

Figure 4.4 depicts flow near the nozzle exit plane for overexpanded and underexpanded conditions. If the flow is highly overexpanded, the potential exists for separation within the nozzle. The separation can occur over just a portion of the circumference (asymmetric separation), or uniformly around the entire circumference. If the flow is mildly overexpanded, the fluid has significant energy to overcome the adverse pressure gradient and the nozzle flows full. Oblique shock waves are observed at the exit plane for this condition.

The perfectly expanded nozzle will have a shockless plume with nearly axial exhaust. Viscous forces (shear layers) at the edges of the plume will cause it to diverge gradually in this case as ambient air is entrained into the high-speed flow. Finally, if the flow is underexpanded, a series of expansion waves are formed at the lip of the nozzle. These expansion waves cause rapid pressure decreases within the plume. Further downstream (2–3 nozzle diameters) a normal shock wave is

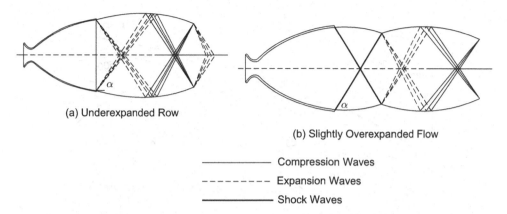

(a) Underexpanded Row

(b) Slightly Overexpanded Flow

————————— Compression Waves

– – – – – – – – Expansion Waves

————————— Shock Waves

Figure 4.4 Nozzle plume characteristics at various exit pressure conditions.

Overexpanded flow, $P_e < P_a$

Vulcain engine, hot-firing at
DLR P5 ground test facility

Perfcetly expanded flow,

$P_e \sim P_a$

RL 10 engine w/o NE

Underexpanded flow, $P_e > P_a$

Saturn 1B with 8 H1 engines,

Apollo 7 mission

Figure 4.5 Nozzle plume characteristics at various exit pressure conditions.

generated to equilibrate plume and ambient pressures. This shock, called a barrel shock or Mach disk, can be repeated a number of times for highly underexpanded nozzles. Figure 4.5 provides a view of actual engine plumes under over, under, and perfectly expanded conditions. The perfectly expanded plume will still show shock and expansion wave structures due to multidimensional effects and interactions with the ambient air.

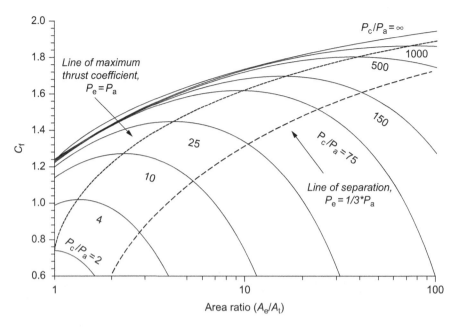

Figure 4.6 Nozzle thrust coefficient behavior ($\gamma = 1.2$).

If $p_a = 0$, we have vacuum conditions at the nozzle exit and the thrust coefficient becomes:

$$c_{fv} = \left[\frac{2\gamma^2}{\gamma - 1} (\frac{2}{\gamma + 1})^{\frac{\gamma+1}{\gamma-1}} (1 - (p_e/p_c)^{(\gamma-1)/\gamma}) \right]^{1/2} + (p_e/p_c)\varepsilon \qquad (4.34)$$

For a given ε, this is the maximum thrust coefficient one can attain. Note that the vacuum thrust coefficient depends only on γ and nozzle size since the ratio p_e/p_c is a function of ε as discussed previously.

A map of thrust coefficient as a function of expansion and pressure ratios is given in Figure 4.6. For a given finite p_c/p_a ratio, a humped-shaped curve is present with points to the left of the maximum represented underexpanded nozzles and points to the right of the maximum representing overexpanded nozzles. As noted above, we can prove that c_f is maximized when the nozzle is perfectly expanded; the locus of $c_{f\max}$ values corresponds to this condition. If the nozzle is highly overexpanded, the flow must work against large adverse pressure gradients and nozzle separation can occur. This line is also noted in Figure 4.6. Finally, if we operate in a vacuum, $p_c/p_a = \infty$ which is the uppermost line of constant p_c/p_a in the figure. If we increase the nozzle size to infinite expansion ratio under these conditions, we can find the ultimate achievable value for c_f. This $c_{f\text{ult}}$ is a function of γ alone and has a value of 2.246 for $\gamma = 1.2$.

For a system operating at a fixed altitude, one can readily use the c_f plot to determine optimal nozzle size to maximize performance. For a launch vehicle system, a very interesting optimization takes place. Assuming a fixed expansion ratio nozzle is used as one ascends through the atmosphere implies that one will travel along a vertical line on the C_f plot. In Section 4.6, we'll show that it is generally desirable to operate on the verge of separation at liftoff so we can gain the benefits of the larger expansion ratio at higher altitudes. We can really see the utility in using the c_f plot (or equivalent tabulated/computed values) to readily decide the optimal nozzle size for a variety of missions.

Values for thrust coefficient can be tabulated as a function of expansion ratio for a given γ value. This can be accomplished because the pressure ratio p_e/p_c is also just a function of expansion ratio for a given γ value. Results for $\gamma = 1.2$, 1.3, and 1.4 can be found on the Purdue Propulsion Web Page (see *Rocket Thrust Coefficient Tables,* at https://engineering.purdue.edu/~propulsi/propulsion/). The value $c_{f_{opt}}$ is the thrust coefficient assuming optimal expansion ($p_e = p_a$), while the quantity c_{fv} is the vacuum thrust coefficient as defined above. We can find the thrust coefficient for an arbitrary freestream pressure by noting that:

$$c_f = c_{fv} - (p_a/p_c)\varepsilon \tag{4.35}$$

which makes the tables very useful. Note the exit Mach number is also provided in the tables (https://engineering.purdue.edu/~propulsi/propulsion/).

Summary of Basic Rocket Relations

At this point, you may be asking, "Why do we go to the trouble of defining these new variables (c^*, c_f) in terms of quantities we already know?" Remember that when we defined c^*, that this quantity was a measure of propellant performance. Therefore, engines with similar propellants should have similar c^* values. We can make use of this fact in determining the efficiency of a given engine with respect to other engines using the same propellants since c^* should be invariant in this case. For example, we may want to test a subscale engine with the same operating conditions and propellant mixture in order to assess combustion characteristics in a smaller and less expensive test bed.

We also stated that thrust coefficient described nozzle performance so that two engines with geometrically similar nozzles should have the same thrust coefficient. Using this notion, we can compare the thrust efficiency of two geometrically similar engines, which might be operating at slightly different chamber pressures. These notions allow propulsion engineers to characterize performance of individual engines within a group of engines with the same design. This procedure is often crucial since overall rocket performance is very sensitive to engine performance and efficiency.

Having defined the new variables c^* and c_f which describe propellant energy and nozzle performance, we can readily compute nozzle mass flow and thrust from very simple relationships:

$$F = c_f p_c A_t; \quad \dot{m} = p_c A_t/c^* = g p_c A_t/c^* \quad \text{(English/Imperial units)}$$

and can easily relate I_{sp} to these fundamental quantities:

$$I_{sp} = \frac{F}{\dot{m}g} = \frac{c_f p_c A_t}{g p_c A_t/c^*} = c_f c^*/g \tag{4.36}$$

Table 4.1 Typical values of rocket performance parameters

Parameter	Ordinary value	High value
T_c	3600–5400 °R, 2000–3000 K	6000–7000 °R, 3300–3900 K
c^*	4000–5000 f/s, 1200–1500 m/s	6000–8000 f/s, 1800–2400 m/s
c_f	1.3–1.5	1.7–1.9
$\mathfrak{M}$	20–25 lb/lb-mol, kg/kg-mol	8–20 lb/lb-mol, kg/kg-mol
γ	1.14–1.18	1.2–1.25

Therefore, we can see that I_{sp} is a direct function of both propellant energy and nozzle performance. As we noted in Chapter 1, I_{sp} is truly the most fundamental rocket performance measure as it reflects the overall efficiency/capability of both the combustion process as well as the nozzle flow process. If we substitute in for c^* and c_f into Eq. 4.36 we can write:

$$I_{sp}g = \sqrt{\frac{2\gamma R_u T_c}{(\gamma - 1)\mathfrak{M}}\left(1 - (p_e/p_c)^{(\gamma-1)/\gamma}\right)}\left[1 + \frac{(p_e/p_c - p_a/p_c)(p_c/p_e)^{1/\gamma}}{\frac{2\gamma}{\gamma-1}\left(1 - (p_e/p_c)^{(\gamma-1)/\gamma}\right)}\right] \qquad (4.37)$$

While this relation is rarely used to calculate I_{sp}, it does demonstrate the very important result that:

$$I_{sp} \propto \sqrt{T_c/\mathfrak{M}} \qquad (4.38)$$

which indicates we want high chamber temperature and low molecular weight of chamber gases to maximize performance. This is the fundamental reason that the liquid oxygen/liquid hydrogen (LOX/LH$_2$) propellant system provides some of the highest I_{sp} values for chemical propulsion systems. The presence of large amounts of hydrogen lowers the molecular weight substantially and thereby increases I_{sp} even though the flame temperature of this combination is similar to many other propellants.

To give the students a feel for the range of certain parameters derived in this chapter, typical values are summarized in Table 4.1. As a practical measure, we don't control γ, it is really an artifact/result of the propellant combination we select. Of course, low values of $\mathfrak{M}$ produce higher performance as reflected in Table 4.1. In many instances, the molecular weight of the fuel is lower than that of the oxidizer. In this case, the parameter $\sqrt{T_c/\mathfrak{M}}$ is maximized by running at fuel rich conditions. This reason explains why we so much afterburning in many rocket plumes as excess fuel is being combusted with atmospheric oxygen.

Example 4.1 Ideal Performance of the Titan IV Stage 1 Engine

The Titan vehicle was used by the US Air Force from the mid 1950s through 2005. The vehicle grew out of the early Titan ballistic missiles by implementation of solid rocket strap-on boosters. The first-stage engine in the core vehicle had a chamber pressure of 700 psi, an expansion ratio of 15 and a throat radius of 7.626 inches. The stage burned nitrogen tetroxide oxidizer with Aerozine

50 fuel at a mixture ratio (oxidizer/fuel mass ratio) of 1.95. Aerozine 50 is a fuel that is 50% hydrazine and 50% unsymmetrical dimethyl hydrazine (rocket fuel in its truest sense). Combustion of these propellants produces a c^* of 5712 f/s with an average γ value of 1.2 and a flame/combustion temperature of 5971 °F. Using these data, you are asked to determine the vacuum I_{sp}, thrust, and propellant flow rate and the sea-level I_{sp}, thrust, and flow rate.

Solution: Vacuum operation

From Appendix B, at $\varepsilon = 15$, for $\gamma = 1.2$ we find that $c_{fv} = 1.790$.

We can compute the vacuum specific impulse from Eq. 4.36:

$$I_{spv} = c_{fv}c^*/g = 1.790 * 5712/32.2 = 317.5 \text{ s}$$

The mass flow (weight-flow in the case of English/Imperial units) can be computed from Eq. 4.30:

$$\dot{m} = gp_cA_t/c^* = 32.2 * 700 * \pi * (7.626)^2/5712 = 721 \text{ lbf/s}$$

The vacuum thrust can then be computed from the flow rate and specific impulse values:

$$F_v = \dot{m}I_{spv} = C_{fv}p_cA_t = 721 * 317.5 = 229,000 \text{ lbf}$$

For sea-level operation, we need to correct our vacuum thrust coefficient to the local ambient pressure using Eq. 4.35:

$$c_f = c_{fv} - (p_a/p_c)\varepsilon = 1.790 - 15 * (14.7/700) = 1.475$$

The resulting sea-level specific impulse is then:

$$Isp = c_fc^*/g = 1.475 * 5712/32.2 = 261.6 \text{ s}$$

The mass flow at sea-level is the same as the result in a vacuum since the nozzle is choked (i.e. since the chamber pressure is high enough to produce supersonic flow when expanded to atmospheric conditions in Eq. 4.5. Choked flow implies that the flow is insensitive to ambient or "back" pressure. Professors love to offer trick questions of this sort. For the sea-level thrust, we can compute from the definition of I_{sp} as above, or use the definition of thrust coefficient:

$$F = c_fp_cA_t = 1.475 * 700 * \pi * (7.626)^2 = 189,000 \text{ lbf}$$

We can see the hefty decrease in both thrust and I_{sp} associated with the ambient pressure operation. As this decrease is entirely attributed to the PA thrust, $(p_e - p_a)A_e$, you can see its importance in the overall mix.

4.3   DESIGNING NOZZLE AERODYNAMIC CONTOURS

An efficient nozzle contour will enable optimal expansion of the high-pressure chamber gases to supersonic conditions without any shocks. Most rocket engines make use of a conventional de Laval nozzle with circular cross-section to accomplish this task. The parameters that describe this

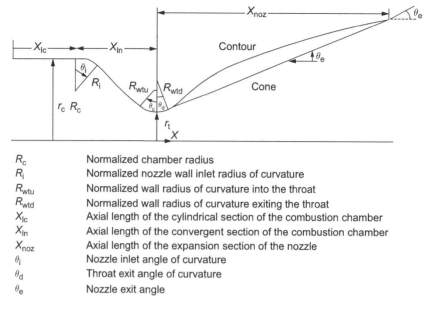

R_c	Normalized chamber radius
R_i	Normalized nozzle wall inlet radius of curvature
R_{wtu}	Normalized wall radius of curvature into the throat
R_{wtd}	Normalized wall radius of curvature exiting the throat
X_{lc}	Axial length of the cylindrical section of the combustion chamber
X_{ln}	Axial length of the convergent section of the combustion chamber
X_{noz}	Axial length of the expansion section of the nozzle
θ_i	Nozzle inlet angle of curvature
θ_d	Throat exit angle of curvature
θ_e	Nozzle exit angle

Figure 4.7 Contour definition for a conventional de Laval nozzle.

contour are given in Figure 4.7. The chamber length, X_{lc}, is set by the combustor designer. The normalized chamber radius, R_c, is also a parameter that is typically set by the combustor designer. The contraction area ratio $A_c = (R_c/R_t)^2$ is an important combustor design parameter as it effectively sets the velocity of gases in the chamber. See Chapters 7 and 8 for more discussion on this matter.

The nozzle entry region in Figure 4.6 is shown as a simple conical surface at *convergence angle* of θ_i. Designs of this region vary considerably depending on volume constraints. For example, a steeper inlet angle can be used to effectively shorten the nozzle and therefore minimize engine length. The nozzle inlet length, X_{ln}, is effectively set by the choice of θ_i, R_{wtu}, and R_i values, where R_i is the radius of curvature of the arc establishing the inlet conical surface.

The throat region is usually formed by circular arcs through the angles θ_i and θ_d for the inlet and exhaust sides of the nozzle. While θ_i values can vary considerably, θ_d is generally restricted to angles between 10 and 25 degrees. The radii of curvature defining upstream and downstream circular arcs are denoted R_{wtu} and R_{wtd}, respectively. In general, the two radii of curvature, which are non-dimensionalized with respect to the throat radius, are not equal. Typical values lie in the range $0.2 < R_{wtu}, R_{wtd} < 4.0$, within many of today's nozzle designs. In some cases, a short cylindrical section is placed between these circular arcs to establish a small constant area section for the throat. This technique is used mainly for machining purposes.

The exit cone geometry can be specified as a simple cone at angle θ_d, or as a series of points defining a contour. In general, *contoured nozzles* can be made shorter (and lighter) than a *conical nozzle* with the same expansion ratio. In addition, a contoured design often can give a lower nozzle exit angle, θ_e, which will be advantageous in reducing 2-D losses as shown in the following section of this chapter. Methods of characteristics or CFD techniques are often utilized to optimize this contour and ensure that no shocks are formed. In the 1950s, Rao (1958) developed an approximate method to determine the exit cone of a contoured nozzle by fitting this region with a series of parabolas.

His approach is still used frequently and in a preliminary design one can specify the shape of the exit contour by a parabola that starts at the end of the R_{wtd} circular arc with angle θ_d and ends at the nozzle exit with angle θ_e. Knowing the starting x location of the parabola and the two angles is sufficient to define the parabolic curve.

The contoured nozzle has the disadvantage of being more difficult to fabricate, although the production complications are only minor when compared to a conical design. Oftentimes, the length reduction of the nozzle is specified with respect to a conical nozzle with the same exit angle. An 80% bell (contoured) nozzle with a given exit angle corresponds to a nozzle that is 20% shorter than the comparable bell design.

Since most solid propellant combustion products contain particles, erosion of insulation and nozzle surfaces can be a concern. Particles tend to impact in the converging portion of the nozzle, leading to high insulation erosion in this region. One technique used to alleviate aft dome erosion problems is to move the nozzle assembly forward, thus "submerging" the nozzle within the aft dome of the motor case. A schematic of such a design is shown in Figure 4.8. Submergence of the nozzle reduces erosion of the insulation in the "reentrant region" above the nosecap and also reduces the overall length of the motor. This reduction in length can be quite beneficial for "upper-stage" motors, which must be stored within the fairing of a launch vehicle or within the bay of the Space Shuttle. Systems of this type are said to be *volume limited* (or volume constrained), and it is desirable to package the propulsion system efficiently for these applications.

Submerged designs are not normally considered for cooled nozzles since the surface area exposed to the flame is much greater than in a conventional design. Effects of submergence on aerodynamic efficiency are quite minor. Results in the literature give conflicting reports;

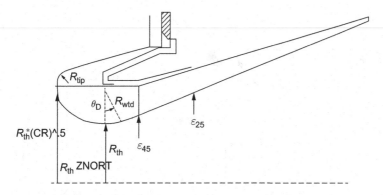

Figure 4.8 Contour definition for a submerged de Laval nozzle.

Landsbaum *et al.* (1980) found that nozzle submergence improved efficiency, while Kordig and Fuller (1967) indicate that nozzle efficiency drops with increasing submergence. Another negative aspect associated with the submerged design is the tendency of partially or completely combusted particles to become trapped in the reentrant region. Trapping of these particles, called *slag*, tends to reduce the overall mass fraction slightly and contributes to a small performance loss.

The internal contour of a submerged nozzle is similar to a conventional nozzle in the throat and exit cone regions. The entry region can be designed with an arbitrary spline or hyperbolic spiral shape. The variable L_{sub} denotes the amount of submergence of the throat in comparison to the aft end of the motorcase. (Note that other references use differing definitions of L_{sub}, such as the percentage of nozzle or motor case length submerged.) The rounded structure forming the forward-most part of the nozzle is referred to as the *nosecap*. The region near the nosecap tangency point is often defined as a circular arc of radius R_{tip}. The design of the contour above the nosecap varies widely and is dependent on the specific motor design to a large extent.

4.4 Non-Conventional Nozzles

For vehicles operating within the Earth's atmosphere over a range of altitudes, the ambient pressure can vary considerably during the course of engine operation. For example, a nozzle used as a first-stage propulsion system for a launch system will be overexpanded initially due to high ambient pressure at sea level. As the rocket rises through the atmosphere, the nozzle will eventually reach a perfectly expanded state and beyond this point the flow will be underexpanded.

To avoid low-altitude performance losses associated with this phenomenon, two types of *altitude compensating* nozzles have been devised. The *aerospike* and *expansion-deflection* nozzles shown in Figure 4.9 both have these features. The aerospike nozzle has two variants in which the base region may or may not have exhaust gases exiting through its surface. As the external pressure changes, the outer flow boundary automatically adjusts and flow remains attached to the inner wall at all times. The expansion-deflection nozzle operates under a similar principle. In this case, a recirculation region is formed behind the centerbody. The size of the recirculation zone shrinks

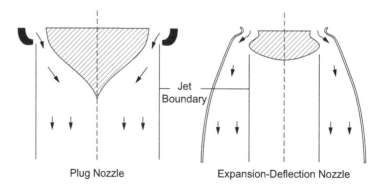

Figure 4.9 Plug (aerospike) and expansion-deflection nozzles (Source: Rao, 1958).

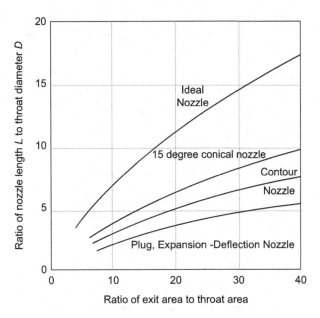

Figure 4.10 Nozzle length comparisons.

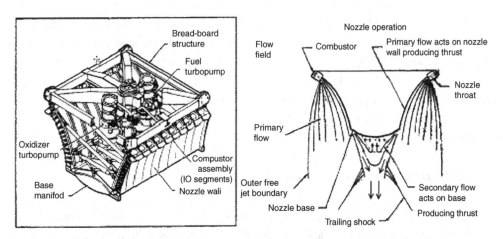

Figure 4.11 (a) Aerojet Rocketdyne Corporation linear aerospike engine. (b) Flowfield produced by aerospike nozzle (Source: images courtesy of Aerojet Rocketdyne).

naturally as the ambient pressure decreases, thus keeping the flow attached to outer walls at all times.

The aerospike and expansion-deflection nozzles also enjoy the advantage of being shorter in length (and possibly lighter in weight) as shown in Figure 4.10, courtesy of Rao (1958). The ideal nozzle would have a zero degree exit angle to maximize axial thrust, while the 15° conical would be

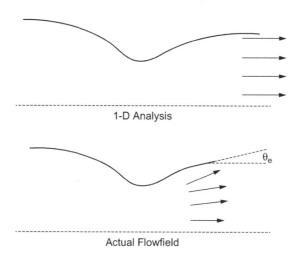

Figure 4.12 Effect of nozzle divergence.

much shorter. Using a contoured nozzle reduces length further since higher expansion angles can be used in the throat region. However, the aerospike and expansion-deflection nozzles are more compact than any of the conventional nozzles.

The difficulties associated with these non-conventional nozzles generally involve the annular throat region. Manufacturing of an annular throat to exacting tolerances is difficult for small engines and next to impossible for larger engines because the throat region is defined as the difference between two large radii. In addition, the annular throat has much higher exposed surface area than in a conventional design so nozzle cooling can be a big problem. For this reason, aerospike engines, which have multiple chambers feeding the nozzle, have been considered. In fact, Rocketdyne has tested aerospikes of this type in both circular and linear designs as shown in Figure 4.11.

4.5 Two-Dimensional Flow Effects

The isentropic flow equations (Eqs. 4.5–4.7) provide the pressure, temperature, and density distribution along an ideal nozzle assuming that the flow is parallel to the nozzle centerline at all locations. In actuality, the flow exits the nozzle at various angles with respect to the nozzle centerline as shown in Figure 4.12. At the nozzle edge, the flow is tangent to the surface with angle θ_e. This phenomena leads to losses in thrust since the flow is completely in the axial direction. Losses due to this mechanism are called *nozzle divergence losses* or two-dimensional (2-D) flow effects.

It is clear from Figure 4.12 that the nozzle divergence loss must depend on the magnitude of θ_e. We can determine the magnitude of the loss by assuming that all the fluid enters the nozzle from a point source as shown in Figure 4.13. It turns out that this "source flow" solution is quite valid as

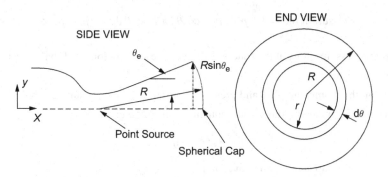

Figure 4.13 Nozzle flow emanating from a point source.

long as the nozzle has an expansion ratio greater than three. The control volume in Figure 4.13 is spherical in nature; therefore the bounding surface at the end of the nozzle is a spherical cap of radius R and angle θ_e. The unit normal vector for this surface is simply $\cos \theta_e$ if we are interested in forces in the x direction as shown in Figure 4.13.

For steady flow, the momentum theorem states that the sum of forces on the control volume is equal to the integral of the momentum flux through the control surface at the end of the nozzle. Let us first look at the momentum flux contribution in the x direction:

$$\text{Momentum flux} = \int v_x \rho (\mathbf{v} \cdot \mathbf{n}) \mathrm{d}s \tag{4.39}$$

where v_x is the velocity component in the x direction. If we define v_s as the velocity vector along the spherical cap then it is clear from Figure 4.13 that $v_x = v_s \cos \theta$. Referencing the end view in Figure 4.13, we can write the incremental area as:

$$\mathrm{d}s = 2\pi r R \, \mathrm{d}\theta = 2\pi R^2 \sin \theta \, \mathrm{d}\theta \tag{4.40}$$

since $r = R \sin \theta$. Making these substitutions into Eq. 4.39 and integrating between $0 < \theta < \theta_e$ gives:

$$\text{Momentum flux} = \pi R^2 \rho_s v_s^2 \sin^2 \theta_e \tag{4.41}$$

where ρ_s is the gas density along the spherical cap. Noting that $\dot{m} = \rho_s v_s A_s$ and that the quantity $\pi R^2 \sin^2 \theta_e = A_e$, we can finally write our momentum flux contribution:

$$\text{Momentum flux} = \dot{m} v_s A_e / A_s \tag{4.42}$$

The sum of forces on the control volume includes the pressure force on the cap and the net thrust of the nozzle. Using our incremental area and unit normal vectors for this situation, we may write (for the x direction):

$$\text{Pressure force} = \int_0^{\theta_e} (p_s - p_a) \, \cos\theta \, 2\pi R^2 \, \sin\theta \, d\theta = (p_s - p_a)A_e \tag{4.43}$$

where p_s is the pressure acting at the spherical cap and the substitution for A_e has been made using the relation shown above.

Having the momentum flux and pressure contributions, the remaining force is simply the thrust of the nozzle. Including this in our momentum equation gives:

$$F = \dot{m}v_s A_e/A_s + (p_s - p_a)A_e \tag{4.44}$$

but we can write the cap area in terms of R and θ_e by noting:

$$A_s = \int_0^{\theta_e} 2\pi R^2 \, \sin\theta \, d\theta = 2\pi R^2(1 - \cos\theta_e) \tag{4.45}$$

so that:

$$A_e/A_s = \frac{\sin^2\theta_e}{2(1 - \cos\theta_e)} = \frac{1 + \cos\theta_e}{2} \tag{4.46}$$

Using Eq. 4.46, the nozzle thrust can be written:

$$F = \lambda_d \dot{m} u_s + (p_s - p_a)A_e \tag{4.47}$$

where $\lambda_d = (1 + \cos\theta_e)/2$ is called the *divergence factor*. Note that from Eq. 4.12, we obtained:

$$F_{id} = \dot{m}v_e + (p_e - p_a)A_e \tag{4.48}$$

Now if the nozzle is long enough (i.e. it has a high enough expansion ratio), the spherical cap lies close to the exit plane at all points so that $v_s \approx v_e$. Assuming that the pressure thrust is negligible provides for a quick estimate of the jet thrust contribution due to nozzle divergence:

$$\frac{F}{F_{id}} \approx \lambda_d = \frac{1 + \cos\theta_e}{2} \tag{4.49}$$

Note that $\lambda_d < 1$ for all $\theta_e > 0$ so that a net loss in thrust is realized due to this effect. While this factor is strictly to be applied to the jet thrust component, it is frequently scaled against the entire ideal thrust (Eq. 4.12). While this isn't strictly correct, in most cases the pressure thrust is small relative to the jet thrust and the overall correction isn't impacted greatly. From a practical standpoint, this simple scaling is only appropriate for preliminary design and the main motivation for sharing the derivation here is to demonstrate the impact of exit angle selection on overall thrust. Despite the rather crude assumptions employed, the result compares well with measurements over a reasonable range of exit angles. Test results at small exit angles are obscured due to large boundary layer losses in these long nozzles. At high exit angles, experiment deviates from theory due to flow separations that occur in this region.

4.6 NOZZLE SHOCKS AND SEPARATION

4.6.1 Treatment of Nozzle Shocks

If the nozzle is poorly designed, there is a possibility that a normal shock could exist within the divergent portion downstream of the throat. While this condition rarely exists in actual nozzles (since we design them to avoid shocks) analysis of this phenomenon is quite straightforward. Since the flow behind the shock is subsonic, it will "sense" the local ambient back pressure and position the shock such that the exit pressure is equivalent to the ambient condition. The following iterative procedure can be used:

1. Assume a shock location (area ratio).
2. Use isentropic flow tables to get the conditions just upstream of the shock; call these conditions $M_1, p_1, p_{c1} = p_c$.
3. Use normal shock relations to get the conditions just downstream of shock (M_2, p_2, p_{c2}).
4. Determine the A^* value (new throat area due to stagnation pressure drop across shock) consistent with M_2.
5. Expand the flow to area ratio A_e/A^* and determine the exit pressure (p_e) from isentropic flow tables.
6. If $p_e = p_a$ assume the shock position is correct; if not return to the first step and assume a new location until iterations converge.

As mentioned above, we rarely see normal shocks within real nozzles. If shocks are present, they are generally oblique in nature due to the presence of boundary layers in real nozzles. Oblique shocks often signal the presence of nozzle separation as discussed in the following subsection.

4.6.2 Nozzle Separation

If the nozzle is sufficiently overexpanded, the possibility of nozzle separation exists. This condition is shown schematically in Figure 4.14. Nozzle separation results from flow in the boundary layer having insufficient energy to overcome the adverse pressure gradients characteristic of overexpanded nozzles. In rocket applications, this phenomenon is usually most prominent at sea level conditions where the ambient "back pressure" is highest. Nozzle designers typically try to avoid having separated flow within the nozzle due to the highly transient nature of the phenomenon. During engine ignition (when the chamber pressure is low), transient separation is observed in many engines. Typically, separation will occur only on one side of the nozzle thereby creating an azimuthal pressure distribution, which can lead to large side forces due to the comparatively higher atmospheric condition on the separated side of the nozzle. Figure 4.15 illustrates the origin of the side force generated from an asymmetric nozzle separation. In practice, unstable situations, such as a rotating separation region, might also be observed.

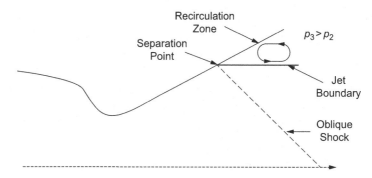

Figure 4.14 Flow separation in a de Laval nozzle.

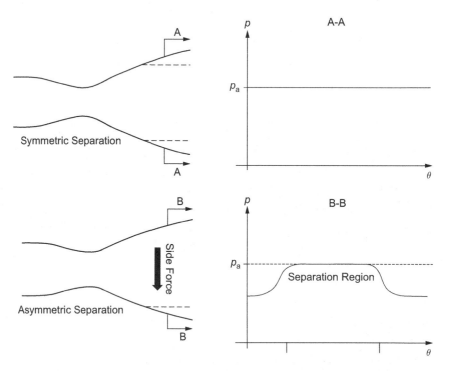

Figure 4.15 Side force generation from an asymmetric nozzle separation event.

Most of the early research on nozzle separation has been experimental in nature, as it has been challenging to predict the separation of a turbulent boundary layer computationally. Our ever-increasing abilities to model turbulent flows will continue to improve our capabilities to predict the exact conditions under which separation will occur. Early experimental efforts showed that the local separation pressure ratio, $p_{\mathrm{sep}}/p_{\mathrm{a}}$, was in the range $0.3 < p_{\mathrm{sep}}/p_{\mathrm{a}} < 0.4$, and as a rough rule of thumb many will claim that we can design a nozzle to be unseparated if we assume an exit pressure of 1/3 of an atmosphere. For preliminary designs, many groups have used the Kalt–Bendall correlation that provides a predicted separation pressure based on the ratio of chamber and ambient pressures (Kalt and Bendall, 1965):

$$\frac{p_{\text{sep}}}{p_a} = 0.667(p_c/p_a)^{-0.2} \tag{4.50}$$

One can use Eq. 4.50 to determine the separation point within a nozzle provided that the nozzle contour $A/A^* = f(x)$ is known (here x is the length along the nozzle centerline measured from the throat). Therefore, for a given x location, A/A^* is known and the local wall Mach number and static/ stagnation pressure ratio p_w/p_c can be determined from 1-D isentropic flow tables. Since $p_w/p_a = (p_w/p_c)(p_c/p_a)$, the separation point occurs when this ratio is equal to the result from Eq. 4.50. The Further Reading at the end of the chapter provides other experimental studies/ correlations for nozzle separation.

To estimate the thrust of a separated nozzle, we can neglect any force contributions within the separated region, as the local pressure there is very near ambient. First determine the separation point and separation pressure using the technique described above. The thrust corresponds to the equivalent force developed, assuming that the nozzle has been shortened to the separation point. Using this approach, we can calculate the thrust coefficient for the abbreviated expansion ratio assuming an exit pressure p_{sep} at this location.

It is interesting to note that the separated thrust actually exceeds the thrust one would calculate assuming that the nozzle was "flowing full." This result is caused by the reduction in the pressure/area contribution to thrust for the full flowing nozzle. While the separated nozzle has a lower exit velocity, the p–A thrust term is less negative than for the full flowing nozzle due to the reduced effective exhaust area and increased pressure, i.e. we are closer to operating the nozzle at its maximum thrust point corresponding to perfect expansion. Designers rarely make use of this benefit since the transient, unpredictable nature of the separation process often forces the nozzle to designs that never separate (at least during steady-state operation). As there are substantial benefits to be gained, there have been a number of research efforts over the years aimed at developing an altitude compensating capability that permits stable operation over a wide altitude range. One could place a series of boundary layer trips (aft-facing steps or other structures) within the nozzle to "lock in" the separation point over a range of altitudes, but these structures will inherently add drag/ viscous losses to the full flowing nozzle.

4.7 Two-Phase Flow Losses

An important phenomenon that can drastically affect the performance of some rockets is caused by the presence of liquid or solid particles within the exhaust stream. This effect is generally restricted to solid rocket motors. Solid propellants generally utilize metals (typically aluminum) as fuels so that the combustion products contain considerable amounts of metallic oxides in either liquid or

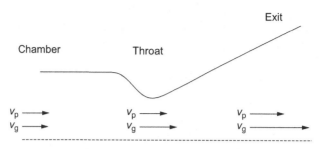

Figure 4.16 Two-phase flow losses caused by "lag" in particle velocity.

solid form. Since these particles do not have infinite drag coefficients, they tend to "lag" gas molecules in their acceleration process through the nozzle. The net effect of this lag is a reduction in exit velocity as compared to the result given in Eq. 4.33. If we let u_p and u_g denote particle and gas velocities respectively, the general situation is depicted in Figure 4.16.

To further investigate this situation, consider the motion of a particle of mass m by writing $F = ma$ on this body:

$$m\frac{dv_p}{dt} = C_D \frac{1}{2}\rho_g A_p (v_g - v_p)^2 \qquad (4.51)$$

where C_D is the particle drag coefficient and A_p is the particle cross-sectional area. If we know the particle drag behavior and the gas flow ρ_g, v_g through the nozzle, we can use Eq. 4.51 to solve for the particle velocity at the nozzle exit plane. Typically, the presence of particles is ignored in calculating the gas-phase flow solution since the particles occupy a very small amount of the available volume.

To express Eq. 4.51 in terms of a spatial distance x, note that $v_p = dx/dt$:

$$mv_p\frac{dv_p}{dx} = C_D \frac{1}{2}\rho_g A_p (v_p - v_g)^2 \qquad (4.52)$$

or

$$\frac{dv_p}{dx} = \frac{\rho_g}{2\beta}\frac{(v_g - v_p)^2}{v_p} \qquad (4.53)$$

where $\beta = m/(C_D A_p)$ is called the particle *ballistic coefficient*. We can see from the relation above that particles with low ballistic coefficients will generate large dv_p/dx values and will therefore closely follow the gas. Particles with high ballistic coefficients will not be affected as much by the drag force and will lag the gas to a greater extent.

Now we need to be reminded of the behavior of the drag coefficient. For spherical particles, C_D is a function of Reynolds number as shown in Figure 4.17. Here the Reynolds number must be based on the velocity relative to the particle. In the chamber, the gas density is high and the Reynolds number is low so that dv_p/dx is large and the particles travel essentially at the gas

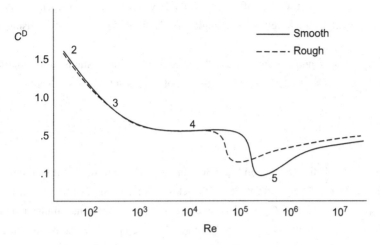

Figure 4.17 Drag coefficients for spherical particles (Source: NASA).

velocity. As the gas accelerates through the throat, the density drops, C_D drops, and a lag begins to develop. In the divergent portion of the nozzle, the low gas density lowers the drag on the particle further, and a significant lag is realized at the exit plane. Since larger particles will have higher Reynolds numbers and lower drag coefficients, these particles will be more susceptible to the phenomenon described above.

For large solid rocket motors, two-phase flow losses can be the major deviation from the ideal performance. Two-phase flow losses can account for a 5–10% loss in thrust for current solid rocket motor propellants employed today. Losses become more important for nozzles with high expansion ratios due to the highly rarefied gas near the exit plane. As the particles lag the flow in both velocity and direction, there can be problems with using a contoured nozzle for a solid rocket motor. The particles may impact the surface that is curving back toward the nozzle centerline and cause substantial erosion (and presumably thrust loss) upon impacting the aft part of the exit cone. For this reason, many solid motors utilize simply conical nozzle designs.

4.8 BOUNDARY LAYER LOSSES

The presence of boundary layers along the walls of the nozzle has two primary effects on the 1-D analysis presented in Chapter 4. Since the presence of fluid viscosity leads to a velocity profile at any point in the nozzle, the average velocity of the real fluid is lower than that calculated using isentropic flow equations. This factor can be taken into account by using a velocity coefficient, C_v, which is defined as the ratio of the actual exit velocity (averaged over the exit plane cross-section) and the ideal result given by Eq. 4.23:

$$C_v = \frac{v_{e-actual}}{v_{e-ideal}} \tag{4.54}$$

In practice, Eq. 4.55 is not used that often since boundary layer, nozzle divergence, and two-phase flow effects are often combined into a single nozzle fluid flow calculation using numerical techniques.

The effect of boundary layers in the throat region has implications for the mass flow rate through the nozzle. The nozzle discharge coefficient, C_d, is introduced to take this effect into account:

$$C_d = \frac{\dot{m}_{actual}}{\dot{m}_{ideal}} \tag{4.55}$$

where $\dot{m}_{ideal}$ represents the result from Eq. 4.30. For an adiabatic flow, the throat radius is reduced by the effective boundary layer displacement thickness which implies that for a given chamber pressure $\dot{m}_{actual} < \dot{m}_{ideal}$ or that $C_d < 1$. However, in most applications, heat transfer at the throat is quite important. Since the wall is cold, the boundary layer temperature can be significantly lower than the outer flow. This effect causes an increase in gas density within the boundary layer, which effectively increases the massflux through this region of the flow in the throat. The net result can exceed the effect of displacement thickness to the point that $C_d > 1$. Actual discharge coefficients typically lie in the range of $0.93 < C_d < 1.15$, but for larger throats they are within a few percent of unity.

4.9 METHOD OF CHARACTERISTICS FOR AXISYMMETRIC FLOWS

While the methods for describing nozzle contours in Section 4.3 should be adequate for preliminary design purposes, they fall short of providing a precise contour that delivers the maximum performance. From a practical perspective, manufacturers will fight to achieve the very highest level of performance and differences/advances as small as 0.1% can be the subject of substantial investigation and effort. The state-of-the-art approach for designing nozzle exit cones assumes an inviscid compressible flow using the technique of Method of Characteristics. This approach is covered in some detail in numerous gas dynamics texts; those interested in a detailed account are referred to Zucrow and Hoffman (1967) or Anderson (2003) at the end of the chapter. For axisymmetric, inviscid compressible flow, the momentum equation can be written:

$$(u^2 - a^2)\frac{du}{dx} + (v^2 - a^2)\frac{dv}{dr} + 2uv\frac{du}{dr} - \frac{a^2v}{r} = 0 \tag{4.56}$$

Where u, x and v, r are axial and radial velocities and coordinates, respectively, and a is the sound speed as before. Additionally, the irrotational flow assumption necessitates that:

$$\frac{du}{dr} - \frac{dv}{dx} = 0 \tag{4.57}$$

These two equations can be algebraically manipulated to give:

$$(u^2 - a^2)C_{+/-}^2 - 2uvC_{+/-} + (v^2 - a^2) = 0 \tag{4.58}$$

where:

$$C_{+/-} = \left(\frac{dr}{dx}\right)_{+/-} = \frac{uv + a^2\sqrt{M^2 - 1}}{u^2 - a^2} \tag{4.59}$$

and M is the Mach number as before. The quantities $C_{+/-}$ are called *characteristics* and their slope is defined by Eq. 4.59. Formally, Eq. 4.59 is actually two equations, one for the slope of the left-running C_+ characteristic and the other for the right-running C_- characteristic. If we write the velocities in terms of flow speed (V) and angle (θ) then we can substitute $u = V\cos\theta$ and $v = V\sin\theta$ into Eq. 4.59 to eliminate the individual velocity components in favor of the flow angle. In addition, if we recall the Mach number is related to the Mach angle, μ, via $\mu = \sin^{-1}(1/M)$, then Eq. 4.59 can be recast into the simple form:

$$C_{+/-} = \left(\frac{dr}{dx}\right)_{+/-} = \tan(\theta \pm \mu) \tag{4.60}$$

A fundamental result from gas dynamics demonstrates that the Mach angle determines the region where information propagates in a supersonic flow, i.e. points at angles greater than $\mp\mu$ cannot affect an axial flow. Since in general the flow can be inclined at an angle θ, the characteristics then describe the uppermost (C_+ characteristic) and lowermost (C_- characteristic) extent that information can be convected downstream. For an accelerating supersonic flow such as in a rocket nozzle, M will increase and lead to corresponding changes in μ; as a result the characteristics will map out curved paths as the flow accelerates downstream.

Equation 4.58 can also be cast in terms of flow angles by once again casting u, v in terms of V, θ to give:

$$d\theta = \mp \sqrt{M^2 - 1}\frac{dV}{V} \pm \frac{dr/r}{\sqrt{M^2 - 1} \mp \cot\theta} \tag{4.61}$$

You may recognize the first term on the right-hand side of this relation as it is identical to the Prandtl–Meyer function, $d\nu$. The Prandtl–Meyer angle, ν, results from integration of this term and represents the angle achieved by an expansion fan in a simple wave region such as flow around a corner. Substituting this result into Eq. 4.61, we arrive at two relations for the flow angles and radial changes along characteristic lines:

$$d(\theta + \nu) = \frac{dr/r}{\sqrt{M^2 - 1} - \cot\theta} \quad \text{(along } C_- \text{ characteristic)} \tag{4.62}$$

$$d(\theta - \nu) = -\frac{dr/r}{\sqrt{M^2 - 1} + \cot\theta} \quad \text{(along } C_+ \text{ characteristic)} \tag{4.63}$$

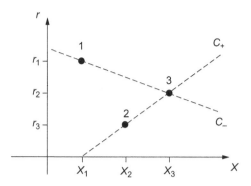

Figure 4.18 MOC application on a discrete mesh. Consider the conditions at points 1 and 2 known and the integration aims to determine the radial location of point 3 (r_3) and the resulting flow conditions at that location.

These two relationships are called *compatibility conditions* as they ensure that the relationship between flow angles along the characteristics satisfy the momentum equation for compressible flow. Recognizing that $v = v(M)$, Eq. 4.60 and Eqs. 4.62–4.63 provide four conditions for incremental changes in r, θ, M, and μ. Mathematically, these relationship represent a set of parabolic nonlinear ordinary differential equations. The equations are nonlinear in that the complex relationships between Mach and Prandtl–Meyer angles with Mach number do not support an analytic solution. The equations are parabolic meaning that they are amenable to "time-stepping" approaches. Here, the flow direction relates to time since we know that no information can travel upstream in a supersonic flow.

The Appendix of this book provides information on various time-stepping algorithms with different levels of accuracy. To illustrate the steps in a numerical MOC approach, let's consider the simplest approach, i.e. the forward Euler scheme. In this method, we argue that time-steps (in this case steps in the axial/main flow direction) are small enough such that derivatives can be estimated using values at the current "time" (i.e. current axial location).

Figure 4.18 outlines the situation where we are stepping in the x direction from points 1 and 2 where all information is currently known. Beginning with the integration of Eq. 4.60, we write:

$$dr_+ \cong\ r_3 - r_1 = \int_{x_1}^{x_3} \tan(\theta + \mu) dx \tag{4.64}$$

The function $\tan(\theta + \mu)$ is a function of x implicitly through the Mach number changes that occur over the incremental distance $x_3 - x_1$. Approximating this function with the value at $x = x_1$ then permits direct integration of the remaining differential:

$$r_3 - r_1 = \tan(\theta_1 + \mu_1)(x_3 - x_1) \tag{4.65}$$

This gives us the relationship between r, x, and angles along the left-running characteristic. Using a similar procedure for the right-running characteristic, we obtain:

$$r_3 - r_2 = \tan(\theta_2 - \mu_2)(x_3 - x_2) \tag{4.66}$$

Proceeding in a similar fashion with Eq. 4.62, we have:

$$\mathrm{d}(\theta + v) \cong (\theta_3 + v_3) - (\theta_1 + v_1) = \int_{r_1}^{r_3} \frac{\mathrm{d}r/r}{\sqrt{M^2 - 1} - \cot\theta} \tag{4.67}$$

Using the forward Euler integration, the differential relationship for this left-running characteristic is then:

$$(\theta_3 - v_3) - (\theta_1 - v_1) = \frac{(r_3 - r_1)/r_1}{\sqrt{M_1^2 - 1} - \cot\theta_1} \tag{4.68}$$

And the difference equation representation of Eq. 4.63 and the right-running characteristic becomes:

$$(\theta_3 + v_3) - (\theta_2 + v_2) = \frac{(r_2 - r_3)/r_2}{\sqrt{M_2^2 - 1} + \cot\theta_2} \tag{4.69}$$

Equations 4.65, 4.66, 4.68, and 4.69 are four conditions for the unknowns r_3, x_3, M_3, and θ_3 assuming we are advancing to a new Prandtl–Meyer angle, v_3. Details on other numerical approaches can be obtained from Zucrow and Hoffman (1976) or Anderson (2003).

FURTHER READING

Altman, D., Carter, J. M., Penner, S. S., and Summerfield, M. (1960) *Liquid Propellant Rockets*. Princeton Aeronautical Paperbacks, Princeton University Press.

Anderson, J. D. (2003) *Modern Compressible Flow with Historical Perspective*, 3rd edn. McGraw Hill.

Elam, S. (2000) "Test Report for NASA MSFC Support of the Linear Aerospike SR-71 Experiment (LASRE)." Tech. rep., NASA Marshall Space Flight Center, NASA/TM-2000–210076.

Engel, C., Bender, R., Engel, B., and Sebghati, J. (1998) "NASA SBIR 95–1 Phase II Aerospike Plug Base Heating Model Development." Final report. Tech. rep., Qualis Corporation, Rept. QTR 017–012.

Frey, M., and Hagemann, G. (2000) "Restricted Shock Separation in Rocket Nozzles," *J. Propulsion & Power*, 16(3).

Hagemann G, Immich H, Dumnov G (2001) "Critical Assessment of the Linear Plug Nozzle Concept," in *37th Joint Propulsion Conference*, AIAA-2001–3683.

Hagemann, G., Preuss, A., Grauer, F., and Krestschmer, J. (2003) "Technology Investigation for High Area Ratio Nozzle Concepts," in *39th AIAA Joint Propulsion Conference*, AIAA-2003–4912.

Hoerner, Ing. S. F. (1958) *Fluid Dynamic Drag*. Published by Author.

Hunter, C. A. (1998) "Experimental, Theoretical and Computational Investigation of Separated Nozzle Flows," in *34th AIAA Joint Propulsion Conference,* AIAA-98–3107.

Kalt, S. and Bendal, D. (1965) "Conical Rocket Performance Under Flow Separated Conditions," *Journal of Spacecraft and Rockets*, 2(3): 447–449.

Kordig, J. W. and Fuller, G. H. (1967) "Correlation of Nozzle Submergence Losses in Solid Rocket Motors," *AIAA Journal*, 5: 175–177.

Landsbaum, E. M., Salinas, M. P., and Leary, J. P. (1980) "Specific Impulse Prediction of Solid-propellant Motors," *Journal of Spacecraft and Rockets*, 17(5): 400–406.

Lary, F. (1967) "Advanced Cryogenic Rocket Engine Program Aerospike Nozzle Concept, Materials and Processes Research and Development." Tech. rep., Rocketdyne, AFRPL-TR-67–278.

Lary, F. (1968a) "Advanced Cryogenic Rocket Engine Program, Aerospike Nozzle Concept: Volume 1." Tech. rep., Rocketdyne, AFRPL-TR-67–280.

Lary, F. (1968b) "Advanced Cryogenic Rocket Engine Program, Aerospike Nozzle Concept: Volume 2." Tech. rep., Rocketdyne, Accession Number: AD0387190.

Morisette, E. L. and Goldberg, T. J. (1978) "Turbulent Flow Separation Criteria for Overexpanded Supersonic Nozzles." NASA Technical Paper 1207, Langley Research Center.

Muss, J. and Nguyen, T. (1997) "Evaluation of Altitude Compensating Nozzle Concepts For RLV," in *33rd Joint Propulsion Conference*, AIAA-1997–3222.

Nasuti, F. and Onofri, M. (1998) "Methodology to Solve Flowfields of Plug Nozzles for Future Launchers," *Journal of Propulsion and Power*, 14(3): 318–326.

Rao, G. V. R. (1958) "Exhaust Nozzle for Optimum Thrust," *Jet Propulsion*, 28: 377–382.

Ruf, J. and McConnaughey, P. (1997) "A Numerical Analysis of a Three Dimensional Aerospike," in *33rd Joint Propulsion Conference*, AIAA-1997–3217.

Schmucker, R. H. (1984) "Flow Processes in Overexpanded Chemical Rocket Nozzles. Part 1: Flow Separation." NASA TM-77396.

Summerfield, M., Foster, C. R., and Swan, W. C., (1954) "Flow Separation in Overexpanded Supersonic Exhaust Nozzles," *Jet Propulsion*, 319–321.

Zucrow, M. J. and Hoffman, J. D. (1976) *Gas Dynamics, Vol. 1*. John Wiley & Sons.

HOMEWORK PROBLEMS

4.1 Consider the liquid rocket engine shown in Figure 4.19. Combustion of fuel and oxidizer at a chamber pressure of 5 MPa produces gases with $T_c = 3000$ K, $\mathcal{M} = 15$ kg/kg-mol, and $\gamma = 1.2$.

    (a) Assuming sea-level operation, 1-D isentropic flow throughout, and that chamber conditions are stagnation conditions, determine the thrust and specific impulse of this engine. Is this nozzle nearly perfectly expanded?

    (b) In our ideal analysis, we typically assume that chamber conditions are stagnation values. How much would the thrust from part (a) change if we relax that assumption and consider the fact that the chamber gases don't have zero velocity as they enter the nozzle contraction?

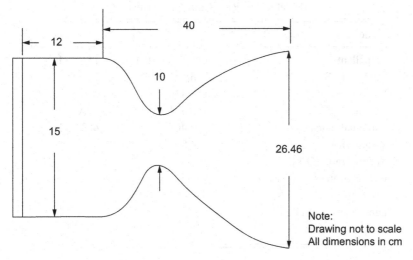

Figure 4.19 Diagram for Problem 4.1.

4.2 Below is an excerpt from a *New York Times* editorial referring to Goddard's rockets and the possibility of space flight. Briefly discuss (in words, without using equations) the errors in the author's logic.

As a method of sending a missile to the higher, and even to the highest parts of the Earth's atmospheric envelope, Professor Goddard's rocket is a practicable and therefore promising device. It is when one considers the multiple-charge rocket as a traveler to the moon that one begins to doubt . . . for after the rocket quits our air and really starts on its journey, its flight would be neither accelerated nor maintained by the explosion of charges it then might have left. Professor Goddard, with his "chair" in Clark College and countenancing of the Smithsonian Institution, does not know the relation of action to re-action, and of the need to have something better than a vacuum against which to react . . . Of course he only seems to lack the knowledge ladled out daily in high schools (*New York Times*, editorial, 1920).

4.3 Table 4.2 summarizes the performance of the aNPO Energomash RD-704 rocket engine (Figure 4.20). This engine is a derivative of the RD-170 engine used in the Russian Zenit and Energia launch vehicles. The engine is unique in that it has a dual mode capability. At liftoff it uses a tripropellant $LO_2/LH_2/RP$-1 and at altitude, the RP-1 flow is terminated and the engine uses LO_2/LH_2 propellants. Additionally, assume that the ratio of specific heats is constant for both modes and equal to a value of 1.2.

(a) Based on this operational scenario, calculate the nozzle throat and exit diameters using Mode 1 information.

For each mode, determine the following:

Table 4.2 RD-704 characteristics

Mode	1	2
Propellants	$LO_2/LH_2/RP$-1	LO_2/LH_2
Thrust (lb) S.L.	386,140	N/A
Vac	441,430	175,560
Impulse (s) S.L.	356/351	N/A
(Nominal worst case) Vac	407/401	452/450
Weight (lb)	5,329	
Mixture ratio (O/F)	4.38	6.0
Chamber pressure (psia)	4,266	1,762
Area ratio	74	
Dimensions (in) Dia.	70.1	
(Single bell) Length	151	

Figure 4.20 The RD-704 rocket engine.

(b) characteristic velocity;

(c) jet and pressure thrust;

(d) flow rates of all propellants;

(e) nozzle exit velocity; and

(f) thrust coefficient.

(g) Why bother with the third propellant, RP-1? What is the advantage of using this fluid at liftoff?

4.4 An arcjet thruster heats hydrogen ($\gamma = 1.4$) to a temperature of 5000 °R in the chamber.

(a) Assuming choked flow, calculated the velocity of the gas at the throat in both English (f/s) and SI (m/s) units. Be sure to describe all dimensions and units conversions required for this calculation.

(b) If the hydrogen entered the engine at 300 K, how much energy (per unit mass) was added by the arc? You may assume constant specific heats ($c_p = 29$ J/g-mol K). Express your answer in BTU/lbm, kcal/kg, ft–lbf/lbm, and KJ/kg noting units and conversions.

(c) Express c_p (from part (b)) in kcal/kg K, J/g K, BTU/lbm °R, and f-lbf/lbm °R. Note conversions and units.

4.5 Intelligence sources indicate that a new missile is under development from another nation. Observations of a test flight of this vehicle reveal the following information:

Nozzle exit diameter = 1.0 m

Nozzle exit pressure (inferred from plume shocks) = 0.8 atm

Initial vehicle mass = 40,000 kg

Burning time = 100 s

Propellant γ and $c^* = 1.2$, 1500 m/s (estimated based on guess of propellant combination)

In addition, trajectory information reveals that the vehicle thrust increased 10% in going from sea level to its burnout attitude of 20 km. Assuming ideal performance and that chamber pressure remains constant for the entire flight, determine:

(a) engine thrust (both liftoff and burnout);

(b) engine chamber pressure;

(c) nozzle throat diameter;

(d) engine mass flow;

(e) engine I_{sp}; and

(f) payload mass assuming that all structural mass is equivalent to 15% of propellant mass.

4.6 A new liquid propellant combination is test fired in a subscale heavyweight test chamber as shown in the schematic in Figure 4.21. The parameters A_t, A_e, p_c, p_e, p_a, $\dot{m}$, and F are all measured for this test.

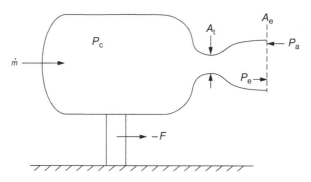

Figure 4.21 Diagram for Problem 4.6.

Outline a procedure to us this information to predict the thrust of a full-scale flightweight engine using the same propellant combination.

4.7 Derive the Mach number–area ratio relation:

$$\frac{A}{A_t} = \frac{1}{M}\left[\frac{2+(\gamma-1)M^2}{\gamma+1}\right]^{\frac{\gamma+1}{2(\gamma-1)}}$$

for steady, isentropic flow through a choked nozzle by using the continuity equation and compressible flow relations.

4.8 Write a computer code (or Matlab routine) to calculate the pressure distribution and thrust within an ideal conical nozzle with the following characteristics:

$p_c = 3$ MPa, $p_a = 0.1$ MPa, $F = 100$ KN,
$c* = 1500$ m/s, $\gamma = 1.2$, $R_{wtu} = R_{wtd} = 0.8$, $\varepsilon = 15$,
Contraction ratio $= 3.5$, $\theta_i = 45°$, $\theta_d = 15°$, $R_i = 0.5$
Use the following steps in your study:

(i) From the geometric information (and thrust level) provided, generate a series of equations for the local radius as a function of distance from the throat ($r = r(x)$). Plot the nozzle contour.

(ii) For a given position in the nozzle, x, use Newton's method to find the local Mach number from Eq. 4.9.

(iii) Using the method from part (b), plot the Mach number and static pressure distribution ($M = M(x)$, $P = P(x)$) for the contour specified above.

(a) Using your code plot c_f vs. ε for $p_c/p_a = 100$, $\gamma = 1.2$ over the range $1 \leq \varepsilon \leq 100$.

(b) For the nozzle defined in Problem 4.8, what range of ambient pressures support normal shock solutions for flow in the exit cone? If a shock exits at $\varepsilon = 5$, what ambient pressure must be present?

(c) Using the Kalt–Bendall separation criteria, predict the point (i.e. x location) where the flow will separate during a sea-level engine test firing.

4.9 An air-launched missile is designed to operate at an altitude of 20,000 feet using a solid rocket propellant which delivers a c^* of 5000 ft/s at $\gamma = 1.2$. The missile design is such that the nozzle expansion ratio is limited to 30 so as to have the nozzle exit fit within the desired envelope. The missile is loaded with 250 lb of propellant and has a 2.0" diameter throat.

    (a) Since the missile will primarily be used to "shoot up" at targets at higher altitudes, we desire for the nozzle to be on the verge of separation at ignition. Assuming a separation pressure of 1/3 the local ambient pressure, determine the chamber pressure required to force this condition.

    (b) Assuming constant chamber pressure operation at 20,000 feet using the value for p_c obtained in part (a), find the motor thrust and burning time.

    (c) Determine the sea-level thrust and burning time for this missile using the p_c value from part (a).

4.10 The solid rocket motor shown in Figure 4.21 is fired on a test stand; both thrust and chamber pressure are measured during the firing. Curvefits of the digital data give the following results for the 50 second burn:

$$p_c(t) = 2 + 0.21t - 0.005t^2 \rightarrow (\text{in MPa})$$

$$0 \le t \le 50$$

$$F(t) = 4 + 0.42t - 0.01t^2 \rightarrow (\text{in KN})$$

The motor is tested in an altitude chamber which is evacuated to an equivalent altitude of 100,000 ft. Prefire measurements indicate a propellant mass of 130 kg. Use these data to determine the following:

    (a) the measured average I_{sp} and corresponding vacuum I_{sp} for this test;

    (b) the characteristic velocity, c^* delivered in this test; and

    (c) the overall efficiency (ratio of average measured to ideal I_{sp}) of this motor assuming that the theoretical c^* for this propellant is 1530 m/s and that the average chamber pressure for this firing was 3 MPa.

4.11 Figure 4.22 shows the TRW Ultra Low Cost Engine (ULCE) currently being marketed for launch vehicle propulsion. The figure dimensions are in inches. The engine has a chamber of 700 psi and uses LOX/LH$_2$ propellants ($c^* = 7800$ ft/s) at a mixture (O/F) ratio of 6.0. Assuming $\gamma = 1.2$ and ideal engine performance, compute:

    (a) total engine propellant flow rate;

    (b) sea level thrust;

    (c) sea level jet thrust;

    (d) sea level pressure thrust;

    (e) vacuum thrust;

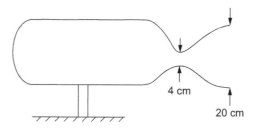

Figure 4.22 Diagram for Problem 4.10.

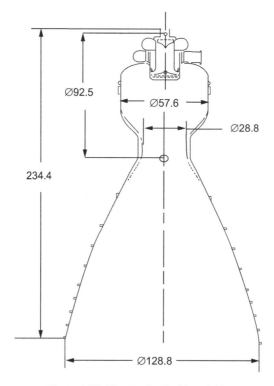

Figure 4.23 Diagram for Problem 4.11.

(f) sea level and vacuum I_{sp}; and

(g) sea level and vacuum nozzle exit velocities.

4.12 Analyze a rocket nozzle with the internal aerodynamic contour given by:

$$r = 1 + 0.435|x| - 0.00365x^2 - .000659|x|^3 \quad |x| \le 10$$

where r is the radial distance to the nozzle wall and x is the axial distance measured from the throat. Suppose that the stagnation quantities feeding this nozzle are:

$$P_c = 500\,\mathrm{psi},\ T_c = 5000\ ^{\circ}\mathrm{R},\ \mathcal{M} = 20\ \mathrm{lb/lb\text{-}mol}$$

(a) What is the nozzle expansion ratio?

(b) What mass flow (in lbm/s) is travelling through the nozzle?

(c) Write a computer code to solve for Mach number, static pressure and static temperature in the nozzle assuming choked flow. Use Newton's method to solve the nonlinear Mach number/area ratio relationship. Plot the following quantities:

   (i) the nozzle contour;

   (ii) Mach number variation as a function of x;

   (iii) static temperature variation as a function of x;

   (iv) static pressure variation as a function of x; and

   (v) velocity variation as a function of x.

(d) Attach a listing of your code from part (c).

(e) What would the ambient (back) pressure have to be to invalidate the assumption of choked flow?

4.13 Consider a rocket nozzle fed by chamber gases with $c^* = 1500\ \mathrm{m/s}$, $\gamma = 1.2$, and $P_c = 5\ \mathrm{MPa}$. If the expansion ratio of this nozzle is 25 and the throat diameter is 10 cm, answer the following questions:

(a) Determine the mass flow rate, thrust, thrust coefficient, and I_{sp} assuming sea-level operation with no nozzle separation.

(b) Determine the mass flow rate, thrust, thrust coefficient, and I_{sp} assuming sea-level operation with nozzle separation.

(c) At what altitude will this nozzle begin to flow full?

(d) What is the thrust and I_{sp} consistent with conditions in part (c)?

4.14 Consider the conical rocket nozzle shown in Figure 4.7 with the following specification:

$$r_t = 1\ \mathrm{ft},\ \varepsilon = 15,\ x_{lc} = 3\ \mathrm{ft}$$

$$R_i = 0,\ R_c = 2.5,\ R_{wtu} = 0.4,\ R_{wtd} = 0.8$$

$$\theta_i = 45^{\circ},\ \theta_d = \theta_e = 17^{\circ}$$

(a) Derive a set of relationships for the nozzle contour $r = r(x)$ using the variables given above. What is the total length of the nozzle?

(b) Plot (to scale) the nozzle contour.

4.15 A rocket with $\varepsilon = 25$ is operated at sea-level with a chamber pressure of 200 psi. Assuming a normal shock lies in the exit cone, determine:

(a) the area ratio of the shock location;

(b) the thrust loss (gain) due to the presence of the shock; and

(c) the shock position (area ratio) assuming an ambient pressure of 25 psi. You may assume $\gamma = 1.4$ for this problem.

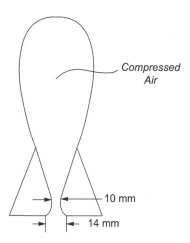

Figure 4.24 Diagram for Problem 4.16.

4.16 Consider the toy rocket running off compressed air as shown in Figure 4.24. A pump is used to force air into the chamber until the pressure is 6 atm; the gas is allowed to equilibrate to the ambient temperature (25 °C).

(a) Assuming ideal performance, compute the initial thrust and I_{sp} of the rocket at liftoff from sea level.

(b) The rocket is observed to fly straight for the first portion of the flight, but begins to go unstable as the pressure in the chamber drops. Assuming the instability is caused by flow separation and that the expansion of gases in the chamber is isothermal, compute the pressure at which separation will occur. You may use the Kalt–Bendall separation criterion.

(c) Suggest a design change to improve the stability of the rocket.

4.17 The RL 10 engine is used as a propulsion system for the Centaur upper stage. This engine utilizes LOX/LH$_2$ propellants which produce c^* of 7670 ft/s at $\gamma = 1.2$. Nozzle throat and exit diameters are 5.14 and 38.8 inches, respectively and the chamber pressure is 400 psi. Assuming vacuum operation, determine:

(a) Engine thrust coefficient;

(b) Total thrust;

(c) Momentum and pressure thrust;

(d) Nozzle exit velocity;

(e) Specific impulse;

(f) Engine mass flow; and

(g) Nozzle exit pressure

4.18 Repeat Problem 4.17 assuming sea-level operation with a full-flowing nozzle.

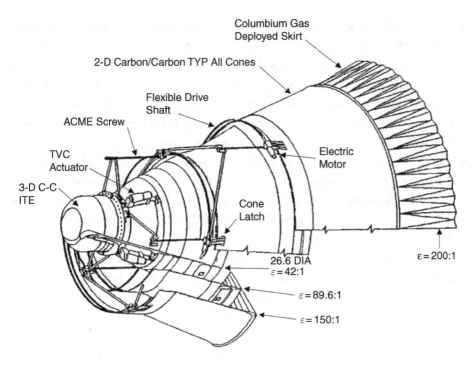

Figure 4.25 Diagram for Problem 4.19.

4.19 You are asked to design a nozzle for the first stage of a new launch vehicle. Your boss has asked you to consider a nozzle with an extendible exit cone (EEC) such as that shown in Figure 4.25 (without the gas deployed skirt). The following information is known:

$$\text{Desired sea level thrust} = 4 \text{ Mn}$$
$$p_c = 9 \text{ MPa}, \quad c* = 1600 \text{m/s}, \quad \gamma = 1.2$$
$$\text{Contraction ratio} = 3, \quad R_{wtu} = 0.5, \quad R_{wtd} = 0.3, \quad x_{lc} = 3$$
$$R_i = 0.3, \quad \theta_i = 60°, \quad \theta_d = 15°$$

You should design the first nozzle extension to be deployed at an altitude of 4 km, while the second extension should be deployed at an altitude of 9 km.

(a) Determine the nozzle throat size and the best expansion ratio for each of the three cases.
(b) Determine the thrust level before and after deployment of the two EEC assemblies.
(c) Make a scale drawing of your nozzle in both deployed and stowed positions. Indicate all important dimensions on your drawing.
(d) Is an EEC really a good idea for this application? Explain why this would (or wouldn't) be a good choice?

4.20 A control system error causes the EEC in Problem 4.19 to deploy after ignition. What happens? Determine the change in thrust and I_{sp} assuming one or both of the EEC subassemblies is deployed.

   Plot thrust coefficient as a function of expansion ratio for the nozzle in 4.10–19 at ignition and just after deployment of each of the EEC subassemblies. To make this plot, modify the computer code from Problem 4.8 above in order to calculate c_f and enclose this calculation in a "Do Loop" in order to recompute at different ε values. Put all three curves on the same plot and indicate operating points at deployment on your curves.

4.21 Consider a missile to be used for an air-launch application at altitudes from 20,000–40,000 ft. Packaging constraints dictate that the nozzle internal diameter be less than (or equal to) 7.0 inches. A minimum total impulse of 12,000 lb-s is required and the solid propellant utilized generates a characteristic velocity of 4900 ft/s. Assuming a constant thrust burn of 10 s is desired determine the maximum allowable nozzle expansion ration and minimum allowable chamber pressure for this propulsion system. You may assume $\gamma = 1.2$ and that the critical pressure for flow separation is 1/3 of the local atmospheric pressure.

4.22 You are asked to evaluate potential propellants for a cold-gas thruster designed to provide attitude control for an upper stage of a launch vehicle. The rocket must provide 1000 N of thrust at an altitude of 30 km. You may assume the desired chamber pressure is 2 MPa and that the gas enters the chamber at a stagnation temperature of 300 K. Assuming hydrogen is used at the "working fluid," determine:

   (a) Characteristic velocity, c^*;
   (b) Nozzle throat area, A_t;
   (c) Nozzle exit area, A_e;
   (d) Thrust coefficient, c_f;
   (e) Specific impulse, I_{sp}; and
   (f) Nozzle exit velocity, V_e
   (g) Repeat parts a–f assuming carbon dioxide gas as a working fluid.

4.23 A solid rocket motor utilizes a grain which produces a parabolic-shaped mass flow history:

$$\dot{m} = a + bt + ct^2$$

   The motor has an I_{sp} of 300 sec, and a c^* of 1500 m/s. The performance engineer desires an initial vacuum thrust level of 10^5 N and a vacuum total impulse of 10^7 N-S. In addition, structural limits prelude p_c values exceeding 5 MPa (i.e. we want peak p_c to equal this value). Finally, we desire a burning time of 100 seconds. Using this information, determine:

   (a) The nozzle throat area required.
   (b) The constants $a, b,$ and c in $\dot{m}$ eq.
   (c) An analytic formula for the vacuum thrust history.
   (d) An analytic formula for the chamber pressure history.
   (e) Plot p_c, F_v, and $\dot{m}$ histories.

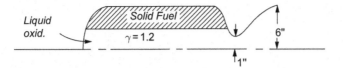

Figure 4.26 Diagram for Problem 4.24.

4.24 Consider the hybrid rocket shown in Figure 4.26. Here, we control the oxidizer flow and maintain a constant flow rate $\dot{m}_{ax} = 10$ lb/s, but the fuel flow is determined by the regression of the cylindrical port and in general is a function of time. In this case, presume the fuel flow is given by:

$$\dot{m}_f = 3 - 0.6(t/t_b) \rightarrow \rightarrow in \; lb_f/s$$

Where t_b is the burning time of 50 seconds. Because the mixture is constantly changing, the characteristic velocity is also a function of time:

$$c* = 4800 + 800(t/t_b) - 800(t/t_b)^2 \quad in \; f/s$$

Assuming ideal performance and that the engine is fired in a vacuum, determine:

(a) The initial thrust and I_{sp} of the engine.
(b) Expressions for the thrust and I_{sp} at arbitrary times $(t \le t_b)$.
(c) The maximum I_{sp} delivered during the firing. When does this maximum occur?

4.25 Design a rocket nozzle to generate 1000 lbf of thrust at sea level conditions. You can assume the propellants deliver a $c*$ value of 5000 ft/s and $\gamma = 1.2$. The chamber pressure is 800 psi.

(a) Determine throat and exit areas to maximize thrust at sea level.
(b) Determine the sea level I_{sp}.
(c) What vacuum thrust and I_{spv} will be delivered by this nozzle?

4.26 You are asked to develop a nozzle for a Mars lander propulsion system. The maximum landing mass of the vehicle is 1000 kg and the chamber pressure for the engine is 1.0 MPa. The propellants deliver a $c*$ of 1500 m/s at $\gamma = 1.2$ with $\mathcal{M} = 18$ kg/kg-mol. Assuming the thrust just balances the max vehicle weight at landing, determine:

(a) The maximum expansion ratio which can be used to avoid flow separation.
(b) The throat and exit diameters of the nozzle.

The ambient pressure and temperature on the system of Mars are 2 kPa and 220 K, respectively and the gravitational acceleration is 38% of Earth's.

4.27 Consider the rocket engine shown in Figure 4.27. The engine used propellants with $c* = 5000 f/s$ and $\gamma = 1.2$ the initial chamber pressure is 800 psi for the sea level firing. Determine:

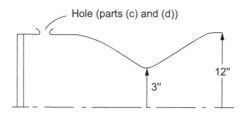

Figure 4.27 Diagram for Problem 4.27.

(a) Engine thrust.

(b) Engine specific impulse,

for the conditions noted above. Toward the end of the firing, a local failure in the chamber wall (as noted in sketch) causes the chamber pressure to drop to 600 psi. Assuming the engine has reached a new steady-state at this pressure and that the propellant flow rates were not affected by the failure, determine:

(c) The cross-sectional area of the hole.

(d) The side force generated by the flow through the hole.

4.28 Consider a first stage of a launch vehicle which operates from sea level to 150,000 ft with constant mass flow. The following information is known:

$$P_c = 900 \text{ psi}, \quad C^* = 5000 \text{ f/s}, \quad \gamma = 1.2$$
$$t_b = 120 \text{ sec}, \quad A_t = 1000 \text{ in}^2$$

You may assume that the altitude of the rocket varies linearly with time, i.e. $h = 150,000(t/t_b)$. Since mass flow is constant, thrust will increase as the rocket ascends due to the decreasing ambient pressure. A subroutine to compute ambient pressure as a function of altitude is available on the Purdue Propulsion Web Page (https://engineering.purdue.edu/~propulsi/propulsion/).

(a) For a given exit area, A_e, (or expansion ratio, ε) write a computer code which will calculate c_f, I_{sp}, and thrust as a function of time for this rocket. Attach a listing of this code to your homework solutions.

(b) Using your code, plot c_f and I_{sp} as a function of time for $\varepsilon = 8$ and 12.

(c) Determine the optimal expansion ratio for this stage. Hint: this is the ε which maximizes total impulse.

4.29 A test engine is operated at sea level on a test stand. Plot the thrust coefficient delivered by this nozzle for chamber pressures between 3 and 100 atmospheres assuming expansion ratios of 10, 50, and 100. You may assume that $\gamma = 1.2$. If we require a nozzle exit pressure greater than 5 psi to avert separation, what is the minimum chamber pressure which can be used for the test for each nozzle?

4.30 An air-launched missile is to be launched at an altitude of 40,000 feet with a chamber pressure of 500 psi. If the propellant used has $\gamma = 1.3$, $c^* = 4800$ f/s, and the desired thrust is 2000 lb, determine:

(a) The nozzle throat and exit diameter.

(b) The nozzle mass flow.

(c) The I_{sp} of the rocket engine.

(d) The thrust coefficient for the nozzle.

(e) Repeat steps (a)–(d) assuming the missile operates at sea level.

4.31 You are asked to determine characteristics of a new missile under development in some foreign country. Intelligence sources have obtained the following information:

- Initial vehicle mass = 60,000 kg
- Initial vehicle acceleration = 1.3g
- Burning time = 140 s
- Final vehicle mass = 15,000 kg
- Final vehicle acceleration = 5.67g
- Burnout altitude = 15 km
- Propellant γ, $c^* = 1.2$, 1550 m/s (estimated)

Assuming ideal performance and that the chamber pressure remains constant for the entire flight, determine:

(a) Engine thrust (both liftoff and burnout).

(b) Engine mass flow.

(c) Engine I_{sp} (both liftoff and burnout).

(d) Engine chamber pressure.

(e) Nozzle throat diameter and expansion ratio.

4.32 A rocket nozzle has the following Mach number progression:

$$M = 1 + 1.8x + 0.55x^2 - 0.35x^3 \qquad -1 \le x \le 2 \text{ ft}$$

Where x is the axial distance as measured from the nozzle throat. In addition, the following information is known:

$$T_c = 6000\ ^\circ R, \quad p_c = 1000 \text{ psi}, \quad \gamma = 1.2, \quad \mathcal{M} = 22 \text{ lb/lb-mol}$$

(a) Plot the nozzle contour $r = r(x)$. What is the nozzle expansion ratio?

(b) Plot Mach number, p/p_c, and T/T_c as a function of x (all on the same curve). What are the exit conditions p_e and T_e?

(c) Plot the thrust coefficient of this nozzle for $0 \le x \le 2$ assuming $p_a = 0$, 5, and 15 psi (all on the same curve).

4.33 A Titan upper-stage engine has the following characteristics:

> Thrust (actual) sea level = 223,500 lb
> Chamber pressure = 809 psia
> Throat area = 183 sq. in
> Expansion area ratio = 15
> Propellant weight flow rate, $w = 880$ lb/s
> Characteristic exhaust velocity, $c^* = 5690$ ft/s
> Ratio of specific heats of combustion products, $\gamma = 1.2$
> Average molecular weight of the combustion products, $\mathcal{M} = 21$ lb/lb-mol
> Determine the following:

(a) Theoretical combustion temperature

(b) Ideal thrust if the stage is started at 25 nautical miles altitude (150,000 ft) $Pa = 0.02125$ psia.

(c) Optimum expansion area ratio, A_e/A^* for conditions in (b) above. Would this be a practical nozzle to put on an engine? Why or Why not?

(d) The ideal thrust for the nozzle A_e/A^* obtained in (c) above.

4.34 A sounding rocket intends to utilize an extendible exit cone (EEC) to improve performance during its ascent through the atmosphere. The following characteristics are known:

> $c^* = 1500$ m/s
> $A_t = 100$ cm^2
> $\gamma = 1.2$
> $p_c = 100$ atm
> $t_b = 100$ s
> $\varepsilon_1 = 10,\ \varepsilon_2 = 25,\ \varepsilon_3 = 50$
> Deployment altitudes: $h_1 = 0$ km, $h_2 = 10$ km, $h_3 = 25$ km
> Burnout altitude = 75 km

(a) Plot the I_{sp} of this rocket as a function of altitude assuming nominal EEC deployments for $0 \le h \le 50$ Km.

(b) Assuming constant mass flow and a constant acceleration trajectory, determine the total impulse delivered under nominal operation.

(c) Repeat parts (a) and (b) assuming the EEC malfunctions and the second and third portions of the exit cone are never deployed.

(d) Suppose the entire exit cone deploys inadvertently just after liftoff. What happens? Determine thrust and I_{sp} for both full-flowing and separated flow conditions. At what expansion ratio does the separation occur? At what altitude would the nozzle begin to flow full?

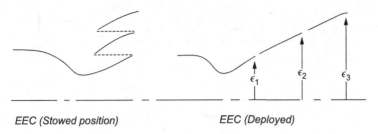

EEC (Stowed position)   EEC (Deployed)

Figure 4.28 Diagram for Problem 4.34.

4.35 You are asked to design a conical nozzle for an upper-stage engine with the following characteristics:

Desired Thrust $=$ 50,000 N
$P_c =$ 6 MP$_a$, $C* =$ 2000 m/s, $\gamma =$ 1.2

Nozzle characteristics:

   Contraction Ratio $=$ 2.5,
   $X_{lc} =$ 3 r$_t$, $R_{wtu} =$ 0.8, $R_{wtd} =$ 0.25, $R_i =$ 0.4, $\theta_i =$ 45°, $\theta_e =$ 16°, $\varepsilon =$ 75

(a) Determine the nozzle throat diameter required for this application.

(b) Make a scale drawing of your nozzle contour.

(c) Designers note that you can shorten the exit cone 20% and maintain the same θ_e value by employing a contoured design. Sketch (in dashed lines) the contoured design on your plot for our part (b).

4.36 The solid rocket motor shown in Figure 4.29 was tested under vacuum conditions. Thermochemical analyses predict the following characteristics for the propellant formulation used in this motor:

   Flame temperature = 6000 °R
   Ratio of specific heats = 1.2
   Molecular weight of chamber gases = 22 lb/lb-mol
   Solid propellant density = 0.0635 lb/in^3

In addition, the measured thrust history was curvefit as a function of time:

$$F = 4000 + 300t - 7t^2 \qquad 0 \leq t \leq 53.5 \text{ s} \qquad F \text{ in lbs.}$$

With this information and assuming ideal behavior, determine:

(a) the maximum thrust during the firing;

(b) the maximum chamber pressure during the firing;

(c) the specific impulse delivered during the firing; and

(d) the amount of propellant expended during the firing.

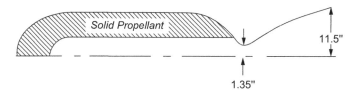

Figure 4.29 Diagram for Problem 4.36.

4.37 A rocket propels an object in a horizontal test flight at an altitude of 50,000 feet. Preflight measurements and post-flight reconstruction of telemetry data reveal the following:

15° conical nozzle

Nozzle diameters = 2.0 in (throat), 8.0 in (exit)

Propellant characteristic velocity = 5800 ft/s

Engine chamber pressure = 750 psi

(a) Determine the engine thrust and specific impulse.

(b) Telemetry data suggest that a nozzle failure occurred during the middle of the test. Thrust dropped by 10% at this time, yet chamber pressure remained constant. You assume that a portion of the end of the nozzle has broken off. Estimate the length of this broken portion.

(You may assume $\gamma = 1.2$ for all calculations.)

4.38 The F-1 engine shown in Figure 4.30 used with the Saturn launch vehicle operates at a chamber pressure of 982 psia. Nozzle geometry gives a throat diameter of 36.0 in and area ratio, ε, of 16/1. A thermochemistry program predicts the molecular weight of the exhaust gases, $\mathcal{M}$, of 22.0 lb/lb mol, $\gamma = 1.23$, and a combustion temperature, T_c, of 5640 °F.

For sea level conditions calculate theoretical values for:

(a) momentum thrust;

(b) pressure thrust;

(c) total thrust;

(d) exhaust velocity;

(e) effective exhaust velocity;

(f) characteristic exhaust velocity;

(g) specific impulse;

(h) propellant weight flow rate; and

(i) thrust coefficient.

Engine tests give an actual I_{sp} value of 265 seconds. Calculate:

(j) specific impulse efficiency.

(You may assume no separation occurs in the nozzle expansion section for all calculations.)

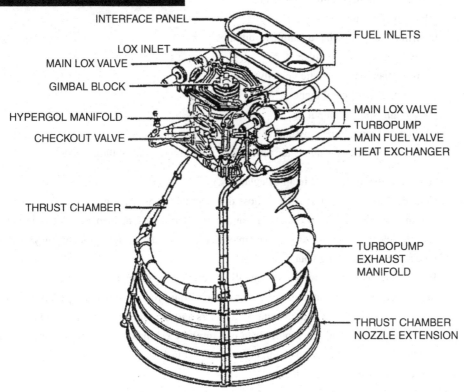

Figure 4.30 Diagram for Problem 4.38.

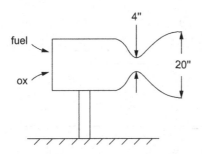

Figure 4.31 Diagram for Problem 4.39.

4.39 The rocket shown in Figure 4.31 was test-fired in a high-altitude test cell. Vacuum pumps were used to lower the ambient pressure in the cell to a simulated altitude of 80,000 feet. The following measurements were obtained from test data:

Average thrust = 18,000 lb

Average chamber pressure = 860 psi

Propellant consumption = 6400 lb

Burning duration = 110 s

Determine the following quantities from these data:

(a) average specific impulse;

(b) propellant characteristic velocity;

(c) average thrust coefficient;

(d) vacuum specific impulse;

(e) chamber temperature, assuming $\mathcal{M} = 20 lb/lb - mol$; and

(f) overall efficiency – this is the ratio of the measured and ideal thrust values

(You may assume $\gamma = 1.2$ in all calculations.)

4.40 Consider three spherical particles whose diameter and density are given in Table 4.3.

If the particles are all exposed to the same nozzle gas flow, rank them in order of decreasing particle velocity at the nozzle exit. You may assume a constant drag coefficient but you should give rationale justifying your ranking.

4.41 Rocket scientists (and engineers) generally determine the efficiency of a rocket based on its ideal performance and measured test data. For a liquid rocket engine, propellant flow rates into the chamber are measured using flowmeters and the thrust is measure using load cells. Internal chamber conditions (other than pressure) are estimated using thermochemistry and combustion codes. Typical data might look like Table 4.4.

From this information, calculate:

(a) ideal thrust (F_{id}) and mass flow from measured pressure;

(b) the ideal c^*_{id} for the calculated propellant conditions;

Table 4.3 Data for Problem 4.40

Particle	Diameter, μm	Density, kg/m^3
1	200	100
2	100	300
3	300	50

Table 4.4 Typical measurement data for Problem 4.41

Calculated parameters	Measured parameters
$T_c = 5800$ °F	$p_c = 1500$ psi, $P_a = 14.7$ psi
$\gamma = 1.2$	$D_t = 6$ in, $D_e = 36$ in
$m = 15$ lb/lb-mol	Thrust = 59,000 lb,
	$m = 205$ lbf/s

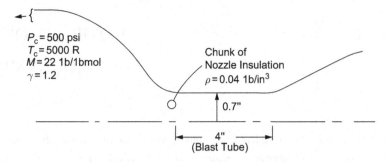

$P_c = 500$ psi
$T_c = 5000$ R
$M = 22$ 1b/1bmol
$\gamma = 1.2$

Chunk of
Nozzle Insulation
$\rho = 0.04$ 1b/in^3

0.7"

4"
(Blast Tube)

Figure 4.32 Diagram for Problem 4.43.

(c) the actual c^* delivered during the test;

(d) the thrust efficiency $\eta_{Cf} = F/F_{id}$;

(e) the c^* efficiency $\eta_{c^*} = c^*/C^*_{id}$; and

(f) the I_{sp} (overall) efficiency, η_o.

4.42 A liquid propellant rocket utilizes liquid oxygen (LOX) and RP-1 (kerosene) as its propellants. This propellant combination produces a characteristic velocity of 5800 ft/s. The chamber pressure is 1000 psi, the throat diameter is 9 inches, and the nozzle expansion ratio is 30. If the ratio of specific heats is 1.2 and the engine is operating at sea level, answer the following questions:

(a) Determine the mass flow rate, thrust, and specific impulse assuming no separation within the nozzle.

(b) Determine the mass flow rate, thrust, and specific impulse assuming that the flow separates.

(c) What expansion ratio would we need if we wanted the flow to be perfectly expanded for this chamber pressure (1000 psi)?

(d) What is the thrust and I_{sp} for the condition described in (c)?

4.43 Some tactical propulsion solid motors utilize a "blast tube" configuration (as shown in Figure 4.32) to improve packaging of electronics and nozzle thrust vector control components. Nozzle development problems have led to conditions where small pieces of nozzle insulator are breaking off and traversing the length of the blast tube. You are asked to perform an analysis to determine the flight history of this debris within the blast tube assuming steady sonic flow throughout the event. You may assume the chamber pressure is unaffected by the blockage caused by the debris and that the debris is spherical with $C_D = 0.7 =$ constant.

(a) Derive analytic expressions for the velocity (V_p) and position (x) of the debris particle in terms of time and gas-phase properties.

(b) Plot V_p vs. t, x vs. t, and V_p vs. x assuming a particle diameter of 0.25 inches.

Chapter 5

COMBUSTION AND
THERMOCHEMISTRY

Many readers of this text may have had some preliminary exposure to combustion processes. For example, in connection with airbreathing engines you may have conducted preliminary analysis of the constant pressure combustion process assuming that the combustion products consisted of water vapor, excess nitrogen and oxygen, and carbon dioxide. We would typically refer to this assumption as corresponding to *complete combustion* in that the final products of oxidation of a hydrocarbon fuel are water and carbon dioxide. However, the high temperatures obtained as a result of energy release, particularly when considering rocket propellants using strong oxidizers, generally leads to *dissociation* of some of the product species. Dissociation refers to the breakup of product compounds due to the high-temperature environment. Due to dissociation reactions, we do not have an a priori knowledge of the product species from any combustion process.

In this chapter, we will investigate techniques to predict both the composition and temperature of combusted gases as a function of the initial temperature and the amounts of fuel and oxidizer present. In Sections 5.1 and 5.2 we present a brief review of the perfect gas law and thermodynamics concepts required for this analysis. In Section 5.3, we will define the concept of *chemical equilibrium* and learn how to determine this state mathematically. In Section 5.4 we will use the notion of chemical equilibrium to determine the *adiabatic flame temperature* for a given rocket fuel combination. We will wrap up this chapter by discussing the implications of thermochemical changes during the expansion processes taking place in a rocket nozzle in Section 5.5.

5.1 Review of Perfect Gases

Since we invoke the perfect gas assumption for the bulk of the material in this text, it is appropriate to provide a brief review of these concepts at this time. The perfect gas law provides a connection between pressure, temperature, and density (or specific volume) of a gas or a collection/mixture of gases. While perfect gas behavior is a special case, most gases (or mixtures of gases) follow the perfect gas assumption for moderate temperatures and pressures. Various forms of the perfect gas law are summarized in Eq. 5.1:

$$pV = nR_{u}T = mRT \quad \text{or} \quad p = \rho RT \tag{5.1}$$

where n is the number of moles present or m is the mass of gas present and V is the volume of gas. Of course, the gas constant, R, is related to the universal gas constant through the molecular weight, $\mathfrak{M}$:

$$R = R_u / \mathfrak{M} \tag{5.2}$$

The universal gas constant may be expressed:

$$R_u = 1545 \frac{\text{ft lbf}}{\text{lb-mol } ^\circ\text{R}} = 8317 \frac{\text{N m}}{\text{kg-mol K}} \tag{5.3}$$

We can show (using basic thermodynamics relations) that gases obeying Eq. 5.1 have specific heats that are a function of temperature alone. As a result of this fact, internal energy and enthalpy changes for a perfect gas may be written:

$$\mathrm{d}u = C_v \, \mathrm{d}T \qquad \mathrm{d}h = C_p \, \mathrm{d}T \tag{5.4}$$

If both C_p and C_v are constant for the temperature range under consideration, then the gas is said to be *calorically perfect*. In this case, the above relations can be integrated directly, say between states "1" and "2":

$$u_2 - u_1 = C_v(T_2 - T_1); \quad h_2 - h_1 = C_p(T_2 - T_1) \quad \text{(calorically perfect gases)} \tag{5.5}$$

For application to rockets (and most airbreathing engines for that matter), the calorically perfect assumption is generally not valid due to the very large temperature changes involved.

Since many of our expressions will be described in terms of R and γ, it is useful to define the relations for these parameters by combining Eqs. 5.1 and 5.4:

$$C_p - C_v = R \tag{5.6}$$

Finally, γ is defined as the ratio of specific heats:

$$\gamma = \frac{C_p}{C_v} \tag{5.7}$$

5.1.1 Mixtures of Perfect Gases

To evaluate the effect of mixing various amounts of differing perfect gases, the reader is first reminded of the definition of 1 mole of gas.

A mole represents the amount of gas which contains Avogadro's number

$(6.02 \times 10^{23}$ molecules/mole) of gas molecules

For this reason, the mass of 1 mole is proportional to the molecular weight of the gas in question. For example, 1 kg-mole of O_2 would have a mass of 32 kg corresponding to the molecular weight of diatomic oxygen. The convention here is that the *mol* is prescribed on a per-gram basis, i.e. 1 mol = 1 gm-mol. For English/Imperial units, we can define a *lb-mol*; for example, one *lb-mol* of O_2 would have a weight of 32 lbf. Any mass or weight can serve as the basis for the mol as in the end the mass/weight of the mol depends on the system of measure and the molecular weight of the substance in question.

Dalton's Law

Another definition we will require for our analysis is the partial pressure of a component making up a mixture of gases:

> *The partial pressure of a component of a mixture of gases is the pressure that component would have stored at the temperature and volume of the mixture.*

Per this definition we may write:

$$p_i V = m_i R_i T = n_i R_u T \tag{5.8}$$

where the subscript "i" refers to a constituent gas making up the mixture. Now if we assume that N constituents are present in the mixture we may write Dalton's law for the mixture as:

$$p = \sum_{i=1}^{N} p_i \quad \text{or} \quad n = \sum_{i=1}^{N} n_i \tag{5.9}$$

Since the fluid pressure is simply the sum of the partial pressures of the constituent gases, a mixture of perfect gases is also a perfect gas. To find the equivalent molecular weight of the mixture we may write:

$$\mathfrak{M} = \frac{m}{n} = \frac{m}{\displaystyle\sum_{i=1}^{N} m_i / \mathfrak{M}_i} = \frac{1}{\displaystyle\sum_{i=1}^{N} y_i / \mathfrak{M}_i} \tag{5.10}$$

where y_i is the mass fraction of species i ($y_i = m_i / m, \sum_{i=1}^{N} y_i = 1$). To obtain the equivalent specific heat of the mixture, we would simply use the weighted average:

$$C_{p_{mix}} = \sum_{i=1}^{N} C_{p_i} y_i \tag{5.11}$$

with an analogous expression holding for $C_{v_{mix}}$. In Eq. 5.11, we presume that the individual specific heat values are prescribed on a per-mass basis such that the mass-weighted average is justified. The ratio of specific heats for the mixture can then be computed:

$$\gamma_{mix} = C_{p_{mix}} / C_{v_{mix}} = \frac{C_{p_{mix}}}{C_{p_{mix}} - R_{mix}} \tag{5.12}$$

As an example of a multi-component system, consider the molecular weight of dry air. Ignoring minor ingredients, air is 21% oxygen and 79% nitrogen on a by volume basis, or 23.2% oxygen and 76.8% nitrogen on a per-mass basis. Using the techniques described above, the molecular weight of air becomes:

$$\mathfrak{M}_{\text{air}} = \frac{1}{0.232/(32) + 0.768/(28)} = 28.8 \text{ kg/kg-mol} \tag{5.13}$$

Note that we could also use partial volumes (i.e. the volume occupied by one component in a mixture) to determine the weightings of the various constituents in the mixture. In other words, another equally valid form of Dalton's law is:

$$pV_i = m_i R_i T = n_i R_u T \quad \text{where } V = \sum_{i=1}^{N} V_i \tag{5.14}$$

Using this approach, we may determine the molecular weight of the mixture:

$$\mathfrak{M} = \frac{m}{n} = \frac{\sum_{i=1}^{N} n_i \mathfrak{M}_i}{n} = \sum_{i=1}^{N} x_i \mathfrak{M}_i \tag{5.15}$$

where x_i is defined as the mole fraction of species "i" (i.e. $x_i = n_i/n$). In this case, the specific heat of the mixture should be computed from individual specific heats that are specified on a per-mole basis:

$$C_{p_{\text{mix}}} = \sum_{i=1}^{N} C_{p_i} x_i \tag{5.16}$$

and then γ_{mix} is computed as in Eq. 5.12. Finally, we note that the mole fraction and mass fractions are related to partial pressure and volume ratios:

$$V_i/V = p_i/p = x_i = y_i(\mathfrak{M}/\mathfrak{M}_i) \tag{5.17}$$

Using the mole fraction approach, and noting that mole fractions correspond to volume fractions as shown in the equation above, we can recalculate the molecular weight of dry air:

$$\mathfrak{M}_{\text{air}} = 0.21(32) + 0.79(28) = 28.8 \text{ kg/kg-mol} \tag{5.18}$$

which is the same result we had in Eq. 5.13.

Mixtures containing fuel and oxidizers are characterized by the *equivalence ratio*, ϕ, or *mixture ratio*, *r*. The equivalence ratio measures the extent to which the reactants deviate from the stoichiometric condition:

$$\phi = \frac{\text{stoichiometric oxidizer mass}}{\text{Actual oxidizer mass}} \tag{5.19}$$

where the stoichiometric condition is the one that provides the minimum amount of oxidizer required to consume all the fuel and oxidizer in a combustion process. Fuel-rich formulations have equivalence ratios greater than unity and fuel-lean mixtures have equivalence ratios less than unity. More commonly, rocket scientists specify proportions of ingredients in terms of the *mixture ratio*:

$$r = \frac{m_{\text{ox}}}{m_{\text{f}}} = \frac{\dot{m}_{\text{ox}}}{\dot{m}_{\text{f}}} \tag{5.20}$$

which is just the mass ratio or mass flow ratio if one assumes steady state flow into the combustor. Using this definition, mixture ratios higher than stoichiometric are lean mixtures, while ratios lower than stoichiometric are fuel rich.

5.2 Thermodynamics Review

As with any thermodynamics discussion, we begin with the First Law of Thermodynamics:

$$dE = \delta Q - \delta W \tag{5.21}$$

Let's discuss each of the three terms in this fundamental equation. The quantity dE represents the total change in energy, measured in calories, BTUs, or joules, of the gases contained within "the system." Remember that the system corresponds to the volume of fluid of interest, or a fixed volume in space with fluid entering and leaving at specified rates. The total energy is usually represented as the sum of internal, kinetic, and potential energy contributions:

$$dE = dU + d(mv^2/2) + d(mgh) \tag{5.22}$$

where U is internal energy (in this discussion), m is the mass of fluid in the system, v is the fluid velocity, g is the acceleration due to gravity, and h is the height measured from some suitable reference location. In Eq. 5.22 we have ignored any electronic or nuclear contributions to the energy of the gas.

Inside the combustion chamber of a rocket engine, velocities are typically small when compared to the large internal energy of the hot gases. In addition, potential energy changes are nearly always negligible in propulsion systems due to relatively small height changes and gas masses. Therefore, in this particular situation:

$$dE = dU \qquad \text{(rocket combustion chamber)} \tag{5.23}$$

which simply states that energy changes of the gases within the combustion chamber can entirely be attributed to changes in their internal energy.

Next, consider the work contribution, δW. Remember that this change in work is in general dependent on the path of the process so that $\delta W \neq dW$ unless additional assumptions are employed. Work changes can be attributed to changes in specific volume or pressure ($p - dV$ work), frictional or viscous forces, or work done through the fluid moving or turning some mechanical device (shaft work). Clearly we won't have to worry about shaft work in our rocket combustor since no mechanical components are being acted upon by the fluid in this region. In addition, viscous work is negligible within the chamber due to the low velocities in this region. By making this assumption, we can not only ignore viscous work, but we can now write:

$$\delta W = dW = p \, dV \qquad \text{(reversible process)} \qquad (5.24)$$

Since no dissipation processes are present and fluid velocities are small, the $p \, dv$ work is accomplished reversibly, and we can write changes in work in terms of the total derivative independent of path.

The other term in Eq. 5.21 represents the amount of heat added or lost from the system, δQ. This quantity is expressed as a non-perfect differential because the amount of heat added is dependent on the temperature at which the reaction occurs.

By combining Eqs. 5.21, 5.22, and 5.24, we can write the form of the First Law valid for a combustion chamber:

$$dE + p \, dV = dH - V \, dp = \delta Q \qquad (5.25)$$

where we have invoked the definition of enthalpy, $H = U + p \, V$. In general, the heat transfer δQ must be computed based on the combustor geometry and operating conditions, but it is often useful to consider the adiabatic case $\delta Q = 0$ as this represents an upper bound for us in estimating the flame temperature of the reaction, i.e. all the energy evolved from combustion remains in the product gases resulting from these chemical reactions. The combustor temperature resulting from this assumption is referred to as the *adiabatic flame temperature*. From a practical perspective, the energy losses/heat transfer from actual devices tends to be quite small and we expect the actual average combustion temperature to be near this ideal limit. One can readily show that if we did a poor job of insulating the walls of the chamber and δQ was large, the walls would heat to temperatures that we can't tolerate and/or we would lose a significant amount of performance (nozzle exhaust velocity), so the adiabatic assumption tends to be quite valid.

Note that if the reaction occurs at constant volume (such as in an idealized internal combustion engine) and under adiabatic conditions, then $dV = \delta Q = 0$ and the First Law result in Eq. 5.25 states that all of the energy interactions are attributed to changes in internal energy, $dU = 0$. Similarly, if the reaction occurs at constant pressure and under adiabatic conditions as is presumed for many rocket combustion processes, then $dp = \delta Q = 0$ and then $dH = \delta Q$. Integrating this equation for the constant pressure case we get:

$$H_2 - H_1 = 0 \qquad (5.26)$$

where subscript "2" refers to the combustion products and "1" refers to the reactants. In words, the fundamental result in Eq. 5.26 might be stated, "in constant pressure, adiabatic combustion the enthalpy of the products equals the enthalpy of the reactants," i.e. $H_2 = H_1$. As we seek to invoke this seemingly simple result, we begin to recognize that the products differ in both composition and temperature from the reactants, so it can be difficult to understand how to compute the enthalpy changes. For this reason, we sometimes find it convenient to think about the process as outlined in Figure 5.1. To get $H_2 - H_1$, we can think of heating or cooling the reactants to some standard temperature, $T°$, then allow the chemical reaction/combustion to occur at this fixed temperature,

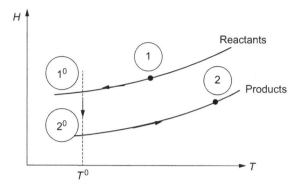

Figure 5.1 *H–T* diagram for constant pressure combustion.

then travel along the products line from the temperature $T°$ to the final temperature of the mixture, T_2. Using this path, we can write the quantity $H_2 - H_1$:

$$H_2 - H_1 = (H_2 - H_2^0) + (H_2^0 - H_1^0) - (H_1 - H_1^0) \tag{5.27}$$

as denoted in Figure 5.1. Here the superscript "0" refers to a standard temperature at which the reaction is assumed to occur (generally 298.16°K).

The quantities $H_2 - H_2^0$ and $H_1 - H_1^0$ represent *sensible enthalpy differences* in changing the temperature of a fixed composition of gases. The quantity $H_2^0 - H_1^0$ is called the *heat of combustion* or *heating value* for the reaction and corresponds to enthalpy changes resulting from combustion at the standard temperature, $T°$. Values for this quantity, often called ΔH^0 or ΔH_f°, are summarized for several compounds and elements in Table 5.1 on a per-mole basis. For example, let's say that we had 3 moles of carbon dioxide present in the reaction. The heat of formation for this molecule is then $\Delta H_f^0 = 3$ moles $* \Delta h_f^0$ where Δh_f^0 is the heat of formation of CO_2 on a per-mole basis which is –94.05 Kcal/mol as is shown in Table 5.1. The Purdue Propulsion Web Page (https://engineering.purdue.edu/~propulsi/propulsion/) contains a list of many other common reactants/propellant ingredients, so it may be convenient for students to bookmark this page.

As an illustrative example of calculating this quantity for a reaction containing a mixture of gases, consider the general case:

$$a_1A_1 + a_2A_2 + \cdots \rightarrow b_1B_1 + b_2B_2 + \cdots \tag{5.28}$$

where a_1 is the number of moles of reactant A_1, b_1 is the number of moles of product B_1, and so forth. In general, then, we may write the total enthalpy of the reactant gases as the sum of all the individual enthalpies of the gases within the reactant mixture:

$$H_1 = \sum_{i=1}^{n} a_i h_{A_i}(T_1) \tag{5.29}$$

where h_{A_i} is the molar enthalpy of species A_i. This quantity is defined:

Table 5.1 Standard heats of formation of selected substances at 298.16 K

Substance	Δh_f^0 (kcal/mole)	Substance	Δh_f^0 (kcal/mole)
O(g)	59.159	C_7H_8(g), toluene	11.9500
O_2(g)	0.000	C_8H_{10}(g), ethylbenzene	7.1200
O_3(g)	34.0	C_8H_{10}(g), o-xylene	4.5400
H(g)	52.089	C_8H_{10}(g), m-xylene	4.1200
H_2(g)	0.000	C_8H_{10}(g), p-xylene	4.2900
OH(g)	10.06	$C_6H_5NH_2$(l), aniline	−4.4510
H_2O(g)	−57.7979	CH_2O(g), formaldehyde	−27.7000
H_2O(l)	−68.3174	CH_3OH(g)	−48.1000
H_2O_2(g)	−31.83	CH_3OH(l)	−57.0360
H_2O_2(l)	−44.84	CF_4(g)	−162.5000
F(g)	19.45	CN_4(c), cyanogen azide	92.6000
F_2(g)	0.000	HCN(g)	31.2000
F_2O(g)	5.5000	$C(NO_2)_4$(l)	5 to +20
HF(g)	−64.2000	CH_5 N(g), methylamine	−6.7000
Cl(g)	29.0120	CH_2N_2(c), cyananiide	9.2000
$Cl_{2(g)}$	0.000	NH_4CN(c)	0.0000
HCl(g)	−22.0630	CH_5N_3(c), guanidine	−17.0000
Cl_2O_7(g)	63.4000	CH_3ON(c), formamide	−61.6000
CIF(g)	−25.7000	CH_3O_2 N(l), nitromethane	−26.7000
Br(g)	6.7100	CH_4ON_2(c), urea	−79.6340
Br_2(g)	7.3400	$CH_5O_4N_3$,(c), urea nitrate	−114.8000
HBr(g)	−8.6600	$CH_2O_2N_4$(c)	22.1400
I(g)	25.4820	$CH_6O_3N_4$(c), guanidine nitrate	−91.1000
I_2(g)	14.8760	C_2H_2(g)	54.1940
I_2(c)	0.000	C_2H_4(g)	12.4960
HI(g)	6.2000	C_2H_6(g)	−20.2360
S(g)	53.2500	C_2H_2O(g), ketene	−14.6000
SO_2(g)	−70.9600	C_2H_4O(g), ethylene oxide	−12.1900
SO_3(g)	−94.4500	C_2N_2(g)	73.6000
H_2S(g)	−4.8150	C_2H_3 N(g), acetonitrile	21.0000
N(g)	112.7500	C_2H_3 N(g), methyl isocyanide	35.9000
N_2(g)	0.0000	C_2H_7 N(g), ethylamine	−6.6000
NO(g)	21.6000	$C_2H_5O_2$ N(l), nitroethane	−30.0000
NO_2(g)	8.0910	$C_2H_5O_2$ N(g), ethyl nitrite	−24.8000
NO_3(g)	13.0000	$C_2H_5O_3$ N(l), ethyl nitrate	−44.3000
N_2O(g)	19.4900	$C_2H_4O_6N_2$(l), glycol dinitrate	58.0000
N_2O_3(g)	20.0000	B(c)	0.0000
N_2O_4(g)	2.3090	B(g)	95.3000
N_2O_5(g)	3.6000	B_2(g)	105.7000

(*continued*)

Table 5.1   *(cont.)*

Substance	Δh_f^0 (kcal/mole)	Substance	Δh_f^0 (kcal/mole)
N_2O_5(c)	−10.0000	B_2H_6(g), diborane	−33.0000
NH_3(g)	−11.0400	B_5H_9(l), pentaborane	−87.0000
N_2H_4(l)	12.0500	$B_{10}H_{14}$(c), decaborane	−195.0000
$N_2H_4H_2O$(l)	−57.9500	BO(g)	−56.5000
P(g)	75.1800	B_2O_3(c)	−342.7600
HNO_3(l)	−41.4040	BF_3(g)	−285.8000
HNO_3H_2O(l)	−112.9600	Li(g)	37.0700
NH_2OH(l)	−25.5000	Li(c)	0.0000
NH_4NO_3(c)	−87.2700	Li_2O(c)	−142.4000
NH_2OHHNO_3(c)	−86.3000	Li_2O_2	−151.7000
NF_3(g)	−27.2000	LiH(g)	30.7000
NH_4Cl(c)	−75.3800	LiH(c)	−21.6100
NH_4ClO_4(c)	−69.4200	LiOH(c)	−116.4500
P(c, III, white),	0.0000	$LiOH-H_2O$(c)	−188.7700
PH_3(g)	2.2100	LiF	−146.3000
C(g)	171.6980	Na(g)	25.9800
C(c, diamond)	0.4532	Na(c)	0.0000
C(c, graphite)	0.0000	Na_2(g)	33.9700
CO(g)	−26.4157	NaO_2(c)	−61.9000
CO_2(g)	−94.0518	Na_2O(c)	−99.4000
CH_4(g)	−17.8890	Na_2O_2(c)	−120.6000
C_2H_6(g)	−20.2360	NaH(g)	29.8800
C_3H_8(g)	−24.8200	NaH(c)	−13.7000
C_4H_{10}(g), n-butane	−29.8120	NaOH(c, II)	−101.9900
C_4H_{10}(g), isobutane	−31.4520	NaF(g)	−72.0000
C_5H_{12}(g), n-pentane	−35.0000	NaF(c)	−136.0000
C_6H_6(g), benzene	19.8200		

Roman numerals following the letter c (c denotes a crystalline compound) identify a specified crystal structure. Heats of formation for a large number of reactants can be found on the Purdue Propulsion Web Page (https://engineering .purdue.edu/~propulsi/propulsion/).

$$h_{A_i} = \int_{T^\circ}^{T} C_{p_{A_i}} dT + h_{A_i}^\circ \tag{5.30}$$

and is measured in BTU/mole or cal/mole depending on the system of units. The first term on the right-hand side of Eq. 5.30 is the *sensible enthalpy of reactant* A_2, while the second term is the *heat of formation* of the species A_i. The heat of formation represents the energy required (or given off) to bring together atoms forming a given compound. Heats of formation are generally quoted at a standard temperature, T°, which is typically 298 K. Basic elements (i.e. C, Ar, Mg, etc.) and diatomic gases (i.e. O_2, N_2, etc.) have $h°$ of zero since they exist in their natural, lowest energy state. Compounds having $h°$ values less than zero give off energy during their formation, while compounds with $h°$ values greater than zero require energy addition in order to be formed.

Of course, the molar enthalpy of products could also be expressed using a relation analogous to Eq. 5.30. Now, using these ideas, we may write the change in enthalpy for the process:

$$H_2 - H_1 = \sum_{i=1}^{n} b_i \left(\int_{T°}^{T_2} C_{p_{B_i}} dT + h_{B_i}° \right) - \sum_{j=1}^{m} a_j \left(\int_{T°}^{T_1} C_{p_{A_j}} dT + h_{A_j}° \right) \tag{5.31}$$

where m reactants and n products are present. Now recall that we previously had written:

$$H_2 - H_1 = (H_2 - H_2^0) + (H_2^0 - H_1^0) - (H_1 - H_1^0) \tag{5.32}$$

Direct comparison of these two expressions clearly shows that the sensible enthalpy differences are given by the integrals in Eq. 5.31, so that the heat of combustion, ΔH_B, is simply:

$$\Delta H_B = \sum_{i=1}^{n} b_i h_{B_i}° - \sum_{j=1}^{m} a_j h_{A_j}° \tag{5.33}$$

or in words we could say:

The heat of combustion of a constant pressure chemical reaction is simply the heat of formation of the product species minus the heats of formation of the reactant species.

Heats of formation are measured using calorimeters and spectroscopic data and are tabulated in many books dealing with combustion. It is likely that many readers are familiar with the review we have presented here. If we know the composition of the reactants (A_i in Eq. 5.28) and the composition of the products (B_i in Eq. 5.28), and we know the initial temperature of the reactants, we can use Eqs. 5.26 and 5.31 and the data in Table 5.1 (or equivalent heats of formation from a reliable source) to compute the temperature of the product gases, T_2. Note that this can be a somewhat arduous process as T_2 lies in the limits of the integral in Eq. 5.31, but in principle the approach can be used to find the temperature of the mixture following the combustion event. The real questions that lie before us are: how do we determine the composition of the product gases? What if dissociation reactions are occurring and we don't necessarily have complete combustion? The answer to these questions lies in the study of chemical equilibrium using the Second Law of Thermodynamics as discussed in Section 5.3.

5.3 CHEMICAL EQUILIBRIUM

5.3.1 The Concept of Chemical Equilibrium

As an example, let us consider the combustion of methane gas with pure oxygen under stoichiometric conditions. From your prior background in complete combustion or a review of the material in

Section 5.2, you would assume that the carbon atoms in the methane are oxidized to form carbon dioxide, and the hydrogen atoms are oxidized to form water, so we would write:

$$CH_4 + 2O_2 \rightarrow CO_2 + 2H_2O \tag{5.34}$$

By calculating the composition in this manner, we can then determine the heat of combustion and the flame temperature for the reaction since the composition of the product species was known from the start.

What is wrong with this process? Well, the basic problem with our procedure is that we have *assumed* that the product species derived in Eq. 5.34 represent the lowest stable chemical state for the pressure and temperature of the reaction species. Like a ball's tendency to roll to the bottom of a big hill, our reaction will proceed in such a manner so as to obtain the most stable end state for the products. For example, the following *secondary reactions* will also be occurring during the course of the combustion process described by Eq. 5.34:

$$
\begin{aligned}
CO_2 &\rightleftharpoons CO + \frac{1}{2}O_2 \\
CO_2 &\rightleftharpoons C + O_2 \\
H_2O &\rightleftharpoons H_2 + \frac{1}{2}O_2 \\
H_2O &\rightleftharpoons 2H + O \\
H_2O &\rightleftharpoons OH + H \\
H_2 &\rightleftharpoons 2H \\
O_2 &\rightleftharpoons 2O
\end{aligned}
\tag{5.35}
$$

Note that the secondary reactions, called *dissociation/recombination reactions*, can proceed in either direction as indicated by the double arrows in Eq. 5.35. Of course, there are many other secondary reactions you could think of, even for this relatively simple example containing only three elements. At the present time, we have no way of knowing which way (and to what extent) each of the secondary reactions proceed, but we do know that Eq. 5.34 certainly doesn't contain the whole story.

Now let us think of our reaction proceeding with primary and secondary reactions occurring. If we assume that there is no outside influence due to temperature changes, mechanical work interactions, or outside heat transfer, the chemical reactions will proceed to a point at which forward and reverse reactions are occurring at exactly the same rate. This is known as the *chemical equilibrium point*, and the resulting product species makeup is known as the equilibrium composition. In view of this fact, we should modify Eq. 5.34 to account for at least some of the other product species which might be present:

$$CH_4 + 2O_2 \rightarrow aCO_2 + bH_2O + cCH_4 + dH_2 + eO_2 + fH + gO + hCO + \cdots \tag{5.36}$$

where the lower case letters *a–h* represent the unknown *stoichiometric coefficients* of the various product species listed.

Note that we have assumed that there may be unburned fuel present in the product species of the equation above. Even the primary reaction (Eq. 5.34) can proceed in both directions, especially when the product species are at high temperature (like in a rocket combustion chamber). Of course, we could consider other species as well, but for the time being let us assume that the eight species we have considered are good enough for our analysis. For this case, we then have eight unknown stoichiometric coefficients. Since we have three elements present in this example, we have three mass balance equations which give us some information:

$$\begin{array}{rl}
\text{carbon :} & 1 = a + c + h \\
\text{oxygen :} & 4 = 2a + b + 2e + g + h \\
\text{hydrogen :} & 4 = 2b + 4c + 2d + f
\end{array} \qquad (5.37)$$

Of course we aren't going to get too far with eight unknowns and only three equations, so we need seek additional information from the Second Law of Thermodynamics and a quantity called "Gibbs free energy."

5.3.2 Thermodynamic Considerations for Equilibrium

Before going too far, we must caution you on a common misconception. The chemical equilibrium process is all about what is happening within the product species. Think of the coming analysis as a way to sort out how the product molecules are coming to some statistically steady state. As we write expressions below they are concentrated on the thermodynamic state of the product species only, i.e. the right-hand side of Eq. 5.36. It is sometimes possible to lose sight of this important fact. With that said, we begin our analysis with the Second Law of Thermodynamics:

$$T dS \geq \delta Q \qquad (5.38)$$

where S is entropy and the equals sign holds if the process is reversible. Substituting for δQ from the First Law and invoking the definition of enthalpy, $H = U + p V$, we may write:

$$dU + p \, dV - T \, dS = dH - V \, dp - T \, dS \leq 0 \qquad (5.39)$$

This equation is essentially a form of the *property relation* or *Gibbs' equation*, which you are likely to have discussed for a reversible process in the context of thermodynamics. The form shown in Eq. 5.39 is valid for any process (reversible or irreversible) and therefore holds during the entire course of the reaction as long as temperature is well defined during the process. Let us now adapt this general form to our specific problem.

Since velocities within rocket combustion chambers are quite low and pressure levels are typically quite high, it is typically valid to assume that the combustion process takes place at constant pressure. In addition, while the temperature of the mixture of product gases will be varying as reactions occur, a steady-state condition will exist near the equilibrium point such that temperature fluctuations will also be negligible. In this case, as the collection of product species approaches equilibrium, $dp = dT = 0$ and Eq. 5.39 can be written:

$$d[H - T \, S] \equiv dF \leq 0 \qquad (5.40)$$

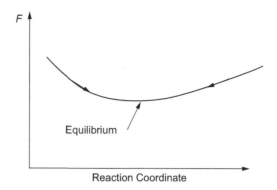

Figure 5.2 Changes in Gibbs free energy during a reaction.

where F is the *Gibbs Free Energy*; a thermodynamic variable made up of the combination indicated in the above equation. Equation 5.40 tells us that the Gibbs free energy must decrease during the irreversible combustion processes, but as equilibrium is reached, we must have $dF = 0$ since all irreversible processes have ceased by the definition of the equilibrium state. Therefore, the reaction process can be monitored by calculating changes in F and we will observe equilibrium when F ceases to change.

This situation is depicted in Figure 5.2, which shows the variation in F with a "reaction coordinate." A reaction coordinate is simply some measure of thermodynamic changes during the combustion process; one example would be the total number of moles in the product species. From Figure 5.2, we see that equilibrium can be found if we can minimize the Gibbs free energy. In some texts, F is referred to as a *thermodynamic potential* because of this behavior. Now we need to express dF in terms of quantities of interest (such as temperature, pressure, and composition) in order to turn this equilibrium criteria into something of practical use.

Since we want to investigate equilibrium for mixtures of perfect gases, let us consider the following reaction:

$$n_a A + n_b B \leftrightharpoons n_c C + n_d D \tag{5.41}$$

where A, B, C, D refer to some chemical species and n_i refers to the number of moles of species "i." Now for a reversible process (near equilibrium), entropy changes for a perfect gas can be written:

$$dS_i = n_i C_{pi} \frac{dT}{T} - R_u n_i \frac{dp_i}{p_i} \tag{5.42}$$

where subscripts "i" refer to an arbitrary reactant or product species, and p_i is the partial pressure for this species. From the definition of free energy we can write:

$$dF_i = dH_i - T \, dS_i - S_i \, dT = n_i \, C_{pi} \, dT - T \, dS_i - S_i \, dT \tag{5.43}$$

But near equilibrium, $dT = 0$ (remember, we are thinking about the product molecules rearranging themselves until they reach a statistically steady state), so use of this fact and combining Eqs. 5.42 and 5.43 gives:

$$dF_i = n_i R_u T \frac{dp_i}{p_i} \tag{5.44}$$

where T = constant. Integration of this result gives:

$$F_i = F_i^0 + n_i R_u T \ln(p_i/p^0) \tag{5.45}$$

where T = constant and F^0 is the free energy of formation; a quantity dependent on the molecular structure of the element or compound being considered. This quantity is analogous to the enthalpy of formation ($h°$) that we discussed above. The free energy of formation is generally specified at a standard temperature, and is at most a function of temperature alone. The pressure integration constant, p^0, corresponds to the pressure at which F^0 values are obtained. Generally, we assume that $p^0 = 1$ atmosphere.

Now we are ready to invoke the equilibrium condition, $dF = 0$, for our general reaction in Eq. 5.41. Remember that this condition must be applied to the entire mixture of gases. For this reason, the equilibrium condition essentially means that the total free energy of the products must equal the total free energy of the reactants, i.e. $F_{products} - F_{reactants} = 0$. Invoking this condition for our example reaction gives:

$$0 = \Delta F^0 + R_u T [n_c \ln p_c + n_d \ln p_d - n_a \ln p_a - n_b \ln p_b] \tag{5.46}$$

where $\Delta F^0 = F_c^0 + F_d^0 - F_a^0 - F_b^0$. Rearranging this result, we obtain:

$$\exp\left(\frac{-\Delta F°}{R_u T}\right) \equiv K_p = \frac{p_c^{n_c} p_d^{n_d}}{p_a^{n_a} p_b^{n_b}} \tag{5.47}$$

where K_p is defined as the *equilibrium constant based on pressure* for this particular reaction. Since ΔF^0 is at most a function of temperature, we can see that K_p is a function of temperature alone. For this reason, K_p values can be tabulated or curvefit as a function of temperature for a number of dissociation/recombination reactions. Table 5.2 contains constant pressure equilibrium constants for a few important reactions as a function of temperature. Curvefits of numerous K_p values are located on the Purdue Propulsion Web Page (https://engineering.purdue.edu/~propulsi/ propulsion/).

For ideal gases, we can substitute for the partial pressures in Eq. 5.47 by using the fact that $p_i/p = x_i$ from Dalton's law (Section 5.1):

$$K_p = \frac{x_c^{n_c} x_d^{n_d}}{x_a^{n_a} x_b^{n_b}} p^{\Delta n} \tag{5.48}$$

Table 5.2   Equilibrium constants for various reactions[1]

T, K	$K_{p,1}$	$K_{p,2}$	$K_{p,3}$	$K_{p,4}$	$K_{p,5}$
298.16	4.8978×10^{-41}	2.4831×10^{-36}	2.8340×10^{-7}	1.1143×10^{40}	1.9187×10^{-60}
300	9.0157×10^{-41}	4.2560×10^{-36}	3.1311×10^{-7}	6.1235×10^{39}	4.6559×10^{-60}
400	5.5335×10^{-30}	1.3459×10^{-26}	2.1419×10^{-5}	1.7418×10^{29}	1.8323×10^{-44}
500	1.7140×10^{-23}	6.9984×10^{-21}	2.6984×10^{-4}	7.6913×10^{22}	4.3351×10^{-35}
600	3.7239×10^{-19}	4.6452×10^{-17}	1.4555×10^{-3}	4.2954×10^{18}	7.8886×10^{-29}
700	4.7315×10^{-16}	2.5351×10^{-14}	4.8362×10^{-3}	3.8282×10^{15}	2.3768×10^{-24}
800	1.0162×10^{-13}	2.9040×10^{-12}	1.1869×10^{-2}	1.9454×10^{13}	5.4828×10^{-21}
900	6.6681×10^{-12}	1.1700×10^{-10}	2.3757×10^{-2}	3.1405×10^{11}	2.2856×10^{-18}
1000	1.9055×10^{-10}	2.2693×10^{-9}	4.1295×10^{-2}	1.1482×10^{10}	2.8708×10^{-16}
1100	2.9703×10^{-9}	2.5852×10^{-8}	6.4834×10^{-2}	7.6015×10^{8}	1.5066×10^{-14}
1200	2.9390×10^{-8}	1.9715×10^{-7}	9.4287×10^{-2}	7.8759×10^{7}	4.0926×10^{-13}
1300	2.0469×10^{-7}	1.1052×10^{-6}	1.2900×10^{-1}	1.1497×10^{7}	6.7143×10^{-12}
1400	1.0824×10^{-6}	4.8596×10^{-6}	1.6899×10^{-1}	2.2060×10^{6}	7.4131×10^{-11}
1500	4.5973×10^{-6}	1.7575×10^{-5}	2.1324×10^{-1}	5.2541×10^{5}	5.9402×10^{-10}
1600	1.6315×10^{-5}	5.4300×10^{-5}	2.6104×10^{-1}	1.4955×10^{5}	3.6787×10^{-9}
1700	4.9911×10^{-5}	1.4710×10^{-4}	3.1180×10^{-1}	4.9204×10^{4}	1.8420×10^{-8}
1800	1.3493×10^{-4}	3.5752×10^{-4}	3.6495×10^{-1}	1.8302×10^{4}	7.7215×10^{-8}
1900	3.2885×10^{-4}	7.9232×10^{-4}	4.2004×10^{-1}	7.5422×10^{3}	2.7861×10^{-7}
2000	7.3350×10^{-4}	1.6233×10^{-3}	4.7664×10^{-1}	3.3931×10^{3}	8.8491×10^{-7}
2100	1.5174×10^{-3}	3.1110×10^{-3}	5.3394×10^{-1}	1.6458×10^{3}	2.5194×10^{-6}
2200	2.9383×10^{-3}	5.6247×10^{-3}	5.9208×10^{-1}	8.5212×10^{2}	6.5283×10^{-6}
2300	5.3753×10^{-3}	9.6627×10^{-3}	6.5041×10^{-1}	4.6677×10^{2}	1.5585×10^{-5}
2400	9.3821×10^{-3}	1.5874×10^{-2}	7.0871×10^{-1}	2.6847×10^{2}	3.4610×10^{-5}
2500	1.5574×10^{-2}	2.5090×10^{-2}	7.6648×10^{-1}	1.6127×10^{2}	7.2161×10^{-5}
2750	4.7424×10^{-2}	6.8250×10^{-2}	9.0910×10^{-1}	5.3272×10^{1}	3.5917×10^{-4}
3000	1.2010×10^{-1}	1.5762×10^{-1}	1.0478	2.0999×10^{1}	1.3515×10^{-3}
3250	2.6381×10^{-1}	3.2048×10^{-1}	1.1788	9.5786	4.2678×10^{-3}
3500	5.1807×10^{-1}	5.8993×10^{-1}	1.3046	4.9295	1.1311×10^{-2}
3750	9.3022×10^{-1}	1.0000	1.4218	2.7498	2.6363×10^{-2}
4000	1.5528	1.5933	1.5315	16.623	5.5361×10^{-2}
4250	2.4416	2.401	1.6351	1.0568	1.0668×10^{-1}
4500	3.6521	3.4602	1.7322	7.0307×10^{-1}	1.9134×10^{-1}
4750	5.2349	4.8029	1.8229	4.9181×10^{-1}	3.2321×10^{-1}
5000	7.2395	6.4506	1.9067	3.5465×10^{-1}	5.1874×10^{-1}

[1] $K_{p,1} = \dfrac{p_O}{\sqrt{p_{O_2}}}$, $K_{p,2} = \dfrac{p_H}{\sqrt{p_{H_2}}}$, $K_{p,3} = \dfrac{p_{OH}}{\sqrt{p_{O_2}p_{H_2}}}$, $K_{p,4} = \dfrac{p_{H_2O}}{p_{H_2}\sqrt{p_{O_2}}}$, $K_{p,5} = \dfrac{p_N}{\sqrt{p_{N_2}}}$

The quantity $\Delta n = n_c + n_d - n_a - n_b$ is the change in the number of moles in the gas mixture as a result of the reaction. The reaction is said to be of *zeroth order* if $\Delta n = 0$, or *first order* if $\Delta n = 1$, etc. Note that while the overall reaction seldom has an integer order, higher-order reactions become more and more sensitive to pressure since the composition is a function of $p^{\Delta n}$. Zeroth-order reactions are completely insensitive to pressure, first-order linear in pressure, etc. As this is an important point, let's consider a couple of the reactions we introduced in the methane/oxygen example above:

$$CO_2 \rightleftharpoons CO + \frac{1}{2}O_2$$
$$H_2O \rightleftharpoons 2H + O$$

The first reaction says that we make 1.5 moles of gases (CO and O_2) from a single mole of CO_2 ($\Delta n = 1.5 - 1 = 0.5$) while the second states that we make 3 moles of gas (atomic hydrogen and atomic oxygen) from a single molecule of water vapor ($\Delta n = 3 - 1 = 2$). Using Eq. 5.46, we could write the K_p values for these reactions:

$$K_{p,1} = \frac{x_{CO}^1 x_{O_2}^{0.5}}{x_{CO_2}^1} p^{0.5} \quad K_{p,2} = \frac{x_O^1 x_H^2}{x_{H_2O}^1} p^2$$

Now let's suppose we consider the mixture at a fixed temperature, this implies that $K_{p,1}$ and $K_{p,2}$ are both constant. What happens then if we, say, double the pressure? Since the left-hand side of both equations is fixed, the amount of this dissociated species CO, and $\frac{1}{2}O_2$ will need to decrease to accommodate the change in pressure. Because of the strong pressure dependence in the second reaction, the amounts of atomic oxygen and hydrogen will drastically reduce with the increase in pressure. This is a general trend for dissociation/recombination reactions: *increasing pressure tends to inhibit dissociations and promote recombinations while decreasing pressure promotes dissociation and inhibits recombination*. It is through the $p^{\Delta n}$ factor that the pressure dependence of the equilibrium composition arises. To illustrate the use of equilibrium constants to find the equilibrium composition and its effect on the temperature of the product mixture, let us consider a simple adiabatic flame temperature calculation.

5.4 CALCULATING THE ADIABATIC FLAME TEMPERATURE

The adiabatic flame temperature is the temperature attained by the product mixture assuming no heat transfer outside of the system ($\delta Q = 0$). This temperature provides a very useful upper bound for rocket combustion processes, i.e. any heat losses through chamber or nozzle walls will lead to a reduction in temperature as compared to the adiabatic limit. As argued previously, most rocket propulsion combustors/nozzles are well insulated out of necessity – the materials of construction we currently have to work with require good insulation from the hottest chamber gases. For this reason, the adiabatic assumption is not only prudent, but is presumably quite accurate (remember that direct measurement of combustion temperature is also not practical because there are no thermocouples that can withstand the temperatures attained in rocket combustion processes). In the laboratory however, we often fire small experimental combustors for short durations using heat-sink cooling (as described in Chapter 6) and in this case heat losses to the wall can be appreciable.

To perform the integration for sensible enthalpies, we need to know the variation of C_p with temperature for each of the species present in the reaction. An alternate approach wou involve the use of sensibleld enthalpy differences $(H - H_B)$, which are tabulated as a function of temperature in many combustion texts. In the 1960s the Joint Army Navy Air Force (JANAF) organization sponsored the development of a standard source of thermodynamic properties for use in both airbreathing and rocket combustion contexts. The JANAF Tables,[1] as they are called, were compiled from known thermochemical data for a large database of reactant and product materials and serve as a well-known standard to this date. The JANAF Tables contain curvefits or tabulated listings of C_p, S, $(F–H^0)/T$, $H–H^0$, ΔH^0, ΔF^0, and $\log(K_p)$ as a function of temperature. Rather than reproduce all these data here (it's a thick book, trust us), we will use C_p curvefits as a function of temperature since these data are easy to incorporate into computer models. A tabulation for some common gases is given in Table 5.3 excerpted from Hamblin's book, *Abridged Thermodynamic and Thermochemical Tables* (1971).

As a simple example, let us consider the combustion of hydrogen and oxygen at a pressure of one atmosphere with the reactants initially at 298 K. We will solve for the adiabatic flame temperature first assuming complete combustion (that all the reactants form water/steam), then repeat the calculation assuming incomplete combustion or chemical equilibrium.

5.4.1 Method 1 – Complete Combustion

In this case, the hydrogen and oxygen combine to form water vapor:

$$2H_2 + O_2 \rightarrow 2H_2O \tag{5.49}$$

The criteria for an adiabatic reaction is simply that $\delta Q = 0$ or that no change in total enthalpy occurs $(H_2 = H_1)$. For this reaction, we may write:

$$2\left[h^\circ + \int_{298}^{T_c} C_p \, dT\right]_{H_2O} = 2\left[h^\circ + \int_{298}^{298} C_p \, dT\right]_{H_2} + \left[h^\circ + \int_{298}^{298} C_p \, dT\right]_{O_2} \tag{5.50}$$

Substituting for the heats of formation and C_p values and performing the one non-vanishing integration gives:

$$-57\,800\,\frac{4.184\,\text{J}}{\text{calorie}} + \left[29.2T + \frac{14.5T}{2}(T/1000) - \frac{2.02T}{3}(T/1000)^2\right]_{298}^{T_c} = 0 \tag{5.51}$$

which can be solved using iterative techniques to give $T_c \approx 5150$ K. While this value lies well outside the range of the C_p curvefit, we shall use it as a basis of comparison.

---

[1] Subsequent to the initial effort on the JANAF Thermochemistry Tables, NASA also joined the organization. The current organization reflects this fact with the acronym JANNAF, i.e. the second "N" refers to NASA.

Table 5.3   Specific heat curvefits for several gases
$$C_p = a + b(T/1000) + c(T/1000)^2 + d(T/1000)^3; \quad \text{in J/g-mole K}$$

Gas or vapor	Range, K	a	b	c	d	% Max. error
Air	273–3700	27.453	6.1839	0.89932	0	1.64
Oxygen	273–3770	28.186	6.3011	−0.74986	0	3.32
Nitrogen	273–3770	27.336	6.2300	−0.95082	0	2.07
Hydrogen	273–3770	26.896	4.3501	−0.32674	0	2.41
Carbon monoxide	273–3770	27.130	6.5565	−0.99939	0	1.86
Carbon dioxide	273–3770	*	*	*	*	2.67
Water	273–3770	29.182	14.503	−2.0235	0	2.27
Hydrogen sulfide	273–1770	29.601	13.096	5.7108	−3.2938	0.69
Ammonia	273–1470	27.578	25.665	9.9072	−6.6909	0.82
Chlorine	273–1470	28.560	23.905	−21.382	6.4770	0.63
Iodine	273–1770	35.605	5.4994	−4.4732	1.3084	0.24
Hydrogen chloride	273–1470	30.329	−7.6200	13.272	−4.3375	0.19
Hydrogen iodide	273–1870	28.060	1.9033	5.0911	−2.0151	1.04
Nitric oxide	273–3770	27.051	9.8725	−3.2259	0.36547	2.32
Nitrous oxide	273–1470	24.108	58.632	−35.621	10.576	0.56
Nitrogen dioxide	273–1470	22.94	57.15	−35.21	7.871	0.46
Methane	273–1470	19.887	50.242	12.686	−11.011	1.34
Propane	273–1470	−4.044	304.76	−157.21	31.74	0.40
Cyclohexane	273–1470	−66.72	688.90	−385.31	80.68	1.59
Toluene	273–1470	−34.39	559.23	−344.57	80.39	0.35
Styrene	273–1470	−24.99	600.97	−383.09	92.24	0.57
Acetylene	273–1470	21.813	92.143	−65.272	18.208	1.46
Methyl chloride	273–1470	12.77	108.69	−52.08	9.63	0.79
Carbon tetrachloride	273–1470	51.25	142.35	−125.39	36.96	1.40
Methanol	273–970	19.050	91.523	−12.184	−8.039	0.17

* $C_{pCO_2} = 75.513 - 0.18732 \times 10^{-3}T - 661.85T^{-1/2}$

5.4.2 Method 2 – Incomplete Combustion, Chemical Equilibrium

Let us consider the case in which a portion of the hydrogen and oxygen do not burn:

$$2H_2 + O_2 \rightarrow aH_2O + bO_2 + cH_2 \tag{5.52}$$

Of course, in the real reaction, many other product species will be present, but let us examine this relatively simple case here. Balancing the total amount of hydrogen and oxygen atoms allows us to remove b and c from the analysis:

$$2H_2 + O_2 \rightarrow aH_2O + \frac{2-a}{2}O_2 + (2-a)H_2 \tag{5.53}$$

so that the total number of moles of products is:

$$n = a + \frac{2-a}{2} + 2 - a = \frac{6-a}{2} \tag{5.54}$$

We must get additional information from equilibrium constants to solve for the unknown coefficient a. We must consider the dissociation reaction:

$$H_2 + \frac{1}{2}O_2 \leftrightharpoons H_2O \tag{5.55}$$

We can write the equilibrium constant for this reaction as:

$$K_p = \frac{x_{H_2O}}{x_{H_2}\sqrt{x_{O_2}}}p^{-1/2} \tag{5.56}$$

But in this case, $x_{H_2O} = a/n = 2a/(6-a)$ and analogous expressions in terms of a can be derived for the other mole fractions. In addition, since the pressure is one atmosphere in this example, we can express K_p as:

$$K_p = \frac{a(6-a)^{1/2}}{(2-a)^{3/2}} \tag{5.57}$$

which effectively gives us another relation for a. The final piece of information involves the enthalpy balance for the reaction assumed in Eq. 5.51. Noting that the enthalpy of formation for hydrogen and oxygen is zero, this balance may be written:

$$a\left[-57800(4.184) + \int_{298}^{T_c} C_p \, dT\right]_{H_2O} + (2-a)\left[\int_{298}^{T_c} C_p \, dT\right]_{H_2} + \frac{2-a}{2}\left[\int_{298}^{T_c} C_p \, dT\right]_{O_2} = 0 \tag{5.58}$$

The solution procedure must be iterative in this case. The approach typically begins by assuming a T_c value that allows us to read K_p from Table 5.2. Knowing K_p, we can solve for a from Eq. 5.57 and substitute the result into the enthalpy balance above. When the enthalpy balance is maintained ($H_2 = H_1$), the correct T_c value has been assumed. The result for the example above gives $T_c = 3500$ K, $K_p = 4.93$, and $a = 1.31$. This result indicates that the high temperatures involved result in only 65% of the maximum water vapor being formed, with a resulting temperature drop of over 1600 K when compared with the case where complete combustion occurs. You can see that our calculations assuming complete combustion were hugely optimistic!

The previous example illustrates the fact that the high temperatures within the chamber can lead to *dissociations* of product species and incomplete combustion, i.e. recombination of some product species into original reactants. Therefore, some of the energy released in the reaction is utilized in forming additional product species and the final temperature of the mixture is lower (sometimes considerably lower) than a calculation assuming complete combustion with no dissociation. Let us now investigate the ramifications of this fact on the performance of a rocket engine.

5.5 ROCKET NOZZLE THERMOCHEMISTRY

Using the notions developed in Section 5.4, we can determine the composition of gases within the combustion chamber of a rocket. Since thrust is dependent on nozzle exhaust conditions (namely the exit velocity), we must consider the changes in composition that occur as a result of expansion of the hot gases through the nozzle. There are two limiting conditions:

1. *Frozen Flow* – This model assumes that the composition of the gases at all points in the nozzle is "frozen" at the pre-determined chamber composition.
2. *Equilibrium Flow* – This model assumes that the composition of gases at any point in the nozzle corresponds to the equilibrium composition for the local pressure and temperature at any particular nozzle location.

Engineers analyzing rocket performance in the 1940s and 1950s used the frozen flow assumption since they only had to compute the composition at chamber conditions. (Remember this was prior to the advent of the computer.) Using this model and ignoring all other loss mechanisms described in the Chapter 4, they observed that larger rockets actually had measured thrust values higher than predictions. In other words, using frozen flow performance gave engine efficiencies greater than unity! Since this could not be physically true, engineers attributed errors to the frozen flow assumption invoked in the calculations.

The advent of high-speed computers in the 1960s permitted the use of equilibrium flow calculations. Thrust predictions using equilibrium flow were significantly higher than the frozen flow results due to the fact that some dissociated species recombined (giving off thermal energy) as the gases expanded to lower pressures and temperatures. Comparison of measured data with equilibrium predictions always gave engine efficiencies less than one.

What really happens? Well it turns out that the frozen and equilibrium flow models provide lower and upper bounds on the ideal performance of an engine. In real life, *chemical kinetics* plays a major role in determining which recombination reactions occur during the short interval of time in which a given element of fluid passes through the nozzle. The frozen flow assumption is consistent with infinitely slow kinetics in that no recombinations are permitted. The equilibrium flow model is consistent with infinitely fast kinetics in that all possible recombinations are assumed to occur. Since the actual performance lies between these two bounds, we may write:

$$I_{sp,frozen} < I_{sp,actual} < I_{sp,equilibrium} \tag{5.59}$$

In today's calculations, engineers start with the equilibrium performance and estimate the loss due to kinetic processes which did not have time to occur for the particular nozzle geometry in question. It turns out that most of the kinetic reactions are very fast, even when compared with the duration of time a given mass of fluid spends in the nozzle. Typically, the chemical kinetics representative of rocket combustion processes occur on a microsecond timescale, while the gas residence time is generally in milliseconds. Even with very high velocities in the nozzle, the local equilibrium assumption is fairly accurate for larger motors/engines. For large engines/motors, kinetics losses are typically less than 1% of the thrust of the device. For micropropulsion systems, however, kinetics losses could be much more important as flow times and chemistry times become much more comparable. We will talk more about kinetics losses in our discussion of solid rocket motor and liquid rocket engine performance in Chapters 7 and 8.

5.6 COMPUTER CODES FOR CHEMICAL EQUILIBRIUM COMPUTATIONS

In the late 1960s, computers started becoming available for those in the propulsion industry. One of the first applications of computational solutions in propulsion problems was in the solution of chemical equilibrium and adiabatic flame temperatures for combustion mixtures. In 1971, NASA researchers Sanford Gordon and Bonnie McBride published a Special Publication (NASA SP-273) that provided the description of a code they had developed to compute chemical equilibrium compositions and rocket performance.[2] This code, sometimes referred to as "the NASA equilibrum code," has served as the basis for more complex software aimed at providing a simulation of processes in rocket combustors and nozzles. The construction of this code and assembly of these data was a major advancement for the rocket community as calculations were transitioning from sliderules to computers in the Apollo era.

     A number of variants exist in today's computational environment; some of which provide graphical user interfaces and even integration into spreadsheets to ease the use of the calculations in more elaborate models. An online version, called CEA (Chemical Equilibrium Analysis), is available for download from NASA websites. A step-by-step tutorial for download and use of this package is provided on the Purdue Propulsion Web Page (https://engineering.purdue.edu/~propulsi/propulsion/). Since readers can access this package, we will amplify here with some general comments on input and output files.

5.6.1 Program Inputs

One of the biggest shortcomings of using a graphical user interface (GUI) or spreadsheet to run CEA lies in the fact that many of the default inputs are somewhat hidden from the user. Interested readers are encouraged to consult the original NASA documentation (SP-273) that provides a complete description. In general, "rocket problems" are modeled as an enthalpy–pressure (HP) problem that assumes constant pressure combustion/reactions. Initial enthalpies at room temperature (or at the fluid boiling point for cryogenics) are generally assumed in the GUI, so if reactants

---

[2]   The code also had the capability to compute equilibrium mixtures of air behind strong shocks as well as computation of combustion products behind detonation waves.

are introduced at other temperatures the reactant information needs to be adjusted accordingly. There are several different ways to prescribe the mixture ratio of oxidizer and fuel, either by weight percentage of each constituent, by the percentage of fuel, as a mixture ratio, or as an equivalence ratio. In general, oxidizer/fuel mass ratio (mixture ratio) is prescribed for hybrid and liquid bipropellant problems, whereas solid propellants are prescribed by inputting the mass fraction of each ingredient.

The combustion (a.k.a. chamber) pressure must also be prescribed as well as an array of area ratios where users desire fluid property and composition information. An option is available to compute local transport properties including mixture viscosity, thermal conductivity, and Prandtl number. One can also stipulate equilibrium or frozen flow options, with the latter corresponding to the case where the composition is "frozen" at the chamber condition.

These basic inputs are generally sufficient for many rocket combustion problems, but readers should be aware of several other inputs that may or may not need to be considered depending on the particular problem of interest. The code is iterating toward a composition and temperature that minimize Gibb's free energy, so there must be some measure of sufficiency that the iteration is converged. Mole fraction values defining convergence are prescribed inputs that are infrequently changed, but may require attention if one is looking for trace ingredients such as pollutants. One is also able to INSERT or OMIT particular species. Sometimes, convergence is upset by small amounts of condensed phase or frozen material (e.g. ice) and the OMIT function can be used to remove these minor species and aid convergence. In general, OMIT inputs should not be used as the user is wanting the algorithm to make its own choices in this regard.

The code draws from a major thermochemical database that was created by JANNAF (Joint Army Navy NASA Air Force) organization that was formed just after World War II. Specific heat and entropy data are curvefit as a function of temperature, typically up to temperatures of 5000 K. Users of the code need to keep this in mind as experience shows that many are happy to believe many more digits in the output than are really significant given the accuracy of information and the iterative nature of the solution. Readers interested in this issue can consult Stull and Prophet (1971) at the end of the chapter or the JANNAF website (https://www.jannaf.org/) . It is possible to add additional species to the database as long as the format requirements are maintained. We have done a bit of this in adding inerted reactants (i.e. ones that don't participate in the combustion, but absorb heat from the reaction) to study their effect on performance.

5.6.2 The Output File

Table 5.4 shows output for combustion of high-concentration hydrogen peroxide with methane. The output begins by echoing the input (reactants, proportions, and combustion pressure). You can see that the mixture ratio of 8.0 corresponds to a fuel percentage of

Table 5.4   Sample output file from rocket thermochemistry calculation

```
theoretical rocket performance assuming equilibrium composition during expansion
0pc = 1200.0 psia
```

		wt fraction	energy	state	temp
chemical formula		(see not (e)	cal/mol		deg k
oxidant h 2.00000 o 2.00000		0.900000	−44880.000	l	298.15
oxidant h 2.00000 o 1.00000		0.100000	−68317.000	l	298.15
fuel c 2.00000 h 4.00000		1.000000	−12700.000	s	298.15

```
o/f=   8.0000   percent fuel= 11.1111   equivalence ratio= 1.0047   phi= 1.0104
```

0	chamber	throat	exit	exit	exit	exit
pc/p	1.0000	1.7350	73.894	610.15	1746.58	2347.28
p, atm	81.655	47.063	1.1050	0.13383	0.04675	0.03479
t, deg k	2791.7	2627.1	1555.8	1071.4	594.3	593.4
rho, g/cc	7.8753-3	4.8531-3	1.9540-4	3.4366-5	2.1704-5	1.6173-5
h, cal/g	-1442.93	-1576.78	-2268.50	-2510.61	-2632.63	-2648.03
u, cal/g	-1694.03	-1811.62	-2405.45	-2604.92	-2684.79	-2700.12
g, cal/g	-9209.35	-8885.41	-6596.80	-5491.14	-4285.85	-4298.83
s, cal/(g)(k)	2.7820	2.7820	2.7820	2.7820	2.7820	2.7820
m, mol wt	22.093	22.230	22.575	22.575	22.638	22.638
(dlv/dlp)t	-1.00777	-1.00574	-1.00001	-1.00000	-1.00000	-1.00000
(dlv/dlt)p	1.1821	1.1429	1.0002	1.0000	1.0000	1.0000
cp, cal/(g)(k)	0.9810	0.9092	0.5253	0.4715	19.5085	15.9440
gamma (s)	1.1368	1.1399	1.2014	1.2296	1.0045	1.0055
son vel, m/sec	1092.9	1058.3	829.7	696.5	468.2	468.1
mach number	0.000	1.000	3.168	4.291	6.739	6.784

0performance parameters						
ae/at		1.0000	10.000	50.000	75.000	100.00
cstar, ft/sec		5285	5285	5285	5285	5285
cf		0.657	1.632	1.856	1.959	1.971
ivac, lb-sec/lb		202.6	290.3	318.3	328.8	330.8
Isp, lb-sec/lb		107.9	268.0	304.8	321.7	323.8

0mole fractions						
co	0.02191	0.01663	0.00209	0.00092	0.00000	0.00000
co2	0.15310	0.15946	0.17674	0.17791	0.17128	0.17246
c3h4(al)	0.00000	0.00000	0.00000	0.00000	0.00268	0.00229
h2	0.01630	0.01276	0.00344	0.00461	0.00000	0.00000
h2o	0.77913	0.79018	0.81771	0.81657	0.81808	0.81887
h2o2	0.00001	0.00000	0.00000	0.00000	0.00000	0.00000
o	0.00065	0.00037	0.00000	0.00000	0.00000	0.00000
oh	0.01524	0.01057	0.00002	0.00000	0.00000	0.00000
o2	0.01251	0.00931	0.00000	0.00000	0.00796	0.00638

11.11% per mass and an equivalence ratio very near stoichiometric. Following these data is a tabular listing of gas properties including pressure, temperature, density (p, t, rho), enthalpy, internal energy, and entropy (h, u, s). Following this is the molecular weight (m) and some thermodynamic derivatives that are related to local composition changes. The local gas specific heat (cp), ratio of specific heats ($gamma$), sonic velocity and Mach number round out the table. Following this is a list of rocket performance parameters including the local expansion ratio (the chamber and throat conditions are always output in addition to the supersonic area ratio "supar" locations we specified in input), c^*, thrust coefficient (cf) and both vacuum and locally perfectly expanded I_{sp} values (ivac, Isp, respectively).

Following the tabular listing of rocket performance parameters is a detailed listing of the local composition at each of the locations specified by the user. In this particular example, the reactants contain the elements hydrogen, oxygen, and carbon. The code will consider all product species that contain these three elements (and only these three elements) that are contained within its thermochemical database including condensed and solid forms of these potential products (e.g. ice and liquid water). Some of these potential products will appear in very small amounts and a threshold limit is considered whereby a given product species may be neglected if its mole fraction falls below this limit. Given this selection process, the resulting composition is computed as outlined in the file below.

The local equilibrium assumption employed in this example allows one to assess the effect of recombination reactions as the flow expands through the nozzle. In the combustion chamber, substantial dissociations are present as evidenced by the non-negligible amounts of such species as CO, O, and OH. As the flow expands into the nozzle, complete combustion products CO_2 and H_2O are formed from recombination reactions involving these species. We'll talk later about the repercussions of the local equilibrium assumption. This is a very useful starting point in that it represents an upper bound on the performance that any system can attain.

In the 2000s, NASA Glenn Research Center developed a web-based version of the tool and a subsequent GUI that provides a modern, windows-based approach toward solving these thermo-chemistry problems. A tutorial on the use of this tool is provided on the Purdue Propulsion Web Page (https://engineering.purdue.edu/~propulsi/propulsion/).

FURTHER READING

Barrere, M., Jaumotte, A., Frais de Veubeke, B., and Vandenkerckhove, J. (1960) *Rocket Propulsion*. Elsevier Publishing Company.

Gordon, S. and McBride, B. J. (1971) "Computer Program for Calculation of Complex Chemical Equilibrium Compositions, Rocket Performance, Incident and Reflected Shocks, and Chapman-Jouguet Detonations," NASA Special Publication, SP-273.

Hamblin, F. D. (1971) *Abridged Thermodynamic and Thermochemical Tables: In S.I.Units (C.I.L.)*. Pergamon Press.

Penner, S. S. (1957) *Chemical Problems in Jet Propulsion*. Pergamon Press.

Siegel, B. and Schieler, L. (1964) *Energetics of Propellant Chemistry*. John Wiley and Sons.

Stull, D. R. and Prophet, H. (eds.) (1971) *JANAF Thermochemical Tables*, 2nd edn. US Department of Commerce, National Bureau of Standards, National Standard Reference Data System, NSRDS-NBS37.

Van Wylen, G.J . and Sonntag, R. E. (1973) *Fundamentals of Classical Thermodynamics*, 2nd edn. John Wiley and Sons.

HOMEWORK PROBLEMS

5.1 Write down balanced chemical reactions assuming complete combustion with 1 mole of fuel for the following propellant combinations:

  (i) LOX-RP-1 at a mixture ratio of 2.2;

  (ii) N_2O_4 – MMH at a mixture ratio of 2.0;

  (iii) stoichiometric combustion of hydrogen peroxide (H_2O_2) and hydrazine; and

  (iv) stoichiometric combustion of wood ($C_{0.513}H_{0.061}O_{0.411}$) in air.

Determine the molecular weight, C_p, γ, and c^* values of the products in the LOX-RP-1 reaction given above assuming a flame temperature of 6000 °R.

5.2 Consider chemical equilibrium combustion of chlorine pentafluoride ClF_5 with MMH at a mixture ratio of 2.8 under vacuum conditions with an expansion ratio of 50. Plot the mole fraction of the four most prominent exhaust species at the nozzle exit plane as a function of chamber pressure for $10 < P_c < 100$ atmospheres. Discuss the behavior of these curves, i.e. why do some species increase with pressure while others decrease? Which dissociation reactions are important to explain this behavior?

5.3 You have been asked to examine the influence of aluminum loading on the performance of a hybrid rocket propellant using 90% aqueous solution of hydrogen peroxide as the oxidizer rocket. The current design specifies chamber and exit pressures of 750 and 14.7 psi, respectively. The fuel is assumed to be hydroxyl-terminated polybutadiene (HTPB), which has a chemical structure $C_{7.3}H_{10.96}O_{0.06}$ and a heat of formation of –250 cal/mole. Determine the optimal mixture ratio (to maximize ODE I_{sp}) assuming the fuel is loaded with 0, 5, 10, and 15% aluminum by weight. Tabulate the optimal mixture ratio, I_{sp}, flame temperature, molecular weight, and propellant bulk density for these four cases. Discuss which of the four formulations you would choose and provide reasons for your results. Remember, two-phase flow losses will increase with aluminum content.

5.4 Determine the adiabatic flame temperature for the reaction:

$$2H_2 + O_2 \rightarrow aH_2O + bO_2 + cH_2 + dOH$$

by including the dissociation reactions:

$$H_2 + \frac{1}{2}O_2 \leftrightharpoons H_2O \ and \ \frac{1}{2}O_2 + \frac{1}{2}H_2 \leftrightharpoons OH$$

Perform calculations assuming the reactants are at 298 K and a pressure of 50 atmospheres. Write down the balanced chemical equation, i.e. determine values for a, b, c, and d as defined above. You **must** include temperature dependent C_p data from Table 5.3.

5.5 Consider the lean, constant pressure combustion of carbon at one atmosphere in which a significant amount of carbon monoxide is produced:

$$3C + 4O_2 \rightarrow \alpha CO_2 + \beta CO + \gamma O_2 \tag{1}$$

Here α, β, and γ represent unknown stoichiometric coefficients. To close the problem, we must consider the dissociation of carbon dioxide:

$$CO_2 \rightleftharpoons CO + \frac{1}{2}O_2 \tag{2}$$

(a) For the reaction in Eq. (1), derive an expression (it need not be simplified) which relates α (and only α) to the equilibrium constant, K_p, for the dissociation reaction in Eq. (2).

(b) A mass spectrometer is used to determine that two moles of CO_2 are present in the products. Determine (within 50 K) the temperature of the product gases.

5.6 Write balanced chemical equations assuming complete combustion for the following reactions (assume 1 mole of fuel is present when writing out reactions):

(a) stoichiometric combustion of Aerozine-50 and nitrogen tetroxide (A-50 is 50% hydrazine, 50% MMH per mass);

(b) combustion of LOX/RP-1 at a mixture ratio of 2.6;

(c) combustion of LOX/RP-1 at a mixture ratio of 4; and

(d) combustion of methane and fluorine at an equivalence ratio of 0.8.

5.7 Calculate heats of combustion for the reactions in Problem 5.6. Consider the combustion of hydrogen and oxygen at a pressure of 100 atm:

$$2.5H_2 + O_2 \rightarrow aH_2O + bOH + cH_2 + dH$$

You may assume that reactants are at a temperature of 298 K and that the following dissociation reactions are important:

$$OH + \frac{1}{2}H_2 \rightleftharpoons H_2O \text{ and } \frac{1}{2}H_2 \rightleftharpoons H$$

Determine the adiabatic flame temperature and equilibrium composition for this reaction. You must use temperature dependent specific heats (Table 5.3 or equivalent).

5.8 The propellant combination LOX/RP-1 is used in many US and Soviet rocket propulsion systems (RP-1 is essentially kerosene). At $P_c = 1000$ psi, $\varepsilon = 50$, determine the optimal mixture ratio for this propellant combination by plotting I_{spv} as a function of this parameter assuming shifting chemical equilibrium flow. Also plot T_c, $\mathcal{M}_{ch}$ and $T_c/\mathcal{M}_{ch}$ as a function of the mixture ratio at the pressure given above. Discuss the locations of the maximum in T_c, I_{sp}, and $T_c/\mathcal{M}$, i.e. are all these functions maximized at the same mixture ratio? What is the stoichiometric mixture ratio for this propellant combination?

5.9 Consider the hybrid rocket propellant combination LOX/HTPB at a chamber pressure of 50 atm.

    (a) Assuming shifting chemical equilibrium flow, determine the mixture ratio that optimizes vacuum performance for this propellant combination at $\epsilon = 50$.

    (b) At the optimal mixture ratio, plot the mass fraction of the five most prominent exhaust species as a function of expansion ratio $1 < \epsilon < 100$. What recombination reactions are occurring within the nozzle?

5.10 The Space Shuttle SRB utilizes the following propellant combination (on a percent mass basis):

Aluminum – 16%
Ammonium perchlorate (NH_4ClO_4) – 69.7%
PBAN ($C_{6.719} H_{9.548} O_{0.410} N_{0.222}$) – 14%
Iron oxide (Fe_2O_3) – 0.3%

The heats of formation for PBAN and Fe_2O_3 are –15,110 and –197,300 cal/gm-mol respectively. The average p_c and ϵ values are 660 psi and 7.48, respectively. If the nozzle throat radius is 27.36 inches, determine:

    (a) sea level mass flow, thrust and I_{sp}; and

    (b) mass flow of HCl and Al_2O_3 as a fraction of the total mass flow.

    (c) Can you predict the solidification temperature of Al_2O_3 using your data?

    (d) Is this an environmentally "clean" propellant?

5.11 Repeat Problem 5.4 using CEA to compute the chamber temperature and exhaust composition. How did your previous answer compare? Draw up a table comparing the mole fractions of the five major products and compare these results (in the same table) with your hand calculation. Explain any differences in the results from the two methods.

5.12 A mixture of hydrogen and fluorine gas, initially at a temperature of 198 K, is combusted at constant pressure of 50 atm per the following reaction:

$$H_2 + 4F_2 \rightarrow aHF + bF_2 + cF$$

where a, b, and c are unknown stoichiometric coefficients. To determine the composition of the products of this reaction, consider the dissociation of hydrofluoric acid:

$$F_2 \leftrightharpoons 2F$$

    (a) Derive an expression relating the stoichiometric coefficient b to the equilibrium constant for the dissociation reaction given above.

(b) Determine the adiabatic flame temperature and equilibrium composition for this combination process assuming that the specific heats are constant, as in the table below:

Constituent	C_p (J/g-mol K)
H_2	31
F_2	25
HF	30
F	30

[Hint: $b < 0.5$.]

5.13 Assuming $b = 0.5$ and $T_2 = T_c = 3000$ K, determine the characteristic velocity for the propellant combustion process described in Problem 5.12. (The molecular weight of atomic fluorine is 19 gm/gm-mol.)

5.14 A mixture of 30% hydrogen, 30% carbon dioxide, and 40% nitrogen (by volume) is stored at ambient conditions in a cylindrical container with a piston at the top. The mixture is heated to a temperature of 200 °C. Determine:

(a) the amount of energy added per kg of gas in the container; and
(b) the amount of motion of the piston (i.e. the percentage increase in length of the cylinder).

5.15 Many liquid rocket engines utilize non-cryogenic (i.e. storable) propellants to simplify operations and provide a means of firing the engine after extended time intervals. One of the more common combinations in use today involves the use of nitrogen tetroxide (N_2O_4) and monomethyl hydrazine (MMH) as oxidizer and fuel, respectively. Assume that these propellants are combined at a mixture ratio of 2.4 in a combustion held at 25 atm. Consider water vapor, carbon monoxide, carbon dioxide, diatomic nitrogen, and oxygen as combustion products and determine:

(a) the adiabatic flame temperature assuming the reactants are gases at 298 K;
(b) the equilibrium composition of the product species;
(c) the molecular weight of the product gases;
(d) the mass fraction of each of the assumed product species; and
(e) the characteristic velocity obtained by this combustion process. [Hint: Consider the dissociation of carbon dioxide in your analysis.]

5.16 Consider the state-of-the-art solid propellant in the following table:

Ingredient	Formula	Weight %
Ammonium perchlorate	NH_4ClO_4	70
Aluminum	Al	18
HTPB	$C_{7.3}H_{10.96}O_{0.06}$	12

Here HTPB is the binder material, and has a heat of formation of –250 cal/mole. Suppose this propellant is combusted in a solid rocket motor (SRM) at a chamber pressure of 500 psi with an expansion ratio of 12. Determine:

(a) sea-level and vacuum I_{sp};
(b) sea-level and vacuum C_f;
(c) c^*; and
(d) $\mathcal{M}$ in the chamber.
(e) Repeat (a)–(d) for $P_c = 100$ psi. Are c^* and vacuum I_{sp} sensitive to p_c?

5.17 Consider the Lox-RP-1 propellant combination combusting at 100 atmospheres with $\varepsilon = 75$. Determine the optimal mixture ratio by plotting T_c, $\mathcal{M}$ (in chamber), and I_{spv} as a function of r for $1.5 < r < 5$. At the optimal mixture ratio, plot $\mathcal{M}$ vs. A/A_t in the exit cone. Explain the behavior of this curve. What recombination reactions are important? Tabulate the **mass fraction** of the six most prominent exhaust species at $\varepsilon = 1, 3, 5, 15, 25, 75$, for the optimal r condition.

5.18 In the combustor of an airbreathing engine, JP-5 (approximately CH_2) is reacted with air at a constant pressure of 10 atm. Sensors in the combustor indicate a temperature of 3000 K. At this elevated temperature, we presume that appreciable amounts of NO are formed (in addition to water vapor and carbon dioxide). From fuel and air flow measurements, the reactants can be expressed as:

$$CH_2 + 2O_2 + 7.5N_2 \rightarrow \text{Products}$$

Determine the composition of the product species for this reaction. How would this result change if we doubled the pressure?

5.19 Consider the combustion of 1 mole of hydrazine with 0.1 mole of gaseous oxygen at a pressure of 5 MPa. Assuming complete combustion, nitrogen gas, hydrogen gas, and water vapor are formed. However, at high temperature, a small amount of ammonia gas (NH_3) can be present. The amount of ammonia is determined by the dissociation reaction:

$$NH_3 \leftrightarrows \frac{1}{2}N_2 + \frac{3}{2}H_2$$

Presume that the reactants are at standard (298 K) temperature. Calculate the equilibrium composition and adiabatic flame temperature for this reaction. Calculate the molecular weight, γ, and c^* values for the equilibrium mixture. Does dissociation of the ammonia increase or decrease c^*, i.e. is some level of dissociation a good thing?

5.20 Consider the combustion of N_2O_4 and hydrazine at a pressure of 4 MPa. Using the NASA code, plot T_c, $\mathcal{M}$, $T_c/\mathcal{M}$, and I_{spv} as a function of mixture ratio for $1.0 < r < 2.0$. What is the stoichiometric mixture ratio for this propellant combination?

5.21 Consider the combustion of 1 mole of methane with 2 moles of oxygen at a pressure of

5 atmospheres. Due to the high combustion temperature of this reaction, we presume significant dissociation of carbon dioxide (to carbon monoxide and diatomic oxygen) will occur. Measurements indicate a temperature of 3000 K associated with this reaction. Determine the equilibrium composition of the product species in this case.

5.22 In our discussion of hydrazine thruster performance, we noted that the decomposition reaction could be written in terms of the fraction of NH_3 (ammonia) dissociation, x:

$$3N_2H_4 \rightarrow 4(1-x)(1+2x)$$

For optimal performance (I_{sp}), we desire $x \cong 0.2$. Suppose we ran a test of the thruster at 500 psi and determined that $x \cong 0.3$ for this condition. Can you predict (i.e. calculate) the chamber pressure required to reduce x to the desired value? You may assume that a negligible variation in flame temperature occurs over this narrow range of x values.

5.23 Ambient pressure air (79% N_2, 21% O_2 by volume) is heated electrically at constant pressure using an arc-heater. The gases are initially at a temperature of 400 K and are heated to 3500 K using the arc-heater. At this high final temperature we expect significant dissociation of the O_2 (into monatomic oxygen) in the container.

(a) Determine the composition of the gases in the heated container.

(b) Assuming the container is well insulated, determine the amount of energy added by the heater per mole of O_2.

For this problem, you may assume constant specific heats as provided in the table below:

Constituent	C_p (cal/gm-mol K)
O_2	8.2
N_2	8.0
O	15.0
NO	10.0

5.24 Consider the decomposition of room temperature hydrazine (N_2H_4) at a pressure of 50 psi. Decomposition products include N_2, H_2, and ammonia gas (NH_3). Using equilibrium constants, determine the composition of the product gases, and the adiabatic flame temperature and c^* value for this reaction. What would the c^* value be if we had assumed complete combustion?

5.25 Pure oxygen is stored in a piston-type container which is to be heated from below, as shown in Figure 5.3. Initially, the oxygen is at room temperature (298 K) and a pressure of one atmosphere. The gases are then heated by the burner to a temperature of 3000 K. At this elevated temperature, we expect substantial dissociation to monatomic oxygen.

(a) Determine the relative proportions (mole fractions) of O_2 and O in the chamber at the end of the heating process.

(b) What is the height of the piston at the end of the heating process?

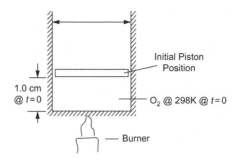

Figure 5.3 Diagram for Problem 5.25.

You may assume that the values of C_p are constant in your analysis: $C_{p_{O_2}} = 35$ J/mol K, $C_{p_O} = 25$ J/mol K.

5.26 Previous rocket systems have been tested with fluorine/oxygen mixtures (called FLOX) as oxidizer. Assume we utilize a 30/70 fluorine/oxygen blend (based on weight percent) and we wish to use MMH as fuel. Write down the balanced chemical equation for complete combustion of these propellants under stoichiometric conditions. Write your results assuming 1 mole of fuel is present in the reactants.

5.27 Hydrazine (N_2H_4) is a monopropellant commonly used in space propulsion applications. This fluid decomposes into nitrogen gas, hydrogen gas, and ammonia (NH_3) depending on the pressure level. Assuming the hydrazine enters the reaction at room temperature and a decomposition at 10 atmospheres, determine:

(a) the equilibrium composition and adiabatic flame temperature using K_p data on the Purdue Propulsion Web Page (https://engineering.purdue.edu/~propulsi/propulsion/); and

(b) the resulting $c*$ of the decomposition gases.

5.28 Repeat the calculations in Problem 5.27 using the NASA thermochemistry code. Attach the relevant output table summarizing chamber conditions. What happens to the exhaust composition as the gases expand through a nozzle?

5.29 Figure 5.4 shows an electrically augmented hydrazine thruster (see Chapter 8 for more discussion). Here, the idea is to use excess electricity from solar arrays to augment the performance (I_{sp}) of the thruster. The electricity flows through a resistance heating coil (just like a toaster) to increase the temperature of the gases entering the nozzle. We would like to model the performance of this device using the NASA thermochemistry code. Assume we know the p_c, propellant flow rate $\dot{m}$, nozzle expansion ratio ε, and power input to the heating element P_h. Using this information, can you devise a way to "trick" the NASA code into simulating the thermochemical performance of this thruster? Briefly explain the approach you would use.

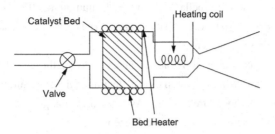

Figure 5.4 Diagram for Problem 5.29.

5.30 Assume you wish to develop a process to make hydrofluoric acid (HF) by first decomposing fluorine gas (F_2) into atomic fluorine, then combining with hydrogen gas. The fluorine is heated electrically as it flows through a tube at essentially constant pressure.

(a) To what temperature would we need to heat the F_2 gas to insure that at least 50% is dissociated when it exits the heater, assuming the process occurs at 20 atm pressure?

(b) Assuming that F_2 enters at room temperature (298 K), how much energy (per mole of F_2) is required to cause the dissociation desired?

(c) You want to increase the F_2 dissociation to 75%, keeping the same amount of heating to the plant. Determine the temperature and pressure of the gases, omitting the dissociations step in this case.

You may assume constant specific heats in this problem: $C_{P_{F_2}} = 0.25$, $C_{P_F} = 0.6$, $C_{P_{H_2}} = 3.4$ (all in cal/gm K).

5.31 (a) What mixture ratio corresponds to a combination of 7 moles of N_2O_4 with 10 miles of MMH?

(b) Determine the stoichiometric mixture ratio for combustion of Lox with RP-1.

(c) Write an equation for the reactant formulation in a Titan launch vehicle liquid engine. The fuel is a 50/50 mixture (by weight) of MMH and UDMH and the oxidizer is N_2O_4. The mixture ratio for these propellants is 2.0 during engine operation.

5.32 Write down balanced chemical formulas for the reactants in the following reactions.

(a) Combustion of 90% hydrogen peroxide (H_2O_2) with polyethylene (C_2H_4) at a mixture ratio of 8. Write down the reactants assuming 1 mole of fuel is present.

(b) Combustion of a solid propellant which is 15% aluminum, 70% ammonium perchlorate, and 15% HTPB ($C_{7.33}H_{10.96}O_{0.06}$) by weight. Write down the reactants assuming a single mole of aluminum is present.

(c) Combustion of 90% hydrogen peroxide (H_2O_2) with polyethylene (C_2H_4) at a mixture ratio of 8. Write down the reactants assuming 1 mole of fuel is present.

(d) Combustion of a solid propellant which is 15% aluminum, 70% ammonium perchlorate, and 15% HTPB ($C_{7.33}H_{10.96}O_{0.06}$) by weight. Write down the reactants assuming a single mole of aluminum is present.

5.33 Air is heated in a container at 10 atmospheres pressure through the use of electric arc discharge. Gas sample probes are then used to determine the concentration of various species present in the high-temperature mixture. As a result of this process, we assume that nitric oxide is created via the following chemical reaction:

$$O_2 + 3.76N_2 \rightarrow 0.9O_2 + 3.66N_2 + 0.2NO$$

(a) Determine the temperature of the heated mixture.

(b) Assuming that the air was originally at room temperature, determine the amount of energy (in joules) added by the electric arc discharge. You may assume there is negligible heat loss to the walls of the container.

(c) Describe what would happen to the products in this reaction if the pressure were doubled to 20 atmospheres.

You may assume that specific heats are constant and equal to the value in the table below:

Gas	C_P (cal/mol K)
Oxygen	8
Nitrogen	7.5
Nitric Oxide	11

5.34 A mixture of gases at 1.0 MPa and a temperature of 350 K is made up of 1 kg of carbon monoxide, 3 kg of carbon dioxide, and 2 kg of water. Calculate the following properties of the mixture: molecular weight, gas constant, specific heat ratio, partial pressure of each component, and gas density. What is the $c*$ of the mixture?

5.35 In a thermally isolated chamber, 8 kg of oxygen and 1 kg of hydrogen, both initially at 298 K, react at atmospheric pressure to form a mixture of combustion products. Write down the balanced equation. How much heat is evolved during the reaction? Approximate the temperature of the final mixture by assuming the heat capacity of the mixture is the heat capacity of the major constituent in the combustion gas (H_2O). At that temperature (it should be pretty high), dissociation occurs, and an equilibrium exists between O_2, H_2O, H_2, H, O, and OH. Calculate the partial pressures of each of these species.

5.36 Write down balanced chemical equations assuming complete combustion of the limiting reactant (which could be either oxidizer or fuel) for the reactions described below. For each reaction assume we begin with 1 mole of fuel.

(i) Stoichiometric combustion of nitrogen tetroxide and monomethyl hydrazine.

(ii) Combustion of hydrogen peroxide and polyethylene at a mixture ratio of 7.0.

(iii) Combustion of Lox with LH_2 at a mixture ratio of 6.0.

(iv) Combustion of a mixture of 75% ammonium perchlorate (NH_4ClO_4) and 15% HTPB by weight.

(a) Determine heats of combustion for each of the reactions given in above.

(b) What would the $c*$ value be for (ii) if we had assumed complete combustion?

[Hints: HTPB $= C_{7.332}H_{10.962}O_{0.058}$ $h° = -250$ cal/mol; polyethylene $= C_2H_4$ $h° = -12{,}700$ cal/mol.]

5.37 Run the NASA CEA code to optimize the performance of the Aerojet LR 87 engine used on the first stage of the Titan IV launch vehicle. This engine has $P_c = 800$ psi, $\varepsilon = 15$, and utilizes N_2O_4 as oxidizer. The fuel is a blend of 50% hydrazine and 50% UDMH (track name, Aerozine-50).

(a) What is the stoichiometric mixture ratio for these propellants?

(b) Using the results from the code, plot T_c, $\mathcal{M}$, and vacuum I_{sp} vs. r. What is the optimal mixture ratio?

(c) Tabulate the mass fraction of the six most prominent species in the chamber and at the nozzle exit at the $r = r_{opt}$ condition.

5.38 Some engineers have proposed the use of hydrogen injection to augment the performance of the Space Shuttle. Plumbing is installed such that LH_2 (from the external tank) is pumped into the head end of each of the SRMs. Since the total mass of all propellants is assumed to be fixed, the liquid rocket engines (LREs) operate at a higher mixture ratio. Using the NASA CEA code, you are asked to investigate this design change. Data for your analysis are tabulated below:

Propulsion system	Space Shuttle main engine (SSME)	Solid rocket booster
Propellants	LOX/LH$_2$	16% Al, 14% PBAN, 69.7% AP, 0.3% Fe$_2$O$_3$
Propellant weight (lb)	0.228×10^6 fuel- 1.362×10^6 oxygen	1.107×10^6 (per SRM)
Avg. p_c (psi)	3000	645
Throat diameter (in)	10.30	53.86

Assuming ideal (ordinary differential equation) performance, use the NASA code to determine the optimal amount of LH_2 to be used in the SRMs. Note that the p_c will change slightly as hydrogen is redistributed; you may neglect this effect on the SRMs, but should consider it on the LREs. Remember, there are three LREs and two SRMs on the vehicle.

(a) Plot SSME I_{sp}, RSRM I_{sp}, and total vehicle impulse as a function of the percentage of LH$_2$ pumped to the SRMs. What is the optimal percentage (to maximize vehicle impulse)? Assume that the SRMs are operating at an altitude of 50,000 feet and the LREs are operating at an altitude of 150,000 feet for the purpose of calculating I_{sp}.

(b) Plot the SSME p_c vs. percentage LH$_2$ pumped to the RSRMs.

(c) Can you speculate in any problems (other than additional plumbing) one might have in implementing this design change?

5.39 (a) Write down balanced chemical equations assuming complete combustion of the limiting reactant (could be either oxidizer or fuel). As a convention, write your reactions in terms of 1 mole of fuel.

  (i) Stoichiometric combustion of chlorine pentafluoride (ClF$_5$) and monomethyl hydrazine (CH$_3$NHNH$_2$).

  (ii) Combustion of LOX and RP-1 at $r = 2.0$.

  (iii) Combustion of LOX and LH$_2$ at $r = 5$.

  (iv) Combustion of aluminum with ammonium perchlorate (NH$_4$ClO$_4$) with 10% fuel by weight.

  (v) Combustion of magnesium with ammonium nitrate NH$_4$NO$_3$ at an equivalence ratio of 0.9.

(b) Determine the heat of combustion for each of the reactions above. Note $h_f^\circ = -60.5$ kcal/mol for ClF$_5$.

(c) Consider the combustion of 1 mole of hydrazine with 0.1 mole of gaseous oxygen at a pressure of 5 MP$_a$. Assuming complete combustion, nitrogen gas, hydrogen gas, and water vapor are formed. However, at high temperature, a small amount of ammonia gas (NH$_3$) can be present. The amount of ammonia is determined by the dissociation reaction:

$$NH_3 \leftrightharpoons \frac{1}{2}N_2 + \frac{3}{2}H_2$$

Presume that the reactants are at standard (298 K) temperature. Calculate the equilibrium composition and adiabatic flame temperature for this reaction.

(d) Calculate the molecular weight, γ, and $c*$ values for the product mixture you obtained in (b).

5.40 Use the NASA thermochemistry code to reproduce the results from Problem 5.39(ii). Assume a nozzle expansion ratio of 50 and tabulate the mass fraction of the five most prominent species in the chamber and at the exit assuming chamber pressures of 10 and 50 atmospheres.

5.41 Consider the combustion of N$_2$O$_4$ and hydrazine at a pressure of 4 MPa. Using the NASA code, plot T_c, $\mathcal{M}$, $T_c/\mathcal{M}$, and I_{spv} as a function of mixture ratio for $1.0 < r < 2.0$. What is the stoichiometric mixture ratio for this propellant combination?

5.42 (a) Write down balanced chemical equations for the following reactions. Assume complete combustion and that 1 mole of fuel is present.

(i) Stoichiometric combustion of chlorine trifluoride and monomethyl hydrazine.

(ii) Combustion of LOX and RP-1 at an equivalence ratio of 1.2.

(iii) Combustion of hydrogen peroxide and hydrazine at a mixture ratio of 2.0.

(iv) Combustion of nitrogen tetroxide and methane at a mixture ratio of 2.5.

(b) Calculate heats of combustion for the reactions above. Note: $h°_f = -60.5$ kcal/mol for ClF_3.

5.43 Many liquid rocket engines utilize non-cryogenic (i.e. storable) propellants to simplify operations and provide a means of firing the engine after extended time intervals. One of the more common combinations in use today involves the use of nitrogen tetroxide (N_2O_4) and monomethyl hydrazine (MMH) as oxidizer and fuel, respectively. Assume that these propellants are combined at a mixture ratio of 2.4 in a combustion held at 25 atm. Consider water vapor, carbon monoxide, carbon dioxide, diatomic nitrogen, and oxygen as combustion products and determine:

(a) the adiabatic flame temperature assuming the reactants are gases at 298 K;

(b) the equilibrium composition of the product species;

(c) the molecular weight of the product gases;

(d) the mass fraction of each of the assumed product species; and

(e) the characteristic velocity obtained by this combustion process. [Hint: Consider the dissociation carbon dioxide in your analysis.]

5.44 Write down balanced chemical equations for the following reactions:

(a) Complete combustion of RP-1 ($CH_{1.9}$) with Lox at a mixture ratio of 2.0. (Assume 1 mole of RP-1 is present.)

(b) A proposed "clean" SRM propellant (no HCl in exhaust) uses ammonium nitrate, AN, ($NH_3 NO_3$) as an oxidizer instead of AP. Suppose a solid propellant uses 2 moles of this oxidizer for each mole of aluminum. A binder (C_2H_4) is added to make a stoichiometric mixture. Write down the balanced equation in this case assuming 1 mole of aluminum and complete combustion.

5.45 As you are probably aware, automotive companies are aggressively pursuing fuel cells using hydrogen and oxygen. Since hydrogen gas is difficult/dangerous to store, one concept involves "cracking" of hydrocarbon fuels to release the required hydrogen. Consider a reaction with methane (CH_4) gas:

$$\text{methane} + \text{heat} \rightarrow \text{hydrogen} + \text{carbon(s)} + \text{residual methane}$$

Suppose we want to have at least 99% efficiency in converting the methane to hydrogen and carbon. Determine the temperature in the feedstock methane to attain this efficiency level.

[Hint: For K_p reactions forming solid species, we can ignore the partial pressure of the solid species and the number of moles of solid in writing the expression for K_p. The solid is assumed to occupy negligible volume in the product gases and therefore exerts a negligible partial pressure. Only moles of gas contribute to partial pressures in this case.]

5.46 Estimate the amount of energy required to crack the methane in Problem 5.46. You may assume that the reactant enters at 30 °C and that the products are at 1000 °C for your analysis. In addition, you may neglect the energy stored in the residual methane and assume constant specific heats: C_p values in J/(g-mol K)

$$C_{p_{H_2}} = 30$$

$$C_{p_{C_{(s)}}} = 40$$

$$C_{p_{CH_4}} = 72$$

5.47 (a) Write down balanced chemical equations assuming complete combustion of the limiting reactant (could be either oxidizer or fuel) for the reactions described below. As a convention, write your results assuming we begin with 1 mole of fuel.

   (i) Stoichiometric combustion of nitrogen tetroxide (NTO) and unsymmetrical dimethyl hydrazine (UDMH).

   (ii) Combustion of hydrogen peroxide and polyethylene at a mixture ratio of 7.0.

   (iii) Combustion of Lox with LH_2 at a mixture ratio of 5.5.

   (iv) Combustion of a mixture of 75% ammónium perchlorate (NH_4ClO_4) and 15% HTPB, and 10% Al by weight. (Assume 1 mole of Al for this case.)

   (b) Determine heats of combustion for each of the reactions given above.

5.48 Consider the decomposition of room temperature hydrazine (N_2H_4) at a pressure of 2 atm. Decomposition products include N_2, H_2, and ammonia gas (NH_3). Using equilibrium constants, determine the composition of the product gases and the adiabatic flame temperature, and c^* value for this reaction. [Hint: $T_c < 2000$ K.]

5.49 What would the T_c value be for Problem 5.49 if we had assumed complete combustion?

5.50 1 kg-mol of carbon dioxide is in an insulated container whose top is a frictionless piston (as shown in Figure 5.5). The gas is initially at standard conditions (298 K, 1 atm). At $t = 0$, a burner is lit and heats the gas to a temperature of 3000 K leading to substantial dissociation to carbon monoxide and diatomic oxygen.

   (a) Determine the equilibrium composition of the gases in the container at the end of the heating process.

   (b) Determine the energy input to the gas (in joules).

   You may assume constant specific heats in this problem; use the values tabulated below:

Constituent	C_p (J/kg-mol K)
CO_2	60
CO	36
O_2	35
O	25

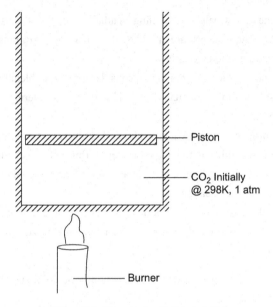

Figure 5.5 Diagram for Problem 5.51.

5.51 Determine the adiabatic flame temperature for the reaction containing hydroxyl radials:

$$2H_2 + O_2 \rightarrow aH_2O + bO_2 + cH_2 + dOH$$

by including the dissociation reactions

$$H_2 + \frac{1}{2}O_2 \leftrightharpoons H_2O \text{ and } \frac{1}{2}O_2 + \frac{1}{2}H_2 \leftrightharpoons OH$$

Perform calculations assuming the reactants are initially at a temperature of 298 K for a pressure of 100 and 200 atmospheres. Write out the balanced chemical equations for both cases and explain differences in your results. You must include temperature dependent C_p data from Table 5.3.

5.52 Determine the c^* values resulting from your calculations in Problem 5.52.

5.53 Liquid oxygen and methane are combusted at a mixture ratio of five.

(a) How many moles of oxygen are required *per mole of methane* for this mixture? Write down the chemical equation for this reaction assuming 1 mole of methane in the reactants and the following product species:

$$\text{product species} = a\text{CO}_2 + b\text{H}_2\text{O} + c\text{H}_2 + d\text{O}_2 + e\text{C}$$

Product species include carbon dioxide, water vapor, hydrogen, oxygen, and carbon, and a, b, c, d, e are unknown stoichiometric coefficients.

(b) Write down the energy balance assuming an adiabatic constant pressure reaction for your equation in (a). The reactants are at 298 K and 50 atmospheres pressure and all specific heats can be assumed to be constant.

(c) Which dissociation reactions should you consider for this problem? Write down relations to aid in the solution for coefficients a, b, c, d, and e by considering equilibrium constants based on pressure.

(d) Assume the products are at a temperature of 2000 K. At this temperature use mass conservation and equilibrium reactions to derive (but not solve) an algebraic system of five equations for a, b, c, d, and e. Note that a, b, c, d, and e should be the **only** unknown quantities in your five equations.

5.54 (a) Write down balanced chemical equations assuming complete combustion of the limiting reactant (could be either oxidizer or fuel). As a convention, write your reactions in terms of 1 mole of fuel.

   (i) Stoichiometric combustion of chlorine pentafluoride (ClF_5) and monomethyl hydrazine (CH_3NHNH_2).
   (ii) Combustion of LOX and RP-1 at $r = 2.2$.
   (iii) Combustion of LOX and LH_2 at $r = 6$.
   (iv) Combustion of aluminum with ammonium perchlorate (NH_4ClO_4) with 15% fuel by weight.
   (v) Combustion of magnesium with ammonium nitrate (NH_4NO_3) at an equivalence ratio of 0.75.

(b) Determine the heat of combustion for each of the reactions above. Note $h_f^\circ = -60.5$ Kcal/mol for ClF_5.

5.55 Many liquid rocket engines utilize non-cryogenic (i.e. storable) propellants to simplify operations and provide a means of firing the engine after extended time intervals. One of the more common combinations in use today involves the use of nitrogen tetroxide (N_2O_4) and monomethyl hydrazine (MMH) as oxidizer and fuel, respectively. Assume that these propellants are combined at a mixture ratio of 2.0 in a combustion held at 50 atm. Consider water vapor, carbon monoxide, carbon dioxide, diatomic nitrogen, oxygen, and hydrogen as combustion products and determine:

(a) The adiabatic flame temperature assuming the reactants are gases at 298 K.

(b) The equilibrium composition of the product species.

(c) The molecular weight of the product gases.

(d) The mass fraction of each of the assumed product species.

(e) The characteristic velocity obtained by this combustion process. [Hint: consider the dissociation of water and carbon dioxide in your analysis.]

5.56 You have been asked to examine two oxidizers for a proposed hybrid rocket. The current design specifies chamber and exit pressures of 350 and 14.7 psi, respectively. The fuel is assumed to be hydroxyl-terminated polybutadiene (HTPB), which has a chemical structure $C_{7.3}H_{10.96}O_{0.06}$ and a heat of formation of -250 cal/mole. Assume the fuel grain is comprised of seven axial ports (with circular cross-section), spaced evenly over the motor cross-sectional area.

Run the NASA Chemical Equilibrium Code for Liquid Oxygen and Hydrogen Peroxide (90% Concentration) and answer the following:

(a) Determine the optimum mixture ratio for each propellant combination.

(b) Plot I_{spv}, T_c, Mol wt, T_c/Mol wt vs. r. Comment. Which propellant would allow large changes in mixture ratio without large changes in the performance?

(c) What is the effect on performance if the desired chamber pressure is increased to 500 psi? 1000 psi? Determine the optimum mixture ratio and plot I_{spv} vs. r for these two pressures.

(d) Someone suggested to you that the performance can be increased if aluminum is added to the fuel. Consider 15% aluminum by weight for each propellant in the baseline case and determine how the performance changes. Tabulate optimal I_{sp} and r values for propellants with and without aluminum. Has adding aluminum to the fuel created two-phase flow through the nozzle? Determine what percentage of the exhaust is not gas at the throat and the exit. Tabulate the mass fractions of the five major product species.

(e) Should you include aluminum in your propellant formulation? Why (or why not)?

5.57 A combustion experiment using hydrogen and oxygen at a mixture ratio of 6.4 is carried out in a constant pressure "bomb" at 2.5 atmospheres pressure. Spectroscopic data reveal that the flame temperature is 3000 K and that the combustion process can best be represented considering the species:

$$H_2 + aO_2 \rightarrow aH_2O + bOH + cH_2$$

where α, a, b, and c are unknown stoichiometric coefficients.

(a) Determine α for the equation above using the specified mixture ratio.

(b) Determine the composition of the products.

5.58 Consider the combustion of aniline and nitric acid at a pressure of 50 atmospheres:

$$C_6H_5NH_{2_{(l)}} + 9HNO_{3_{(l)}} \rightarrow 6CO_2 + 4N_2 + 7H_2O + 2OH + 2O_2 + 2NO$$

The reactant liquids are initially at a temperature of 273 K. For the gaseous products given above, a fellow student has estimated an adiabatic flame temperature of 2298 K. Determine whether or not this estimate is correct. If the estimate is incorrect indicate whether it is too high or too low.

For this problem you may assume that specific heats are constant and equal to the values given in the table below:

Constant pressure specific heats, cal/mol K

Gas	C_p
Aniline	45.0
Nitric acid	27.0
Oxygen	8.1
Nitrogen	7.8
Carbon dioxide	18.0
Water vapor	10.0
Nitric oxide	11.2
Hydroxyl group	4.4

5.59 Consider an electrically augmented hydrazine thruster with a throat diameter and expansion ratio of 0.185 in and 25, respectively. The ODE code was used to assess the unaugmented performance of this thruster; output from this run (at 100 psi chamber pressure) is shown in Table 5.5.

(a) What mass flow (in kg/s) is passing through the thruster under the conditions noted in the ODE run?

(b) Assume we turn on the electric heater and that it provides 10 kg of power in heating chamber gases. What is the resulting increase in chamber temperature and vacuum I_{sp} assuming that negligible changes in specific heats and molecular weight occurs as a result of the heating.

(c) Can you think of a way to "trick" the ODE code into making computations for the electrically augmented case? Describe how to do this, i.e. which inputs would you modify and how would you change them?

5.60 Consider a mixture of perfect gases, as shown in the table below, at a temperature of 298 K.

Constituent	Mass fraction, %
Ammonia	35
Hydrogen	10
Carbon Dioxide	30
Nitrogen	25

Determine the following properties assuming constant composition:

Table 5.5   Theoretical rocket performance assuming equilibrium composition during expansion

```
 pc =  100.0 psia
                                wt fraction     energy     state     temp
chemical formula                (see not(e)    cal/mol               deg k
   fuel  n  2.00000  h  4.00000     1.000000    12050.000      1    298.10

o/f=  0.0000      percent fuel= 100.0000
```

	chamber	throat	exit	exit	exit	exit	exit
pc/p	1.0000	1.8685	1.7968	2.5620	273.51	632.96	1026.65
p, atm	6.8046	3.6417	3.7871	2.6560	0.02488	0.01075	0.00663
t, K	871.4	739.6	747.1	682.9	353.0	329.1	316.9
rho, g/cc	1.0199-3	6.4523-4	6.6405-4	5.1116-4	1.0981-5	5.2708-6	3.4430-6
h, cal/g	376.03	283.04	288.42	241.64	-137.71	-181.38	-204.59
u, cal/g	214.46	146.35	150.30	115.81	-192.58	-230.78	-251.21
g, cal/g	-2971.65	-2558.14	-2581.57	-2382.06	-1493.95	-1445.88	-1421.95
s, cal/(g) (k)	3.8417	3.8417	3.8417	3.8417	3.8417	3.8417	3.8417
m, mol wt	10.717	10.752	10.749	10.785	12.786	13.242	13.507
(dlv/dlp)t	-1.00328	-1.00645	-1.00616	-1.00942	-1.10654	-1.11309	-1.11511
(dlv/dlt)p	1.0245	1.0557	1.0527	1.0870	2.7092	2.9249	3.0237
cp, cal/(g) (k)	0.7162	0.7585	0.7538	0.8127	4.8354	5.4733	5.7807
gamma (s)	1.3670	1.3607	1.3617	1.3486	1.1486	1.1383	1.1332
son vel,m/sec	961.3	882.1	887.1	842.6	513.5	485.0	470.2
mach number	0.000	1.000	0.965	1.259	4.038	4.453	4.688

Transport properties (gases only)

  with equilibrium reactions

	chamber	throat	exit	exit	exit	exit	exit
cp, cal/(g) (k)	0.7162	0.7585	0.7538	0.8127	4.8354	5.4733	5.7807
conductivity	0.5275	0.4833	0.4852	0.4735	1.2145	1.3344	1.3920
prandtl number	0.4608	0.4771	0.4755	0.4941	0.6621	0.6323	0.6136

  with frozen reactions

cp, cal/(g) (k)	0.6821	0.6698	0.6704	0.6647	0.5739	0.5566	0.5468
conductivity	0.5135	0.4512	0.4548	0.4232	0.1940	0.1708	0.1588
prandtl number	0.4509	0.4512	0.4511	0.4520	0.4918	0.5022	0.5089

Performance parameters

ae/at		1.0000	1.0010	1.0500	25.000	50.000	75.000
cstar, ft/sec		3974	3974	3974	3974	3974	3974
cf		0.728	0.707	0.875	1.712	1.783	1.820
ivac, lb-sec/lb		156.1	156.1	158.8	222.7	230.0	238.8
isp, lb-sec/lb		90.0	87.3	108.1	211.4	220.2	224.8

Mole fractions

h2	0.66391	0.66118	0.66143	0.65857	0.50252	0.46691	0.44625
nh3	0.00331	0.00659	0.00629	0.00971	0.19698	0.23971	0.26450
n2	0.33278	0.33224	0.33229	0.33171	0.30050	0.29338	0.28925

Additional products that were considered but whose mole fractions were less than 0.50000E-05 for all assigned conditions
h h2n2 n nh nh2 n2h4 n3

(a) $\mathcal{M}$, C_p, γ at $T = 298$ K.

(b) $\mathcal{M}$, C_p, γ at $T = 2000$ K.

(c) Calculate th enthalpy difference between $T = 298$ K and $T = 2000$ K (use variable values of C_p in the calculation).

5.61 Consider the reaction:

$$CH_4 + \frac{3}{2}O_2 \rightarrow aH_2O + bCO_2 + cO_2 + dH_2 + eCO$$

Assuming the products are at a pressure and temperature of 10 atm and 1000 K, determine the composition of the product species, i.e. find the stoichiometric coefficients a–e.

6   HEAT TRANSFER IN CHEMICAL ROCKETS

A detailed discussion of heat transfer is clearly beyond the scope of this text as this is a subject unto itself and the topic of many other textbooks. Despite this fact, we must emphasize the critical importance of this topic in chemical propulsion devices as heat transfer considerations place severe limitations on many systems and end up being the source of numerous development problems. For this reason, students who are interested in developing competency in rocket propulsion are encouraged to take additional coursework in this subject as there is much to learn. In this chapter, we focus on the most critical elements that pertain to rocket combustors/nozzles, but many topics will be introduced in a very cursory manner. The Further Reading at the end of the chapter has numerous other resources for the interested reader.

Chemical rocket propulsion systems are subjected to some of the highest heat fluxes of any man-made device, with values in excess of 8 kW/cm^2 in modern high-pressure engines/motors. For those just learning the subject, this heat "flow" per unit area may be difficult to put into perspective, so let's consider the following example from everyday student life.

Say you are preparing your dinner of macaroni and cheese or raman noodles by boiling a quart (about a liter) of water on the top of the stove. In this case, we have:

Mass of water ≈ 1 kg
Area of burner $\approx \pi \, (10 \text{ cm})^2 \approx 300 \text{ cm}^2$
Time to boil water ≈ 2 minutes $= 120$ s

The energy input required to bring the water to a boil (i.e. to its *saturation temperature*) is simply its specific heat (1.0 kcal/kg °C) times the 75 °C temperature change required to heat the fluid to its saturation temperature/boiling point of 100 °C:

Total energy input $= [(100–25)(1.0)] \times 1$ kg $= 75$ kcal ≈ 300 kJ $= 300$ kW-s

where we have used the fact that there are 4.184 kJ per kcal. Now we have all the information we need to compute the heat flux, $\dot{q}$, since this is a heat flow per unit area per unit time:

$$\dot{q} = Q/(A \times t) = 300/(300 \times 120) \simeq 0.008 \text{ kW/cm}^2$$

Comparing this to the 8 kW/cm^2 value at rocket conditions, we can see that the rocket heat flux is *1000 times* larger than that produced by a burner on your stove! In fact, if we exposed your pot of water to the rocket level heat flux, we could bring it to a boil in:

Table 6.1   Nomenclature and units for many common variables encountered in heat transfer
problems

Variable	Name	English/Imperial units	Metric/SI units
Q	Total heat load	BTU	Cal, joules
$\dot{Q} = dQ/dt$	Total heat transfer rate	BTU/s	Cal/s, watts
$q = Q/A$	Heat load/unit area	BTU/in^2	Cal/cm^2, J/cm^2
$\dot{q} = \dot{Q}/A$	Heat flux	BTU/in^{2}s	$\frac{Cal}{cm^2 s}, \frac{kw}{cm^2}$
k	Thermal conductivity	$\frac{Btu}{ins^\circ R}$	$\frac{w}{m\,K}, \frac{cal}{cm\,s\,K}$
$\alpha = k/(\rho\,C_p)$	Thermal diffusivity	f^2/s	m^2/s
$h_g = \dot{q}/\Delta T_g$	Gas film coefficient, or heat transfer coefficient	$\frac{Btu}{sf^2\,R}$	$\frac{cal}{scm^2\,K}, \frac{w}{m^2 K}$
h_ℓ	Liquid film coefficient	$\frac{Btu}{sf^2\,R}$	$\frac{cal}{scm^2\,K}, \frac{w}{m^2\,K}$

$$t = Q/(\dot{q}A) = 300/(300(8)) = 1/8\text{th second}$$

Now that makes for some speedy mac and cheese! This simple example demonstrates the daunting problem for cooling rocket combustors. In fact, the heat fluxes in rocket environments are roughly an order of magnitude higher than those in nuclear reactors, so the thermal management problem is one of the most difficult among all manmade devices.

    With this motivation, we describe some of the pertinent approaches to solving rocket propulsion heat transfer problems in this chapter. Section 6.1 provides an overview of nomenclature and dimensionless variables related to heat transfer. Section 6.2 provides a summary of cooling techniques used in rocket engines/motors and Section 6.3 reviews fundamentals of conductive, convective, and radiative heat transfer. Section 6.4 describes approaches used for scaling heat transfer processes to unique conditions and Section 6.5 provides a detailed analysis of liquid rocked engine (LRE) regenerative cooling systems.

6.1 INTRODUCTION

6.1.1 Heat Transfer Nomenclature

Since heat transfer is a new subject for many readers of this text, we include some description of the major variables involved in Table 6.1. In the example above, we related the total heat load, Q, to the heat flux for a case where we specified SI units. English/Imperial units are still used quite heavily in the US with the British Thermal Unit (BTU) being the amount of heat required to raise 1.0 lbm of water one degree Fahrenheit (Rankine). The calorie is the metric equivalent, representing the amount of energy required to raise 1.0 g of water one degree Celsius (kelvin). The joule is the SI

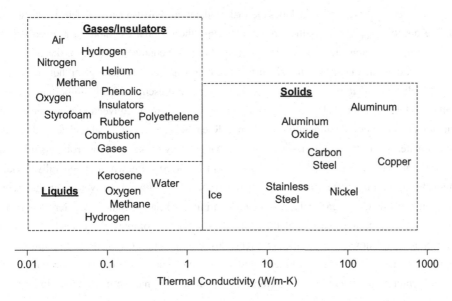

Figure 6.1 Thermal conductivity values for gases, liquids and solids in SI units.

measure, representing 4.184 calories or 0.000948 BTU. The watt (or kilowatt, kW), defined as 1.0 J/s, is the most universally accepted measure of heat flow. In many propulsion system thermal management problems, the heat flux, $\dot{q}$, is the principal quantity of interest as it prescribes the local amount of heat flow per unit area of surface exposed to said heat. As we will see in subsequent discussion, knowing the heat flux allows us to relate temperatures at various points in our system, i.e. to relate the hot gas wall temperature to the combustion or "flame" temperature for a wall that is directly exposed to combustion gases. Knowing the local heat fluxes in a system allows us to add up the total heat transfer rate ($\dot{Q}$) by integrating $\dot{q}$ over the area exposed to heat.

The *thermal conductivity*, k, and *thermal diffusivity*, α, are critical properties of the fluid that influence heat transfer rates. In general, gases are poor conductors of heat (low k values), while liquids and metals are much more conductive, due mainly to the fact that the spacing of individual atoms/ molecules generally decreases as the density of the material increases. Physically, k represents the rate energy can pass through a material of a given thickness with a given exposed temperature gradient, i.e. the material's inherent ability to transfer heat. In propulsion problems, we desire good insulators (low k values) to impair the transfer of heat and good conductors (high k values) to promote heat transfer. Figure 6.1 summarizes approximate k values for several materials in order to give the reader a feel for the typical values we encounter. Here we can see that k values for conductors can be several orders of magnitude higher than those of insulators. It is sometimes difficult to understand how we might want a higher heat transfer value in a rocket chamber cooling problem, but if the hot gas/ combustion temperature is fixed, the higher the heat transfer, the lower the hot gas wall temperature as heat is wicked away rapidly from the combustion zone. For this reason, in rocket applications there are homes for both insulators and highly conductive materials such as copper.

The *thermal diffusivity*, α, includes thermal conductivity, but also material density and specific heat. The quantity ρC_p measures the amount of energy the material can store in a given volume. We will see that this parameter governs the behavior of transient conduction problems that address the question regarding the rate at which wall surfaces heat up when exposed to a given hot gas source.

Finally, the *heat transfer* or *"film" coefficients* represent critically important physical variables in heat transfer problems as they describe the amount of convective heat transfer that occurs for a given imposed temperature difference. Recall that convective heat transfer is defined as energy transfer due to fluid motion – we will see that in many rocket cooling problems this is the chief heat transfer mechanism. If h (h_g for gas or h_l for liquid) is known, it is generally trivial to compute the heat transfer rate from this simple definition i.e. $\dot{q} = h\Delta T$. However, given that h_g or h_l depend on the flow of the gas or liquid, they are intimately related to the character of boundary layers near heated/cooled walls and therefore dependent on complex fluid mechanics of the Navier–Stokes equations with turbulence effects. For this reason, it can be a daunting problem to compute/estimate heat transfer coefficients and a major element of heat transfer research over the past two centuries has focused on developing correlations or analytic expressions to compute/predict h_g or h_l for various fluid flow situations.

6.1.2 Dimensionless Variables Related to Heat Transfer

Since we have introduced new physical parameters such as k, α, and h it should not come as a surprise that there are dimensionless quantities that relate these new variables to length and velocity scales in fluid and heat flow problems. You should be familiar with the Reynolds number that characterizes the ratio of inertial and viscous forces and is a prime variable for consideration of boundary layers in fluid flow systems:

$$\text{Reynolds number: } \text{Re} = \frac{\rho VL}{\mu} = \frac{\rho L}{v}$$

Because convection problems involve fluid flow, the Reynolds number remains as a prime consideration in these applications. However, another dimensionless quantity, the Prandtl number, Pr, also becomes important in many cases. Physically, the Prandtl number represents the ratio of the thickness of the momentum boundary layer to the thermal boundary layer:

$$\text{Prandtl number: } \text{Pr} = \frac{\mu Cp}{k} = \frac{v}{\alpha}$$

From its definition, you can see that Pr represents the ratio of molecular diffusivity, v, to thermal diffusivity, α, and hence represents the relative ability of the fluid to transport momentum and heat via molecular collisions in viscous/thermal layers. Figure 6.2 illustrates this situation for a fluid flowing adjacent to a wall that is being heated or cooled. For many gases, the Prandtl number is near unity thereby implying that thermal and momentum layers are of similar thickness. In the heat transfer literature, numerous analytic solutions exist for this special case because the thermal and

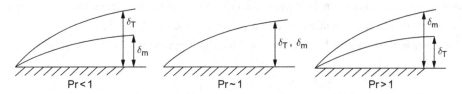

Figure 6.2 Illustration of thermal and momentum boundary layer thicknesses near a wall being heated/cooled by flowing fluids.

momentum layers are similar in shape and height. If Pr < 1 the momentum layer is thinner than the thermal layer, a situation applicable to high-conductivity fluids such as liquid metals. If Pr > 1 the opposite is true and more viscous fluids such as oil tend to fall into this category. In general, the Prandtl number looms as a consideration in all convective heat transfer problems.

If:

Pr $\approx$ 1. Many gases: the thermal and momentum layers are of similar size.

Pr < 1. Liquid metals, high-conductivity fluids – the momentum layer is thinner than thermal layer.

Pr > 1. Water, low-conductivity fluids, oils: the thermal layer is thinner than the momentum layer.

Prandtl numbers for combustion gases typically lie in the range 0.7–1.0. For fuels used as coolants (kerosene, liquid hydrogen, methane or perhaps oxygen) values typically lie in the range 1–3. The Appendix references several websites with properties data for a large number of fluids/materials.

In high-speed compressible flows, the kinetic energy in the gases becomes an important contributor and the Eckert number is often used to characterize heat transfer in this regime:

$$\text{Eckert number} = \text{Ec} = \frac{v^2}{C_p(T_w - T_\infty)}$$

where v is fluid velocity, and T_w and T_∞ are the wall and freestream temperatures, respectively. Physically, Ec represents the ratio of kinetic to thermal energy in the flow. In general we don't see this too often in rocket problems but Ec appears frequently in the hypersonics field.

Perhaps the most common dimensionless variable appearing in convective heat transfer problems is the Nusselt number:

$$\text{Nusselt number} = \text{Nu} = hL/k$$

Physically, Nu represents the ratio of convective to conductive heat transfer rates. Because fluids can transport hot fluids more readily than heat can conduct through a solid media Nu > 1 in virtually all convective heat transfer problems. In rocket applications we can see values in the hundreds. One can also think of Nusselt number as a dimensionless heat transfer coefficient as h can immediately be computed once Nu is known. For this reason, many heat transfer correlations focus on describing Nu in terms of other dimensionless fluid flow parameters such as Re and Pr.

Closely related to the Nusselt number is the Stanton number:

$$\text{Stanton number} = \text{St} = \frac{h}{\rho v C_{\text{p}}} = \frac{\text{Nu}}{\text{Re} \times \text{Pr}}$$

The Stanton number serves the same purpose as the Nusselt number and represents a dimensionless h value. In some problems this dimensionless group appears when considering a specific set of assumptions regarding the fluid flow and it becomes a preferred non-dimensionalization to Nu as a result.

6.2 Cooling Techniques used in Rockets

There are several cooling/insulation techniques for protecting chamber and nozzle walls from excessive heating. In many cases, several of these techniques are used in concert given the magnitude of the problem at hand. This section provides a summary of the most common approaches.

6.2.1 Regenerative Cooling

This elegant approach has been used in numerous engines that employ liquid propellants. Fuel is generally employed as coolant and flows through a series of tubes or channels that form the wall of the chamber itself. Cryogenic fuels such as liquid hydrogen or methane are particularly good coolants and provide potential to cool walls at high heat flux levels. In addition, the heat that goes into the fuel serves to augment performance since it largely remains within the propellants. The main disadvantages of regenerative cooling include a complex chamber design and fabrication as well as substantial pressure drop in the cooling passages that must be compensated for by the turbomachinery feeding the chamber. In Section 6.5 we go into details of the temperature profiles within the walls and cooling channels of a regenerative cooling jacket.

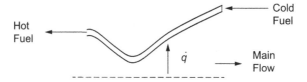

6.2.2 Dump Cooling

This technique is similar to regenerative cooling, but the fuel is dumped at some location along the chamber wall rather than pumped all the way to the injector. This approach has the advantage of a reduced feed pressure requirement and the dumped fuel serves to insulate the chamber wall from the hottest combustion gases in the core of the chamber. However, there is typically a performance (c^*) loss associated with the dumped fuel as it is not completely combusted when it exits the nozzle.

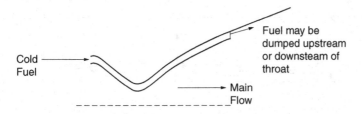

6.2.3 Film Cooling

This technique is nearly always applied in bipropellant liquid engines and can be used in concert with other cooling concepts within this list. A fuel-rich region is injected along the wall in order to insulate the metal against the hottest combustion gases in the "core" region as described schematically in the diagram below. Here rings of fuel injectors are denoted "F" and rings of oxidizer injectors are denoted "O." Typically fuel film cooling is specified as a percentage of the overall fuel flow, i.e. 10% film cooling implies that 90% of the fuel is injected within the core region. While there is a small performance decrement due to poor combustion of the fuel film, it is often a necessary approach to ensure the thermostructural integrity of the chamber/nozzle walls. The effectiveness of the film drops with distance from the injection point due to mixing with hot combustion gases and it becomes a practical performance issue to estimate the local effectiveness as a result. In fuel-rich staged combustion applications, fuel-rich turbine exhaust might also be used as a film coolant. In low-pressure space engines, film cooling may be the only cooling mechanism utilized for thermal management. A related technology is known as *bias layer cooling*. In this case, the mixture ratio near the wall is biased to be fuel rich, but some oxidizer may be included in order to enhance combustion performance while still maintaining acceptable wall environments.

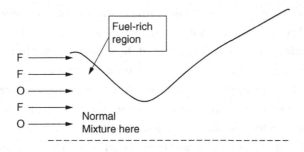

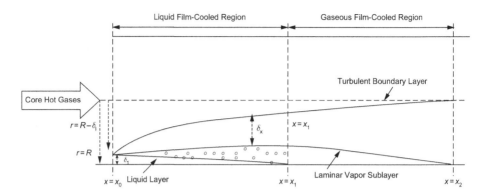

Figure 6.3 Flowfield near the wall where liquid film coolant is introduced.

Figure 6.3 shows the profile near a film-cooled wall. Near the injection point a liquid film will be present. As the film interacts with hot gases in the core region, it will shed droplets (or filaments of supercritical fluid if such conditions exist), and the layer will eventually dry out as noted as location x_1 in Figure 6.3. In this region, the liquid film itself is cooling the wall. Cooler vaporized liquid will occupy a sublayer near the film and near the wall downstream of the x_1 location. In this region, the wall is cooled by the gaseous film comprised of these vaporized liquid constituents. Eventually these vapors will mix out with core region gases as noted in Figure 6.3 at the location x_2.

6.2.4 Transpiration Cooling

Also known as "sweat" cooling, this is an elegant approach wherein the chamber wall is made from a porous material that permits a small fuel flow through the wall itself assuming the appropriate pressure differential can be applied. Cooling effectiveness is excellent in principal as the heat of vaporization of the fuel also contributes to absorbing some fraction of the heat input from chamber gasses. While in principal this technique could yield superior performance, there are two practical challenges. Fabrication of a wall that maintains uniform porosity and can tightly control fuel flow for a given imposed pressure drop is highly challenging – a local region with low porosity will be starved of fuel/coolant flow and can fail as a result. In addition, the pressure in the combustor/ nozzle drops as gases accelerate into the nozzle so if there is a uniform fuel feed pressure for the transpiration jacket there will naturally be more flow where the internal pressure is lower. This is the opposite effect we typically desire as higher pressure regions impose higher heat loads as we will show later in the chapter. A series of manifolds in the fuel feed system could be employed to better control local fuel pressure drops and transpiration flows, but this does add substantial complexity.

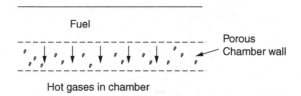

Fuel

Porous
Chamber wall

Hot gases in chamber

All these techniques (regenerative, dump, fuel film, and transpiration) are typically employed with fuel as the coolant. Oxidizers tend to present challenges as coolants; as they heat up they become very corrosive (and even combustible) with the wall materials and as a result much less temperature rise can be tolerated. Local hot spots can always be a problem. Nevertheless, virtually all liquid bipropellant systems tend to optimize at oxidizer/fuel mixture ratios greater than unity and in most application there is, in principal, more cooling capacity available in the oxidizer than the fuel. Use of oxidizer-resistant coatings on cooling passages is an approach that has been developed with some degree of success. The references in this chapter point to a number of studies where liquid oxygen has been employed as a coolant and this technology does remain an option if one can verify coolant properties are within acceptable limits over all operating conditions within all locations in the flowpath.

6.2.5 Ablative Cooling

In solid and hybrid rocket motors, liquids may not be available for cooling chamber and nozzle surfaces. In addition, the high cost (or perceived high cost) of building a regeneratively cooled chamber has caused some liquid engine proponents to utilize ablatively cooled chambers. In this case, an insulating material is placed along chamber and/or nozzle walls to provide thermal integrity. Typical ablative materials employ a rubber base material loaded with particles or fibers of a refractory material that has a very high melting point. For example, EPDM (ethylene propylene diene monomer) rubber can be loaded with silica or carbon phenolic fibers as a modern-day material. Early motors employed asbestos fibers, but their toxicity has precluded use in current devices.

Phenolic materials, comprised of high melting point fibers in a polymer matrix, are used extensively for ablatively cooled parts. Options here include silica, fiberglass (glass), carbon, or graphite. Carbon and graphite phenolics sport the best heat resistance, but are also the most expensive so in general their usage is limited to highest heat flux regions in the throat. Other options for throat insulators are carbon/carbon materials that are formed from interwoven sheets of the high melting point material. Carbon/carbon, graphite, and silicon carbide materials have wonderful strength at high temperatures, but they are thermally conductive so when they are used a backup insulator (typically phenolic) is generally employed. Figure 6.4 shows the design of a typical ablatively cooled liquid rocket thrust chamber noting the wide use of numerous phenolic and structural materials.

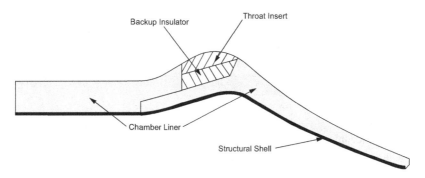

Figure 6.4 Construction of ablative liner for liquid rocket application.

When exposed to the heat from the combustion gases, the ablative material begins to decompose as carbon–hydrogen and carbon–carbon bonds in the hydrocarbon rubber are broken. This process causes the rubber material to pyrolyze and the lighter hydrocarbons tend to outgas and mix with combustion gases in the chamber. The blowing effect of the pyrolysis products tends to temporarily reduce the heat flux to the wall during this stage of the process. Eventually, the insulation material exposed to the flame becomes completely charred and the pyrolysis process tapers away. In this phase of the process the embedded fibers play their role to retain the char in place as it has very little structural stiffness or tensile strength. By retaining the char for some time, the fibrous material reduces the ablation rate. Eventually the char layer builds to a thickness that can no longer be retained under motor vibration and acceleration conditions and this layer flakes off and exposes a new layer of virgin material thus initiating a new pyrolysis process.

Typical ablation rates are measured in mil/s (a "mil" is 1/1000th of an inch) so that a 100 second firing might lead to a fraction of an inch (or about 1 cm) of ablation. Ablative cooling is advantageous in that it does not require any liquids and is viewed to be lower in cost than most other alternatives. However, there is a performance loss due to ablation in the throat region as this leads to a reduction in nozzle expansion ratio with time. For very high-pressure liquid engines, ablation rates become too large to give a practical solution and the techniques described above tend to be better choices. See Chapter 7 for additional discussion on this cooling technique.

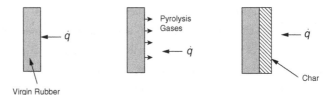

6.2.6 Radiation Cooling

Radiation is often employed as a cooling technique for nozzles of space motors/engines. In the exit cone, the heat fluxes from combustion gases drops to the point where refractory metals can be cooled by radiating the incoming heat to the vacuum of space. Niobium (formerly known as columbium) is typically the refractory metal of choice due to its high melting point of 4476 °F or 2469 °C. Upon ignition of the engine, combustion gases heat the wall until a thermal balance is attained (incoming heat flux is equivalent to radiation heat transfer). At this condition, the wall reaches its steady-state temperature. On-board camera footage of this process has recently become available on the SpaceX Falcon 9 launch vehicles with ignition and heat-up of the MVac (Merlin Vacuum Engine). For example, see www.spacex.com/news/2013/10/14/upgraded-falcon-9-mission-overview.

While radiative cooling provides a simple and elegant solution for cooling exit cones of space engines, the flame temperatures of most of our rocket propellant combinations preclude its usage in chamber and throat regions. In addition, for systems operating in the atmosphere, this option is not terribly attractive as the radiative heat transfer is not as efficient and pressure loads imposed by the atmosphere itself would necessitate thick and heavy walls. For these reasons, the approach has been used almost exclusively on space engines.

6.2.7 Heat Sink Cooling

For short duration ground tests, heat sink cooling is often employed. In this case, we provide sufficient thermal mass to conduct away heat from the combustion gases and simply allow the wall to heat up during the process. Copper, or copper alloys are typically employed for these tests since the transient response of the wall must be such that the metal temperature at the hot gas interface remains well below the melting point. The high thermal conductivity of this material (k = 400 W/(m-K)) permits a rapid "wicking" of the heat from the hot gas interface into the depth of the chamber/nozzle wall. However, as the wall begins to heat up, the interface temperature will also increase and therefore there is a limited duration over which this approach is viable. Heat sink cooling is only really viable for short duration ground testing and we have extensively employed this approach in research at Purdue. The chambers tend to be quite heavy with thick walls in order to provide thermostructural integrity for firing durations typically less than 1–2 seconds. High-speed instrumentation provides ample data on combustion processes during this short duration.

6.2.8 Insulate

In some situations, it is possible to use a material that has a very high melting point that can withstand the heating imposed by the chamber environment. The materials community continues to work to develop very high-temperature materials that can be used in this manner. In some situations, thermal barrier coatings (TBC for short) can be applied to chamber/nozzle walls to reduce the heat flux into the structural materials on the backside. Ceramic materials such as yttria-stabilized zirconia are applied in a thin coat on chambers walls. Sporting a melting point near 2700 °C, these materials can survive environments in lower pressure bipropellant engines or monopropellant devices. Typically, use of these coatings can reduce liner wall temperatures by roughly 50 °C and can decrease strains in the liner wall by 20–30%. However, as with all coatings, their lifetime may be short due to powerful oxidation from hot oxidizer exposure and thermal stresses imposed from startup/shutdown events. The mismatch in thermal expansion coefficients between the liner and the wall liner induces strains on the bondline that can eventually lead to TBC failure (physical loss of material in affected sections). In the highest chamber pressure bipropellant engines, the heat fluxes are so high that wall temperatures exceed the material's capability and only schemes such as those discussed above can be employed.

6.3 HEAT TRANSFER FUNDAMENTALS

Recall that there are three mechanisms for transferring heat:

      (i) conduction – diffusion of heat through a solid;
     (ii) convection – heat transfer due to fluid flow; and
    (iii) radiation – Energy emitted by matter at finite temperature.

In the following subsections, we will address the fundamental relationships defining heat transfer by each of these mechanisms. We should point out that this is a very cursory treatment and several of the texts referenced at the end of this chapter should serve as more elaborate discussion of these classical fundamentals.

6.3.1 Conduction

Conduction processes are governed by Fourier's law:

$$\dot{q} = -k\left(\frac{dT}{dx}\right) \tag{6.1}$$

Figure 6.5 depicts the heat flow through two faces of a cube of dimensions dx, dy, dz. Considering the heat flow through all six faces of the cube, we can relate these fluxes to the rate at which the cube's temperature is changing. Here, the energy balance can be described:

Net energy flux + rate of change of energy in control volume = heat sources

The rate of change of energy in the control volume depends on the mass ($\rho dx\, dy\, dz$), the specific heat of the material, and the rate at which its temperature is changing. Since there are no heat sources inside a solid material, the overall energy balance becomes:

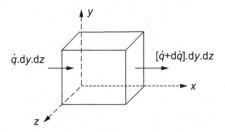

Figure 6.5 Energy balance on a cube of solid material exposed to transient heating.

$$\frac{d\dot{q}}{dx} + \frac{d\dot{q}}{dy} + \frac{d\dot{q}}{dz} + \rho C_p \left(\frac{\partial T}{\partial t}\right) = 0 \tag{6.2}$$

Using Fourier's law, and employing vector notation, we may write:

$$\rho C_p \left(\frac{\partial T}{\partial t}\right) = \nabla \cdot (k\nabla T) \tag{6.3}$$

where ∇ is the divergence operator. Equation 6.3 is called the transient conduction equation and it describes how a material changes temperature when exposed to heating/cooling. In problems we typically care about, the temperature differences can be very large, and since the thermal conductivity is a function of temperature, terms involving spatial derivatives of this parameter generally are important. However, if we are willing to assume that we can use a suitable average conductivity such that k = constant, then Eq. 6.3 becomes:

$$\frac{\partial T}{\partial t} = \alpha \Delta^2 T, \text{ where } \alpha = \frac{k}{\rho C_p} = \text{thermal diffusivity} \tag{6.4}$$

Equation 6.4 is often called the *heat equation* and it shows that the diffusion of heat through a solid of constant thermal conductivity is governed by a lone physical parameter, the thermal diffusivity, α. Mathematically, the heat equation is a partial differential equation and a well-posed formulation for solution of this equation requires an initial condition (at $t = 0$) and two boundary conditions for each dimension in the problem (i.e. some knowledge of the left, right, forward, back, top, and bottom faces in the cube).

If we think about a simpler one-dimensional problem, then we can think about four potential options for boundary conditions:

1. Constant wall temperature, $T(0,t) = T_w$. This condition might be appropriate on an interface adjacent to a large thermal mass.

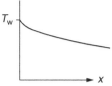

2. Constant heat flux, $dT/dx|_{x=0} = \dot{q}/k$. This condition implies that the gradient of the temperature is prescribed. In practice, this is difficult to realize because as the wall heats up (at least in our case) the heat flux changes as this happens.

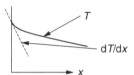

3. Adiabatic wall, $dT/dx|_{x=0} = 0$. This condition is really a subset of the constant heat flux case, but is quite useful to bracket problems where insulators are being employed. If we are applying this adjacent to a chamber wall, this condition implies that the wall has heated the local flame temperature; at this point there is no temperature gradient to support heat flow through the wall itself, but it must be able to withstand such temperature for use in a practical system.

4. Convection surface condition. This is the most useful, but most complex, boundary condition because the temperature derivative can have a complex behavior. In the general case, the heat transfer coefficient (h) and the wall thermal conductivity can vary with time:

$$\frac{dT}{dx}\bigg|_{x=0} = \frac{-h(T_\infty - T(0,t))}{k}$$

T_α ⟨ $T(0,t)$

$dT/dx)_{x=0}$

Equation 6.4 has a number of analytic solutions that are generated via the use of similarity variables or using Laplace transforms. Transient conduction into a cylinder, a plane, or a sphere are several examples of the types of simple geometries amenable to analytic treatments. The book by Carslaw and Jaeger (1986) is a classic that is comprehensive in its presentation of these types of solutions, but virtually all heat transfer texts provide at least some classical results.

When we relax the assumption of constant thermal conductivity, we must use the most general form of the heat equation in Eq. 6.3. Formally, this is a nonlinear partial differential equation and is not amenable to analytic solution due to the presence of terms involving dk/dx, dk/dy, etc. However, using numerical methods, Eq. 6.3 is readily solvable and numerous thermo-structural modeling codes such as Abaqus, ANSYS, and NASTRAN have embedded capabilities for solving this problem on the most complex geometries. Given the substantial computational power available in the current environment, even the most complex conduction problems are quite solvable with our current software.

One-dimensional Transient Conduction in Chamber or Nozzle Wall

A fundamental result that illustrates physics associated with the heat equation and the behavior of the transient conduction process is the transient heating of a wall from convective heat flux. This situation is the one we would encounter at ignition of a rocket chamber where the wall is initially cool and the convective heat flux is coming from chamber gases that are formed from the ignition event. An analytic solution does exist for the one-dimensional (1-D) problem (conduction in the direction x, locally perpendicular to the wall), if we are willing to assume constant properties, i.e. constant k, C_p, α. In addition, we shall assume a semi-infinite solid, which implies that the wall is thick enough such that heat is never conducted to the backside over the time interval of interest (as illustrated in Figure 6.6). This approximation would typically restrict us to consideration of a fairly small timeframe since any chamber wall is obviously of finite thickness. If we let T_i be the initial wall surface temperature at $t = 0$, this condition is then expressed: $T = T_i$ as $x \to \infty$.

Using a similarity variable that scales as $x/\sqrt{\alpha\,t}$, Eq. 6.4 can be transformed to an ordinary differential equation in this parameter that has an analytic solution. The solution is written in terms of a dimensionless temperature related to the maximum available temperature difference between the hot gas temperature (T_∞) and the initial wall temperature (T_i):

$$\frac{T(x,t) - T_i}{T_\infty - T_i} = erfc\left(\frac{x}{2\sqrt{\alpha t}}\right) - Exp\left[\frac{hx}{k} + \frac{h^2 \alpha t}{k^2}\right] erfc\left[\frac{x}{2\sqrt{\alpha t}} + \frac{h\sqrt{\alpha t}}{k}\right] \tag{6.5}$$

where

$$erfc(w) = \text{complimentary error function} = 1 - erf(w) \tag{6.6}$$

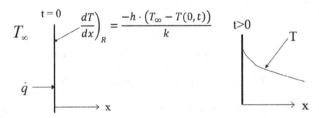

Figure 6.6 Transient conduction due to convective heat flux into a semi-infinite solid.

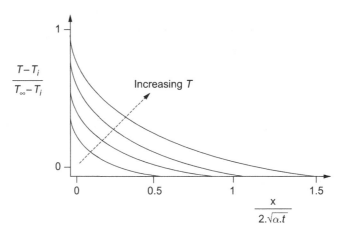

Figure 6.7 Transient conduction into a semi-infinite solid showing heat propagation with increasing time.

and:

$$\operatorname{erf}(w) = \frac{2}{\sqrt{\pi}} \int_{0}^{w} e^{-y^2} dy$$

For given T_i, T_∞, h, α, and k values, Eq. 6.5 can be plotted at various times in order to show how heat soaks back into the part. Figure 6.7 shows the behavior plotted against the similarity variable $x/(2\sqrt{\alpha\,t})$. This similarity variable has great importance in transient conduction problems. Note that α has dimensions of L^2/t so that $\sqrt{\alpha\,t}$ is a distance – a scaled conduction depth if you will. This behavior is general; if you wish to get a rough understanding of the distance heat can penetrate in a given time, compute this quantity. The constant of proportionality is not necessarily 1.0, but you will get a rough idea of conduction depth (to within a factor of maybe 2 or 3) as a result. This result shows also implies that the heat affected depth increases as $t^{1/2}$, so if we impose a heat source for twice as long the heat affected depth will only increase by $\sqrt{2}$, or about 40%. The reason for this behavior is that as the wall exposed to the flame heats up dT/dx at the surface gets smaller and the heat flux into the part diminishes.

Steady Conduction in Chamber or Nozzle Wall
Steady conduction is really just a degenerate case from the transient result. All we need to do is set $\partial T/\partial t = 0$ in Eq. 6.3:

$$\nabla \cdot (k\nabla T) = 0 \qquad\qquad (6.7)$$

For a 1-D situation with a constant thermal conductivity, this result reduces to:

$$d^2 T/dx^2 = 0 \text{ or } dT/dx = \text{constant} = \dot{q}/k \qquad\qquad (6.8)$$

where the latter part of the equation comes from the definition of conductive heat flux. Equation 6.8 is very useful to get the steady-state temperature distribution through the wall material. In general, as the wall heats up, both wall temperatures and heat flux change with time, but at the steady condition, the heat flux through any part of the wall structure must be constant. For example, if the wall has a thermal barrier coating, we can use this condition to compute the temperature on both sides of the coating, as well as on the backside of the wall. Making this assumption permits a quick estimate of surface temperatures assuming an estimate of the heat flux exists.

6.3.2   Convection

Heat transfer due to convection can be written in terms of h and the temperature difference between the hot gas at the edge of the thermal layer (T_∞) and the wall (T_w):

$$\dot{q} = h\,(T_\infty - T_w) \tag{6.9}$$

where h is h_g for gas or h_l for a liquid. Recall that h is most typically called the *heat transfer coefficient*, but is also termed *film coefficient* or *film conductance* in the heat transfer community to emphasize the fact that fundamentally it is derived from properties of the thermal and boundary layers near the wall.

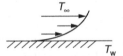

While Eq. 6.9 is deceptively simple, the real problem in convection is that "h" is an unknown. Fundamentally, we can determine h from the energy balance at the wall (convective inflow heat flux = conduction heat flux into the wall):

$$h = \frac{-k\left[\frac{dT}{dx}\right]|_{x=0}}{T_w - T_\infty} \tag{6.10}$$

Determining the temperature gradient at the wall necessitates solution of the Navier–Stokes equations and the energy equation. In gas flow problems, the momentum layer (boundary layer) is coupled to the thermal layer due to compressive heating of the gas, i.e. the Mach number at the wall is zero while the Mach number at the edge of the boundary layer can be supersonic. Liquid thermal and boundary layers also have strong coupling because the liquid density and viscosity can be a strong function of temperature. One representation of the equation set to consider (focusing on gaseous boundary layers) would look like this:

$$\frac{D\vec{v}}{Dt} = -\frac{1}{\rho}\nabla P + \upsilon \cdot \nabla^2 \vec{v}$$

$$\rho \cdot C_p \cdot \frac{DT}{Dt} = \frac{DP}{Dt} + \nabla \cdot (k\nabla T) + \Phi \tag{6.11}$$

where Φ is a complex term involving velocity derivatives referred to as viscous dissipation and $D()/Dt$ denotes a substantial derivative including advection. This is a coupled nonlinear system in which turbulence and variable properties are important. In combustion chambers, we may also have chemical reactions occurring within the boundary/thermal layers as well as significant mass transfer/mixing between fuel films and hotter combustion gases. This fluid flow problem represents one of the most complex challenges to the CFD community and while strides are being made for solution to a limited range of problems, no general treatment or capability is currently in place. Even given the very substantial computational power that can be dedicated to these problems, run times are very long in the current environment such that quick turnaround computations simply don't exist. For these reasons, the community still finds great utility in empirical methods/correlations for estimating h values.

Correlations for h_g

Correlations for h are typically expressed in dimensionless form using the Nusselt number or Stanton number. Correlations for h_g historically are derived from experiments involving flow of hot or cold gas in a pipe where the local wall temperatures are measured such that the temperature gradient in Eq. 6.10 can be inferred. While the analogy from a pipe flow to flow in a rocket combustion chamber and nozzle is tenuous, very substantial efforts would be required to improve on the situation given the range of conditions and propellant combinations of interest. In the end, heat transfer experts must fold in substantial conservatism given all the unknowns, yet estimates are desperately required at all design stages in order to assess thermal margins.

Because the heat transfer is related to local boundary layer thicknesses, the Reynolds number is the prime parameter for correlating this effect. The Prandtl number also appears to reflect the local thickness of thermal layers. A few of the more popular correlations are summarized below.

McAdams pipe flow

$$\text{Nu} = \frac{h_g D}{k} = 0.026 \cdot \text{Re}^{0.8} \cdot \text{Pr}^{0.4} \tag{6.12}$$

In Eq. 6.12, the Reynolds and Prandtl numbers are defined based on average fluid properties or conditions at the edge of the thermal layer and the diameter of the pipe, D:

$$\text{Re} = \frac{\rho_\infty V_\infty D}{\mu_\infty} \text{ and Pr} = \frac{\mu_\infty C_{p\infty}}{k_\infty}$$

While the McAdams correlation is really not suitable for rocket applications, it is used sometimes due to its simplicity.

Bartz equation   D. R. Bartz was employed at JPL in the 1950s during the earliest days of chemical rocket development in the United States. He was tasked with thinking specifically about the rocket heat transfer problem and recognized that property variations across thermal layers that have temperature differences of thousands of degrees are first-order considerations. His derivation and methodical treatment of these effects has stood the test of time and still remains as the primary approach most folks take when designing modern combustors even today. There are several versions of the Bartz equation; we report the one from Bartz (1957) here:

$$h_{\mathrm{g}} = \frac{0.026}{D^{0.2}} \left(\frac{\mu^{0.2} C_{\mathrm{p}}}{\mathrm{Pr}^{0.6}} \right)_0 (\rho_\infty V_\infty)^{0.8} \left(\frac{\rho_{\mathrm{am}}}{\rho_\infty} \right)^{0.8} \left(\frac{\mu_{\mathrm{am}}}{\mu_0} \right)^{0.2} \tag{6.13}$$

where $()_0$ denotes property evaluation at the local stagnation temperature, and $()_{\mathrm{am}}$ denotes a property evaluation at the arithmetic mean temperature ($T_{\mathrm{am}} = (T_\infty + T_{\mathrm{w}})/2$). Bartz's original paper provides this relationship as well as a relation similar to that more commonly used in Huzel and Huang (1992). The two relations give similar, but not identical results although many users opt with the Huzel and Huang version.

Ambrok's equation   This expression (see Kays and Crawford, 1980) has the advantage that it attempts to take the wall shape ($R = R(x)$) into account so that boundary layer thinning due to flow acceleration is modeled. However, we should point out that it has received much less attention than the more popular Bartz equation. Here the heat transfer coefficient is written in terms of the Stanton number:

$$\mathrm{St} = \frac{h_{\mathrm{g}}}{\rho_\infty V_\infty C_{\mathrm{p}\infty}} = 0.0287 \, \mathrm{Pr}^{-0.4} \cdot \frac{R^{0.25}(T_\infty - T_{\mathrm{w}})\mu^{0.2}}{\left[\displaystyle\int_0^x R^{1.25}(T_\infty - T_{\mathrm{w}})\rho_\infty V_\infty \mathrm{d}x \right]^{0.2}} \tag{6.14}$$

where $\mathrm{Pr} = \mu_\infty C_{\mathrm{p}\infty}/k_\infty$. For $\mu^{0.2}$ one should probably use the mean value, although Kays and Crawford (1980) point out that there isn't much sensitivity here due to the small exponent. Ambrok's equation has been used quite successfully in solid rocket heat transfer problems. It probably should be the preferred form to use since it includes the influence of wall shape. However, it is the most difficult to evaluate.

Comments   Readers must be cautioned that while these expressions have been or are currently being used to model rocket heat transfer there are hefty error bars. In particular, mixing processes associated with fuel film cooling are not accounted for, nor are any chemical reactions (endothermic or exothermic) that may be occurring in the boundary layer. In the end, we must view these as a starting point and gain experience (presumably via subscale testing) that enhances confidence or provides empirical correction. Both Ambrok's and Bartz equations require prediction of T_{w}, which makes them inherently iterative as T_{w} is presumably a main outcome from this analysis. Ambrok's equation is said to fail for strongly accelerated flows while that is precisely the condition we have in the nozzle region.

   Another level of analysis capability exists in the US industry with a large code developed in the 1970s and 1980s. The Two-Dimensional Kinetics (TDK) code was written under NASA and DoD sponsorship and provides a boundary layer module that solves integral method boundary layer equations for the actual wall shape. The TDK code and its sister, Solid Rocket Performance Prediction (SPP) code, are more comprehensive analysis capabilities but

Table 6.2   Recovery factors for various types of flow

Flow	Recovery factor, r
Laminar couette	Pr
Laminar free boundary layer	$\sqrt{Pr}$
Turbulent free boundary layer	$Pr^{1/3}$

are restricted to US firms and persons for usage. While not a full-blown multidimensional CFD analysis, these tools provide enhanced resolution of many of the engineering performance prediction relative to the simple introductory approaches employed in this text. Chapters 7 and 8 touch on this issue a bit more for the case of solid rocket motors and liquid rocket engines, respectively.

Recovery Temperature and its Role in Compressible Boundary Layers
An important final caveat regarding gas-side heat transfer in rocket nozzles pertains to the compressibility of the boundary layer itself. Compressible boundary layers are subject to viscous dissipation robs some of the kinetic energy from fluid. As a result, rather than having the stagnation temperature of the gas "driving" the heat transfer, a parameter called *the recovery temperature*, T_r serves this role. Viscous dissipation dictates that the recovery temperature is always somewhat less than stagnation temperature in fluid. Recovery temperature is defined:

$$T_r = T_\infty \cdot \left(1 + \frac{\gamma - 1}{2} \cdot r \cdot M_\infty^2\right) \tag{6.15}$$

where r is called the *recovery factor* $(r \leq 1)$. The value of r depends on the type of boundary layer, but general is function of Prandtl number. Table 6.2 summarizes recovery factor values for several types of flow.

For all but the smallest (lowest Reynolds number) chemical rockets the turbulent free boundary layer result is the most applicable. Since $Pr < 1$ for gases in the nozzle, $r < 1$ and $T_r < T_{stag}$. Using the recovery temperature, we define the gas-phase convective heat flux:

$$\dot{q} = h_g(T_r - T_w) \tag{6.16}$$

where T_w is the wall temperature adjacent to the hot gas. Equation 6.16 should always be employed for gas-side heat transfer. In the chamber where the Mach numbers are low, T_r is near the gas static temperature, but in the nozzle region it will be substantially greater than the local static temperature.

Correlations for h_l
As with gases, the convective heat flux in liquids is determined by the boundary and thermal layer profiles near the wall. Here we must be aware that boiling can occur. The nuclear power industry

takes advantage of enhanced heat transfer created by bubble formation and resulting turbulence, but in rocket applications we typically are at pressures too high to take advantage of this mechanism. In addition, use of boiling could be disastrous in practice since there may be a local spot in the wall that "dries out" and resulting heat transfer to the vapor in that region would be poor, leading to greatly increased wall temperatures. Figure 6.8 shows the critical heat flux (CHF) value for water as a function of the driving temperature difference. In rocket applications, we must maintain heat fluxes below these values to avoid dryout of the heated wall, with disastrous consequences.

Cryogenics propellants are often at pressures such that supercritical conditions exist. Liquid oxygen is typically injected at supercritical pressure and subcritical temperature – as the fluid heats in the pre-combustion zone it becomes a supercritical fluid. Here the physics are interesting and challenging as surface tension, and hence the presence of physical droplets, ceases to exist for supercritical fluids. Strong variations in properties occur near the critical point; for example, specific heat values tend to peak in a narrow region near the critical temperature. In this case, the mixing process is represented by gas-type mixing of fluids of differing densities. Figure 6.9 highlights some of these physics and a phase diagram for a typical fluid.

The Seider–Tate correlation is the one most frequently applied to rocket cooling systems:

$$\text{Nu} = \frac{h_1 \cdot D}{k} = a \cdot \text{Re}^m \cdot \text{Pr}^n \cdot \left(\mu / \mu_w\right)^b \qquad (6.17)$$

The constants a, b, m, and n vary between liquids. The parameter μ_w is the fluid viscosity evaluated at the local wall temperature. The initial Seider–Tate equation gave $b = 0.114$, which indicates that

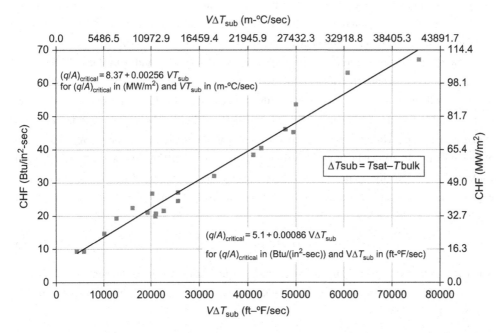

Figure 6.8 Critical heat flux (CHF) values to avoid film dryout in coolant passages. Data are for water.

Table 6.3 Empirical constants for liquid Nusselt number correlations in Eq. 6.15

Fluid	a	m	n
Methane	0.0023	0.8	0.4
Propane	0.005	0.95	0.4
Kerosene	0.023	0.8	0.4
Liquid hydrogen	0.062	0.8	0.3

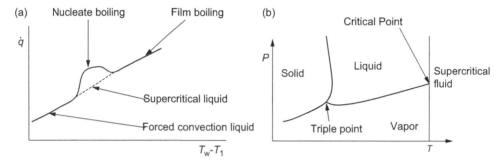

Figure 6.9 Boiling (a) and supercritical flow (b) effects in liquid systems can be important. Although heat flux can be improved, we typically don't take advantage of boiling in rocket applications. Cryogenic fluids will often be at supercritical or trans-critical conditions, i.e. transitioning from a liquid to a supercritical fluid.

viscosity variations are not terribly important due to this low exponent value. For this reason, the viscosity ratio is often ignored as correlations are developed for specific fluids. Table 6.3 shows a, m, and n values for several propellants used as coolants in rocket applications.

Properties in the Reynolds and Prandtl numbers in Eq. 6.17 should be evaluated at the mean temperature, which is reflective of the average between the wall and the bulk fluid. Once again, note the $Re^{0.8}$ influence that is predominant in these internal flows. Here, as we think about an incompressible liquid as coolant, we achieve higher Reynolds numbers by increasing velocities in cooling passages. This comes at the expense of higher pressure drops, of course. We also should point out that, with cryogenic fluids, density variations can be hugely important. Liquid hydrogen coolant changes density by nearly an order of magnitude when heated in cooling passages in engines. Finally, for cooling passages that are not circular, one should use the *hydraulic diameter* in Eq. 6.17

$$D_h = 4A/\text{Per} \tag{6.18}$$

where A is the cross-sectional area of the cooling passage and Per is its perimeter.

A modified Dittus–Boelter correlation similar to Seider–Tate has been employed in rocket cooling jacket applications:

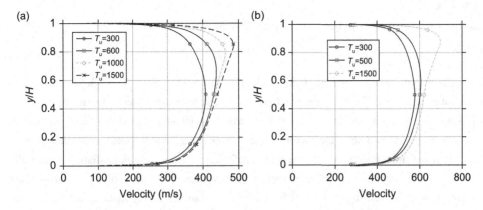

Figure 6.10 Computed velocity profiles in a 2-D channel 100 channel heights downstream from inlet. Entering fluid temperature 150 K, lower wall temperature 300 K, with different upper (heated) wall temperatures T_u. (a) Re = 1.0E05 and (b) Re = 1.0E06.

$$\mathrm{Nu} \ = \ \frac{h_1 \cdot D}{k} = 0.023 \cdot \mathrm{Re}^{0.8} \cdot \mathrm{Pr}^{0.4} \left(\frac{T_{wl}}{T_1}\right)^{-0.3} \tag{6.19}$$

In Eq. 6.19 average properties should be employed to reflect changes across the thermal layer. In general, these equations are applicable for Re > 10,000, 0.7 < Pr < 160 and channel lengths greater than $10D_h$. The modified Dittus–Boelter expression reflects the high temperature differences between the hot wall and the fluid, but neither this equation nor the Seider–Tate correlation take into account the fact that the wall passage is not straight, but curved to form the nozzle shape. Meyer (1997) considered corrections to account for this factor, as well as wall roughness and entrance region effects. Jung (2009) has recently compared a number of the existing correlations with CFD studies and recommends a value of –1.0 for the temperature ratio term as opposed to –0.3 in Eq. 6.19.

Coolant flow simulations

Since there is no combustion occurring in the coolant channel, the channel flow is more amenable to computations than the hot gas side. Here, the main challenge results from the large length (relative to the channel dimensions), the curvature of the passage as it forms the nozzle contour, and the drastic changes in properties of the coolant due to temperature changes across the thermal layer. Figure 6.10 shows computed results for liquid/transcritical nitrogen for a two-dimensional (2-D) channel with different heated wall temperatures. Wall heating creates a lower density region near that wall and the boundary layer becomes distorted as a result. Since wall friction depends on the velocity gradient at the interface, large changes in friction can result from the heating process. Section 6.5 provides additional discussion on this issue.

6.3.3 *Radiation*

Radiative heat transfer is governed by the Stefan–Boltzman law:

$$\dot{q} = \varepsilon\sigma(T_\mathrm{w}{}^4 - T_\mathrm{a}{}^4) \tag{6.20}$$

where ε is the emissivity ($\varepsilon = 1$ for a "black body"), $\sigma = 5.67 \times 10^{-0.8} \mathrm{W/m^2/K^4}$ is the Stefan–Boltzman constant, and T_a is the ambient temperature. Here the main difficulty lies in computing or estimating the emissivity, as it depends on the type of material emitting energy, the wavelength of the emitted energy, and its view factor to the surroundings. Inside combustion chambers of chemical propulsion systems, convective heat transfer tends to dominate and radiation is generally not considered as a primary mechanism, mainly because most gases have low emissivity, i.e. are poor emitters. However, in plumes that have a view toward aft-facing surfaces on the vehicle, radiation plays a significant role. For example, in solid rockets that use aluminized propellants, the aluminum oxide combustion products can be a strong emitter for aft surfaces. The Al_2O_3 particles can also exhibit higher temperatures in the plume than the expanding gases due to their own thermal inertia.

As we mentioned in Section 6.2, we typically employ radiation cooling in the nozzle exit cone of space engines. Figure 6.11 shows a typical arrangement for these systems. Radiation-cooled skirts have used on liquid rocket engines, but have not seen service in hybrid or solid motor applications. The emissivity of the niobium skirt material is high and a value of 0.95 is a good estimate in most applications.

We can find the equilibrium (steady-state) temperature in metal by balancing the convective heat flux into the part with the radiative heat flux leaving the part as illustrated in Figure 6.12. As the skirt thickness is small, it can be characterized with an average bulk wall temperature, T_w. Given this assumption, at steady-state conditions, $\dot{q}_\mathrm{out} = \dot{q}_\mathrm{in}$, we get a simple relationship for the local wall temperature:

$$\varepsilon\sigma T_w^4 = h_\mathrm{g}(T_\mathrm{r} - T_w) \tag{6.21}$$

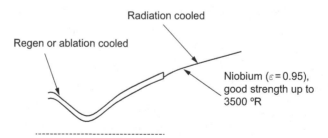

Figure 6.11 Radiation-cooled nozzle extension (skirt) employed on some space engines. The skirt attachment point is determined by the local convective heat flux; as this quantity drops with the expanding and cooling gases, a columbium skirt becomes a viable option. The skirt tends to be lighter in weight and lower in complexity than the adjacent cooling jacket or ablative liner.

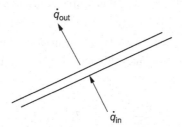

Figure 6.12 Energy balance on radiation-cooled skirt.

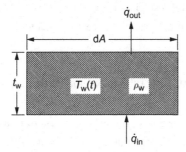

Figure 6.13 Lumped parameter energy balance to get transient response of a wall material.

We can get T_w from this equation.

$$\dot{q}_{out} = \varepsilon \sigma T_w^4 \leftarrow T_a = 0$$
$$\dot{q}_{in} = h_g(T_r - T_w)$$

6.3.4   *Lumped Parameter Transient Analysis*

The presumption that the wall can be characterized by a single bulk temperature in Eq. 6.21 is often referred to as a *lumped parameter* approach. Lumped parameter methods permit rapid evaluation of the average conditions during unsteady processes. In Chapters 7 and 10, we employed these methods for characterization of chamber pressure histories in SRMs or HREs and in Chapter 8 we use this approach to get the average tank pressure in LRE systems. In heat transfer, one can always get the leading order wall response under the constraint that we are characterizing the wall with a single bulk temperature. As an example, consider the transient heating of the radiation-cooled skirt. Figure 6.13 provides a schematic of this situation for an incremental area along the skirt (dA), where the local wall thickness is t_w. The rate of change in energy in a local section of the skirt depends on the mass of material in that skirt (m), its specific heat, and the relative importance of incoming and outgoing heat fluxes:

$$m \cdot C_p \cdot \frac{\partial T_w}{\partial t} = (\dot{q}_{in} - \dot{q}_{out}) dA \tag{6.22}$$

We can calculate the transient behavior (for a thin wall) by considering the energy balance on a unit volume of material (see Figure 6.13).

However, the mass of the element can be written in terms of its density, $m = \rho_w t_w dA$, and if we substitute for $\dot{q}_{in}$, $\dot{q}_{out}$ we get an ordinary differential equation for the local wall temperature as a function of time:

$$C_p \cdot \rho_w \cdot t_w \left(\frac{\partial T_w}{\partial t} \right) = h_g \cdot (T_r - T_w) - \varepsilon \cdot \sigma \cdot T_w^4 \tag{6.23}$$

In general, Eq. 6.23 must be solved numerically since h_g may be a function of T_w. This approach can be applied to other situations in order to get bulk material transient response. For example, in the wall of a cooling jacket that we consider in Section 6.5, we would have a competition between convective heat flux from chamber gases into the wall material with convective cooling from coolant flow in the liner.

6.4 Scaling of Convective Heat Transfer Processes

Rocket environments represent difficult problems with regard to measurements and oftentimes the only data that exist are for conditions that are significantly different to those of interest. For this reason, we often are placed in a position of having to scale results to new conditions (e.g. higher pressure) in order to have some estimate of the heat transfer rates for the system of interest. As you know, convective heat after processes (McAdams, Bartz, Seider–Tate) all show influence of turbulent boundary layer development via the Reynolds number or massflux parameter raised to the power of 0.8, i.e. $\text{Re}^{0.8}$ or $(\rho v)^{0.8}$. The figure of 0.8 results from the character of a turbulent boundary layer. Since the gas density in the Reynolds number or massflux scales directly with pressure, this implies that convective heat transfer in rockets scales as $p^{0.8}$. This is a very fundamental result that not only indicates the difficulties of cooling at high pressure, but also provides a mechanism to compare local heat transfer rates within the flowpath, i.e. along the nozzle. We can see how things scale along the nozzle if we recognize that $\rho v = \dot{m}/(\rho A) = 4\dot{m}/(\rho \pi D^2)$. This implies that $\text{Re} \sim 1/D$ such that the maximum heat transfer will occur at the throat of the nozzle.

There are numerous situations where a scaling of the heat transfer to different conditions is useful. In fact, this may be the only alternative due to the practical difficulties in measuring heat transfer rates in operating combustors. In general, the convective heat transfer correlations depict the $\text{Re}^{0.8}$ scaling:

$$\text{Nu} = \frac{hD}{K} \alpha \text{Re}^{0.8} \tag{6.24}$$

For gas-phase heat flux, we can write the Reynolds number in terms of local Mach numbers using the definition of M:

$$\mathrm{Re} = \frac{\rho V D}{\mu} = \frac{p M \sqrt{\gamma R T} D}{R T \mu} \tag{6.25}$$

Using this result, we can assess influence of p, D, or heat transfer levels by combining Eqs. 6.21 and 6.22 to give:

$$h = \frac{\mathrm{Nu} \times k}{D} = \text{constant} \times \frac{k \times \mathrm{Re}^{0.8}}{\mathrm{D}} = \text{constant} \times \left[\frac{kpMD}{\sqrt{RT/\gamma\mu}}\right]^{0.8} \tag{6.26}$$

From isentropic flow, we know that the Mach number is a function of the local area ratio A/A_t, therefore the Mach number progression only depends on the shape of the nozzle and does not vary as we change either the pressure or the scale (size) of engine. Similarly, if subscale or different pressure engines incorporate the same propellants, then the gas properties μ, T, γ, k all remain the same. In this case, all these factors can be lumped into the constant in Eq. 6.26, and we get the fundamental result:

$$h \propto p^{0.8}/D^{0.2} \tag{6.27}$$

Rocket combustor designers know this result all too well as it speaks to the difficulties in attaining high-pressure performance. The heat transfer goes up nearly linearly with increased pressure. Moreover, if we consider the engine thrust level (and hence mass flow) as fixed, then when we increase pressure, we also decrease throat size since $\dot{m} = p_c A_t/c* = $ constant implies that $pD^2 = $ constant or $D^{-0.2} \sim p^{0.1}$. In this case, the scaling in Eq. 6.27 becomes:

$$h \propto p^{0.8}(\text{ fixed thrust result}) \tag{6.28}$$

so doubling the chamber pressure on a given thrust engine design nearly doubles the difficulty in cooling the engine. This result points to just one of the difficulties in developing very high-pressure systems. Structural loads and sealing issues created by high pressure are other challenges that must be met in order to successfully develop systems to perform under these arduous conditions.

6.5   Regenerative Cooling System Analysis

One of the most fundamental elements of cooled LRE thrust chamber designs is the analysis of the performance of a regenerative cooling system. This approach is employed in most large engines and represents an elegant and effective solution to manage thermal environments. A collateral benefit comes from the fact that the heat that is dumped into the coolant (typically fuel) does not leave the system and therefore nearly all the energy of combustion gets imparted to fluids exiting the nozzle.

The "regen" jacket analysis lies at the heart of LRE system design because the pressure drop in the coolant system must be provided by the fuel turbopump. Typically the highest pressure in the entire engine cycle is the value required for the coolant distribution manifold that distributes the flow to individual coolant tubes/channels. The heat transfer in the coolant channels is improved as

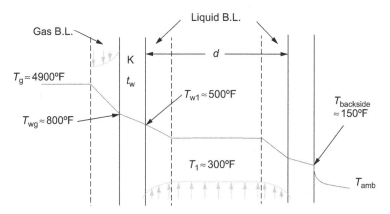

Figure 6.14 Cross-section of a chamber wall and cooling passage showing thermal/boundary layers and typical temperature progression under steady-state conditions.

velocities increase in the channel (remember the $Re^{0.8}$ influence in the Seider–Tate equation), so this benefit lies in direct competition with the pressure drop, which also increases at higher channel velocity. Additionally, the fuel exit condition (temperature, pressure) becomes the initial condition for analysis of injector and thrust chamber.

Figure 6.14 highlights the chamber wall region with a given cooling passage. Typical temperatures are included to give the reader an overall view of the changes through the system. The assumptions for the analysis include:

- neglect backside wall heat transfer – all energy goes to the liquid and gas-side wall;
- T_1 is the bulk temperature in the liquid; and
- steady-state has been achieved.

By neglecting backside heat transfer we obtain an upper bound on the fuel temperature. Note that Figure 6.14 does show some temperature drop in the backside material that would be present if backside heat transfer is included. In principle, it is straightforward to relax this assumption and consider free or forced convection on the outer wall of the jacket, but in practice there is very little heat lost here and results typically don't differ much as a result. A thermal layer is assumed to exist on the heated side of the coolant channel, but the overall energy balance in the liquid is used to compute the local bulk/average temperature, T_1 at each axial station in the jacket. The steady-state assumption dictates that the heat flow at any point in the system must be the same, i.e. conduction through the heated wall and convection into the coolant must be the same as the driving convection coming from the hot gases:

$$\dot{q} = \ h_g(T_r - T_{wg}) = k\left(\frac{T_{wg} - T_{wl}}{t_w}\right) = h_l(T_{wl} - T_1) \tag{6.29}$$

Figure 6.15   Photo of copper-based cooling jacket liner using milled channels to form the coolant flowpath. A nickel-based closeout would typically be bonded to the backside of these channels to form the final part.

where we have approximated the wall temperature gradient and assumed a linear distribution in order to estimate dT/dx, which is driving conduction. Given the fact that the wall is thin and that it supports a large temperature difference, this is typically an excellent assumption. The result in Eq. 6.29 is very important in the design process. Given rough estimates of the gas heat transfer coefficients, bulk coolant temperature, and a CEA results for the combustion gas temperature, Eq. 6.29 provides estimates of T_{wg} and T_{wl} so we can immediately assess whether or not the wall material will be able to sustain the predicted heat flux. A similar energy balance can be constructed in the case where a thermal barrier coating is applied and the material temperatures of the TBC can be evaluated in similar fashion.

Heat transfer folks refer to this as a conjugate gradient heat transfer problem due to the coupling of convection and conduction via the shared wall temperatures T_{wg} and T_{wl} on the heated/cooled surfaces respectively. Mathematically, Eq. 6.29 provides two conditions for three unknown temperatures, T_{wg}, T_{wl}, and T_l. The main challenge here is that, in general, h_g and h_l depend on T_{wg}, T_{wl}, which are not known a priori.

6.5.1   Tube/Channel Design

Many of the early regen chambers were fabricated from individual tubes that were bent into the desired shape to form the thrust chamber, then brazed together in a large braze furnace. In this case, the number of tubes, N, is normally set by the heat flux in the throat region. For a given throat diameter D_t, we have:

$$N = \frac{\pi[D_e + 0.8(d + 2t_w)]}{(d + 2t_w)} \qquad (6.30)$$

where the factor of 0.8 accounts for the fact that tubes are brazed in a circle, rather than along a line. More modern regen chambers are being manufactured from milled channels that are cut into the back wall of a pre-formed chamber/nozzle shape. Here the main limitation is the precision with which these passages can be cut. The integral walls (called "lands") between adjacent channels serve as fins that help to wick heat from the chamber to these surfaces and eventually into the coolant. The limitations of the milling process set a lower bound on the land thickness between channels and the overall depth of the channel for a given width. Structural constraints may limit the ratio of channel width to land thickness or the ratio of channel width to face thickness. Figure 6.15 illustrates the design of a modern cooling jacket comprised of a copper/chromium/zirconium alloy. Copper alloys are used almost exclusively for jacket liners due to the excellent thermal conductivity of the material. Alloys of interest display excellent strength at temperature. Specific alloys are developed for rocket applications, for example the Narloy Z material was developed for the SSME thrust chamber.

Most recently, the prospects of additive manufacturing of intricate jacket designs using direct metal laser sintering/melting has permitted yet another method of fabrication. Here, the passage size is limited by the particular machine and process being used, but in general a number of manufacturers are marketing capabilities to manufacture sub-millimeter sized passages. In this case, the wall roughness is substantial as it is an artifact of the powder size being melted/sintered.

Regardless of the manufacturing approach being employed, one can obtain the individual tube/passage mass flow based on the total coolant flow (presumed to be fuel here) and the total number of tubes/channels:

$$\dot{m}_{chan} = \frac{\dot{m}_f}{N} \qquad (6.31)$$

assuming all the fuel goes through the jacket.

The tube or land thickness is set by structural constraints. Equation 6.29 shows that we desire the thinnest possible tube (smallest t_w possible) in order to maximize heat transfer. The thermostructural challenges are highlighted in Figure 6.16 as both mechanical and thermal stresses are imposed on the jacket.

In general, a multidimensional thermostructural analysis must be carried out for all the most stressful operating conditions in order to verify integrity of the design. There are some approximate analytic results from NASA SP-125 that highlight the important physical parameters. Tangential stresses are created by the internal pressure in the coolant passage and its relationship to the local gas pressure as well as due to thermal growth of the material on the heated side of the passage:

$$\sigma_t = \frac{(p_l - p_g)D}{2t_w} + \frac{Ea\dot{q}t_w}{2(1-v)k} \qquad (6.32)$$

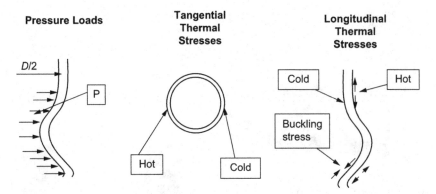

Pressure Loads   **Tangential Thermal Stresses**   **Longitudinal Thermal Stresses**

Figure 6.16 Thermostructural loads imposed on coolant tubes/channels. Mechanical (pressure loads) work to explode the chamber while thermal stresses create buckling loads on the cold side of passages.

where E is Young's modulus (tension), v is Poisson's ratio, and a is the coefficient of thermal expansion of the tube wall material. Physically, the first term in Eq. 6.32 is the hoop stress and the second term is the thermal stress induced by the heat transfer. Due to the tremendous heat flux imposed in regen jackets, thermal stresses can easily be as large or larger than the mechanical/hoop stresses. The hot-gas wall in a milled channel has a similar stress balance as the wall is supported by the channel lands on either side and is being pushed inward toward the engine centerline due to the high internal pressure. In this case, the tangential stress can be written:

$$\sigma_t = \frac{(p_1 - p_g)}{2} (w/t_w)^2 + \frac{Ea\dot{q}t_w}{2(1 - v)k} \tag{6.33}$$

where w is the internal width of the channel.

The longitudinal stresses result from the tendency of the heated side of the passages to grow in length:

$$\sigma_l = Ea\Delta T \tag{6.34}$$

where $\Delta T = T_{hot} - T_{cold}$ in tubes.

Huzel and Huang (1992) suggest that one should keep $\sigma_l \leq 0.9\sigma_t$ for good design, but using modern tools we can of course determine the detailed stresses at all locations and look at how these compare to allowable levels. Thrust loads from the nozzle create forces to work to buckle the jacket in the axial direction. This buckling stress can be estimated via:

$$\sigma_b = \frac{EE_c{}^{t_w}/_D}{(\sqrt{E} \pm \sqrt{E_c})^2 \sqrt{3(1 - v^2)}} \tag{6.35}$$

where E_c is the modulus of the wall material in compression. While Eqs. 6.32–6.35 are replaced with a thermostructural finite element analysis in the modern era, they do elucidate the basic loads on the structure. In many cases, thermal stresses exceed mechanical stresses – a non-intuitive

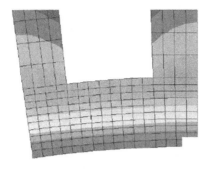

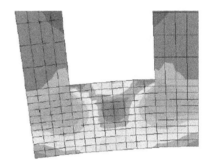

Plastic Strain Distribution

Temperature Distribution

Figure 6.17 Thermostructural analysis results for Space Shuttle main engine regenerative cooling jacket.

revelation given the high pressures/mechanical loads we typically work with in rocket thrust chambers. Figure 6.17 provides a typical modern thermostructural analysis result for the SSME regen jacket. Note that the temperature distribution is predominantly radial, but the strain distribution contains concentrations near the corners of the channel and in the region between the channel and the heated wall.

There is an obvious negotiation between the structural analyst and the thrust chamber designer on these issues and the overall analysis and design process is inherently iterative. If thermostructural loads are excessive, additional fuel film cooling might be required. Results are very sensitive to the wall thickness due to the central role it plays in Eq. 6.26. As illustrated in Figure 6.17, the thin web between adjacent tubes/channels is subject to high stresses and is generally a life-limiting feature in the jacket. Locally, the web will buckle and lead to "dog-house" cracks that eventually will fail altogether. Figure 6.18 provides an image of dog-house cracks in a used liner. A single failure in a jacket is not catastrophic as this simply dumps additional coolant into the failed region, thereby limiting propagation. However, numerous failures would significantly alter injector flows and core mixture ratios and would eventually become catastrophic. In addition, if liquid oxygen is used as coolant, the tolerance for leaks becomes more tenuous as combustion may occur in a near-wall region designed to be fuel-film cooled. This is but one of the many challenges in attempting to use oxidizer as coolant.

The local pressure drop in an incremental length Δx along the coolant flowpath can be estimated based on a friction factor, c_f:

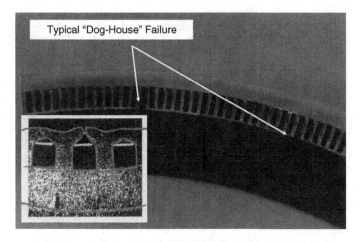

Figure 6.18 Dog-house cracks in used cooling jacket.

$$\Delta p = 4c_f \left(\frac{\Delta x}{d}\right)\left(\frac{1}{2}\right)\rho_1 V_1^2 \tag{6.36}$$

where d is the local hydraulic diameter of the coolant passage. The friction factor depends on the boundary layer characteristics within the passage and is therefore a function of the Reynolds number and the relative roughness of the wall. The relative roughness is computed as the ratio of the wall surface height variation non-dimensionalized by the passage diameter. The famous Moody diagram provides values of c_f as a function of Reynolds number and relative roughness – a representation can be found on the engineering toolbox website (www.engineeringtoolbox.com). For smooth pipes/walls, curvefits have been developed:

$$
\begin{aligned}
c_f &= 16/\mathrm{Re} & \mathrm{Re} < 2100 \ (\text{laminar flow}) \\
&= 0.046/\mathrm{Re}^{0.2} & 5000 < \mathrm{Re} < 200,000 \\
&= 0.0014 + 0.125/\mathrm{Re}^{0.32} & 3000 < \mathrm{Re} < 3E06
\end{aligned} \tag{6.37}
$$

Because the passages are small, wall roughness effects can be very important. In addition, "printed" passages using laser melting/sintering approaches produce high roughness values. In our application, heating of the wall causes significant changes in the boundary layer profiles as noted in Figure 6.10 above. For these 2-D CFD results, computed c_f values are shown in Figure 6.19 below. The transcritical nature of cryogenic coolant creates compressibility effects as the Mach number becomes appreciable as the fluid is heated. Compressibility effects alone can change friction factors significantly. The distortion of the velocity profiles under heated conditions as shown in Figure 6.10 leads to enhanced friction factors. For these reasons it is difficult to use existing literature (Moody diagram, etc.) for accurate representations in rocket cooling jackets. However, this is often the best starting point if we are exploring a new design.

(a)

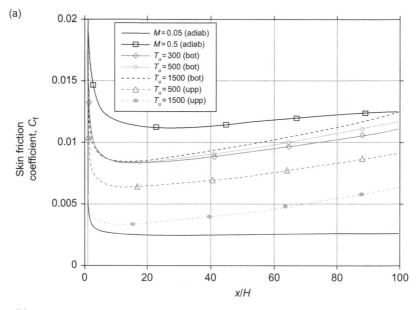

(b)

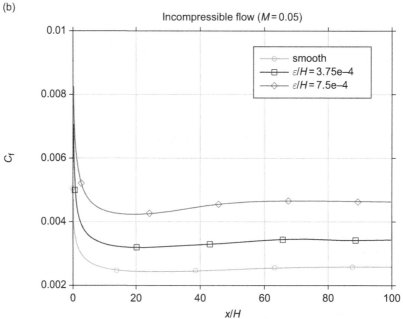

Figure 6.19 Computed friction factors for a 2-D channel at Re = 1.0E06. (a) Effect of upper wall temperature, T_u (in K) and Mach number, M. (b) Surface roughness effects at a fixed Mach number of 0.05.

Since the mass flow rate in the channel is $\dot{m}_{chan} = \rho_1 v_1 \pi d^2/4$ we can write the result in Eq. 6.36 as:

$$\Delta p = 32 c_f \left(\frac{\Delta x \, \dot{m}_{chan}^2}{\rho_1 \pi^2 d^5} \right) \qquad (6.38)$$

In Eq. 6.38 we can see the sensitivity of the channel pressure drop to mass flow and hydraulic diameter. Doubling the mass flow in a given channel will quadruple the pressure drop. The $1/d^5$ sensitivity implies that only a 5% reduction in internal dimension can increase the overall jacket pressure drop by 30%. As mentioned previously, the density drop with heating of the passage also has great impact on the overall pressure drop and this feature needs to be considered in order to get reasonably accurate results.

6.5.2   Regen Jacket Analysis Procedure

Equations 6.30–6.38 provide the basic elements of the steady-state regen jacket analysis. All inlet conditions and fluid properties must be known in order to proceed. We let x be the coordinate locally parallel with the coolant passage and d represents the local hydraulic diameter. In general, d can be a function of x, i.e. we can change channel dimensions along the length. Since the engine internal diameter is changing to form the nozzle, and since heat flux is a maximum at the throat, there is a tendency to want to decrease d in the throat region. Some chamber manufacturing approaches allow for this feature while others do not. Figure 6.20 summarizes the notation and coordinates for the analysis. The channel flow rate and the inlet pressure and temperature (p_{11}, T_{11}) must be provided as well as all gas-phase conditions (presumably from CEA or alternate computation) along the length of the nozzle. The channel can be discretized into m segments and a typical segment "i" is represented in Figure 6.20. Given these inputs, the overall procedure for integrating along the length of the channel is outlined below.

Procedure

1. Pick a system design:
   - Single pass/double pass, start/finish points, etc. We discuss this issue in more detail in Chapter 8.

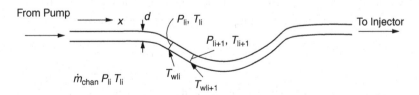

Figure 6.20 Nomenclature and coordinate system for regen jacket analysis.

- Prescribe tube/channel inflow conditions (p_{11}, T_{11}), cross-section and geometry ($d = d(x)$, $t_w = t_w(x)$), Set $i = 1$ at start of integration along passage length.

2. Calculate N, $\dot{m}_{chan}$ from Eq. 6.30 (or equivalent).
3. Begin stepping down tube/channel $i = i + 1$.
4. Guess the wall temperature on gas side, T_{wgi}.
5. Calculate h_g then $\dot{q} = h_g(T_r - T_{wgi})$. Here, we presume that we have 1-D isentropic flow or multidimensional CFD results to give the local chamber gas conditions (T, p, etc.) that feed into correlations for h_g.
6. Calculate T_{wli} from conduction through the wall, $\dot{q} = k(T_{wgi} - T_{wli})/t_w$. Note that $\dot{q}$ is the same value as computed in step 5 under the steady-state assumption.
7. Compute h_l based on $T_{1(i-1)}$ (temperature in previous segment).
8. Compute heat flux $\dot{q} = h_l(T_{wli} - T_{1(i-1)})$, and check whether equal to step 6.
9. If yes, continue; if no, return to step 4.
10. Now we have correct $\dot{q}$, h_g, h_l, T_{wg}, T_{wl} values for segment "i." Obtain a new liquid temperature in the ith segment by considering the amount of heat dumped in over the incremental length Δx:

$$T_{1(i+1)} = T_{1i} + \frac{1}{\dot{m}_{tube}c_{pl}} \int_{x_i}^{x_{i+1}} \dot{q}\, dA_i \qquad (6.39)$$

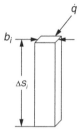

where $dA_i = b_i \Delta x_i$ and b_i is the circumferential span of the channel in question. The coordinate x is assumed to be aligned with the local wall such that area can be computed in this fashion. Note that heat enters the entire tube/channel cross-section including tube walls or channel lands so that heat traveling up the lands also has to enter coolant passages sharing those lands. For a tubular design this would be the azimuthal length to the outer ends of the tube, for milled channels b_i would be measured from the center of the land on either side of the channel.

Given the new liquid temperature we can now update all liquid properties (ρ_l, μ_l, etc.) that are functions of this parameter. From Eq. 6.36, the new liquid pressure is:

$$P_{1(i+1)} = P_{1i} - c_{fi}\left(\frac{\Delta x_i}{d_i}\right)2\rho_l v_l^2$$

where c_{fi} is computed based on the current Reynolds number in the passage and the velocity is computed from the channel mass flow and cross-sectional area.

11. At this point, we have all properties in the current segment. Now return to step 3 to move onto the next segment. At the end of the jacket or at intermediate manifolds in the jacket, we typically assume all the dynamic pressure in the flow is lost so the manifold pressure will be equivalent to the local static pressure of the last segment feeding that manifold.
12. After completing integration, check structural integrity at all points using Eqs. 6.32–6.35 or more commonly a detailed thermostructural finite element analysis.

Comments

- Typical overall $\Delta T = 100$–200 °F (50–100 °C) in jacket.
- The T_{wg} iteration is not sensitive – it converges rapidly. Clearly the best first guess is the T_{wg} value obtained in the previous segment. If one uses at least 100 points to define the contour, the segment-to-segment changes in T_{wg} are modest.
- The critical region is the throat as this is where heat transfer is maximum.
- Another important aspect are *"finning effects"* or "extended surfaces" that result from lands between cooling channels or adjacent tube walls, as touched on in connection with Eq. 6.39. Since the conductivity of the wall material is higher than that of the coolant, heat from the chamber wicks along these extended surfaces and augments the overall heat transfer relative to the conduction region through the channel/tube wall. Corrections for these effects can be applied using analytic techniques described in the Bergman *et al.* (2011). At some level, the three-dimensionality of the situation becomes important and detailed thermostructural tools are then required to achieve this next level of resolution.

FURTHER READING

Bartz, D. R. (1957) "A Simple Equation for Rapid Estimation of Rocket Nozzle Convective Heat Transfer Coefficients", *Jet Propulsion*, 27(1): 49–51.

Bergman, T. L., Lavine, A. S., Incropera, F. P., and Dewitt, D. P. (2011) *Fundamentals of Heat and Mass Transfer*, 7th edn. New York: Wiley.

Bucchi, A. E. (2005) "Transpiration Cooling Performance in LOX/Methane Liquid-Fuel Rocket Engines," *AIAA Journal of Spacecraft and Rockets*, 42(3).

Carslaw, H. S. and Jaeger, J. C. (1986) *Conduction of Heat in Solids*, 2nd edn. Oxford Science Publications.

Computer Program, Vol. I, *Engineering Methods*, Software and Engineering Associates, Inc., NASA Contract NAS8-39048, 1993.

Dunn, S., Coats, D., Dunn, S., and Coats, D. (1997) "Nozzle Performance Predictions Using the TDK 97 Code," in *33rd Joint Propulsion Conference and Exhibit*, p. 2807.

Edwards, D. K., Denny, V. E., and Mills, A. F. (1974) *Transfer Processes*. Wiley.

Huzel, D. and Huang (1992) *Design of Liquid Propellant Rocket Engines*. Washington, DC: AIAA.

Jankovsky, R. S., Smith, T. D., and Pavli, A. J. (1999) "High Area Ratio Rocket Nozzle at High Combustion Chamber Pressure – Experimental and Analytical Validation." NASA TP-1999–208522.

Jung, H. (2009) "Conjugate Analysis of Asymmetric Heating of Supercritical Fluids in Rectangular Channels," Ph.D. Dissertation, Purdue University.

Jung, H., Merkle, C., Schuff, R., and Anderson, W. (2007) "Detailed Flowfield Predictions of Heat Transfer to Supercritical Fluids in High Aspect Ratio Channels," in *43rd Joint Propulsion Conference and Exhibit*. AIAA.

Kays, W. M. and Crawford, M. E. (1980) *Convective Heat and Mass Transfer*, 4th edn. McGraw Hill.

Keller, H. B. and Cebicci, T. (1972) "Accurate Numerical Methods for Boundary Layer Flows II: Two-Dimensional Turbulent Flows," *AIAA Journal*, 10(9): 1193–1199.

Martinez-Sanchez, M. (2005) MIT Open Course Ware Lecture 7: Convective Heat Transfer: Reynolds Analogy.

Meyer, M. L. (1997) "The Effect of Cooling Passage Aspect Ratio on Curvature Heat Transfer Enhancement," NASA TM 107426.

Nickerson, G. R., Berker, D. R., Coats, D. E., and Dunn, S. S. (1993) *Two-Dimensional Kinetics (TDK) Nozzle Performance Computer Program Volume 11, Users' Manual*. Prepared by Software and Engineering Associates, Inc. for George C. Marshall Space Flight Center under contract NAS8-39048, March 1993.

Rouser, D. C. and Ewen, R. L. (1976) Aerojet Liquid Rocket Company, "Combustion Effects on Film Cooling," NASA Contractor Report CR-135052.

Schuff, R., Jung, H. Anderson, W., and Merkle, C. (2007) "Experimental Investigation of Asymmetric Heating in a High Aspect Ratio Cooling Channel with Supercritical Nitrogen," in *43rd Joint Propulsion Conference and Exhibit*. AIAA.

Suslov, D., Woschnak, A., Sender, M., and Oschwald, M. (2003) "Test Specimen Design and Measurement Technique for Investigation of Heat Transfer Processes in Cooling Channels of Rocket Engines under Real Thermal Conditions," in *39th Joint Propulsion Conference and Exhibit*. AIAA.

Wennerberg J., Anderson, W., Haberlen P., Jung, H. and Merkle, C. (2005) "Supercritical Flows in High Aspect Ratio Cooling Channels," in *41st Joint Propulsion Conference and Exhibit*. AIAA.

Wennerberg J., Jung, H., Schuff, R., Anderson, W., and Merkle, C., (2006) "Study of Simulated Fuel Flows in High Aspect Ratio Cooling Channels," in *42nd Joint Propulsion Conference and Exhibit*. AIAA.

Woschnak, A., Suslov, D., and Oschwald, M. (2003) "Experimental and Numerical Investigations of Thermal Stratification Effects," in *39th Joint Propulsion Conference and Exhibit*. AIAA.

Young, W. C. and Budynas, R. G. (2002) *Roark's Formulas for Stress and Strain*, 7th edn. New York: McGraw-Hill.

Homework Problems

6.1 The small test rocket shown in Figure 6.21 is to utilize heat sink cooling of the nozzle throat by

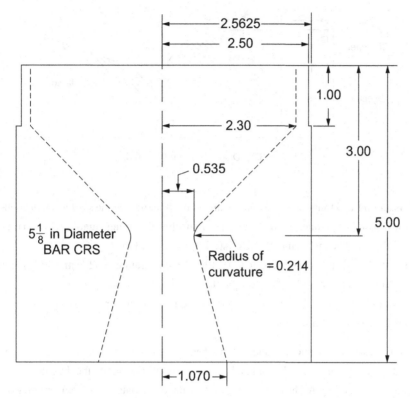

Figure 6.21 Thruster design for Problems 6.1, 6.2 and 6.7.

making this part out of a thick steel billet. The device will utilize 85% hydrogen peroxide and methyl alcohol as propellants at a mixture ratio of 3.0 and a chamber pressure of 500 psi.

(a) Use the Bartz equation to determine the gas-phase heat transfer coefficient at the throat.

(b) Assuming 1-D transient conduction and that the thickness of the steel is much greater than the heat affected zone, estimate the maximum allowable firing duration to avoid melting of the throat. (The melt temperature of steel is 2500 °F/1370 °C.)

(c) Let the time in part (a) be defined as t_m. Plot the temperature as a function of distance from the surface at $t = t_m/4,\ t_m/2,\ 3t_m/4,\ t_m$.

(d) Is your analysis conservative? That is, is the real t_m greater or less than the value you predict? Is the infinite solid assumption valid?

6.2 The small test rocket shown in Figure 6.21 is to utilize heat sink cooling of the nozzle throat by making this part out of a thick steel billet. The device will utilize Gox and RP-1 as propellants at a mixture ratio of 2.5 and a chamber pressure of 200 psi.

(a) Use the Bartz equation to determine the gas-phase heat transfer coefficient at the throat.

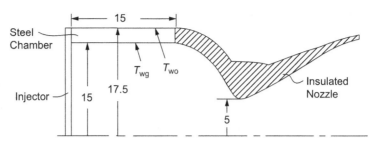

Figure 6.22 Diagram for Problem 6.3.

(b) Assuming 1-D transient conduction and that the thickness of the steel is much greater than the heat affected zone, estimate the maximum allowable firing duration to avoid melting of the throat. (The melt temperature of steel is 2500 °F/1370 °C.)

(c) Let the time in part (b) be defined as t_m. Plot the temperature as a function of distance from the surface at $t = t_m/4$, $t_m/2$, $3t_m/4$, t_m.

(d) Is your analysis conservative? That is, is the real t_m greater or less than the value you predict? Is the infinite solid assumption valid?

6.3 A liquid rocket engine injector is tested in a heavy-walled chamber as shown in Figure 6.22. While the throat region is well-insulated, the chamber itself relies on the thermal mass of steel as a means of cooling. At the end of the test, thermocouple data record temperatures of 900 K and 600 K on inner and outer surfaces of the chamber (T_{wg} and T_{wo} in the figure). In addition, you may assume that the temperature varies linearly through the thickness of the metal and that negligible heat is transferred from the outer surface of the chamber wall to the surroundings. The following information is known:

Gas properties	Steel properties
Gas temperature, $T_g = 2000$ K	Initial temperature, $T_0 = 300$ K
Specific heat, $C_p = 0.02$ J/kg K	Specific heat, $C_s = 477$ J/kg K
Conductivity, $k = 0.027$ W/m-K	Conductivity, $k_s = 14.9$ W/mK
Ratio of specific heat, $\gamma = 1.2$	Density, $\rho_s = 7900$ kg/m^3

(a) Estimate the gas film heat transfer coefficient for the chamber gases.

(b) Estimate the duration of the test firing. Carefully describe the assumptions required to make this estimate.

6.4 Prove that convective heat transfer is a maximum at the throat of a rocket nozzle. You may assume negligible temperature dependence of μ, C_p, and k for your proof.

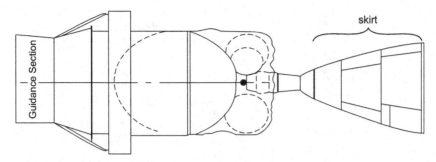

Figure 6.23 Radiation-cooled skirt design for Problems 6.5 and 6.8.

6.5 A columbium skirt (emissivity at 0.9) is to be used as a nozzle extension on a space engine as shown in Figure 6.23. The skirt attaches to an ablatively cooled chamber at an expansion ratio of 7:1 and extends to 60:1. The engine utilizes LOX/RP-1 as propellants with fuel-film cooling. We estimate a mixture ratio of 2.2 near the wall in the skirt region. The engine has a chamber pressure of 20 atm and a throat diameter of 15 cm. You are asked to perform a thermal analysis to verify the integrity of the skirt.

(a) To avoid extra work, you only want to check thermal integrity at one point. Which point should you choose?

(b) Determine the maximum temperature in the skirt. (Use the Bartz equation to estimate h_g.)

(c) Assume the engine starts instantaneously and that radiation heat transfer can be neglected as the wall heats up. How long will it take the 200 mil thick skirt to heat up to 90% of the temperature calculated above? (Assume the skirt is at 550 °R prior to ignition.)

Note: $\rho = 8750 \text{ kg/m}^3$, $C_p \cong 300 \text{ J/kg K}$ for niobium/columbium.

6.6 Consider the flow of a constant property gas through a rocket nozzle.

(a) Prove that heat transfer is a maximum at the nozzle throat.

(b) How does doubling the chamber pressure affect the heat transfer?

(c) How does doubling the throat radius affect the heat transfer?

6.7 A heavy-walled steel nozzle shown in Figure 6.21 is used in our hybrid rocket research. Assuming a 15 second test firing, we like to know if the throat will survive. We estimate an average h_g of 10,000 W/m² K and for steel, we have $k = 60.5$ W/m-K and $\alpha = 17.7 \times 10^{-6}$ m^2/s. The test will be conducted at $p_c = 300$ psi with an oxidizer of 85% aqueous solution of H_2O_2 and HTPB as fuel at a mixture ratio of 7.0. Plot the temperature profile at the throat through the thickness of the part at $t = 2, 5, 10,$ and 15 seconds. Clearly state all assumptions in your analysis. Will the steel melt? (The melt temperature of steel is 2500 °F/1370 °C.)

6.8 The nozzle extension (skirt) on the second stage of the Delta launch vehicle is made of columbium and is radiation cooled (see Figure 6.23). The skirt attaches to an ablatively cooled chamber at an expansion ratio of 5.9:1. The engine operates at a chamber pressure of 105 psi

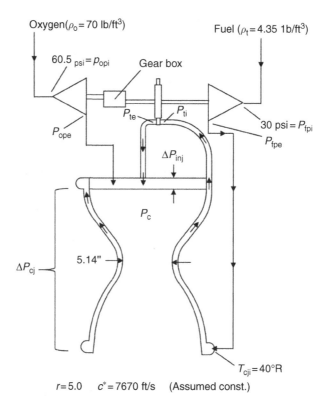

Figure 6.24 Diagram for Problem 6.9.

and has a throat diameter of 7.48 inches. The fuel is a 50/50 blend of Hydrazine and UDMH and the oxidizer is nitrogen tetroxide. These propellants are combined at a mixture ratio of 2.0.

(a) What is the highest temperature in the skirt?

(b) Assuming the engine starts instantaneously $(P_c|(t = 0) = P_c = 105 \ \text{psi})$ and that radiation heat transfer can be neglected during the heat-up of the wall, how long will it take the 200 mil thick skirt to reach 95% of the temperature calculated in part (a)? (Assume an ambient skirt temperature of 530 °R prior to ignition.)

6.9 Consider the LOX/LH$_2$ expander cycle engine shown in Figure 6.24. The oxidizer pump is geared so as to allow for the proper flow rate to insure a mixture ratio of 5.0 for all engine operating conditions. The following pressure drops are known as functions of fuel and oxidizer flow rates $(\dot{m}_f, \dot{m}_o)$:

$$\Delta p_{\text{injf}} = 93 \left(\frac{\dot{m}_f}{5.63}\right)^2 \quad \Delta p_{\text{injo}} = 197 \left(\frac{\dot{m}_o}{28.16}\right)^2$$

$$\Delta p_{\text{cj}} = 292 \left(\frac{\dot{m}_f}{5.63}\right)^2$$

where mass flows are in lb/s and Δp is in psi. The hydrogen coolant has an average C_p of 3.254 BTU/lb °R within the jacket (which includes latent heat of evaporation). For the turbine, assume a C_p of 3.5 BTU/lb °R. Finally, we estimate an average wall heat flux of:

$$\dot{q} = 4(p_c/400)^{0.8}(\text{in BTU/in}^2 \text{ S})$$

acting over a wall area of 3680 in². The overall efficiency of the turbopump system is $\eta_t\eta_p = 42\%$. You are asked to determine the behavior of the power system for various p_c values and to determine the limiting p_c for this engine.

(a) Develop an algorithm to solve for all pressures and temperatures noted in the schematic starting with an initial p_c value.

(b) Determine the maximum flow rates allowed ($\dot{m}_{f\ max}$, $\dot{m}_{o\ max}$) which will still support a power balance. What p_c corresponds to this value?

(c) Plot p_{fpe}, p_{ope} vs. p_c for $100 \leq p_c \leq p_{cmax}$. Plot turbine Δp vs. p_c for $100 \leq p_c \leq p_{cmax}$. Plot turbine power vs. p_c for $100\ 100 \leq p_c \leq p_{cmax}$.

(d) Discuss why the engine has an upper bound in p_c. What physical properties lead to this limit? Why is gearing of the LOX pump required? What would happen to the p_c limit if we increased the nozzle length?

6.10 Hydrogen peroxide/hydrocarbon engines are receiving significant attention these days due to the high density impulse and non-toxic, storable nature of the propellants. Some concepts envision the use of high-concentration hydrogen peroxide (HP) as a regen-jacket coolant. Consider the engine shown in the schematic – a small part of the HP flow enters a manifold in the exit cone at an expansion ratio of 6.0 and flows forward through a series of square channels 30 mil on each side. You can assume that the channels remain the same dimensions along the length of the jacket. At the throat, the channels are spaced 60 mils apart, i.e. there is a 30 mil "land" between the adjacent channels. The channels are to be painted blue. The majority of the HP flows proceeds through a catalytic bed and decomposes into hot steam and oxygen. RP-1 fuel is then sprayed in downstream of the catbed exhaust. The liquid exiting the jacket is injected and decomposed in the chamber downstream of the fuel injectors. You are asked to analyze and size the coolant system assuming that the HP enters the manifold at room temperature with a given mass flow and sufficient inlet pressure in order to permit a 100 psi drop across the orifices exhausting the liquid into the chamber. You may assume a chamber pressure of 1200 psi for the study.

A 1-D isentropic flow analysis was conducted to assess the flow conditions in the nozzle. Results of this analysis are included on the Purdue Propulsion website. Two datafiles will include the x, r coordinates (in inches) of the nozzle wall and a table of x, M values where M is the local Mach number computed from the Mach number/area ratio relation. The end of this file represents the axial location of the HP injection in the exit cone.

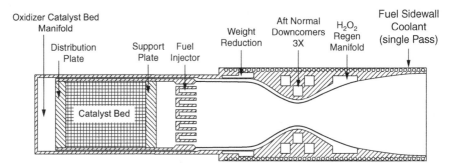

Figure 6.25 Diagram for Problem 6.10.

(a) Find the definition of density impulse, a "mil," and identify the useless information that has been provided to you. Determine the optimal mixture ratio for maximum vacuum I_{sp} assuming equilibrium flow. Compute the respective fuel and oxidizer flows at this condition. Determine the engine thrust and I_{sp} under vacuum conditions assuming ideal performance.

(b) Verify with your TA or instructor that you all have the correct answers in part (a) – you're on your own for this one as these concepts should be familiar to you at this point. If you don't have the correct answers – return to part (a).

(c) Assume that the film coolant provided from the HP entering the chamber at the top of the jacket effectively creates a mixture ratio of 10 in the near wall region. From the ODE thermochemistry calculations, add a capability to compute static pressure, static temperature, mixture viscosity, specific heat, and Prandtl number at the edge of the gas boundary layer (these can be thought of as additional columns in the table).

(d) Construct an algorithm to read in this data for a given HP flow rate, compute the heat flux to the wall using Bartz equation, and compute the local HP temperature in the channel using the approach outlined in Section 6.5.1. Use the following information for the specific heat (C_p), thermal conductivity (k) and density (ρ) of the HP in the jacket:

$C_p = C_{pref}(w_{ref}/w)^{2.8}$, where $w = 0.1745 - 0.0838T_r$, T_r is the reduced temperature $(T_r = T/T_{crit})$, $T_{crit} = 457\Delta°C$ is the critical temp, and $C_{pref} = 20.62$ cal/(g-mol K) at 25 °C, $k = 0.339$ BTU/(hr ft °F), $\mu = 1.26$ centipoise, $T_r < 0.65, \rho(T) = 14.297 - 0.0051$ T, where T is in °R and ρ is in lb/gal.

(e) Assume a jacket flow of 0.1 lb/s and compute the properties in the channels, the heat flux, and the gas and liquid side wall temperatures, T_{wg}, and T_{wl}, respectively. Plot these temperatures as well as T_g and T_l as a function of x. Also plot the heat flux and pressure in the regen channels as a function of x (on separate plots). You may assume the liquid enters the jacket at a relative pressure of 0 psi for the latter plot, i.e. the last entry (at $x = 1$) will be the pressure at which the coolant enters the injection orifices to the chamber.

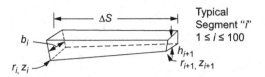

Figure 6.26 Diagram for Problem 6.11.

What pressure is required at the entry to the jacket in order to permit adequate injection into the chamber?

(f) The combustion chamber designer is asked to keep the wall temperature below 900 °F at all points in the jacket in order to meet life requirements. By varying the mass flow of coolant, find the optimal flow which produces this condition.

(g) Ideally, your cycle will balance if the coolant passage injection pressure is about the same needed by the catbed for injection of the remainder of the HP. Is this condition met with your results? What might you change in order to meet this condition?

6.11 In this problem, you will analyze the performance of a regenerative cooling system for the first stage engine of the titan launch vehicle. A schematic of the double pass system, including chamber geometry, is given in Figure 6.26. The following engine and fuel characteristics are known:

$$P_c = 827 \text{ psi}, \quad \dot{m}_f = 288 \text{ lb/sec}, \quad r = 2.0$$
$$T_c = 5971°\text{R}, \quad \gamma = 1.2, \quad \mathfrak{M} = 23 \text{ lb/mol}$$
$$Pr = 0.64 \qquad \mu = 0.183 \times 10^{-5}(T/5640)^{0.7} \text{ in lb} - \text{s/f}^2$$
$$\kappa = 0.03 \text{ lb/sec} °\text{R}$$

Fuel enters the fuel torus at a pressure and temperature of 1100 psi and 80 °F, respectively. From here the fuel enters 64 "downcomers" through a ¾-inch opening as shown in View A. At the end of the jacket, the downcomers dump into the 64 "upcomers" as shown in View B. After the end of the second pass, the fuel in the upcomers enters the injector manifold through a ¾-inch slot. We will assume the rectangular cross-section of thickness 0.025 inches for all 128 tubes in the jacket.

The Assignment – Part I

(a) Modify your code from Chapter 4, Homework Problems 4.8 and 4.12 to account for the new chamber geometry.

(b) Let S represent a coordinate aligned parallel with the chamber/nozzle wall at all points. Determine the total length along the chamber. Divide the length of the chamber into 100 segments of constant length ΔS as shown below. Determine the "b" tube dimension based on the local circumference at point "i" and assume a constant tube cross-sectional area of

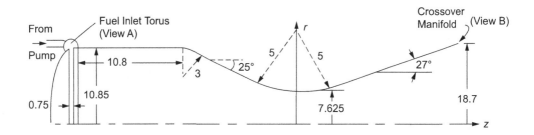

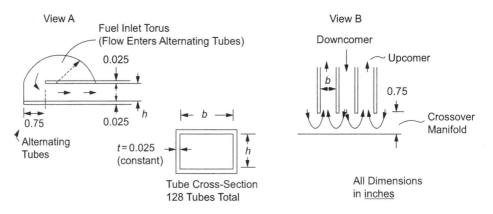

Figure 6.27 Diagram for Problem 6.11.

0.2 in^2 in order to calculate the tube height, h. Store all geometric information in arrays and print out the following table by solving the internal flow at the center of each segment.

$$\text{Segment } z \ r \ b \ h \ M \ p \ T$$

where M, p, T are the Mach number and static pressure temperature values of the chamber gases.

(c) Solve for the jacket pressure drop neglecting heat transfer from the chamber. You will need separate calculations to account for losses in each manifold. For the "upcomers" you will just need to loop backward through your arrays containing geometric information. Print out the following table:

$$\text{Segment } z \ r \ q \ \text{Re}_l \ p_l$$

Indicate which friction factor correlation (as a function of Reynolds number) was used. Assume a fuel density of 0.03 lb/in^3 and viscosity of $0.3 \times 10^4 \text{ lb/in-s}$ in your calculations.

(d) Repeat part (c) assuming a square tube cross-section, i.e. $b = h$. In this case tube area will not be constant.

The Assignment – Part II

The following liquid fuel properties were curvefit as a function of temperature for use in the analysis:

$$\rho_1 = 0.0426 - 1.82 \times 10^{-5} T_l \sim \text{lb/in}^3$$
$$C_{pl} = 0.285 + 8.38 \times 10^{-4} T_l \sim \text{Btu/lb}_m{}^{\circ}\text{R}$$
$$K_l = 5.57 \times 10^{-6} - 3.32 \times 10^{-9} T_l \sim \text{Btu/(in-s-}^{\circ}\text{R)}$$
$$\mu = 3.16 \times 10^{-4} - 7.67 \times 10^{-7} T_1 + 4.86 \times 10^{-10} T_1^2 \sim \text{lb/(in-s)}$$

These curvefits assume the liquid temperature, T_l, is in $^{\circ}$R and that T_l lies between 540 and 74 $^{\circ}$R (30 and 280 $^{\circ}$F).

(e) Using the procedure in Section 6.5.1, determine the jacket pressure drop and liquid temperature rise for the case where the tube cross-sectional area is a constant 0.2 in^2. Tabulate the following quantities for both downcomer and upcomer tubes:

$$\text{Segment } z \ r \ p \ \dot{q} \ h_g \ h_1 \ T_1$$

Plot $\dot{q}$, h_g and h_1 as a function of z for the downcomer tubes. You may neglect heat transfer in the fuel torus and crossover manifolds in your study. You may use McAdams correlation to obtain h_g and Eq. 3.15 (forced connection relation) for h_1. To account for fuel film cooling on the gas side, you will need to (rather arbitrarily) reduce the h_g values by a factor of 1000. Remember, Reynolds numbers in these relations are in terms of local hydraulic diameter! Also, take some time to insure that units work out *before* you start coding; it may save you many headaches later.

(f) Congratulations! You now have a working model of a regen jacket. Investigate the effects of tube design on the jacket pressure drop and temperature rise characteristics. For example, assume we want a temperature rise of no more than 100 °F and look at the behavior of the pressure drop for different assumed tube height variations. For example, $h = ar^b a$, b = constants to be chosen or determined. Tabulate your results as before and explain any trends you have observed.

(g) Organize both parts of assignment in a rational manner to facilitate grading. Include explanations of what you did and attach a listing of your code.

6.12 Consider the solar thermal propulsion concept shown in Figure 6.28. Here, the basic energy flux from the sun ($\dot{q}_s$) is focused by a collector mirror to a high-temperature absorber plate. The cross-sectional areas of the collector and absorber plate are A_c and A_a, respectively. The absorber must be kept at or below a temperature, T_a, to insure structural integrity of this element. The concept shown employs an injector in which cold hydrogen gas (at temperature T_i) impinges on the absorber plate. Heat is transferred to the gas convectively, presume that the heat transfer coefficient, h_g, is known. As a result of the heat imparted to the fluid, it reaches a temperature T_o at the outlet of the injector.

(a) Assuming A_c, $\dot{q}_s$, T_i, and T_a are known, derive an expression for the absorber area, A_a.

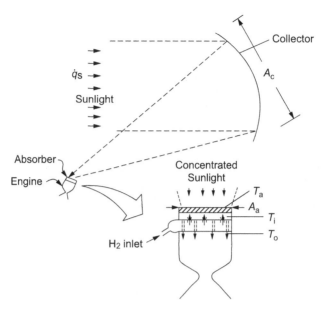

Figure 6.28 Diagram for Problem 6.12.

(b) Given your result from part (a), determine the temperature T_0 assuming the hydrogen rate, $\dot{m}$ is known.

(c) Outline how you would determine the performance (thrust and I_{sp}) of the engine assuming you choose an expansion ratio and chamber pressure.

(d) Derive an expression for the overall efficiency of the device in terms of known information. Here, the overall efficiency measures the power output relative to the power input to the device.

6.13 Thermocouple probes are placed in a regenerative cooling tube to measure wall temperature on gas and liquid sides (T_{wg} and T_{wl}) at the throat of a large liquid rocket engine. After the wall has heated up to steady conditions, the probes measure:

$$T_{wg} = 800 \text{ K}, \ T_{wl} = 300 \text{ K}$$

In addition, the following gas-phase chamber conditions are known:

$$p_c = 6.8 \text{ MPa} \quad T_c = 3800 \text{ K} \quad Pr = 0.48 \quad \gamma = 1.15$$
$$\mu_g = 0.1 \times 10^{-6} \ \text{Ns/m}^2 \quad K_g = 1.4 \times 10^{-3} \ \text{J/(cm K s)}$$

The tube wall is 0.04 cm thick and has a conductivity of 15.0 W/(mK). From this information, estimate the gas-phase heat transfer coefficient at the throat of this engine.

6.14 A student has an idea of determining an average heat transfer coefficient, $\bar{h}_g$, in a rocket engine by dropping a series of steel balls of different diameters through the engine. Neglecting any temperature gradients through the ball itself, the student is able to analyze

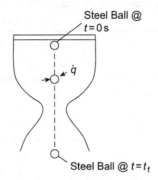

Steel Ball @
$t = 0\,$s

$\dot{q}$

Steel Ball @ $t = t_f$

Figure 6.29 Diagram for Problem 6.14.

bulk temperature changes in the steel for an assumed heat flux input of $\dot{q} = \bar{h}_g g(T_c - T)$, where T_c = chamber temperature and T = steel temperature, i.e. recovery factory of unity is assumed. The idea is to drop balls of different diameters through the engine as shown and measure the time at which they appear. Very small balls will melt and break-up, but at some critical size, the steel ball will emerge intact (presumable very near its melting temperature.

(a) Write down the differential equation governing the change in steel temperature (T) with time in terms of ball radius (R), steel density and specific heat (ρ, C_s) and $\bar{h}_g$ and T_c.

(b) Integrate the equation over the flight time, t_f, and derive a relation between initial and final ball temperatures (T_i, T_f) and physical parameters given in part (a).

(c) Suppose the student finds the critical ball size to be 1 cm at a flight time of 0.5 s. Determine $\bar{h}_g$ in this case given

Steel properties $T_i = 25°C$, $\rho = 7900$ Kg/m^3, $C_s = 477$ J/KgK
Melt temperature = 1670 K
Chamber temperature $T_c = 3000$ K
Calculate $\bar{h}_g$ for kw/m^2 K.

(d) Explain why this information wouldn't be terribly useful for inferring wall heat fluxes in the engine.

6.15 A regenerative cooling jacket has been instrumented in the chamber region of an engine in order to assess performance (see Figure 6.30). At a given instant in time, wall temperatures are reasonably constant over the length of duct shown in the figure. The wall is made of Inconel, and the coolant is hydrazine. The following properties are given:

Wall properties	Hydrazine properties
Density = 8510 kg/m^2	Density = 1020 kg/m^3
Specific heat = 439 J/(kg K)	Specific heat = 0.75 Cal/(kg K)
Thermal conductivity = 11.7 W/(m-K)	Thermal conductivity = 0.57 W/m-K

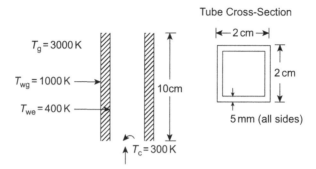

Figure 6.30 Diagram for Problem 6.15.

The heat transfer group estimates gas and liquid side heat transfer coefficients at 250 and 3000 W/m^2 K, respectively for the region shown in the sketch. Finally, the liquid temperature at the entry to the section is measured at 300 K.

(a) Has the process reached steady state? Explain your answer, i.e. are we looking at a startup or shutdown condition?

(b) If the steady-state liquid temperature in the tube really is 300 K, what wall temperatures would you expect to measure on each side of the tube in this region?

(c) At the equilibrium heat flux from part (b) how long would it take to boil the hydrazine ($T_{\text{boil}} = 386$ K) if it were left stagnant in this region of tube assuming constant heat flux over this period?

6.16 Wall temperature sensors have been mounted on surfaces of a regenerative cooling system under steady-state operation. You are asked to analyze the system taking into account back-side heat transfer. The liquid properties can readily be measured, but gas-phase properties are not well known. You may assume that the wall temperatures are constant over the length ΔX in Figure 6.31 and that the tube has a square cross-section with side length S as shown below. Finally, wall thicknesses (t_{wh}, $t_{\text{w(c)}}$) and thermal conductivities (K_{wh}, $K_{\text{w(c)}}$) differ from hot to cold sides as indicated in the figure. In terms of the variables noted below, determine:

(a) The fraction of heat flux from the chambers that actually enters the liquid.

(b) The liquid film heat transfer coefficient.

(c) The change in liquid temperature (ΔT_l) over the interval ΔX.

Liquid Channel

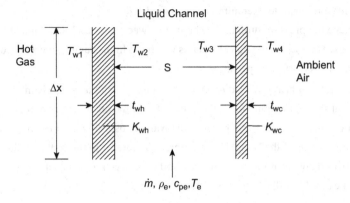

Figure 6.31 Diagram for Problem 6.16.

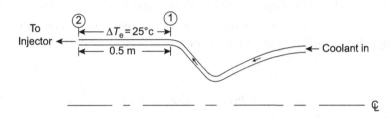

Figure 6.32 Diagram for Problem 6.17.

6.17 A regeneratively cooled liquid rocket engine plans to use film cooling in the chamber to minimize the heat flux to the wall. Tests on a heavy weight chamber reveal a gas-phase heat transfer coefficient of 10.0 W/m^2 K in this low-velocity region. The liquid temperature rise has been measured to be 25 °C under a mass flow of 2 kg/s (per tube). Details regarding liquid, wall, and gas properties (assumed to be constant) are tabulated below and Figure 6.32 shows the chamber and tube geometry for the test.

Liquid (RP-1) properties	Gas (chamber) properties
$\rho = 700 \text{ Kg/m}^3$	$T_c = 2700 \text{ K}$
$C_p = 1.88 \text{ j/Kg K}$	$\dot{m} = 200 \text{ Kg/s}$
$K = 0.133 \text{ w/m K}$	$\text{Pr} = 0.6$
$\mu = 0.55 \text{ centipoise}$	$\rho_c = 1.3 \text{ Kg/m}^3$

Wall properties (stainless steel)

$K = 40 \text{ w/m K}$
$C_p = 434 \text{ J/Kg K}$
$\rho = 8130 \text{ Kg/m}^3$

Tube cross-section

(a) Calculate the wall temperature on the gas-side, T_{wg}.

(b) Estimate the pressure drop in the channel between stations 1 and 2 noted below.

(c) Suppose the T_{wg} value you calculate is much too high. Suggest a way to lower T_{wg} with minimum performance penalties.

6.18 When we test five heavy-walled copper chambers, we tend to get a chamber pressure trend such as that shown in Figure 6.33. We believe the initial chamber pressure, p_{ci}, is lower than the final chamber pressure, p_{cf}, because the wall is absorbing a lot of energy as it heats up during early stages of the firing. If we assume heat transfer to the wall only during this initial heat-up period and an adiabatic wall afterward we can estimate the ratio p_{cf}/p_{ci}. Assume that the heat flex to the wall during the initial phase is:

$$\dot{q} = -h_g(T_{ci}-T_w)$$

where h_g is the film coefficient, T_{ci} the initial chamber flame temperature, and T_w is the average wall temperature during the heat-up phase. Assuming constant properties, the enthalpy changes in the combustion process can be written:

$$H_2-H_1 = \overline{C}_p(T_C-T_B) + \Delta H_B$$

where $\overline{C}_p$ is the average specific heat, ΔH_B is the heat of combustion, and $T_c = T_{ci}$ or T_{cx} depending on which phase of the process is to be analyzed. Further, assume the chamber mass flow, $\dot{m}$, is constant and the effective wall area being heated, A_w, is known. Given this information, derive an expression for the pressure ratio p_{cf}/p_{ci}. [Hint: Remember that c^* will change with chamber temperature.]

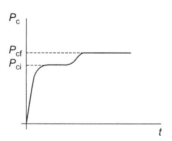

Figure 6.33 Diagram for Problem 6.18.

7   SOLID ROCKET MOTORS

7.1 INTRODUCTION

One of the more common rocket propulsion systems in use today is the solid rocket motor, or SRM. Solid rocket motors have a large number of historical applications, including:

- strap-on boosters or stages for space launch vehicles;
- upper-stage propulsion systems for orbital transfer vehicles (OTVs);
- numerous types of missile propulsion systems; and
- gas generators for starting liquid engines, pressurizing tanks, or for use in attitude control.

In fact, even the technology employed in developing early versions of airbags for automobiles was essentially derived from SRM fundamentals. Advances in launch vehicle technology over the past two or three decades has greatly reduced the need for solid-based OTVs but solid propulsion is still used on a number of deep space missions to provide for escape from Earth's gravity field.

The main advantages of an SRM include its simplicity (very few moving parts), ability to be used in a quick-response situation (no need to fill tanks with liquid), and the ability to deliver very high thrust levels without significant design compromises. The SRM offers the maximum volumetric impulse (impulse per unit volume) of any chemical propulsion system, and for that reason it is attractive for volume-limited applications. While these advantages typically translate to high mass fractions when compared to hybrid or liquid propulsion alternatives, the specific impulse of solid propellant systems tends to be lower than these other chemical systems. While throttling has been demonstrated in SRMs by using a movable pintle that adjusts throat area, in general the SRM is more difficult to throttle than LRE or HRE systems. The nature of the solid propellant combustion is such that it is impractical to test an SRM for adequate combustion prior to usage and for this reason solid motor manufacturers rely on strict manufacturing and handling protocols to ensure reliability. Finally, because both propellants are intimately mixed in an SRM, handling procedures must reflect this fact and tend to complicate logistics a bit when compared to LRE or HRE systems. Despite these disadvantages, the SRM still has a firm place in quick-response and volume limited approaches as well as applications where loading propellant tanks is not practical.

In this chapter, we will describe the major components of an SRM and discuss design/analysis techniques for predicting the performance of these devices. Section 7.2 presents ballistic analyses aimed at predicting the motor chamber pressure history and other aspects of performance are addressed in Section 7.3. Component technologies are presented in Section 7.4 and Section 7.5 provides a discussion of solid propellants. Finally, thrust vector control alternatives are discussed in Section 7.6 .

7.2 SRM Internal Ballistics

The field of modeling or computing the chamber pressure and thrust history for a given SRM design is referred to as internal or *interior ballistics*. The ballistician has the responsibility of assessing the propellant burn rate, the exposed propellant grain surface area, the nozzle throat and exit areas, as well as the amount of energy in the propellant (c^*, T_c, $\mathfrak{M}$) in order to make a performance prediction or reconstruction. The propellant burning rate, r_b, has been determined both theoretically and experimentally to be a function of the motor chamber pressure. In many cases, we may write:

$$r_b = a p_c^n \tag{7.1}$$

where a is called the *burning rate coefficient* and n is called the *burning rate exponent*. Equation 7.1 is generally referred to as St. Robert's law, even though it is essentially a curvefit of experimental data. Typical solid propellant burn rates lie in the range 0.05–2 in/s (0.1–5 cm/s), with most lying in the range 0.25–0.5 in/s (0.6–1.2 cm/s). The constants a and n are determined experimentally as the intercept and slope of a line on a plot of $\ln r_b$ vs. $\ln p_c$, i.e. if we take the log of Eq. 7.1 we get an equation for a line $\ln r_b = \ln a + n \ln p_c$. Students need to be aware of the strange units that result for the burning rate coefficient: in English/Imperial units a appears in in/s/psi^n while in SI units it is reported in cm/s/MPa^n.

When port velocities are high, the propellant burning rate can be augmented substantially above that predicted by St. Robert's law. This response is typically referred to as *erosive burning*, although the community still debates if physical erosion is at play in these instances. While higher burning rates might be advantageous in some applications, the reproducibility of the effects has been a concern for exploitation of these physics in actual designs. Real motors likely do undergo erosive effects early in the firing where the port cross-sectional area is a minimum, but as the propellant recedes from the center of the motor, port velocities generally drop. While it is a bit beyond the scope of this text, numerous references are included at the end of this chapter for those interested in further reading on the topic.

If we presume that the burning rate is known, we should be able to determine the rate of mass entering the chamber at a given instant. Recalling that for steady flow, $\dot{m} = \rho v A$, we may write:

$$\dot{m}_{in} = \rho_p\, r_b\, A_b \tag{7.2}$$

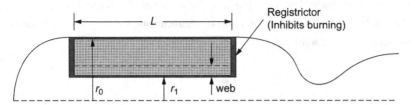

Figure 7.1 Sample grain geometry for a typical solid rocket motor.

where ρ_p is the propellant density, the velocity corresponds to the burn rate, r_b, and A_b is called the propellant *burning surface area*. From this expression, we can see that the amount of mass entering the chamber is directly proportional to the burn surface area currently exposed to the flame and the instantaneous burning rate.

Consider the simple tubular grain shown in Figure 7.1. The burning surface is simply the surface area of a cylinder with changing radius. The propellant is assumed to regress in a direction that is perpendicular to the local surface, so in this simple example the motion is entirely in the radial direction. The *web distance*, w, measures the extent to which the surface has moved relative to the original surface location. As the propellant burns, the web distance increases, and the instantaneous burning surface depends solely on this parameter. From the definition of burning rate, we can write:

$$dw/dt = r_b \text{ or } w = \int_0^t r_b \ dt \qquad (7.3)$$

so that the instantaneous web distance is simply the integral of the burning rate history.

In our simple example in Figure 7.1, the ends of the cylinder are assumed to be restricted from burning by bonding a rubber *restrictor* or *inhibitor* on these faces. For the geometry shown, we can see that the initial surface area is simply $2\pi r_i l$ while the surface area just prior to motor extinction is $2\pi r_o l$. Since $r_o > r_i$ we see that the burn surface in this case would increase with time. Motors of this type are called *progressive burning* motors. If the burn surface is reasonably constant with time the motor is said to be *neutral burning*, while a motor with a burn surface that decreases with time is said to be *regressive burning*.

Since we will see that both chamber pressure and thrust follow the burn surface trends, by designing the proper grain geometry one can obtain the desired thrust history for the motor. Generally, we desire either neutral or regressive thrust histories for real SRMs since a progressive history would give the highest thrust when the vehicle is lightest, which generally leads to structural problems. A neutral pressure trace will provide the minimum inert weight design since structural components must be designed to withstand maximum pressure loads and the maximum and average pressures are very nearly the same for a neutral trace.

Let us assume that we know the geometry of the burning surface so that A_b and $\dot{m}_{in}$ can both be determined. We know from our definition of characteristic velocity (c^*) that the mass flow exiting the nozzle may be written:

$$\dot{m}_{out} = \frac{g\, p_c A_t}{c^*} \tag{7.4}$$

7.2.1 Quasi-Steady Lumped Parameter Method

The internal ballistics of the motor are simply a reflection of the instantaneous mass balance, i.e. the extent to which Eqs. 7.2 and 7.4 agree. In general, there can be non-negligible changes in the chamber pressure as we progress through different portions of the burning propellant grain. In this case, the burning rate in Eq. 7.2 would be spatially dependent and we would need to evaluate the local inflow contributions and add them all up to get the global mass flow entering the grain. We will discuss this case in Section 7.2.2.

However, many motors are designed such that there are minimal changes in the chamber pressure as one traverses the grain. In this case, we can characterize the pressure in the chamber by a single value. Ballistic calculations performed under this assumption are referred to as *lumped parameter* methods, and these techniques are the simplest approaches to compute the chamber pressure and thrust histories for a given configuration. We introduced this topic in Chapter 6, but there are a wide range of applications for a lumped parameter approach across the propulsion field. In this case, the lumped parameter assumption implies that the inflow in Eq. 7.2 becomes a function of just a single pressure. The outflow in Eq. 7.4 is always a function of a single pressure as this parameter is dependent on the stagnation pressure feeding the nozzle, i.e. $\dot{m}_{out}$ is a function of the pressure entering the nozzle.

If one makes a further assumption that the mass flows are slowly varying (since the burning surface is changing shape slowly over the course of the firing), then we arrive at the simplest ballistics scheme, the *quasi-steady lumped parameter* approach. In this case, we can equate the mass flow entering the chamber with the mass flow leaving ($\dot{m}_{in} = \dot{m}_{out}$) and using St. Robert's law (Eq. 7.1) we can solve for the instantaneous pressure:

$$p_c = \left[\frac{a\rho_p A_b c^*}{gA_t}\right]^{1/(1-n)} \tag{7.5}$$

Equation 7.5 gives us a lot of information as to how solid rocket motors work. We can see that as we increase the burning rate coefficient, propellant density, or burning surface area, or decrease the throat area, the pressure will tend to increase. For a given propellant formulation (a, ρ_p, c^*, n all constant) the chamber pressure history will mimic the burn surface profile, i.e. $p_c \sim A_b^{1/(1-n)}$.

Equation 7.5 also indicates the importance of the burning rate exponent, n, in determining the behavior of a given propellant formulation. Consider a plot of the individual mass flows $\dot{m}_{in}, \dot{m}_{out}$ as a function of pressure as shown in Figure 7.2. In Figure 7.2(a), we have assumed

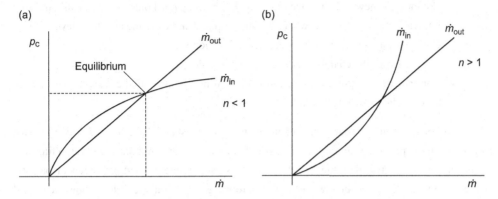

Figure 7.2 Equilibrium operating point for $n < 1$ (a) and $n > 1$ (b).

$n < 1$ as is the case with solid propellants in use today. The equilibrium point is where the two mass flows are equal. If there is a mass flow excursion, for whatever reason, we can see that the situation in the left graph is stable. For example, if we have a perturbation to a mass flow above the equilibrium value, $\dot{m}_{out}$ exceeds $\dot{m}_{in}$ in this region and the pressure will fall toward the equilibrium point. If the mass flow drops below the equilibrium value, then $\dot{m}_{in}$ exceeds $\dot{m}_{out}$ thereby causing the pressure to climb once again toward the equilibrium point.

Figure 7.2(b) shows a case where the burning rate exponent is greater than unity. In this case a perturbation in flow above the equilibrium point leads to a situation where $\dot{m}_{in}$ exceeds $\dot{m}_{out}$, which causes the pressure to climb even more without bound in a divergent manner. Formulations with this unstable behavior are called *explosives* for obvious reasons. As a burning rate exponent gets larger and approaches unity, the overall stability of the motor becomes less and less tenuous. We will see more on this issue as we discuss dynamic treatments in the following section.

Given a burning surface vs. web distance (A_b vs. w) profile, propellant properties (a, ρ_p, c^*, n), and nozzle design (A_t, ε), it is very straightforward to construct a quasi-steady lumped parameter ballistics code. In general, one can construct the analysis such that one takes incremental steps in either web distance or time. Stepping in time allows for a more general treatment for any input A_b vs. w behavior; generally it is sufficient to use 100–200 time-steps to provide adequate accuracy, but the user should determine this by assessing the sensitivity to time-step in the particular problem at hand. The following steps would be taken in the process:

1. Input grain and nozzle geometry information, propellant characteristics, and time-step, Δt. Set initial time and web distances equal to zero, $w_1 = t_1 = 0$, where the subscripts indicate array position.
2. Calculate/look up initial burn surface area, A_{b1} at $w_1 = 0$.
3. Compute the initial chamber pressure (p_{c1}) using Eq. 7.5.
4. Begin time-stepping: $i = i + 1$; $t_{i+1} = t_i + \Delta t$.
5. Evaluate the current burning rate $r_b = a p_c^n$ (or equivalent) using the current chamber pressure.

6. Compute the new web distance assuming this burning rate holds over the small time-step interval, $w_{i+1} = w_i + r_b \Delta t$ where the subscript $i+1$ indicates the new time level.
7. Compute/look up the new burn surface at location w_i.
8. Compute the new chamber pressure (p_{ci}) using Eq. 7.5.
9. Continue time-stepping until the propellant burns out, i.e. until the current web distance exceeds the total web available.

Although the lumped parameter technique is the simplest of the alternatives available to the ballistician, it can provide a suitably accurate solution for a number of motors. However, if transient effects become important then higher-fidelity models such as the lumped parameter or ballistic element methods discussed below are more appropriate. The example below shows a typical lumped parameter calculation at motor ignition time.

Example: Quasi-steady Lumped Parameter Ballistics Calculation

Given:
$$r = 0.5\, P_c^{0.35} \quad P_c \sim Mpa \quad r \sim cm/s$$
$$\rho_p = 1700 \text{ kg/m}^2$$
$$c^* = 1500 \text{ m/s}$$
$$P_c = \left[\frac{a P_p A_b c^*}{A_t} \right]^{\frac{1}{1-n}}$$
$$A_b = 2\pi (10^2 - 2^2) + 2\pi (2)\, 50 = 392\pi$$
$$A_t = \pi 1.5^2 = 2.25\pi$$
$$A_b / A_t = 174$$

$$t=0 \left\{ \begin{array}{l} P_{c_i} = \left[6.5\ 1700\ 174\ 1500 \ \ \underbrace{\frac{cm}{sMPa^{0.35}} \frac{kg}{m^3} \frac{m}{s} \frac{m}{100cm} \frac{MPa}{10^6\,pa}} \right]^{\frac{1}{1-n}} \\[1em] \qquad\qquad\qquad\qquad\qquad\quad \frac{P_a}{MPa^{0.35}} \overline{100} \\[0.5em] = 3.4 MP_a \\[1em] F_{v_i} = 1.854\ 3.4\text{x}10^6 \ \ \frac{N}{m^2}\ 2.25\pi \text{ cm}^2\ \frac{m^2}{10{,}000cm^2} \ = 4.46\text{x}10^3\ N = 4.46 \text{ kN} \end{array} \right.$$

$$t=t_b \left\{ \begin{array}{l} A_b = \pi\,(20)\,34 = 680\pi \qquad P_c \propto A_b^{\frac{1}{1-n}} \qquad P_{c_f} = 3.4 \left(\frac{680}{392} \right)^{\frac{1}{0.65}} = 7.93\ MPa \\[1em] F_{v_f} = F_{v_i}\ \frac{P_{c_f}}{P_{c_i}} = 4.46\ \frac{7.9}{3.4} = 10.3\ kN \end{array} \right.$$

7.2.2 Unsteady Lumped Parameter Method

The conservation of mass states that the rate of change of mass of gas (within the chamber) is equal to the difference between the mass entering the chamber and the mass leaving through the throat:

$$\frac{dm_c}{dt} = r_b \rho_p A_b - g p_c A_t / c^* \tag{7.6}$$

where we can express the mass in the chamber (m_c) as:

$$m_c = \frac{p_c V_c \mathfrak{M}}{R_u T_c} \tag{7.7}$$

To find the rate of change of m_c, we just need to differentiate Eq. 7.7. Before we do this, let us investigate which terms in this differential might be important (i.e. in general, p_c, T_c, $\mathfrak{M}$, and V_c can vary with time). What we are trying to do is obtain an equation for changes in chamber pressure in terms of grain (and throat) geometry, and propellant properties. In Chapter 5, we showed that propellant properties (T_c, $\mathfrak{M}$) are quite insensitive to changes in pressure under equilibrium conditions. In this case, pressure is also varying, so it is not obvious that variations in these parameters are negligible. Heister and Landsbaum (1985) showed that we can safely neglect variations of temperature and molecular weight unless a very large pressure disturbance is present.

Therefore, differentiating Eq. 7.7 we obtain

$$\frac{dm_c}{dt} = \frac{m_c}{p_c}\frac{dp_c}{dt} + \frac{m_c}{V_c}\frac{dV_c}{dt} \tag{7.8}$$

but since $dV_c = r_b A_b$, we can combine Eqs. 7.6 and 7.8 to give:

$$\frac{m_c}{p_c}\frac{dp_c}{dt} = r_b \rho_p A_b - g p_c A_t / c^* - m_c r_b A_b \tag{7.9}$$

Finally, by substituting for m_c on the right-hand side of this relation we can write:

$$\frac{m_c}{p_c}\frac{dp_c}{dt} = r_b A_b(\rho_p - \rho_c)p_c^n - g p_c A_t / c^* \tag{7.10}$$

where ρ_c is the density of the gases in the chamber. Typically, $\rho_p/\rho_c \approx 1000$ so that the ρ_c term is usually neglected. Under this assumption, the dimensionless form of Eq. 7.10 can be written:

$$\tau \frac{d\hat{p}}{dt} = k_b \hat{p}^n - k_t \hat{p} \tag{7.11}$$

where $k_b = A_b/A_{bo}$ and $k_t = A_t/A_{to}$ represent dimensionless burn surface and throat areas, respectively, and $\hat{p} = p_c/p_{co}$. The subscript $()_o$ in these definitions refers to the initial undisturbed state before the transient process begins. The quantity τ in Eq. 7.11 is a very important quantity in problems of this type; it is referred to as the *motor time constant*. Physically, τ represents the time to exhaust the current mass of gas in the combustion chamber at the equilibrium flow rate:

$$\tau = \frac{\rho_{\mathrm{g}} V}{\dot{m}_{\mathrm{o}}} \tag{7.12}$$

where, $\dot{m}_{\mathrm{o}} = \frac{g p_{\mathrm{co}} A_{\mathrm{to}}}{c*} = a p_{\mathrm{co}}^n \rho_{\mathrm{p}} A_{\mathrm{bo}}$ is the equilibrium flow rate, ρ_{g} is the gas density (based on equilibrium pressure), and V is the chamber volume at the start of the disturbance. The motor time constant is reflective of the time scale over which the combustion chamber will respond so processes that take a long time relative to τ can be thought of as quasi-steady, while those occurring on a time scale near τ are inherently unsteady. Typical values for τ lie in the range 0.01–0.5 s; in general, τ tends to increase over the duration of the firing as the gas volume in the chamber (and hence gas mass in chamber) increases with time. See Humble (1995: Ch. 6) for more information and solutions to Eq. 7.11.

7.3 Specific Impulse, Mass flow, and Thrust Predictions

Engineers make use of test data to determine the efficiency of the motor in terms of the ideal analyses we discussed in Chapters 4 and 5. Since the mass of propellant loaded into a test motor is carefully measured, we can introduce a $c*$ efficiency to give the proper expended mass:

$$\eta_{c*} = \frac{1}{m_{\mathrm{p}}} \int_0^{t_{\mathrm{b}}} \frac{g p_{\mathrm{c}} A_{\mathrm{t}}}{c_{\mathrm{th}}^*} \mathrm{d}t \tag{7.13}$$

where c_{th}^* is the theoretical value calculated from one-dimensional equilibrium (ODE) codes as discussed in Chapter 5. Of course the integration in Eq. 7.13 is for the actual pressure history measured in the test (with an appropriate prediction for the throat area as a function of time).

Typical η_{c*} values lie between 0.96 and 1.0, which indicates a very high combustion efficiency in solid motors. The actual, or *delivered* $c*$ is then simply:

$$c^* = \eta_{c*} c_{\mathrm{th}}^* \tag{7.14}$$

which could be used in predicting the pressure for future motors.

The average delivered vacuum I_{sp} can be determined from the test data by dividing the total impulse by the loaded propellant mass:

$$I_{\mathrm{spv}} = \frac{1}{m_{\mathrm{p}}} \int_0^{t_{\mathrm{b}}} (F + p_{\mathrm{a}} A_{\mathrm{e}}) \mathrm{d}t \tag{7.15}$$

Based on this measurement, the overall efficiency of the motor can be defined as:

$$\eta_o = \frac{I_{\mathrm{spv}}}{I_{\mathrm{spvth}}} \tag{7.16}$$

where I_{spvth} is the vacuum theoretical I_{sp} calculated using an ODE code. Typical η_o values lie in the range 80–87% for small motors and 88–96% for large motors.

In Chapter 6, we introduced readers to US industry standard codes Two Dimensional Kinetics (TDK) and Solid Motor Performance Prediction (SPP). These tools are used by the US industry to perform more detailed design and analysis of liquid and solid rockets, respectively. In the case of the SRM, analysts like to be able to determine the composition of η_o in order to refine predictions and to be able to make accurate forecasts for new motors. For this reason, η_o is normally broken down into the following components:

$$\eta_o = \eta_{kin}\eta_{2D2P}\eta_{bl}\eta_{eros} \tag{7.17}$$

where these four efficiencies address losses (when compared to ODE values) attributed to kinetics, 2-D and two-phase effects, boundary layer drag, and throat erosion. The SPP code has modules to predict the magnitude of each of these terms. The kinetics efficiency, η_{kin}, estimates the effects of incomplete combustion due to lack of residence time. In most SRMs this term is very near unity as the flame temperatures are so high that chemical kinetics rates are very large. The two-dimensional, two-phase flow losses are accounted for in the η_{2D2P} term. Two-phase flow losses are the dominant loss mechanism in motors utilizing metalized propellants, and we touched on the physics here in Section 4.7. Boundary layer losses are measured in the η_{bl} term using an integral method boundary layer treatment described briefly in Chapter 6 to estimate viscous drag losses in the fluid. This loss mechanism is not very important in larger SRMs, but can become significant as motor size shrinks since Reynolds numbers are also getting smaller in this case. The η_{eros} term measures losses due to throat erosion. Because SRMs employ ablative throat materials, there is some surface regression during the firing such that the nozzle expansion area ratio tends to reduce in time. Eroded parts also tend to have higher surface roughness and this can also feed into this loss mechanism.

Using the efficiencies developed above, the thrust of the motor is simply:

$$F = \eta_o I_{spth}\dot{m} \tag{7.18}$$

where the predicted mass flow contains c^* efficiency as defined in Eq. 7.14. Chamber pressure predictions for new motors will include this factor as well as evidenced by the c^* dependence in Eq. 7.14. By predicting chamber pressure using current burn rate data for the motor under consideration, mass flow and thrust follow immediately using the notions discussed in this subsection. As pointed out in Chapter 4, propulsion engineers typically report vacuum thrust and I_{sp} since these conditions can readily be corrected to an arbitrary ambient pressure.

7.4 SOLID ROCKET MOTOR COMPONENTS

Figure 7.3 shows a typical SRM with its major components highlighted. The pressure vessel enclosing the propellant charge is referred to as the *motor case*. The motor case is generally bolted to the nozzle assembly through a fitting referred to as a *polar boss*. Most motor cases contain polar bosses at both ends although Figure 7.3 only indicates the aft boss. The motor case generally has a cylindrical extension called the *thrust skirt*, which enables the motor to be attached to the upper

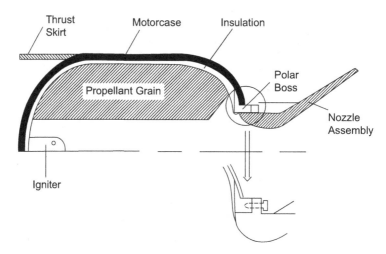

Figure 7.3 Components in a typical solid rocket motor.

stage or payload of the rocket. Motors may contain either zero, one, or two skirts (forward and aft) depending on the application.

Burning within the chamber is initiated by the *igniter*, while *internal insulation* shields the motor case from hot propellant gases. The *nozzle assembly* converts the thermal energy of propellant gases to thrust. Nozzles can be quite complex and contain numerous parts depending on whether or not they have the capability to gimbal in order to provide thrust vector control. The mass of propellant is generally called a *propellant grain* to highlight the fact that it is cast into a mold of specified geometry. The term propellant grain actually has been carried along from the gun ballistics field in which the amount of propellant (gun powder) is usually specified in grains. Let us now discuss in more detail each of these components in a typical SRM.

7.4.1 Motor Cases

The motor case acts as a pressure vessel containing the high-pressure combustion processes occurring within the bore of the motor. Cases may be constructed from metal materials, from composite fiber materials or from a combination of a metal "liner" with a composite overwrap. Table 7.1 provides a summary of some of the critical properties of various candidate materials, where the variables ρ, F_{tu}, and E represent the density, ultimate tensile strength, and Young's modulus for the material.

Many lower-cost SRMs will use metal designs for simplicity especially if the motor is quite small or has to be reusable (such as the Space Shuttle SRB). Many small motors used in air-launched missiles employ steel cases using either the 4130 alloy or the higher strength D6AC alloy. Small and mid-size space motors have frequently employed titanium cases for motors designed in

Table 7.1 SRM motor case materials

Material	ρ,lb/in^3	F_{tu}, ksi [b]	E, Msi [b]	F_{tu}/ρ,min
Aluminum	0.101	82.0	10.0	0.812
Titanium	0.161	178.0	15.0	1.10
D6AC steel	0.283	220.0	29.0	0.777
4130 steel	0.283	125.0	29.0	0.442
Graphite [a]	0.056	340.0	10.8	6.07
Kevlar [a]	0.050	190.0	10.0	3.80
Fiberglass [a]	0.072	167.0	xx.x	2.32

[a] Composite fiber materials using an epoxy resin matrix for structural stability.
[b] ksi = kilopounds per square inch, Msi = megapounds per square inch.

the 1960s and 1970s. Aluminum is rarely an attractive choice for a case material due to its low strength, and it has been included in Table 7.1 mostly as a basis of comparison.

As highlighted in Table 7.1, composite materials are very attractive case materials due to their high strength-to-weight ratio. Motor cases designed with these materials can be 20–30% lighter than conventional metal cases. Some of the weight savings is lost due to the fact that these materials are not as stiff (i.e. they have lower modulus values) than titanium or steel, so that large deflections are possible at high strain levels. For example, the composite motor case, which was tested for the Space Shuttle SRB, actually grew more than three inches under pressurization. Oftentimes, stiffness concerns of this type override strength considerations thus increasing the weight of the composite case above its minimum value to retain internal pressure alone. Designs of this type are said to be *stiffness critical* rather than *strength critical*, considerations that nearly always prevail in metal designs.

The motor case must be able to withstand the maximum chamber pressure incurred under any possible operating condition for the SRM. Generally, this condition occurs at the upper limit of the SRM operating range and includes worst-case grain geometry, propellant burn rate, and nozzle geometry allowed by the motor specifications. The pressure calculated via this process is called the *maximum expected operating pressure*, or MEOP, for the motor. The case will be designed to burst at a pressure p_b equal to:

$$p_b = f_s \text{ MEOP} \tag{7.19}$$

where f_s is the *factor of safety* applied to account for case material variations, possible inaccuracies in the MEOP calculation, and overall safety margin for the design. Historically, safety factors are generally 1.25 for unmanned systems, 1.4 for manned systems, and as high as 2.0 for small pressurization bottles or gas generators.

Using these ideas, the case thickness can be calculated for a given case material, burst pressure and case size. For a cylindrical case, thin-walled pressure vessel theory gives:

$$t_{cs} = p_b r_{cs} / F_{tu} \qquad (7.20)$$

where t_{cs} and r_{cs} are the case thickness and radius, respectively. This thickness can be used to estimate case weight provided that the additional geometry (i.e. case length and dome geometry) is known.

Oftentimes, the parameter pV/W is used in the preliminary design of motor cases. In this expression, p refers to burst pressure, V the enclosed case volume, and W the actual weight of the case. The parameter is a fundamental measure of case performance since it measures the mass of material required to maintain the total structural load (pV). Typical values lie between 300,000 and 600,000 for metal cases and from 900,000 to nearly 2 million inches for composite cases. Knowledge of this parameter can permit rapid estimation of case weight.

The case volume is of course a function of the propellant volume required to accomplish the mission. Oftentimes, preliminary estimates can be obtained by introducing a parameter called the *volumetric loading efficiency*, η_v. This parameter measures the percentage of enclosed case volume utilized by propellant:

$$\eta_v = V_p / V_{cs} \qquad (7.21)$$

where V_p and V_{cs} represent propellant and case volumes respectively. Typical η_v values lie in the range $0.6 < \eta_v < 0.92$. Of course, we would like η_v to be as close to unity as possible, but grain design and thrust history requirements usually dictate the actual value obtained in practice.

Thrust Skirts and Polar Bosses
The critical design factor influencing the thickness and weight of thrust skirts is the compressive stress induced by thrust loads, which must be transmitted through this member. Since composite materials have lower strength in compression, skirts (and the region where the skirt is joined to the case) are usually thicker than the case material itself.

Polar bosses are designed to accommodate *blowout loads* from the internal chamber pressure and any torques provide by the nozzle during vectoring operations. Note that at the aft boss location, thrust loads through the nozzle provide a relieving force since the tendency is to have the boss pushed outward by the chamber pressure. Bosses are nearly always made of metal (usually aluminum) and are generally minor in weight as compared to the case itself (roughly 10–20%) on many designs.

7.4.2 Igniters

Igniters can be divided into two general categories. In *pyrogen igniters* the igniter itself operates as a small SRM in order to provide hot gases and high pressures for ignition of the main propellant

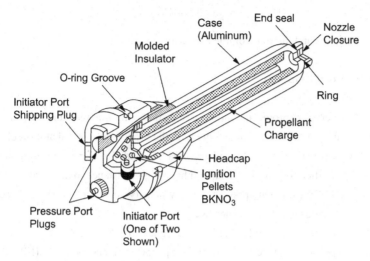

Figure 7.4 Standard pyrogen igniter used on Star 20 space motor.

grain. This type of igniter is used most frequently in larger SRMs that require a substantial amount of hot gas to ensure ignition. One typical example of a pyrogen igniter is shown in Figure 7.4. An electrical signal initiates a small charge, which ignites the boron potassium nitrate pellets at the head end of the igniter. Gases from the $BKNO_3$ charge initiate combustion in the main igniter grain.

Smaller SRMs may utilize *pyrotechnic igniters* to start combustion. In the pyrotechnic design, just the pellets are present. Ignition of the pellets via electrical signal begins the combustion process.

Designing an igniter is similar to designing a small rocket motor. The igniter case (or housing) must be able to withstand the predicted internal pressure while the igniter is burning. In addition, igniters must be well insulated since they normally must withstand the chamber environment during the entire SRM firing. Consumable igniters have been developed to alleviate this requirement and improve motor mass fractions slightly.

The design (and weight) of the igniter is often neglected in preliminary designs since the igniter mass is typically a few percent of the total inert mass. The propellant charge in pyrogen designs is usually counted in the total useful propellant load since it does eventually exit the nozzle and contribute to thrust. One method of estimating nozzle weight is based on the free volume within the main chamber at ignition. We can estimate this quantity based on the volumetric loading efficiency:

$$V_{\text{free}} = V_{\text{p}}\left(\frac{1}{\eta_{\text{v}}} - 1\right) \tag{7.22}$$

where V_{free} is the free volume and V_{p} is the volume of propellant in the main chamber. A simple correlation for igniter weights developed in US industry using existing devices relates this parameter to the free volume that must be pressurized by igniter gases:

$$m_{ig} = 0.0664 V_{free}^{0.571} \qquad\qquad (7.23)$$

where m_{ig} is in pounds and the free volume is in cubic inches.

7.4.3 Internal Insulation

Internal insulation serves to protect the motor case from the high-temperature environment within the chamber. The insulator must be compatible with the propellant selected and should have good strain capability to ensure that the propellant remains bonded under case deformations due to pressurization. In addition, the insulation should have high resistance to heat in order to minimize the thickness required for a given application. Low density is also a desirable characteristic for insulation since this tends to minimize inert weight.

These requirements generally lead the designer to select rubber-based materials with fibrous fillers to improve resistance to heat. Ethylene propylene diene monomer (EPDM) or natural butadiene rubber (NBR) are popular rubber materials used in many SRM applications. These rubbers are normally mixed (loaded) with fibrous materials such as asbestos, silica, or Kevlar pulp.

Most insulations react to the heat of the chamber through a sublimation process referred to as *ablation*. Rubber material exposed to the flame decomposes to form *pyrolysis gases* leaving a *char layer* composed mostly of carbon near the surface. The filler materials are instrumental in increasing the strength of the char, which is normally quite brittle and easily removed. As the char layer builds up it provides excellent insulation for virgin rubber lying behind it due to its very low thermal conductivity. As a result of this situation, superior insulators are normally those whose filler material does a good job of retaining the char near the surface under stress and vibration conditions of the firing. The process of char loss is highly variable, which normally leads to reasonably large variations in ablated depth for a given controlled test condition. This result must be incorporated into the design, normally through a safety factor on the design thickness.

Design of the insulation thickness required at any location within the motor is a largely empirical process. The required thickness is a function of the *exposure time*, which is the amount of time a particular place on the insulator is exposed to the chamber environment. (Remember, the propellant itself will shield the insulator in many locations for portions of the burn.) In addition, the thickness is proportional to the *ablation or erosion rate*, $\dot{e}$, and a safety factor (f_s):

$$t_i = t_{exp}\dot{e}f_s \qquad\qquad (7.24)$$

where t_i is the insulation thickness.

There are numerous ways to define the safety factor in the design of the insulator. Using a factor of twice the average char depth at a given location (as measured from test data) or 1.25 times the maximum depth at a given axial station has commonly been employed. Oftentimes, initial tests of a new design will employ large amounts of insulation to ensure thermal protection. Use of ablation data from these tests will enable the designer to generate a *flight weight* insulation design to be validated on later tests.

Oftentimes, subscale ablation test data will be available for the designer to utilize in developing the full-scale design. To scale these data to actual motor conditions, we assume that the erosion rate is proportional to the convective heat transfer coefficient, h_g. In Chapter 6, we presented a typical correlation of h_g:

$$h_g = \frac{\kappa}{D} \text{Re}^{0.8} \text{Pr}^{0.4} \tag{7.25}$$

where κ, D, Re, and Pr are the thermal conductivity, local internal diameter, Reynolds number (based on D), and Prandtl number of the gas flow, respectively.

Let us presume that we have conducted a subscale test (subscript s) and measured the erosion rate $\dot{e}_s$. If the test uses the same propellant as the full-scale article, then the thermal conductivity and Prandtl number are matched exactly. Presuming that the erosion is directly proportional to h_g as given above produces the following scaling relation:

$$\dot{e} = \dot{e}_s (\text{Re}/\text{Re}_s)^{0.8} (D_s/D) \tag{7.26}$$

Now we can write the Reynolds number as:

$$\text{Re} = \frac{\rho V D}{\mu} = pMD\sqrt{\frac{\gamma}{RT\mu^2}} \tag{7.27}$$

The quantities under the square root in Eq. 7.27 are gas-phase properties that are dependent on propellant type alone. If the subscale test uses the same propellant, the Reynolds number ratio may be written:

$$\frac{\text{Re}}{\text{Re}_s} = \frac{pMD}{p_s M_s D_s} \tag{7.28}$$

which implies that Eq. 7.26 can be written:

$$\dot{e} = \dot{e}_s \left[\frac{pM}{p_s M_s} \right]^{0.8} (D_s/D)^{0.2} \tag{7.29}$$

This relation would permit us to scale ablation data from subscale measurements. Further simplifications can be made if the subscale test article is geometrically similar to the full-scale design. In this case, we have the same area ratio variation with distance as in the full-scale article and since Mach numbers are dependent on area ratio alone for a 1-D choked flow, Mach numbers must also match for the two cases. In addition, if we presume 1-D flow, all pressures are scaled relative to the chamber (or stagnation) pressure. Under these assumptions, we may write:

$$\dot{e} = \dot{e}_s \left[\frac{p_c}{p_{cs}} \right]^{0.8} (D_s/D)^{0.2} \tag{7.30}$$

so that erosion rates scale as $p_c^{0.8} D^{-0.2}$. This is a fundamental result used in many preliminary heat transfer analyses.

While the scaling notions described above are often used in practice, they have several deficiencies. First of all, the assumption of 1-D flow is not generally valid in most realistic nozzles. In addition, the assumption that convective heat transfer is the major contributor to erosion is not always valid. In regions near the throat, particle impingements are known to influence erosion rates; this phenomenon is not replicated in a subscale test article since actual physical dimensions and particle sizes govern this process. In regions of very low-speed flow (such as the head end of the chamber), radiation heat transfer can be an important contribution. Finally, one should recognize that the heat transfer correlation itself was developed from pipe-flow data and may not be applicable to nozzle regions where high wall angles are typically present.

Predicting insulation weights for preliminary design purposes is also a challenge since, in this case, the propellant grain design is not known. Most preliminary design codes utilize correlations (regressions involving known weights from previous motors) in order to estimate insulation weight. One such correlation, based on data from space motors utilizing silica loaded EPDM insulation, is shown below:

$$m_{in} = 1.7 \times 10^{-6} \times m_p^{-1.33} \ t_b^{0.965} (L/D)^{0.144} L_{sub}^{0.058} A_w^{2.69} \tag{7.31}$$

where m_{in} and m_p are insulation and propellant weights in pounds, t_b is burn time, L/D is for the motor case, L_{sub} is the percentage of case length in which the nozzle is submerged, and A_w is the exposed wall area inside the motor case. This regression reproduces insulation weights within approximately 15% on the SRMs within the database. This type of accuracy is typical for insulation correlations since various motors have insulation thicknesses optimized to varying degrees.

7.4.4 Nozzles

It is safe to say that a large percentage of the design, analysis, and fabrication time associated with an SRM lies in development of the nozzle. The nozzle is responsible for converting the high thermal energy in chamber gases to kinetic energy and thrust. For this reason, the highest velocities, heat fluxes, and pressure gradients exist in this portion of the motor. Studies of historical SRM failures indicate that roughly 50% of all failures occur due to problems in the nozzle area. The high failure rate in this region is attributed to the severe environment and the fact that nozzles can contain many parts and may have a vectoring requirement to provide vehicle control.

Most of today's SRM nozzles are comprised of composite materials. *Phenolic* materials are composites using an epoxy resin with a fiber reinforcement, which is subjected to high temperatures and pressures in an autoclave "curing" process. Graphite and carbon/carbon materials utilize woven fibers as well, but in this case negligible resin is present after the very high-temperature graphitization (or carbonization) process.

Table 7.2 SRM nozzle insulator and structural materials

Material	ρ, lb/in^3	C_p, BTU/lb °R	κ, BTU/hr ft °R	F_{tu}, ksi	$\dot{e}*$, mil/s
Pyrolytic graphite	0.079	0.50	0.034	15.0	4.4
Polycrystalline graphite	0.063	0.60	15.0	7.0	4.4
2-D carbon/carbon	0.052	0.54	8.0	16.0	–
3-D carbon/carbon	0.069	0.50	18.2	27.0	4.4
Carbon/phenolic	0.052	0.36	0.58	10.5	7.2
Graphite/phenolic	0.052	0.39	0.92	7.6	11
Silica/phenolic	0.063	0.30	0.32	7.6	53
Glass/phenolic	0.07	0.22	0.016	60.0	60
Paper/phenolic	0.044	0.37	0.23	22.0	76

* At $p_c = 1000$ psi, $D_t = 1$ ft, $T_c = 5460°$R. Note that values are scaled using the convective heat transfer correlation.

A summary of insulator and structural materials used in today's nozzles is presented in Table 7.2. Since properties are dependent on both temperature and orientation for many of these materials, the quantities tabulated should be viewed as approximations. In the throat region, we desire materials with low erosion to minimize the loss in expansion ratio during the course of the firing. Graphite and carbon/carbon materials are most often used in this application. Here we note that actual erosion rates were scaled to the condition noted at the bottom of Table 7.2 using the notions developed in the insulation section of this chapter. If the heat flux is proportional to the quantity $p^{0.8}T^{0.4}D^{-0.2}$ it is easy to show that the highest heat fluxes are present in the throat region.

Typically, we require an insulating material behind the throat insert due to the high cost and thermal conductivity associated with graphite and carbon/carbon materials. Low density (weight) and thermal conductivity are desirable characteristics for these components. Carbon, graphite, glass, and silica phenolics are typically used in this application. Glass phenolic has excellent insulating properties, but use is generally minimized since it is heavier (note the higher density) than the other insulators. Finally, a paper phenolic material has been proposed as a low-cost alternative for some applications but is mainly reserved for hobby motors and the like.

In the exit cone region, the thermal environment is less severe and some of the insulators (i.e. carbon and silica phenolics) can be used as primary gas side materials. Both 2-D and 3-D carbon/carbon materials have also been used in this region, both alone and in conjunction with insulators. The thickness of exit cones is typically set by thrust (pressure) loads placed on the structure and this region can be very thin in many designs (as low as 30–50 mils).

In addition to the so-called "plastic parts" used as nozzle insulators and throat and exit cone materials, most nozzles include some metal components to serve as structural members. Aluminum, steel, and titanium are the most common metals used in these applications.

Figures 7.5–7.7 present nozzle designs for the Space Shuttle SRB, the Castor IV motor, and the Star 48 space motor. The Space Shuttle SRB nozzle uses carbon phenolic (with glass phenolic backup) in regions of highest heat flux and transitions to the less expensive silica phenolic where heat fluxes are lower. The entire nozzle has a metallic housing backup structure to facilitate reusability in the design. All plastic parts were replaced after each mission for this nozzle.

The Castor IV motor is an expendable motor used as first-stage propulsion early versions of the Delta launch vehicle. Since the nozzle has no TVC requirements, it is very simple in design using graphite as a throat insert and a combination of graphite and carbon phenolic in regions upstream and downstream of the throat, respectively. This nozzle also uses a metallic backup structure made of an aluminum alloy.

The Star 48 nozzle, shown in Figure 7.7, uses a 3-D carbon/carbon throat insert with carbon phenolic backup insulators and primary insulators in regions of lower heat flux. The exit cone is

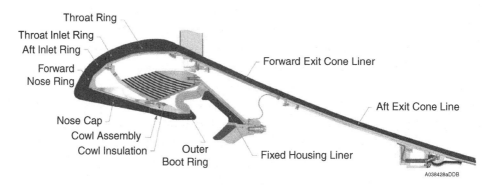

Figure 7.5 Space Shuttle SRB nozzle design (Source: Orbital ATK).

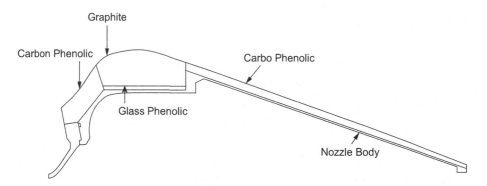

Figure 7.6 Castor IV SRM nozzle design (Source: Orbital ATK).

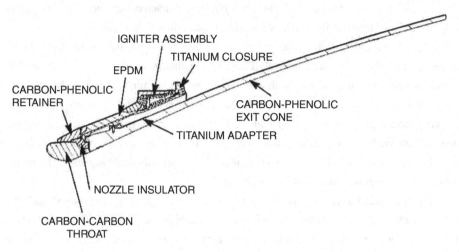

Figure 7.7 Star 48 Space motor nozzle Design (Source: Orbital ATK).

overwrapped with a glass epoxy to provide hoop strength. Another version of the nozzle, using a carbon/carbon exit cone, has also been used in flight. A minimum amount of metallic structure is used to minimize the inert weight in this upper-stage motor.

It is difficult to predict the weight of the nozzle for preliminary design purposes. This difficulty arises from the fact that the nozzle contains many parts, that the design is influenced heavily by incorporation of a TVC system, and that historical designs have varying degrees of conservatism in material thicknesses. Nozzles built for ground launched applications may be heavier by virtue of the fact that buckling loads (due to low pressure within the cone) at ignition may drive the design. For these reasons, it is difficult to develop a correlation of historical nozzle weights to any degree of accuracy.

With these caveats in mind, we present a correlation capable of predicting nozzle weights:

$$m_{noz} = 2.46 \times 10^{-4} \left[\frac{(m_p c^*)^{1.2} \varepsilon^{0.3}}{p_c^{0.8} t_b^{0.6} (\tan\theta_e)^{0.4}} \right]^{0.917} \tag{7.32}$$

where m_p is the propellant mass in lbm, c^* is in ft/s, p_c is in psi, and t_b is in seconds. This equation was developed using space motors and strap-on boosters and is accurate to only about 20% for the database employed. As mentioned above, it is more desirable to actually consider the detailed design of the nozzle (even at the preliminary design stage) in order to obtain a more accurate weight estimate.

7.5 SOLID ROCKET PROPELLANTS

The propellant in an SRM must not only burn at the proper rate so as to maintain the desired chamber pressure, but it must also have enough structural continuity to withstand pressure and acceleration loads introduced during flight. Most physical characteristics (including burning rate) cannot be predicted analytically so heavy use of test data is normally required.

Solid propellants fall into two general categories. In *double base* propellants, fuel and oxidizer are mixed on a molecular level. These propellants generally contain a nitrocellulose material dissolved in nitroglycerine with additional amounts of minor additives. Nitrocellulose, also known as *gun cotton*, was developed in the early 1800s. Both of these primary ingredients are monopropellants/explosives containing both fuel and oxidizer (oxygen) within their molecular structures. Double-base propellants have been used most frequently in military applications (primarily in ballistic and tactical missiles) but their use is currently declining due to an increased emphasis on safety of munitions devices by the military.

The other general type of solid propellant is called a *composite propellant* to highlight the fact that it is a heterogeneous mixture of fuel, oxidizer, and binder ingredients. The binder serves as a fuel in the propellant, but many composite propellants use metallic powders as additional fuel sources to augment performance. Oxidizers are generally crystalline materials, while the binder is a rubber-based material that serves as a means to hold the powder/crystal mixture together in a cohesive grain.

7.5.1 Fuels

Table 7.3 summarizes typical fuels used in composite propellants in ascending order of perform-ance capability. The most common fuel in use today is aluminum. Current aluminized propellants utilize anywhere from 2–21% Al. Magnesium has also been proposed as a fuel and is currently being used in some modern "clean" formulations with reduced exhaust emissions. The beryllium fuels are the most energetic available but the exhaust products (BeO_x) are highly toxic. For this reason, beryllium fuels are considered seriously only for space-based applications such as space interceptor weapons for ballistic missile defense.

Table 7.3   Fuels used in composite propellants

Fuel	Chemical symbol	Molecular weight	Density, g/cm^3	Remarks
Zirconium	Zr	91.22	6.4	Low performance but high ρ
Magnesium	Mg	24.32	1.75	Clean propellant applications
Aluminum	Al	26.98	2.7	Low cost, common fuel
Aluminum hydride	AlH$_3$	30.0	1.42	Tends to decompose
Beryllium	Be	10.81	2.3	Toxic exhaust products
Beryllium hydride	BeH$_2$	11.03	≈ 0.65	Toxic exhaust products

7.5.2 Oxidizers

Table 7.4 gives descriptions of typical oxidizers used in composite propellants. Ammonium perchlorate, or AP, is by far the most common oxidizer in use today due to its moderately low cost, moderate performance, and good processability. The one problem associated with AP is that the chlorine in this compound combines with hydrogen to yield hydrochloric acid in the exhaust products. The HCl cloud from a large solid motor can combine with water vapor to form acid rain in the area near the launch pad.

Of the remaining oxidizers in Table 7.4, ammonium nitrate (AN) has the most applications. While AN is not as energetic as AP, it is even cheaper and provides a "clean" exhaust (at least in terms of HCl). Problems with AN include its inherently low burning rate as well as the fact that the compound is susceptible to a crystalline phase change at about 30 °C. Phase-stabilized AN has been developed to mitigate the phase change issue. Sodium nitrate has also been proposed as a prospective oxidizer for some clean propellant applications. Potassium perchlorate sees some uses in high-powered hobby motors, and ammonium dinitramid (ADN) is a more recent entry developed in Russia and Europe that is proposed for usage some uses in tactical motors due to the fact that it does not produce smoke in its exhaust.

7.5.3 Binders

The role of the binder is to hold the entire formulation in a structurally sound grain capable of withstanding temperature variations and the acceleration loads both prior to and during flight. It is desirable to have high density and energy of combustion in binders. In addition, the best binder

Table 7.4   Oxidizers used in composite propellants

Oxidizer	Chemical symbol	Oxygen content, %	Density, g/cm^3	Remarks
Ammonium perchlorate	NH_4ClO_4	34.0	1.95	Moderate performance and cost
Ammonium nitrate	NH_3NO_3	20.0	1.73	Moderate performance, low cost
Sodium nitrate	$NaNO_3$	47.1	2.17	Lesser used in SRMs
Potassium perchlorate	$KClO_4$	46.2	2.52	Low r, used in hobby SRMs
Potassium nitrate	KNO_3	39.6	2.11	Low cost and performance
Ammonium dinitramid	$NH_4 N(NO_2)_2$	51.6	1.81	Smokeless propellant

Table 7.5   Binders used in composite propellants

Binder chemical	SRM applications
Historical binders with limited current use	
Polysulfide (PS)	JATOs, older rockets
Polyether polyurethane (PEPU)	Polaris stage 1
Polybutadiene acrylic acid (PBAA)	Older rockets
Polybutadiene acrylonitrile (PBAN)	Titan and Shuttle SRMs
Nitrocellulose (plasticized) (PNC)	Minuteman and Polaris
Carboxy-terminated polybutadiene (CTPB)	Minuteman stages 2, 3
Current and potential future binders	
Hydroxy-terminated polybutadiene (HTPB)	IUS, Peacekeeper, ASRM
Nitrate ester polyether (NEPE)	Peacekeeper, SICBM
Glycidal azide polymer (GAP)	High energy binder
Hydroxyl terminated polyether (HTPE)	Insensitive munitions binder
DiCyclopentadiene (DCPD)	High strength/burning rate binder

materials are the ones that can provide the essential structural integrity using a minimum amount of available propellant volume, as binders tend to be less energetic than the fuel and oxidizer ingredients.

Binder materials used in SRM composite propellants are summarized in Table 7.5. The first three compounds described in this table are generally considered to be older technology, while PBAN, CTPB, and particularly HTPB are used in many of today's applications. The GAP binder has been used in missile and explosives applications; it offers high performance but is also quite costly as compared to other current binders. There is great interest in the military to develop insensitive munitions that are more forgiving in terms of potential events such as inadvertent drops of hardware, bullet impact, and immersion in a fire (slow and fast "cookoff" events). The HTPE binder has received some attention as it may provide improved insensitivity. Finally, the DCPD binder has been evaluated for a number of solid and hybrid propellant formulations. This material is best characterized as a high-strength plastic whereas the other binders are more rubber-type materials.

Binders are generally long chain polymers capable of retaining the powders and crystals in the propellant in place through a curing process called *crosslinking*. The crosslinking process takes place by adding a *curative* to the propellant just prior to *casting* (pouring) the mixture into the motor. In HTPB-based propellants, crosslinking takes place between hydroxyl (OH) groups that stem from the main backbone of the polymer. In PBAN, acrylonitrile groups (CH_2CHCN) are the crosslinking agents. The cast grain is cured at high temperature and pressure to promote the crosslinking process. The DCPD binder undergoes a slightly different type of polymerization called "ring-opening metastasis

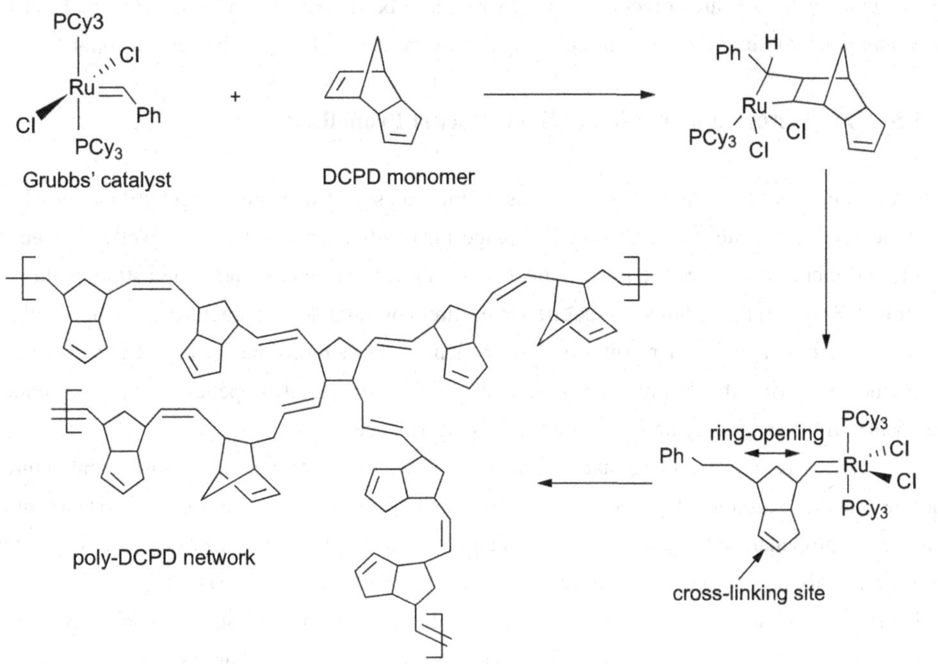

Figure 7.8 Polymerization process in the DCPD binder showing both ring opening and cross-linking.

polymerization." Figure 7.8 highlights the elements of this process and also illustrates cross-linking that is more common in other binder types.

7.5.4 Minor Ingredients

We mentioned above that a small amount of *curative* is required to promote the polymerization process. A *cure catalyst* is often included with the curative to increase the rate of chemical reaction cause by addition of this ingredient. The HTPB binder requires an isocyanate-based curative, while PBAN uses an epoxy-based curative. Some propellants also utilize a *plasticizer* to improve low-temperature physical properties. Plasticizers act to improve the strain capability of the material and to reduce its glass transition temperature (the temperature at which it becomes brittle) to values below the required operational temperature range. Many double-base and some composite propellant formulations require a *darkening agent* (such as carbon black) to make the translucent propellant formulation darker. The darkening is required to avoid excessive radiation through the propellant to a backside (motor case or insulation) that could overheat and cause ignition in an undesired location.

Since the propellant must be "tailored" to achieve the desired burn rate, *burning rate catalysts or modifiers* are often required in small amounts. These ingredients, such as iron oxide or copper chromite, tend to increase the burning rate of the propellant. Oftentimes, *antioxidants* are also required to prevent binder oxidation reactions in the propellant over long storage periods. Small amounts of *liquid processing aids* are sometimes used to reduce mix viscosity. Finally, some propellants use a *bonding agent* to improve the molecular bond between the oxidizer and binder materials.

7.5.5 Manufacture and Testing of Solid Rocket Propellants

Solid propellants have unique requirements as they must not only be energetic, but must have sufficient structural integrity to survive all storage and handling prior to launch as well as the launch loads and vibrations. Since the propellants cannot undergo structural and combustion evaluation prior to use, careful procedures are required to ensure reproducible manufacturing and the resulting mission performance. As a result, solid propellant manufacturers have very strict procedures implemented during the manufacturing and carry out a number of propellant testing procedures to ensure structural quality and proper ballistic performance.

There are two top-level manufacturing approaches for propellants. Most manufacturers utilize batch processing of propellants by mixing individual batches and casting one or more batches of propellant into the motor. In contrast, continuous processing may be applied such that propellant mixing occurs in an auger or screw-type machine that is capable of continuously delivering propellant to motor cases on an assembly line. In general, the continuous processing would be most attractive if one were using very large amounts of propellant as it would reduce the infrastructure requirements, but most applications today don't demand these large volumes, and most manufacturers use batch processing.

Figure 7.9 shows a sketch of a typical batch-type mixer, a lab-scale unit that we operate here at Purdue University. Most mixers are of the "dual planetary" type in which the blades rotate and the location of the blade rotation is also rotating (as two planets rotate about the sun and each have their own individual rates of rotation about their respective centers). Mixers are jacketed to allow water to flow through passages around the mix bowl such that the mix temperature can be controlled. Using heated water in the jacket keeps the binder warmer and reduces its viscosity. In addition, these mixers are capable of mixing under vacuum conditions, as the use of vacuum in mixing and casting operations ensures that no air bubbles are in the cast propellant. As air bubbles would introduce additional surface area when encountered by the burning front, they could alter the ballistic performance substantially.

Propellant manufacture begins with careful weighing of all ingredients. Fuel powder and all liquids (except cure catalyst) are mixed in what is called the fuel premix. Oxidizer powder, which makes up the bulk of the volume of the mixture, is then added in steps. In general, two or more particle sizes are used for the oxidizer ingredient as the use of large and small particles provides for a better packing efficiency and higher effective density of the manufactured propellant. The larger particles (coarse particles) are generally added first, as the large surface area associated with the smaller particles (fine particles) makes mixing more difficult due to the increased friction. High surface area mixtures become highly viscous to stir/mix, and managing the viscosity of the propellant through the mixing process is a major concern.

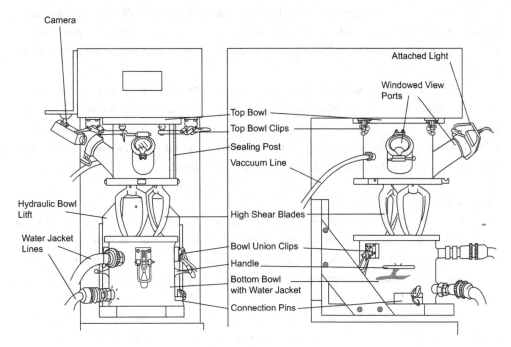

Figure 7.9 Jacketed dual planetary mixer at Maurice Zucrow Lab (Source: Tim Manship).

The propellant can be evaluated during the latter stages of mixing in order to ensure proper burning rate. Liquid strands can be drawn into straws and burned to ensure that the propellant lies within burning rate targets. Additional burning rate catalyst can be added if the liquid strand burning rate (LSBR) is below a target level (we must always approach this problem from below as it is impossible to remove burning rate catalyst from the mixed propellant). This level of burning rate control is typically required only for strap-on boosters where it is of paramount importance that both motors burn for the same duration; otherwise large yaw forces would be generated if one motor were to burn out earlier than the other.

When the propellant is adequately and completely mixed, the cure catalyst is added to the mixture. At this point, the curing (binder polymerization) process is initiated, and the clock starts ticking relative to the time available to mix in the catalyst and cast the propellant before it becomes a solid. We call this time availability the "pot life" of the propellant. By controlling the amount of the cure catalyst and the temperature of the mixture, pot life can be controlled over some range.

The mixed propellant is then cast into the motor by pouring or valving from the mix bowl directly. A mandrel is used to provide the shape of the initial exposed surface, and this device is pulled from the assembly (a ticklish operation for very large motors) once the propellant has been fully cured. As air bubbles would likely be trapped in the propellant, vacuum casting is typically employed to alleviate this potential problem. A slit plate is used to form small ribbons of propellant, thereby allowing any trapped air bubbles to be released as the propellant falls into the motor case. Propellant samples are cast into small containers for evaluation of structural and/or ballistic properties. Small ballistic test motors may also be cast to obtain burning rate data on the mix in question.

After casting, the loaded motor is typically transported to a curing oven and the curing process is allowed to run to completion in this controlled temperature environment. Typically, propellants are cured at elevated temperatures of the order of 120–150 °F and cure cycles can last several days for large motors. The propellant generally shrinks a bit due to the curing process and provisions must be made for this in the grain and insulation designs.

Propellant Structural Evaluation

Formally, solid propellants are classified as viscoelastic materials. Viscoelastic materials have complex structural behavior in that the stress–strain curve that characterizes their tensile performance is dependent on the rate at which strain is applied. This complex behavior is also termed non-Newtonian by the structures community; it implies that we have a whole family of stress–strain curves for a variety of different straining rates. The modulus (Young's modulus) of the propellant generally is reduced as the strain rate is reduced; at high rates the behavior becomes largely insensitive to the strain rate. Tensile testing is conducted by cutting dog-bone shaped pieces of propellant as highlighted in Figure 7.10. Procedures are similar to those employed with metallic samples (as you may have done in a strength of materials course).

Typical stress–strain behavior under high strain rate conditions is shown in Figure 7.11. In general, solid propellants can withstand peak tensile stresses of the order of a few hundred psi, which makes them a poor structural material measured against aluminum or steel, for example. It is this relatively low strength that can make it quite challenging for the grain designer as the use of fins, star points, or other appendages that provide large surface area also provide the potential for cracking and structural failure under the acceleration and vibration loads of flight.

Propellant Ballistic Evaluation

When the main motor is cast, the manufacturer may also choose to cast some of the propellant into small ballistic test motors in order to get data on the burning characteristics of that particular batch of propellant. These motors are generally 1–2 inches in diameter and are cartridge loaded into a heavy case as shown in Figure 7.12. By carefully choosing the L/D of the simple cylindrical propellant grain (see Homework Problem 7.18), one can obtain an approximately neutral pressure history such that the average burning rate of the motor is simply the total web thickness divided by the measured burning time. The small motors can be temperature conditioned in order to assess the temperature sensitivity of the propellant as well.

Another alternative is to cut samples from a carton of propellant that was cast from the parent batch. The samples are generally cut into rectangular strands that are inhibited on all sides except the very top. These strands are then burned at constant pressure conditions in a high-pressure

Figure 7.10 Typical tensile strength test for a propellant "dogbone."

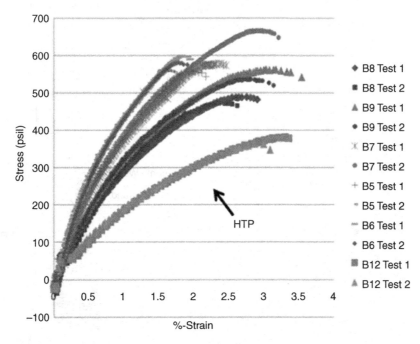

Figure 7.11 Typical high rate stress–strain curves for HTPB and a series of experimental DCPD formulations.

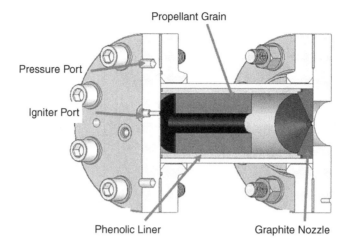

Propellant Grain

Pressure Port

Igniter Port

Phenolic Liner

Graphite Nozzle

Figure 7.12 Typical ballistic test motor design.

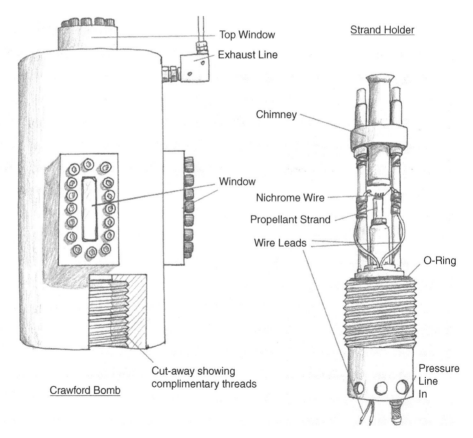

Top Window

Exhaust Line

Strand Holder

Chimney

Window

Nichrome Wire

Propellant Strand

Wire Leads

O-Ring

Pressure Line In

Cut-away showing complimentary threads

Crawford Bomb

Figure 7.13 Crawford bomb (strand burner) at Maurice Zucrow Lab (Source: Tim Manship).

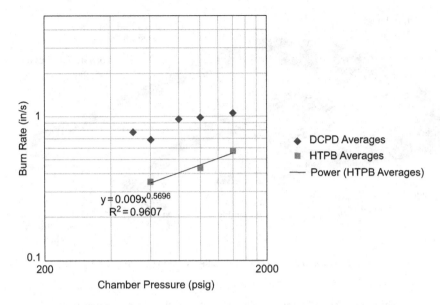

Figure 7.14 Strand burner data for two different propellant binder systems.

combustion bomb, typically referred to as a Crawford bomb. The bomb is typically charged to the desired combustion pressure using an inert gas such as argon. Figure 7.13 provides a sketch of Purdue's Crawford bomb; it contains a window in order to visually assess the rate of surface regression at a given pressure and an internal part that contains the strand mounting, ignition, and gas venting provisions.

Burning rate data attained from small motors or strands can then be plotted and fit to St. Robert's law (or a comparable burning rate model). Figure 7.14 outlines this process for some strand burns conducted in Purdue's labs. Recent research is showing that DCPD propellants display burning rates about double that of comparable HTPB formulations as indicated in the example figure. In general, strand data tend to be less reliable than small motor data, but they are much easier to attain as well. Most manufacturers prefer small motors in order to provide the best assessment of burning characteristics. However, strand burners are still utilized in many laboratories and within industrial R&D organizations.

7.6 THRUST VECTOR CONTROL AND THROTTLEABLE SYSTEMS

Most SRMs contain thrust vector control (TVC) systems to aid in the guidance and control of the rocket during the time period in which the solid motor is operating. Inclusion of a TVC tends to complicate the nozzle design considerably since this component of the motor normally must be capable of handling the side loads introduced when the thrust vector is moved off the motor axis. Some of the more popular TVC approaches are summarized as follows:

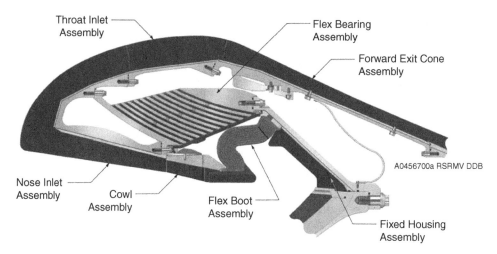

Figure 7.15 Space Shuttle booster flexseal design (Source: Orbital ATK).

- **Flexible bearing (Flexseal)** – This is probably the most popular TVC scheme in use today. A schematic of an application of this system in the Space Shuttle SRB is shown in Figure 7.15. The flexseal is comprised of alternating layers of *shims*, which provide the structural stability for the bearing, and *pads*, which are made of an elastomeric material that has low shear strength. By pushing on the side of the nozzle with a hydraulic actuator, the pads will permit a few degrees (generally 4–6°) of nozzle rotation.

- **Techroll seal** – This system was developed by Chemical Systems Division of United Technologies Corporation in the 1980s for use in the Inertial Upper Stage (IUS) orbital transfer motors. A schematic of the system is shown in Figure 7.16. The Techroll seal was developed to provide TVC with relatively low actuator torques (as compared to the Flexseal, which has relatively high torque requirements). The forward part of the nozzle actually "rides" on a rubber-lined, oil-filled reservoir. When the actuator torques the nozzle to one side, the oil rushes around the annulus of the reservoir permitting nozzle movement. While the system does provide lower torques than a flexseal, it is more complex.

- **Liquid injection TVC (LITVC)** – In this system, a liquid is injected into the exit cone at various circumferential locations to provide the required control moments. Side forces are developed by the wall pressure distribution associated with the shock structure formed by the injectant. A schematic of an LITVC system is shown in Figure 7.17. The injected fluid causes a shock over some fraction of the nozzle circumference and raises wall pressures in that region. By using energetic liquids (such as a liquid oxidizer), the injectant can actually augment the axial thrust a small amount during operation. LITVC systems require no actuators, but are complicated by the liquid manifolding and numerous valves within the system. The Titan launch

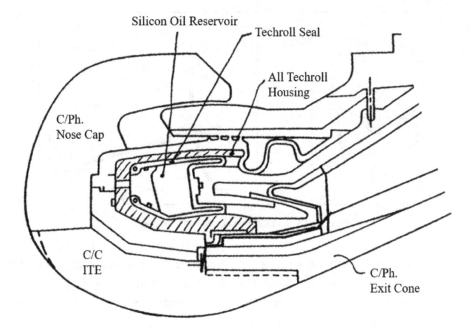

Figure 7.16 IUS Nozzle with Techroll seal. The nose cap and exit cone are carbon/phenolic (C/Ph) while the integral throat entrance (ITE) is carbon/carbon (C/C). The Techroll seal is formed by an elastomeric tube filled with silicon oil.

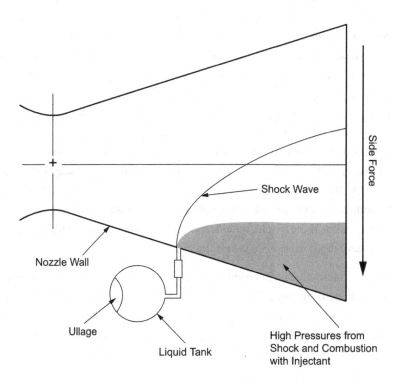

Figure 7.17 Liquid injection TVC system.

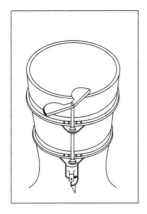 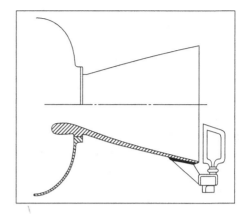

Figure 7.18 Jet tab/vane nozzle TVC concepts.

vehicle employed this system for vehicle steering during ascent under power of the core liquid stage and two strap-on solid boosters. As the core stage used nitrogen tetroxide as oxidizer, this fluid was used as injectant. Since solid motors tend to optimize with fuel rich stoichiometry for optimum I_{sp}, the injected oxidizer afterburned with fuel-rich exhaust to augment control authority.

- **Jet vanes, tabs, and pintles** – These devices generate TVC forces by physically deflecting the flow within the nozzle without moving the nozzle itself. Schematics of the jet vane/tab concepts are shown in Figure 7.18. As one can see from Figure 7.18, both concepts require moving parts within the hot nozzle flow, which generally limits their application to motors with short burning times. It is obvious that both concepts introduce additional frictional losses even when they are not being operated. In many applications, no materials yet exist that can withstand the full duration flow environment; for this reason, these systems are not as common as those described above.

- **Throttleable systems** – As we have emphasized in Section 7.2, by use of features in the grain design, a pre-programmed thrust shape can be provided. Excellent examples of this strategy include the large solid rocket boosters used on Titan, Space Shuttle, and Ariane vehicles. By use of clever segmented designs with large surface area forward closures, these motors are able to provide the highest thrust at liftoff, and then "throttle back" thrust during max q. These are just one example of the abilities to create progressive, neutral, or regressive thrust behavior using the grain design.

However, to react to current conditions, a true throttling capability is required to react to instantaneous conditions. In this case, variable geometry (mainly in the nozzle throat area), can alter the instantaneous chamber pressure, propellant flow, and hence thrust. Advances in computer control, electromechanical actuator size and power-to-weight ratio, and materials are enabling new SRM designs that are truly throttleable. Thrust changes are enabled by a movable pintle that can

Figure 7.19 Orion launch abort attitude control motor (ACM) firing with eight individually controlled nozzle throats (Source: Orbital ATK).

change the net throat area of the rocket motor on command. The best current example of this technology is the NASA Orion Launch Abort Attitude Control Motor (ACM) made by Orbital ATK. This motor incorporates an array of eight electromechanically controlled, hot gas valves around the circumference of the unit. These valves control both the pressure and thrust of the unit to generate pitch and yaw moments for the system to direct the crew capsule and provide the proper thrust direction to control the flight path in an abort scenario. Figure 7.19 shows the variable thrust between the eight valves in a static test from an early development test in 2009.

In situations when extinguishment is needed, the hot gas valve and pintle sizing for either single or an array of valves needs to be compared with the propellant ballistic characteristics to achieve a depressurization fast enough to extinguish the propellant. This is easier to accomplish against a vacuum back pressure than a sea-level back pressure, and a means of re-ignition must be provided through multiple igniters or other means.

FURTHER READING

Bluestone, S., Austin, B., Heister, S., and Son, S. (2010) "Development of Composite Solid Propellants Based on Dicyclopentadiene Binder," in *46th AIAA Joint Propulsion Conference*, Nashville, TN. AIAA.

Green, L. (1954) "Erosive Burning of Some Composite Solid Propellants," *Jet Propulsion*, 24(9): 8–15.

Hasegawa, H., Hanzawa, M., Tokudome, S., and Kohno, M. (2006) "Erosive Burning of Aluminized Composite Propellants: X-Ray Absorption, Measurement, Correlation, and Application," *Journal of Propulsion and Power*, 22(5): 975–983.

Heister, S. D. and Landsbaum, E. M. (1991) "Analysis of Ballistic Anomalies in Solid Rocket Motors," *Journal of Propulsion and Power*, 7(6): 887–893.

Humble, R. ed. (1995) *Space Propulsion Analysis and Design*. McGraw-Hill Publishing.

Jackson, T. L. (2012) "Modeling of Heterogeneous Propellant Combustion: A Survey," *AIAA Journal*, 50(5): 993–1006.

Landsbaum, E. M. (2003) "Erosive Burning Revisited," in *39th AIAA/ASME/SAE/ASEE Joint Propulsion Conference and Exhibit*. AIAA.

Lawrence, W. J., Matthews, D. R., and Deverall, L. I. (1968) "The Experimental and Theoretical Comparison of the Erosive Burning Characteristics of Composite Propellants," in *ICRPG/AIAA 3rd Solid Propulsion Conference*. AIAA.

Lenoir, J. M., and Robillard, G. (1957) "AMathematical Method to Predict the Effects of Erosive Burning in Solid Propellant Rockets," in *Proceedings of the 6th Symposium (International) on Combustion*. Reinhold Publishing Corporation, pp. 663–667.

Moss, J. M. (2008) "Development of Tools for Experimental Analysis of Erosive Burning In/Near Slots in Solid Rocket Motors," M.S. Thesis, School of Aeronautics and Astronautics, Purdue University, West Lafayette, IN.

Moss, J., Heister, S. D., and Linke, K. (2007) "Experimental Program to Assess Erosive Burning in Segmented Solid Rocket Motors," in *43rd AIAA/ASME/SAE/ASEE Joint Propulsion Conference and Exhibit*. AIAA.

Nickerson, G. R. *et al.* (1987) The Solid Rocket Motor Performance Prediction Computer Program (SPP), Version 6.0, AFAL-TR-87–078.

Razdan, M. K. and Kuo, K. K. (1979) "Erosive Burning Study of Composite Solid Propellants by Turbulent Boundary-Layer Approach," *Journal of Propulsion and Power*, 17(11): 1225–1233.

Rettenmaier, A. K. and Heister, S. D. (2013) "Experimental Study of Erosive and Dynamic Burning in Polybutadiene-Based Composite Propellants," *AIAA Journal of Propulsion and Power*, 29(1): 87–94.

Slaby, J., Moss, J., and Heister, S. D. (2009) "Experimental Program to AssesErosive Burning in Segmented Solid Rocket Motors," in *45th AIAA/ASME/SAE/ASEE Joint Propulsion Conference and Exhibit*. AIAA.

McGrath, D., Williams, J., Porter, M., Dominick, T., and Paisley, J. (2014) "Orion Launch Abort System (LAS) Attitude Control Motor Progress and Future Activities," in *50th AIAA Joint Propulsion Conference*. AIAA.

Homework Problems

7.1 Consider the solid rocket motor shown in Figure 7.20. The propellant has a characteristic velocity of 5000 ft/s and the initial chamber pressure (p_c) is 500 psi. Assume a burning rate law of $r_b = aP_c^{0.4}$ in units of in/s.

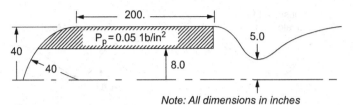

Note: All dimensions in inches

Figure 7.20 Diagram for Problem 7.1.

(a) Determine the value for the constant a in the expression given above.

(b) Using this value, determine the chamber pressure at a web distance of 10 inches.

(c) Estimate the amount of time required to reach the 10 inch web distance.

7.2 You have been asked to optimize the burn time and throat size of an SRM preliminary design under the assumption that the chamber pressure and nozzle expansion ratio are held fixed at 750 psi and 50, respectively. The propellant weight is 500 lb and inert weights due to motor case, bosses, skirts, and igniter can be assumed to be 25 lb (and constant) in your study. However, insulation and nozzle masses will vary. Your preliminary designer predicts the following functional dependence for these parameters:

$$M_{in} = 10 + 0.01 t_b$$
$$M_{noz} = 15 + 0.2 A_t$$

where M_{in} and M_{noz} are weights in lb, t_b is burn time in seconds, and A_t is throat area in square inches. Finally, the following propellant characteristics are known:

$$I_{sp} = 300s \ , \ c^* = 5000 \ f/s, \ \gamma = 1.2$$

With this information, determine the optimal burn time and throat diameter (in inches) for this SRM. Determine the ideal Δv of the rocket under these conditions assuming a payload of 100 lb.

7.3 A subscale SRM firing is conducted for establishing the insulation design on a full-scale SRM development. The 1/10 scale SRM utilizes the same propellant to be used in the full-scale design and the average chamber pressure in the test was the same as that desired for the full-scale unit. The measured erosion rate at a subsonic area ratio (see Figure 7.21) of 3.0 was 0.13 mm/s in the subscale test. Using this information, predict the erosion rate at a subsonic area ratio of 2.0 in the full-scale unit.

7.4 In this assignment, you will generate a preliminary design for an apogee kick motor (AKM) capable of placing 2270 kg of payload into a geosynchronous orbit from an initial 185 × 35,000 km geosynchronous transfer orbit. Most of the data and assumptions for your analysis are included in the table overleaf:

Figure 7.21 Diagram for Problem 7.3.

Assumed motor performance and design characteristics

Performance parameters	Motor case
Average chamber pressure = 5.17 MPa	Graphite Composite Construction
Burning time = 80 s	Overall L/D = 3
Propellant	$P_{c-max}/P_{c-avg} = 1.4$
18% Al, 71% AP, 11% HTPB	Safety factor = 1.25
Density = 1800 kg/m^3	**Insulation**
Burn rate exponent = 0.3	Silica loaded EPDM formulation
Nozzle	Density = 1100 kg/cu m
State-of-the-art materials	**Igniter**
Flexseal TVC	Aft-end flame stick design
20% submergence	
18° Effective divergence angle	

(a) Determine the velocity requirement for your mission.

(b) Using the NASA Thermochemistry Code determine the theoretical I_{sp} of the propellant selected for the motor. Assume an overall efficiency of 88% and an expansion ratio of 80 to estimate delivered I_{sp}.

(c) Estimate the amount of propellant required to accomplish the mission by assuming a motor mass fraction, λ of 0.9. (Design is an iterative process by its very nature, we have to start somewhere.)

(d) Assume the grain design shown in Figure 7.22 and estimate the volumetric loading efficiency of the motor. Using this and the propellant volume, determine the motor case and propellant grain dimensions.

(e) Determine the case thickness and weight based on predicted burst pressure, geometry and material characteristics. Assume that polar bosses (and associated material build-ups) amount to 30% of case weight and estimate their mass.

Motor Case and Grain Design

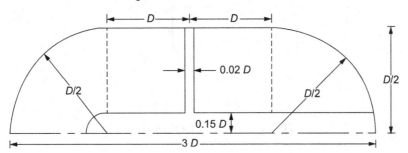

Figure 7.22 Diagram for Problem 7.4.

(f) Assume the thrust skirts are equal in thickness to case material and that both skirts are one case radius in length. Estimate their weight.

(g) Size the nozzle throat to exhaust all propellant in the required burn time. Assume the exit diameter is 80% of the case diameter and calculate weight using the correlation given in class. Assume the igniter weight is included in this mass since we have a head-end web grain design. Assume that the TVC system is 30% of nozzle weight and estimate its mass.

(h) Determine the insulation weight using the correlation given in Eq. 7.31. Determine the thickness of the insulation, assuming that it is uniformly distributed over the inner case wall.

(i) Write a computer code to calculate chamber pressure and thrust histories for the grain geometry shown above, assuming quasi-steady operation at all times. Adjust the burn rate coefficient to give the correct burning time and calculate the average chamber pressure as well. Plot the predicted thrust and chamber pressure histories.

(j) Sum all inert masses and calculate motor mass friction. Calculate the actual expansion ratio from your nozzle design and recalculate I_{sp} accordingly. Estimate ΔV for this situation using I_{sp} from part (b).

(k) Congratulations! You have just completed one design iteration of a space motor. Make a scale drawing of your design. Describe what you would do at this point since your "new" ΔV (part (j)) doesn't agree with the required ΔV (part (a)).

7.5 Consider the cylindrical port solid rocket motor shown in Figure 7.23. To produce the desired thrust history an inhibitor (restrictor) has been bonded to the aft propellant face as indicated in the figure. The propellant properties are:

$$\text{Density} = 0.06 \text{ lb/in}^3$$
$$\text{Burn rate} = p_c^{0.4} \text{ (in inches/s)}$$
$$\text{Characteristic velocity} = 5000 \text{ ft/s}, \gamma = 1.2$$

(a) Determine the initial chamber pressure and vacuum thrust of this motor.

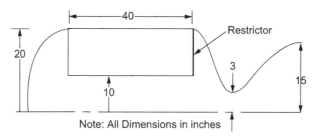

Figure 7.23 Diagram for Problem 7.5.

(b) A static test firing of the motor reveals a chamber pressure 15% higher than your prediction from part (a). You speculate that the restrictor may have unbonded a certain depth from the bore to provide the additional surface area. Calculate the depth of the unbond required to match the observed pressure assuming constant depth 360° around the circumference.

7.6 Consider the solid rocket motor shown in the sketch below. In addition to the dimensions shown in the figure (all given in cm), the following information is known regarding the propellant formulation:

Composition by weight

Propellant constituent	Formula	h_f°, cal/mol	wt. %
Aluminum	Al	0	19
HTPB	$C_{7.3}H_{10.96}O_{0.06}$	−2970	10
AP	NH_4ClO_4	−70730	69
DDI	$C_{26}H_{50}O_4$	−344210	2

Propellant density, $\rho_p = 0.048 \text{ kg/cm}^3, c^* = 1550 \text{ m/s}, T_c = 3333 \text{ K}$
Propellant burning rate, $r_b = 1.1 \, (p_c/6.9 \text{ MPa})^{0.4} \text{(incm/s)}$
             $\gamma = 1.2 \text{ in nozzle}$

Finally, independent estimates from flowfield analysis codes predict efficiencies of 98% and 92% in c^* and I_{sp}, respectively. Thermal analysts predict an average throat erosion rate of 0.13 mm/s.

(a) Using this information, write a computer code to predict the vacuum ballistic performance of this motor. You may use an average c^* value assuming a chamber pressure of 1000 psi.

(b) You must then develop (from the given geometry) relationships for A_b as a function of web distance. You may assume constant throat erosion rate in your calculations and that equilibrium conditions exist in the chamber at all times. Use the following approach:

1. Set initial conditions including pressure and thrust.

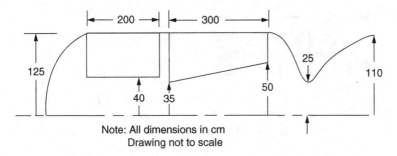

Note: All dimensions in cm
Drawing not to scale

Figure 7.24 Diagram for Problem 7.6.

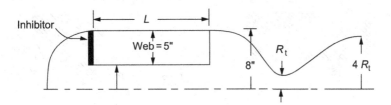

Figure 7.25 Diagram for Problem 7.7.

2. Begin time-stepping $t = t_o + \Delta t$.

3. Calculate change in web $\Delta w = r_b \Delta t$.

4. Calculate new burn surface area at new web.

5. Calculate new throat area.

6. Calculate chamber pressure at new time level.

7. Calculate thrust and store or print results.

8. If web < web/max go to (2).

(c) Attach a listing of your code and a tabulation of web, p_c, A_b, and F versus time.

(d) Plot A_b vs. web, p_c vs. time and F vs. time.

(e) Estimate the motor case, nozzle, insulation, and igniter weights presuming D6AC steel as a case material for the rocket in Figure 7.24. Assume the actual inert weight is 25% greater than the sum of these components and determine the motor mass fraction. State assumptions in your calculations.

7.7 The solid rocket shown in Figure 7.25 utilizes a cylindrical-shaped grain which is inhibited on the forward end. The propellant characteristics are:

$$\gamma = 1.2, \ c^* = \ 5000 \ \text{ft/s}, \ \rho_p = \ 0.063 \ \text{lb/in}^3$$
$$r_b = \ 0.03 p_c^{0.35} (r_b \ \text{in inches/s}, \ p_c \ \text{in psi})$$

We want to develop a vacuum thrust of 3000 lbf at a chamber pressure of 500 psi just after ignition of the motor.

(a) Under these constraints, determine the dimensions L and R_t as shown in Figure 7.25.

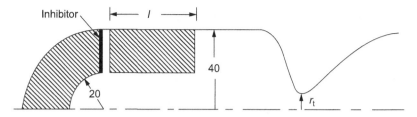

Figure 7.26 Diagram for Problem 7.8.

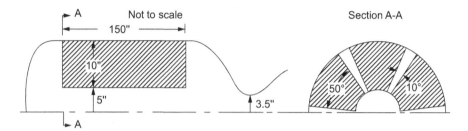

Figure 7.27 Diagram for Problem 7.9.

(b) Determine the chamber pressure and thrust of the motor just prior to burnout.

(c) Estimate the overall burn time of the rocket using data developed in parts (a) and (b).

7.8 The solid rocket motor shown in Figure 7.26 is to be designed for a space mission.
The following propellant data are known:

$$\rho_p = 1700 \text{ kg/m}^3, \ c^* = 1550 \text{ m/s}, \gamma = 1.2$$
$$r_b = 0.6p_c^{0.3}(p_c \text{ in MPa}, r_b \text{ in cm/s})$$

(a) What length of the cylindrical segment, l, would give a neutrally burning behavior for this motor?

(b) For the value of l in part (a), estimate the burning time, assuming a chamber pressure of 4 MPa.

If $l = 75$ cm and $\epsilon = 75$, determine:

(c) the throat radius, r_t, to give an initial chamber pressure of 5 MPa; and

(d) the ideal ΔV the SRM would provide to a 1000 kg payload assuming $\lambda = 0$.

(f) Is the grain design in part (c), progressive, neutral, or regressive? Why?

7.9 Consider the solid rocket motor shown in Figure 7.27 utilizing a six fin grain design. The propellant formulation is the same as in Problem 7.6 and the density and burn rate are given by:

$$\rho_p = 0.06 \text{ lb/in}^3$$
$$r_b = 0.015p_c^{0.35}(p_c \text{ in psi}, r_b \text{ in in/s})$$

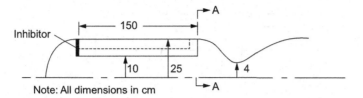

Note: All dimensions in cm

Figure 7.28 Diagram for Problem 7.10.

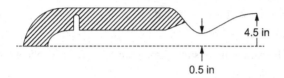

4.5 in

0.5 in

Figure 7.29 Diagram for Problem 7.11.

You are asked to develop a steady-state ballistics analysis of this motor. Use the following steps in your analysis:

(a) Using the initial burning surface areas, calculate the initial chamber pressure. Use this value in the NASA code to get $c*$. You should assume that this c^* value is accurate for the entire burn.

(b) Develop a set of equations for the burning surface area as a function of web distance based on the grain geometry shown in this schematic. Plot A_b vs. web distance for this SRM.

(c) Write a computer code, spreadsheet macro, or Matlab procedure to generate the chamber pressure history given the propellant information and the A_b–web relation from part (b). Plot p_c vs. time. Attach a listing of your code.

7.10 Consider the solid rocket motor shown in Figure 7.28. During a test firing of this motor, the initial (just after ignition) and final (just prior to burnout) chamber pressures were 3 and 7.5 MPa, respectively. The propellant used in this test has a density of 1800 kg/m³ and a characteristic velocity of 1500 m/s. Assuming the propellant burns according to St. Robert's law ($r_b = aP_c^n$), determine the burning rate coefficient (a) and exponent (n).

7.11 The solid rocket motor shown in Figure 7.29 has a chamber pressure history given by:

$$p_c = 300 + 30t - 0.75t^2, \quad 0 \le t \le 45 \text{ s} \ (p_c \text{ in psi})$$

In addition, the following propellant characteristics are known:

$$c^* = 4800 \text{ ft/s}, \rho_p = 0.06 \text{ lb/in}^3, \gamma = 1.2$$

The motor is to be used to launch a space-based missile. Telemetry data reveal a velocity increment of 5000 ft/s during a horizontal flight test of the device. Finally, the motor mass fraction (λ) is known to be 0.8. Using this information, determine:

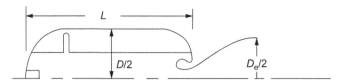

Figure 7.30 Diagram for Problem 7.12.

(a) the mass (weight) of propellant in the SRM;

(b) the maximum thrust of the SRM; and

(c) the payload of the missile.

7.12 Consider the solid rocket motor shown in Figure 7.30. The propellant used in this motor has the following characteristics:

$$r_b = 1 \text{ cm/s}, \rho_p = 1700 \text{ kg/m}^3, c^* = 1500 \text{ m/s}$$

In addition, the following characteristics are known:

$$80\% \text{ web fraction}, \eta_v = 0.9, \theta_e = 16°, D_e = 90\%D$$

$\bar{P}_c = 5$ MPa, $p_b = 6.5$ MPa, 10% nozzle submergence

The motor case will incorporate a graphite composite construction with the following characteristics:

$$F_{tu} = 1.3 \text{ Gpa}, \rho_c = 1550 \text{ kg/m}^3, f_s = 1.25$$

You may also assume that the motor case employs spherical domes and that polar bosses and skirts have a weight corresponding to 30% of the motor case.

Using this information, you are asked to predict the motor mass fraction (λ) for this SRM as a function of propellant mass (M_p) for motor case $L/D = 1$, 4, and 8. Plot λ vs. ln M_p for 10 kg $< M_p < 100{,}000$ kg for each of L/D values on the same curve. For $L/D = 4$, plot nozzle, motor case, insulation, and igniter mass as a function of M_p for the same range of M_p values. Discuss the "scaling effects" important to each of these motor components, i.e. how do these masses behave with increasing M_p?

7.13 Consider the solid rocket motor shown in Figure 7.31. The following propellant and nozzle information is known:

$r_b = 0.015 \, p_c^{0.35}$	r_b in in/s, p_c in psi
$c^* = 5000$ ft/s	$\rho_p = 0.06 \text{ lb/in}^3$
$A_t = 5 \text{ in}^2$	$I_{sp} = 300 \text{ s} = \text{constant}$
$T_f = 6000$ °R	$\gamma = 1.2$

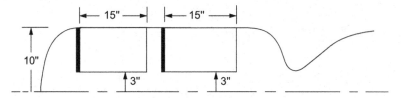

Figure 7.31 Diagram for Problem 7.13.

The motor employs the two-segment grain design with inhibited faces shown in the figure. Determine:

(a) the chamber pressure in the motor just after ignition;

(b) the maximum chamber pressure which occurs during the firing; and

(c) the total impulse of the SRM.

7.14 Consider the solid rocket motor in Figure 7.32. The following propellant and dimensional data are known:

$r_b = 0.025\, p_c^{0.3}$	r_b in in/s, p_c in psi
$\rho_p = 0.06$ lb/in^3	$c^* = 5000$ ft/s
$A_t = 2.0$ in^2	$w = 5$ in
$L = 15$ in	$x = 10$ in

(a) Derive an analytic relationship defining the burning surface (A_b) as a function of web distance for $0 \leq$ web $\leq W$. (You may neglect the tiny sliver formed at the center of the grain.) Plot A_b vs. web.

(b) Write a short computer code to predict the chamber pressure history for this SRM.

(c) Plot the resulting chamber pressure history and attach a listing of your code.

(d) Give reasons why the grain design used in this motor might not be too practical.

7.15 A designer chooses a "rod in tube" grain design for the SRM shown in Figure 7.33 in order to maintain a completely neutral burning behavior. The propellant used in the motor has the following characteristics:

$$\rho_p = 1700 \text{ kg/m}^3, \quad c^* = 1550 \text{ m/s}, \quad \gamma = 1.2$$
$$r_b = 0.6 P_c^{0.35} \,(p_c \text{ in MPa, } r_b \text{ in cm/s})$$

In addition, we desire the motor to have a chamber pressure of 5 MPa, an expansion ratio of 25, a burning time of 10 s, and a vacuum total impulse of 100,000 N-s. Using this information, determine:

(a) the vacuum I_{sp} of the motor;

(b) the propellant mass necessary to meet impulse requirements;

(c) the throat radius required to maintain the desired chamber pressure; and

(d) the grain dimensions r_i and l shown in Figure 7.33.

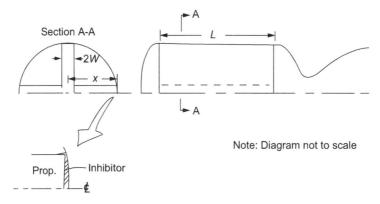

Figure 7.32 Diagram for Problem 7.14.

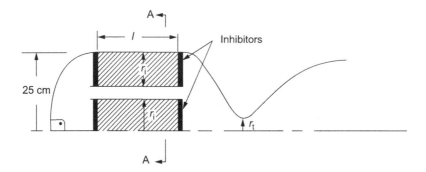

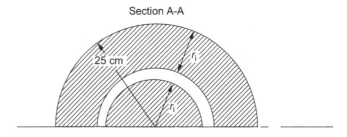

Figure 7.33 Diagram for Problem 7.15.

7.16 Consider the solid rocket motor shown in Figure 7.34. The following propellant, performance, and motor material characteristics are known:

Assumed motor performance and design characteristics

Performance parameters	Motor case
Initial chamber pressure = 2.5 MPa	D6AC steel construction
Propellant	Hemispherical domes
18% Al, 71% AP, 11% HTPB	Safety factor = 1.25
Density = 1800 kg/m^3	**Insulation**
Burn rate exponent = 0.3	Silica loaded EPDM formulation
Characteristic velocity = 1550 m/s	Density = 1100 kg/m^3
Nozzle	**Igniter**
State-of-the-art materials	Aft-end flame stick design
Flexseal TVC	
Expansion ratio = 70	
15° effective divergence angle	

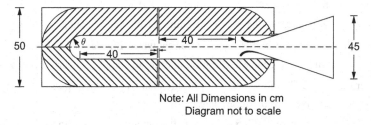

Note: All Dimensions in cm
Diagram not to scale

Figure 7.34 Diagram for Problem 7.16.

(a) Determine the burn rate constant (in cm/s/MPa$^{0.3}$) required to achieve the initial chamber pressure.

(b) Determine the chamber pressure at web distances of 4, 8, 12, and 16 cm.

(c) Presume that MEOP is 30% higher than the largest pressure determined in part (b). Determine the case thickness and weight per this assumption.

(d) Estimate the weight of the thrust skirts assuming a thickness twice that of the motor case.

(e) Assume an average insulation thickness of 0.8 cm and determine the insulation weight.

(f) Determine the nozzle weight using the correlation provided in class. Presume that the igniter weight is included in this estimate.

(g) Estimate the inert weight of the SRM. Increase the weight of your motor case from part (c) by 25% to account for polar bosses and attachment fittings.

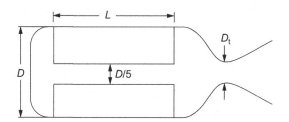

Figure 7.35 Diagram for Problem 7.17.

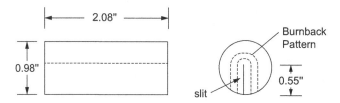

Throat Diameter, $D_e = 0.175"$, Expansion Ratio = 3

Figure 7.36 Diagram for Problem 7.18.

(h) Determine the propellant mass fraction for this SRM.

(i) Estimate the ideal ΔV of the motor assuming a 1000 kg payload and vacuum conditions.

7.17 Consider the solid rocket ballistic test motor shown in Figure 7.35. The small motor uses a cylindrical grain with a center perforation in order to achieve a nearly neutral burn.

(a) Determine the L/D ratio which will provide identical chamber pressures at ignition and just prior to burnout of the propellant grain.

(b) Assume we want a total of 0.5 kg of propellant in the test motor and we desire a chamber pressure of 3 MPa for the test. Determine the dimensions L, D, and D_t to produce these results. The propellant characteristics are:

Burn rate = 1 cm/s, density = 1700 kg/m³, $c^* = 1500$ m/s, $\gamma = 1.2$

[Hint: If you can't get an answer in part (b), you may assume $L/D = 2$ to work this part of the problem.]

7.18 Perform a ballistics & performance analysis on the Aerotech F-50 solid rocket motor shown in Figure 7.36. The propellant formulation is:

Aluminum – 10%

HTPB – 20%

Ammonium perchlorate (AP) – 70%

Density = 0.06 lb/in³

$$r_b = 0.035 p_c^{0.4} \ (p_c \text{ in psi}, r_b \sim \text{in/s})$$

Assume that the slot can be approximated by an axial slit running the length of the grain as shown in the figure. In addition, assume that the end faces are inhibited such that burning occurs only on the periphery of the slit.

(a) Compute the initial p_c from the information given and run the NASA thermochemistry code to get ideal I_{spv} and c^* for this propellant.

(b) Derive equations for the burn surface area, A_b, as a function of web distance burned. What is the total web distance for this grain design?

(c) Using your results from parts (a) and (b) write a short Matlab routine to solve for chamber pressure and thrust histories. Use 100 steps through the total web. Attach a listing of your code. Plot p_c, vacuum thrust, and r_b vs. time and A_b vs. web distance.

7.19 Consider a solid rocket motor to be optimized for a space propulsion mission to impart a ΔV to a 1000 lb payload. The motor contains 5000 lb of propellant, but its inert mass will vary depending on the burn time we choose. Suppose the inert mass can be written:

$$M_{in} = 100 + M_c + M_{ins} + M_{noz} \text{ (in lb)}$$

The case (M_c), insulation (M_{ins}), and nozzle (M_{noz}) masses are all functions of burn time:

$$M_c = 100 + 500/t_b$$

$$M_{ins} = 50 + 0.2t_b$$

$$M_{noz} = 200 + 0.01t_b$$

(all masses in lb, t_b in seconds)

(a) Assuming the nozzle contour (i.e. A_t, ϵ values) and I_{sp} are fixed, find the t_b value which maximizes performance (ΔV) of this motor.

(b) Explain why the case, insulation, and nozzle masses have the functional dependence on t_b as noted in the equations above. In other words, what physical mechanisms lead to this behavior with t_b?

7.20 Solid propellant ignition systems frequently employ cylindrical-shaped pellets such as those shown in Figure 7.37. Assume the pellet density (ρ_p), characteristic velocity (c^*), and geometric dimensions L and D as shown in the figure are all known. Furthermore, you may assume that pellet combustion obeys St. Roberts' law ($r = ap_c^n$) and that the burn rate coefficient and exponent (a, n) are both known.

(a) For a given throat size, A_t, determine how many pellets, N, would be required to obtain an initial chamber pressure of p_{ci}?

(b) Derive a relationship for the ratio of final and initial chamber pressures, p_{cf}/p_{ci}, where the final p_c value occurs just prior to burnout of the pellets. Assume $L = D/4$ in your derivation (your result should only depend on n). Show that this result provides a highly regressive pressure history for a typical n value of, say 0.5.

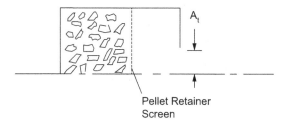

Pellet Design

Figure 7.37 Diagram for Problem 7.20.

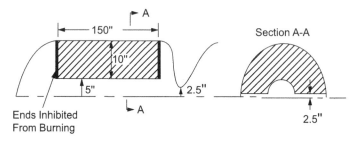

Figure 7.38 Diagram for Problem 7.21.

(c) Suggest a change to the pellet geometry to create a more neutral pressure trace. Draw a sketch of the new pellet and discuss the implications of your change on the design of the system. Would your change effect the size of the pellet container?

7.21 Consider the solid rocket motor in Figure 7.38 utilizing a slotted tube grain design. The propellant formulation Problem 7.6 and the density and burn rate are given by:

$$\rho_p = 0.06 \text{ lb/in}^3, \quad r_b = 0.015 p_c^{0.35} (p_c \text{ in psi}, r_b \text{ in in/s})$$

You are asked to develop a steady-state ballistics analysis of this motor, as follows:

(a) Using the initial burning surface areas, calculate the initial chamber pressure. Use this value in the NASA code to get c^*. Presume that this c^* value is accurate for the entire burn.

(b) Develop a set of equations for the burning surface area as a function of web distance based on the grain geometry shown in this schematic. Plot A_b vs. web distance for this SRM.

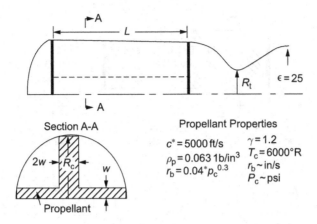

Figure 7.39 Diagram for Problem 7.22.

(c) Write a computer code, spreadsheet macro, or Matlab procedure to generate the chamber pressure history given the propellant information and the A_b–web relation from part (b). Plot p_c vs. time. Attach a listing of your code.

7.22 Consider the SRM with a cruciform grain design as shown in Figure 7.39. The cruciform design has the advantage that it provides a nearly neutral pressure and thrust history. Suppose we desire an initial p_c of 800 psi, a burn time of 20 s, and a total vacuum impulse of 40,000 lb-s for our motor. Using these requirements and the propellant information shown below, determine:

(a) the total web distance, w [hint: since the grain design is neutral, p_c is always near 800 psi];
(b) the grain length L shown in the figure assuming $R_c = 7.0$ inches; and
(c) the throat radius R_t shown in the figure.
(d) Comment on the overall desirability of using the grain design, i.e. are there any foreseeable problems?

7.23 Consider the high-powered solid rocket motor shown in Figure 7.40. The propellant grain is a simple perforated cylinder with inhibitors (restrictors) placed on the ends. As the name implies, the inhibitors inhibit burning on the surfaces to which they are applied. The baseline configuration employs this simple tubular geometry and we expect a progressive burning behavior since the radius of the cylinder increases as the propellant recedes. It is decided to modify the grain by making either one or two longitudinal cuts in order to expose more burn surface and obtain a more regressive behavior. These alternate geometries, along with propellant properties, are shown in Figure 7.40.

To reduce the environmental impact of our solid rocket motors, we have decided to use a "clean" propellant that uses ammonium nitrate (AN) instead of AP as oxidizer. AN propellants contain no chlorine, so we don't have to worry about acid rain showers

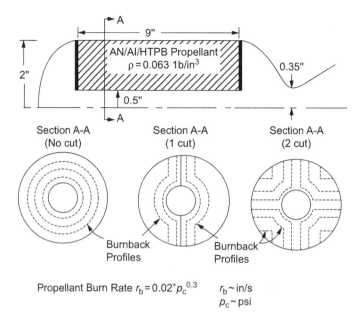

Figure 7.40 Diagram for Problem 7.23.

after we launch it. The down side regarding AN is that it burns slower and has lower performance than AP.

You are required to predict the ballistic performance of the modified rockets for the zero, one, and two cut grain geometries. You may assume that negligible propellant is lost in the cutting process.

(a) Run the NASA code for the *AN* propellant formulation assuming a chamber pressure of 300 psi and a propellant composition: 70% AN, 15% Al, 15% HTPB. Determine the vacuum I_{sp} and c^* for an expansion ratio of 4.0 and attach a listing of the relevant portion of the output (summary tables only).

(b) Develop analytic expressions for the burning surface as a function of web distance for each of the three grain configurations ($n_{cut} = 0, 1, 2$). Note down the definitions of any angles/lengths required in the derivations.

(c) Perform a quasi-steady ballistics analysis for each of the grain geometries and plot the thrust, p_c, and burn rate histories, as well as A_b vs. web.

(d) Attach a listing of your code and summarize your results.

7.24 The solid rocket motor shown in Figure 7.41 is to be designed for a space mission. The following propellant data are known:

$$\rho_p = 1700 \text{ kg/m}^3 \text{ kg/m}^3, \ c^* = 1550 \text{ m/s}, \gamma = 1.2$$
$$r_b = 0.6 P_c^{0.4} (p_c \text{ in MPa}, r_b \text{ in cm/s})$$

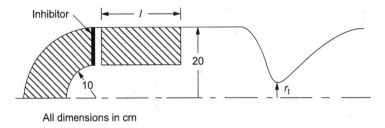

All dimensions in cm

Figure 7.41 Diagram for Problem 7.24.

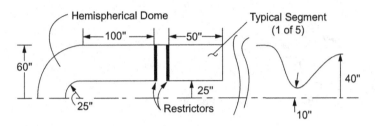

Figure 7.42 Diagram for Problem 7.25.

(a) What length of the cylindrical segment, l, would give a neutrally burning behavior for this motor?

(b) For the l value in part (a), estimate the burning time assuming chamber pressure of 4 MPa.

(c) If $l = 35$ cm and $\epsilon = 75$, determine:

    (i) the throat radius, r_t, to give an initial chamber pressure of 5 MPa; and

    (ii) the ideal ΔV the SRM would provide to a 1000 kg payload assuming $\lambda = 0.8$.

(d) Is the grain design in part (c) progressive, neutral, or regressive? Why?

7.25 Perform a ballistics and performance analysis on the solid rocket booster shown in Figure 7.42. The propellant formulation is:

Aluminum – 18%

HTPB – 12%

Ammonium perchlorate (AP) – 70%

Density = 0.0635 lb/in^3

$r_b = 0.04 p_c^{0.3}$ (p_c in psi, r_b in in/s)

The grain geometry contains five identical segments and a forward closure. Rubber is bonded on the forward face of each segment and on the end of the forward closure to restrict burning on these surfaces.

(a) Assume an average p_c of 1000 psi and run the NASA thermochemistry code to get ideal I_{spv} and c^* for this propellant.

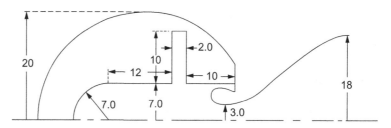

Note: All dimensions in inches

Figure 7.43 Diagram for Problem 7.26.

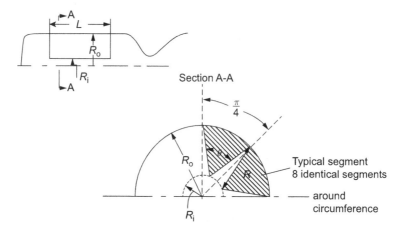

Figure 7.44 Diagram for Problem 7.27.

(b) Derive equations for the burn surface area, A_b, as a function of web distance burned. What is the total web distance for this grain design?

(c) Using your results from (a) and (b) write a short Matlab routine to solve for chamber pressure and thrust histories. Use 100 steps through the total web. Attach a listing of your code. Plot p_c, vacuum thrust, and r_b vs. time and A_b vs. web distance.

7.26 Consider the SRM shown in Figure 7.43. The propellant used in this motor has the following characteristics:

$$\rho_p = 0.063 \text{ lb/in}^3, \ c^* = 4900 \text{ ft/s}, \ \gamma = 1.2$$
$$r_b = 0.06 p_c^{0.25} \ (p_c \text{ in psi}, \ r_b \text{ in in/s})$$

Determine the initial chamber pressure and vacuum thrust of this motor, assuming steady-state operation.

7.27 Develop a ballistic prediction (vacuum F and p_c histories) for the SRM using the eight-point inverted star grain design shown in Figure 7.44. The following information is known:

$$\rho_p = 0.06 \text{ lb/in}^3, \ c^* = 5000 \text{ ft/s}, \ \gamma = 1.2$$
$$A_t = 4.0 \text{ in}^2, \ A_e = 40.0 \text{ in}^2$$
$$R_o = 10 \text{ in}, \ R_i = 2.0 \text{ in}, \ L = 15 \text{ in}$$
$$r_b = 0.025 p_c^{0.3} \ (p_c \text{ in psi}, \ r_b \text{ in in/s})$$

(a) Develop a series of analytic equations for the burn surface area (A_b) as a function of web distance, w. What are the values for R and θ in this grain design? What is the maximum web distance, w_{max}, for this grain design?

(b) Write a short computer code (or Matlab routine) to predict chamber pressure and thrust histories.

(c) Plot p_c vs. t, F_{vac} vs. t, and A_b vs. w.

(d) Attach a listing of your code.

CHAPTER 8   LIQUID ROCKET ENGINES

In this chapter, we will introduce the reader to the fundamentals of liquid rocket engines. A description of engine components and some basic terminology is included in Section 8.1. Section 8.2 provides a discussion of monopropellant systems, while bipropellant engine cycles are discussed in Section 8.3. Section 8.4 discusses propellant tanks and Section 8.5 provides information on thrust chamber design. Sections 8.6 and 8.7 describe LRE injectors and performance calculations, respectively. Section 8.8 provides an outline of lumped parameter systems analyses used for engine control and operations and Section 8.9 a brief note on additive manufacturing advances in the LRE community.

Readers should be aware that numerous other chapters touch on LRE system technology, with Chapter 5 describing performance analysis/thermochemistry aspects, Chapter 10 dedicated to turbomachinery, and Chapter 9 providing background on liquid propellants. In addition, Chapter 6 provides discussion of regenerative cooling systems and Chapter 12 covers feed system and combustion dynamics.

8.1 Introduction: Basic Elements of an LRE

In this chapter, we will investigate analytic and design techniques applicable to liquid rocket engines. As you are probably aware, the liquid rocket engine is one of the more common rocket propulsion systems in use today. Liquid engines enjoy performance advantages (I_{sp}) over solid motors and are more flexible in that many engines have a throttling capability to permit thrust changes as a function of time. On the down side, liquid engine systems are much more complex than solid systems and generally have significantly lower mass fractions.

Figure 8.1 highlights the basic features of a liquid rocket engine propulsion system. The propellant system can be characterized as being one of two basic types. In a *monopropellant* system, a single liquid decomposes (reacts) in the presence of a catalyst material to create hot gas. Monopropellant systems generally have moderate performance with typical vacuum I_{sp} values lying between 200 and 230 seconds. These systems are most attractive in applications where simplicity is highly desirable, and total impulse requirements are not large. For this reason, monopropellant systems are currently in use as control rockets in many of today's satellites and spacecraft. Small engines used in this application are normally called *thrusters* to distinguish them from larger counterparts used in launch vehicles.

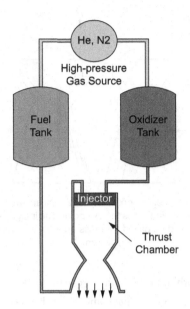

Figure 8.1 Basic features of a liquid rocket propulsion system.

Most larger rockets are *bipropellant* systems in which fuel and oxidizer are stored in separate tanks and then combined in the thrust chamber. Bipropellant systems are attractive in situations where the total impulse requirements are large. These systems can be further categorized into *storable* and *cryogenic* propellant combinations. Storable propellants are those that are liquid at ambient conditions, while cryogenics (usually liquid oxygen (LOX) or liquid hydrogen (LH$_2$)) must be stored at low temperatures to avoid boiling. The specific impulse of most bipropellants using hydrocarbon fuels lie in the 250–350 second range while LOX/LH$_2$ cryogenic systems lie in the 425–500 second range. The additional performance of the cryogenic system is somewhat offset by the additional complications (i.e. insulation, etc.) required in using these liquids.

Some bipropellant combinations (such as liquid oxygen/liquid hydrogen) must be ignited, while others are *hypergolic* in that the combustion occurs instantaneously as fuel and oxidizer combine. Some engines are designed only for a single thrusting operation, while others include a multiple *restart* capability for certain missions.

The *feed system* is also classified by two primary types. In a *pressure-fed* or *blowdown* system, a high-pressure gas (usually N$_2$ or He) stored in a separate tank is used to force propellants into the thrust chamber. The volume (and mass) of pressurant gas required is proportional to the total impulse requirement for the system. For this reason, blowdown systems can get quite large and heavy if the total impulse is substantial.

Applications demanding large impulse typically utilize a *pump-fed* configuration in which turbopumps are used to force propellants into the combustion chamber. The turbine in a turbopump system requires a hot gas source and that can be provided using a number of different alternatives as

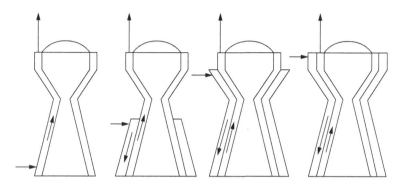

Figure 8.2 Regenerative cooling jacket options. The horizontal arrows show where the fuel enters the jacket, and the vertical arrow at the top indicates the exit point. The image on the left is a single-pass design while the two middle options would be called 1½-pass options. The image on the right is a two-pass design.

described in Section 8.3. The turbine provides the torque to spin the pump and provide the high exit pressure required to drive propellants into the combustion chamber.

The *thrust chamber* is typically a metal structure designed to contain the high-pressure gases, and exhaust them through the nozzle. Many large LREs utilize *regeneratively cooled* chambers where the fuel is pumped through a network of tubes (which comprise the chamber wall) in order to keep wall temperatures at acceptable levels. Several options for regenerative cooling passages are noted in Figure 8.2 and a detailed analysis of a "regen" jacket is described in Chapter 6. If regenerative cooling is not chosen, the walls must be able to remain intact at the high combustion temperatures. *Ablative liners* are often used in this case. In addition, film cooling techniques (as we discussed in accordance with turbine blades) are often utilized to reduce wall temperatures.

As in the design of any chemical rocket propulsion system, the chamber pressure is an important parameter. High chamber pressure is desirable for liquid engines since for a given flow rate (i.e. thrust) requirement a higher pressure engine will be smaller ($A_t = \dot{m}c^*/gp_c$). Figure 8.3 highlights the dramatic change in engine size as a function of chamber pressure. Of course smaller engines will in general yield lighter weight systems, which is the main driver in most rocket designs. The negative trades involved in using high chamber pressure come in through the design of the feed system, which grows in weight as a function of the chamber pressure. Obviously, this tradeoff is made during the preliminary design of any liquid rocket engine.

The *injector* is designed to atomize and mix propellants to promote efficient combustion in the chamber. Injector design is still approached in an empirical fashion due to the highly complex two-phase flow, combustion, and heat transfer processes in this region. The injector must promote mixing and smooth combustion and wall temperatures must also be maintained at acceptable levels.

Performance of bipropellant systems can be characterized in terms of the *mixture ratio, r.* This quantity is expressed as a ratio of oxidizer and fuel flow rates:

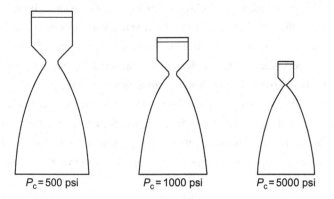

Figure 8.3 The effect of increasing chamber pressure on engine size (all engines designed to achieve same thrust level).

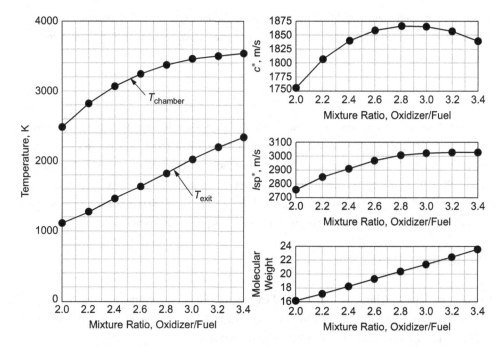

Figure 8.4 Typical bipropellant performance changes with mixture ratio. This example is for liquid oxygen (LOX)/liquid methane (LCH$_4$) propellants at a chamber pressure of 1000 psi.

$$r = \dot{m}_{\text{ox}}/\dot{m}_{\text{f}} \tag{8.1}$$

The typical engine performance is characterized as shown in Figure 8.4 for the liquid oxygen (LOX)/liquid methane (LCH$_4$) propellant combination. The chamber temperature, T_c, is maximized near the stoichiometric mixture ratio of 4.0 as shown on the curve. However, the molecular weight increases steadily with mixture ratio since the oxidizer typically has a higher molecular weight than the fuel. For this reason, the characteristic velocity is maximized at a fuel-rich condition near a mixture ratio of 3. Remember, we want to maximize $\sqrt{T_c/\mathfrak{M}}$ to get the highest c^*. The maximum I_{sp} occurs at a slightly higher mixture ratio of 3.2 because this parameter not only reflects c^* behavior but also c_f behavior that is a strong function of the ratio of specific heats, γ.

If we chose a mixture ratio, the propellant characteristic velocity can be determined. This information, along with the associated nozzle geometry, permits calculation of engine specific impulse. If the desired thrust level is known, the total propellant flow rate can be calculated. We can express fuel and oxidizer flow rates in terms of this mass flow and the selected mixture ratio:

$$\dot{m}_f = \frac{\dot{m}}{1+r} \tag{8.2}$$

$$\dot{m}_{ox} = \frac{r\dot{m}}{1+r} \tag{8.3}$$

Another important variable is the propellant bulk density, ρ_b, which represents the ratio of the total mass of propellant and the total volume it occupies. Bulk density can be written in terms of fuel and oxidizer ratios, as well as the mixture ratio:

$$\rho_b = \frac{1+r}{1/\rho_f + r/\rho_{ox}} \tag{8.4}$$

By knowing this parameter, we can determine the total volume of tankage required. High bulk density is desirable since a smaller volume of tankage will be required to house a given mass. This is the primary problem with the use of LH_2 as a fuel. While LOX has relatively high density (71 lbm/f³, 1140 kg/m³), LH_2 density is only (4 lbm/f³, 64 kg/m³). At a mixture ratio of 6, this cryogenic propellant combination has a bulk density of 21 lbm/f³, which is only about one-third that of water. Therefore, large tanks are required to house this propellant combination.

8.2 Monopropellant Systems

A monopropellant is defined as a single fluid capable of undergoing exothermic reaction to yield gaseous products. Generally, we refer to this reaction as a *decomposition process*. Monopropellant systems enjoy several advantages over bipropellant systems:

- Only a single tank and propellant feed system is required.
- Injection is simplified in that there is no need to bring two separate fluids into contact.
- Monopropellant systems are less sensitive to changes in temperature. Differences in density changes with temperature in bipropellant systems can lead to an *outage* situation in which one propellant is expended prior to the other due to a higher density sensitivity to temperature. This factor also leads to shifts in the mixture ratio.

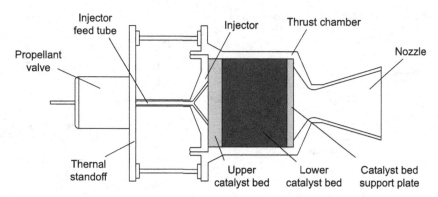

Figure 8.5 Features of a typical monopropellant thruster.

- Field operations (and on-orbit operations for satellite systems) are simplified since only a single propellant is present.

Figure 8.5 highlights some of the features of a monopropellant thruster utilizing hydrazine (N_2H_4) as propellant. Liquid propellant decomposes into hot gases with the aid of a *catalyst*. The function of the catalyst is to promote rapid decomposition of the liquid by reducing the activation energy required to initiate chemical decomposition reactions. The catalyst material is generally applied as a coating and housed within a *catalyst bed*. The "catbed" is composed of high surface area features (either granular material or a series of finely woven screens) in order to rapidly expose liquid molecules to the bed surface and to initiate the exothermic reaction for the relatively cold propellant entering the chamber.

Flow processes within the catalyst bed are poorly understood as a variety of two-phase flow conditions (from bubbly flow that is mainly liquid to dispersed flow that is mainly gaseous with just a few drops) and the process of thruster development tends to be highly empirical. The main parameter governing the flow processes is the massflux or *bed loading*, $G = \dot{m}/A$, where A is the cross-sectional area of the catbed. Here G has units of lbm/in^2-s or kg/cm^2-s. Higher G values are desirable as this implies that we can achieve higher propellant flows within the same cross-sectional area. However, raising G also raises the velocities in the bed and reduces residence time, so there is a threshold beyond which the bed will become flooded and the conversion from liquid to gas will become incomplete. Generally, this limit can only be determined empirically due to the complexity of the two-phase flows within the bed itself. The other aspect of increasing the bed loading is that the pressure drop of the fluid tends to increase due to higher frictional losses within the tortuous path. The pressure drop characteristics are important at a system level as the tankage or fluid delivery system must provide sufficient pressure to overcome this loss and sustain the desired thruster chamber pressure.

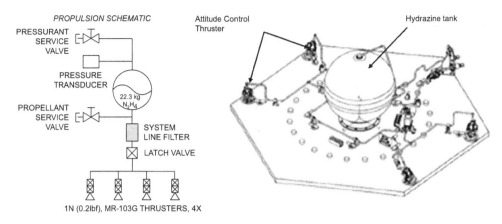

Figure 8.6 Satellite installation of hydrazine propulsion system for NASA EO-1 spacecraft (Source: Aerojet Rocketdyne).

The thruster shown in Figure 8.5 contains both a coarse and fine catalyst bed to optimize the decomposition process. Other designs use a single catalyst that is appropriate for the mission requirements. This thruster also utilizes a *catalyst heater*, which is a simple electrical coil used to preheat the catalyst prior to thruster operation. Because they are often located in outboard locations on spacecraft to maximize the moment arm, many monopropellant thrusters also employ valve heaters (not shown in Figure 8.5). Preheating the catbed minimizes thermal shock to the catalyst particles during the ignition phase and enhances the decomposition rate, allowing for higher bed loading than a cold system. Despite these measures, the large thermal and mechanical loading of the beds at rocket bed loading conditions does lead to erosion of the catalyst material over time and the designer must consider lifetime requirements as a result.

The perforated structure shown in Figure 8.5 is called a *thermal standoff*, and its function is to minimize the temperature increase near the valve during long duration operation of the thruster. Tubes running through the standoff region carry propellant to the first catalyst bed. The *heat shield* (wraps around outer portion of chamber, but not shown in Figure 8.5) serves as a protection barrier for satellite structure in the region near the thruster. This feature also improves the function of the catalyst heater to minimize external heat losses when the heater is operating.

As mentioned above, the relative simplicity of monopropellant systems makes these devices good candidates for satellite and spacecraft attitude control systems. A typical installation of a hydrazine system in a spinning satellite is shown in Figure 8.6. Redundancy in thrusters is sometimes included to ensure that no single point failures are present. In general, two thrusters are required for each axis of control (pitch, roll, yaw). Older spacecraft employed spin stabilization as this reduced the number of control functions (and thrusters), but most modern systems employ full three-axis stabilization with pitch, roll, and yaw thruster requirements. Multiple tanks offer some level of redundancy but also provide a mechanism to maintain the vehicle center of gravity in the desired location. If a single tank is employed, as in the EO-1 system shown in Figure 8.6, it is placed

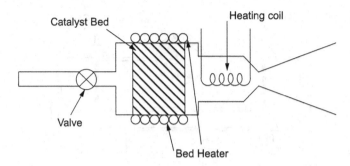

Figure 8.7 Electrothermal augmented hydrazine thruster. The heating coil imparts additional energy into the decomposition gases to raise temperature and specific impulse.

very near the center of gravity (c.g.) of the vehicle so that the c.g. shifts very little with propellant utilization.

In many satellite applications, propulsion maneuvers (called *stationkeeping*) can be scheduled at times of low power usage. Taking advantage of this fact, the *augmented hydrazine thruster*, shown in Figure 8.7, was developed. In the augmented thruster, electrothermal energy (from satellite solar panels) is used to further heat the decomposition gases in order to augment performance. Using this technique, the effective I_{sp} of a hydrazine thruster has been increased from 230 seconds (unaugmented) to 320 seconds by applying 6.2 watts of electrical energy per pound of thrust. These devices have been used on several satellites currently in orbit. The increase in chamber temperature (ΔT_c) can be simply estimated by equating electrical energy (P_e) added to the mixture to the change in thermal energy:

$$\Delta T_c = P_e/(\dot{m}c_p) \tag{8.5}$$

The associated increase in I_{sp} can then be estimated recognizing that this parameter scales with $\sqrt{T_c/\mathfrak{M}}$.

8.3 BIPROPELLANT SYSTEMS AND ENGINE CYCLES

As noted in the introduction to this chapter, most of the larger engines utilize bipropellant systems to achieve higher performance levels than those attainable from monopropellants. In fact, bipropellant systems are even finding applications in the small thruster arena for attitude and control functions and for divert systems for exo-atmospheric interceptors. As noted in the monopropellant discussion, the performance advantages of a bipropellant system must be weighed against the additional complexity associated with two propellants. In general, separate tankage, lines, and feed systems are required for fuel and oxidizer thus increasing system complexity. In addition, injector design, combustion, and ignition processes are all more complex for the bipropellant engine.

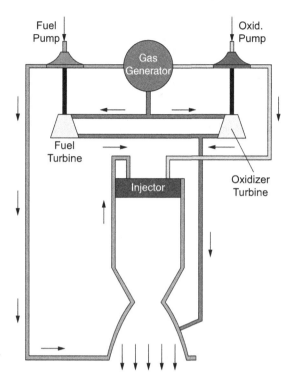

Figure 8.8 Gas generator liquid rocket engine cycle.

In spite of these shortcomings, performance advantages have led many designers to adopt this option, and it is clear that bipropellant systems will be in use well into the future.

Since most of the applications for bipropellant engines have large total impulse requirements, the systems are usually pump fed. In addition, most large engines are regeneratively cooled, which implies that the walls of the chamber are made of a collection of tubes or channels in which fuel flows to act as a coolant. As discussed in Chapter 6, regenerative cooling is a very efficient concept since energy lost through chamber walls serves as a preheating source for the propellant. Fuel is usually used as the regenerative coolant due to the fact that oxidizer in these regions is normally thought of as a potential ignition hazard. Oxidizer cooling has been safely demonstrated in research rigs and represents a tantalizing alternative given the fact that most of the flow is oxidizer (i.e. mixture ratios are greater than unity) in bipropellant combinations in use today.

Because of the variety of approaches to provide power to pumping systems, there are several different *engine cycles* that can be used for these systems. A common cycle in use today is the *gas generator cycle*, which utilizes a small combustion chamber (called a *gas generator*) to provide energetic gases to drive turbomachinery. Figure 8.8 provides a schematic representation of the gas generator (or "gg") cycle, assuming that fuel flows through a regenerative cooling jacket prior to entering the injector. Alternate configurations are possible in which fuel enters the injector directly, for example if an ablative thrust chamber were employed. Temperature limits usually set the gas generator to a fuel-rich mixture ratio and as a result "sooting" within the turbine can be a problem when one uses hydrocarbon fuels such as kerosene. The hot gases from the gas generator drive both the oxidizer and fuel turbines in the configuration shown. Alternately, a single turbine may be employed to drive both pumps. Upon exiting turbines, the gases may be "dumped" into the nozzle exit cone (as shown in Figure 8.8), or a separate exhaust nozzle may be built to optimally expand these gases. In some applications the gg exhaust exits through a moveable nozzle that can provide vehicle roll control. Current engines utilizing the gg cycle include the Merlin engine used on SpaceEx vehicles, the RS-68 used on the Delta vehicle, and the Vulcain engine used on the Ariane vehicle (to name a few).

One variant on the gas generator cycle utilizes the hot chamber gases themselves as a thermal energy source to drive turbomachinery. This cycle, known as the *combustion tap-off*, bleeds off combustion gases near the injector face (where the temperature is perhaps as low as half of the chamber value) in order to power turbomachinery. The main advantages of this approach include the elimination of the gas generator and the wide range of throttling capability afforded by the tap-off feature. Elimination of a combustion device is a major consideration since a large amount of engineering is required to develop each combustor. The design of the tap-off is crucial as it is possible to ingest chamber gases that are too hot for the turbine to tolerate. Fuel is typically mixed with the tap-off gases to reduce their temperature. Downstream of the tap-off location, the hottest chamber gases could reattach to the wall and create undue thermal stresses, so a delicate balance is required to manage the heat flux to the chamber wall. The J-2S engine developed as an upgrade to the original J-2 used as the second stage of the Saturn V utilized this cycle. The Blue Origin BE-3 engine uses this cycle with oxygen and hydrogen propellants.

Another variant on the gas generator cycle utilizes evaporated coolant to provide energetic gases to power turbomachinery. This cycle, called the *coolant bleed or coolant tap-off cycle*, uses coolant that has passed through the regenerative cooling jacket as a working fluid to drive turbines. Normally, this cycle can only be used for cryogenic propellants since evaporation of the coolant (liquid hydrogen) must take place. Because of the required evaporation, the energy level of gases entering the turbines is set by material properties in the main chamber, which tends to limit this cycle to low to moderate chamber pressures. Like the combustion tap-off cycle, this approach requires no gas generator.

The gas generator and tap-off cycles share the attribute of having the turbine exhaust dumped into the exit cone (or a separate nozzle) rather than back into the main combustion chamber. Cycles of this nature are referred to as *open cycles*. In a *closed cycle* engine, pressures of turbine exhaust gases are maintained at pressures high enough to reintroduce these gases into the main combustion chamber. Closed cycle engines enjoy efficiency advantages over the open cycle engines in that turbine exhaust gases are expanded to higher velocities. For this reason, the I_{sp} of a closed cycle engine will be higher than the open cycle alternative if all other design parameters are kept fixed.

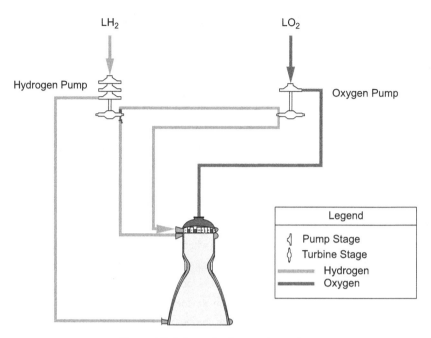

Figure 8.9 Schematic of expander cycle.

One type of closed cycle alternative shown in Figure 8.9 is the *expander cycle*. This cycle is similar to the coolant bleed cycle in that vaporized coolant is used to drive turbomachinery, but in this case the exhaust gases are fed directly into the combustion chamber. The expander cycle is utilized only with cryogenic propellants (generally utilized liquid hydrogen as fuel) and is limited to low-to-moderate chamber pressures due to the energy of gases driving the turbines.

In this cycle, the heat addition from the regen jacket provides the thermal power for the turbine, which in turn determines the maximum pressure rise in the pumps. For this reason, there tends to be a chamber pressure limitation wherein the chamber wall environments become too severe for the materials employed. From a practical standpoint, this limits chamber pressures in expander cycle engines to somewhere in the neighborhood of 1000 psi. In spite of this fact, it is a simple, high-performance cycle that typically provides a lightweight engine. In addition, since liquid hydrogen will vaporize at ambient engine start conditions, this engine is "self-starting." The RL 10 engine used on the Centaur and the Vinci engine used on Ariane use this engine cycle for space engine applications. The expander cycle is arguably the supreme upper-stage engine cycle as it provides a relatively simple engine with a single combustion device and closed cycle performance. The fact that the engine is limited in pressure is not a concern for the space environment since the engine operates in a vacuum and has essentially an infinite pressure ratio on the nozzle as a result.

The *staged combustion cycle* shown in Figure 8.10 is the most advanced of the closed cycle alternatives. Energy to drive turbopumps is obtained from a *preburner*, which combusts a portion of

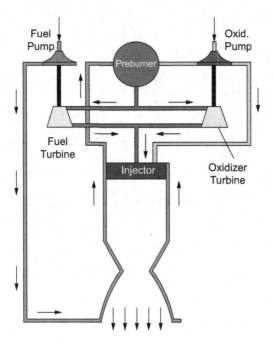

Figure 8.10 Schematic of staged combustion engine cycle. The preburner produces warm gas at high flow rate as a power source for the turbines and can be operated at fuel-rich or oxidizer-rich conditions to control the temperature of gases entering the turbine.

one propellant with essentially all of the other propellant. Because the preburner flow is much higher than from a typical gas generator, staged combustion engines can generate huge amounts of power to turbomachinery that can be translated to very high pump pressure rises and result in very high chamber pressures. For this reason, the staged combustion alternative provides the highest performance for space launch engines. While the high preburner power and chamber pressure translate to high performance, they also introduce engineering challenges and costs associated with containing the high pressures in difficult oxidizer- or fuel-rich combustion gas environments.

In fuel-rich staged combustion (FRSC), all of the fuel is combusted with a small amount of oxidizer, while in oxidizer-rich staged combustion (ORSC), the reverse is true. The Space Shuttle Main Engine (SSME) (now the RS-25 to be used on NASA's Ares launch vehicle as expendable engine) and the Japanese LE-7 engine used on the H-II vehicle make use of FRSC using hydrogen fuel. This cycle is challenging for using kerosene fuels due to the potential for coking within the heavily fuel-rich preburner exhaust stream. However, liquid methane or liquefied natural gas (LNG) are less prone to this issue.

As noted above, significant fluid power can be generated by vaporizing the oxidizer stream as it represents the majority of the flow in propellant combinations employed in bipropellant engines. For this reason, there is an opportunity to develop substantially more turbine power (and hence pump pressure rise and chamber pressure) using the ORSC approach. Here, the

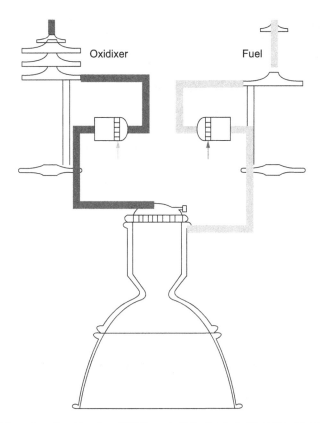

Figure 8.11 Full-flow staged combustion (FFSC) cycle featuring fuel-rich and oxidizer-rich preburners.

principle challenge lies in working with warm/hot oxidizer gases produced by the preburner since warm oxidizers are highly reactive and can regard flowpath walls as "fuel." A key technology for these engines is oxidizer-resistant metal alloys or coatings that inhibit flammability at the aggressive conditions emanating from the preburner and entering the turbine. In the 1970s and 1980s, the Soviet Union conquered these challenges and developed a number of engines (NK-33, RD-180, RD-170, RD-181) that incorporate the ORSC technology. Currently, the Russian RD-180 is used as first-stage propulsion on the Atlas V launch vehicle but a number of US firms are seeking to develop engines based on this technology, including the Aerojet Rocketdyne AR-1 and Blue Origin BE-4 engines.

Another challenge that stems from staged combustion cycles stems from the fact that a fuel- or oxidizer-rich turbine gas shares the same shaft as a pump that is delivering the opposite fluid. In this case, there is an inter-propellant seal that is a critical item as leakage in this seal could bring the two propellants together in small cavities within the turbopump. One way to mitigate this issue is to develop two preburners, one that is fuel-rich and one that is oxidizer-rich, in order to drive the respective fuel and oxidizer pumps, as shown schematically in Figure 8.11. This alternative is referred to as full-flow staged combustion (FFSC). The FFSC cycle is arguably the supreme bipropellant engine cycle as the total flow (or nearly so) of each of the propellants is producing turbine power and hence pressure rise in the respective pumps. The large power generated by the high preburner flows permits operation at lower temperatures, thereby simplifying the environments in at least some parts of the engine. The main combustor operates with gas/gas injection and is believed to be more stable than combinations involving a liquid propellant. However, very little data exist to corroborate this theoretical advantage.

The downside of this cycle stems from the complexity of operating three separate high-power combustion devices at the same time. Also, because the cycle affords high-pressure operation, the demands of working preburner and main chamber environments at very high-pressure levels is always a challenge. NASA first demonstrated the efficacy of this cycle in the integrated powerhead demonstrator (IPD) program in the 1990s. Currently, the SpaceX Raptor engine is to employ this cycle using oxygen and methane propellants.

8.3.1 Engine Power Balance

The power balance lies at the heart of the preliminary engine sizing as the requirement that turbine power output equals pump power demands provides a crucial relationship between turbine and pump operating conditions. Assuming adiabatic flow, the turbine power output (P_t) can be expressed in terms of the mass flow flowing through the device:

$$P_t = \dot{m}_t \eta_t (h_{01} - h_{02}) = \dot{m}_t \eta_t c_p (T_{01} - T_{02}) \tag{8.6}$$

where $()_1$ and $()_2$ represent inlet and outlet conditions, respectively, and η_t is the turbine efficiency, which represents the ratio of delivered to ideal power output. As mentioned in Chapter 10, typical values are in the range 20–70%, depending on the type of turbine and the number of stages employed. Using isentropic flow relations, Eq. 8.6 can be expressed in terms of the stagnation pressure ratio p_{01}/p_{02}:

$$P_t = \dot{m}_t \eta_t (h_{01} - h_{02}) = \dot{m}_t \eta_t c_p T_{01} [1 - (p_{02}/p_{01})^{(\gamma-1)/\gamma}] \tag{8.7}$$

so that larger pressure ratios, T_{01}, or turbine mass flows can be used to generate higher levels of power. Generally we require that the turbine flow be choked so that power output becomes insensitive to ambient pressure variations. Since, a sonic isentropic flow has a pressure ratio $p_{02}/p_a = [(\gamma + 1)/2]^{\gamma/(\gamma-1)}$, this ratio has a value of 1.8–1.9 depending on the γ value of the turbine exhaust. To avoid elaborate cooling, turbine inlet temperatures are generally limited to roughly 1500 °F (800 °C). The turbine inlet pressure is bounded by the pressure level in the combustion device feeding the turbine. As a result, the engine designer really utilizes $\dot{m}_t \eta_t$ to control turbine power level with flow rate being the most obvious parameter to adjust.

The pump power (P_p) demand similarly scales with the enthalpy changes across the device as noted from the First Law of Thermodynamics:

$$P_{\mathrm{p}} = \dot{m}_{\mathrm{p}}(h_{01} - h_{02})/\eta_{\mathrm{p}} \tag{8.8}$$

where we divide by the pump efficiency, η_{p}, as inefficiency in the pump increases the power required to achieve a given enthalpy rise. Typical η_{p} values are in the range 50–80% and this accounts for frictional and other aerodynamic losses in the process. Pumps are also subject to a small amount of leakage that would generally need to be taken into account in a detailed system power balance. For the incompressible fluids being pumped, the stagnation enthalpy rise can be written in terms of the pressure rise using the Second Law of Thermodynamics:

$$h_{01} - h_{02} = (p_{01} - p_{02})/\rho \tag{8.9}$$

so that the net pump demand is:

$$P_{\mathrm{p}} = \dot{m}_{\mathrm{p}}(p_{01} - p_{02})/(\rho\eta_{\mathrm{p}}) \tag{8.10}$$

It becomes obvious that higher power is demanded at higher flow rates, higher pressure rises, or lower liquid densities. Equation 8.10 makes it apparent that pumps operate as volumetric flow devices in that the power scales directly with volumetric flow, $\dot{V}_{\mathrm{p}} = \dot{m}_{\mathrm{p}}/\rho$. It is this factor that makes liquid hydrogen a difficult fluid to pump since its density is only about 1/15th that of water. Finally, we note here that the pump power is negative, i.e. it implies the work input/power demand of this level.

Example: Gas generator cycle power balance

To illustrate the application of a shaft power balance, consider the gg cycle described in Figure 8.12. Since the role of the gas generator is solely to provide the turbine gas source, we should be able to determine all its operating characteristics ($\dot{m}_{\mathrm{fgg}}, \dot{m}_{\mathrm{oxgg}}, p_{\mathrm{cgg}}, T_{\mathrm{cgg}}$) by reflecting on the power balance and turbine conditions we desire. As opposed to finding individual gg flow rates, it is more convenient to solve for the total flow rate $\dot{m}_{\mathrm{gg}} = \dot{m}_{\mathrm{fgg}} + \dot{m}_{\mathrm{oxgg}}$ and mixture ratio, $r_{\mathrm{gg}} = \dot{m}_{\mathrm{oxgg}}/\dot{m}_{\mathrm{fgg}}$. Since we have a single turbine driving two pumps in this case, the turbine power output must equal the sum of the two pump power demands. In addition, the pumps must not only provide fluids for the main combustion chamber, but also the flows going into the gas generator. In this case, using Eqs. 8.9 and 8.10 the applicable power balance can be written:

$$
\begin{aligned}
&\dot{m}_{\mathrm{gg}}\eta_{\mathrm{t}}c_{\mathrm{p}}T_{\mathrm{cgg}}[1 - (p_{\mathrm{te}}/p_{\mathrm{cgg}})^{(\gamma-1)/\gamma}] \\
&= \left(\dot{m}_{\mathrm{ox}} + \dot{m}_{\mathrm{gg}}\frac{r_{\mathrm{gg}}}{1+r_{\mathrm{gg}}}\right)\frac{\Delta p_{\mathrm{ox}}}{\rho_{\mathrm{ox}}\eta_{\mathrm{pox}}} + \left(\dot{m}_{\mathrm{ox}} + \frac{\dot{m}_{\mathrm{gg}}}{1+r_{\mathrm{gg}}}\right)\frac{\Delta p_{\mathrm{f}}}{\rho_{\mathrm{f}}\eta_{\mathrm{pf}}}
\end{aligned} \tag{8.11}
$$

From the definition of turbine efficiency, we can also relate the turbine exit temperature to the pressure ratio:

$$\frac{T_{\mathrm{te}}}{T_{\mathrm{cgg}}} = 1 - \eta_{\mathrm{t}}[1 - (p_{\mathrm{te}}/p_{\mathrm{cgg}})^{(\gamma-1)/\gamma}] \tag{8.12}$$

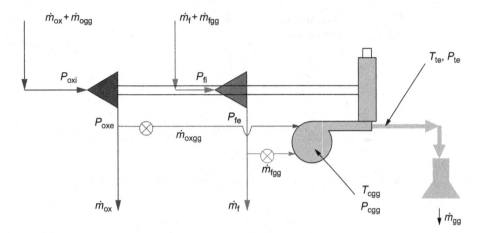

Figure 8.12 Power balance on a gas generator cycle.

We first need to select a gas generator operating pressure. Here, we are restricted to the minimum of the two pump exit pressures and then must account for any losses in lines, valves and across the gg injector, i.e.:

$$p_{cgg} = \min(p_{oxe}, p_{fe}) - \Delta p_{valve} - \Delta p_{line} - \Delta p_{gg\text{-inj}} \qquad (8.13)$$

Guidelines for injector pressure drops are provided in Section 8.5 below. Line losses are generally small and valve losses are typically less than 10% of the overall pressure level. The mixture ratio in the gas generator is set to provide an acceptable gas generator exhaust/turbine inlet temperature by skewing the mixture to very fuel-rich operation as illustrated in Figure 8.13. As an example, for LOX/RP-1 propellants the optimal mixture ratio occurs at r_{gg}=2.3, but a gas generator may run as low as r_{gg} = 0.4 to achieve the desired turbine inlet temperature. With hydrocarbon fuels, sooting can be a major concern, with the excessive amounts of kerosene. While there is another solution in an oxidizer-rich mixture ratio, this result is typically not employed due to the difficulties/hazards associated with working with hot oxidizer-rich gas.

Given p_{cgg} and r_{gg} from this process, one can solve Eq. 8.11 for the required gas generator flow rate and Eq. 8.15 for the turbine exit temperature. Here we stipulate a p_{te} value to ensure a choked exit flow as mentioned previously. Knowing this pressure, we can design a turbine nozzle to return the maximum amount of thrust from this stream as we would do with any nozzle flow. The resulting I_{sp} from the turbine can then be computed from the theoretical exit velocity and pressure thrust as described in Chapter 10. Similar results can be generated for the main combustion chamber, where a better mixture ratio and a higher overall pressure ratio will permit a much higher I_{sp} value. An engine level performance can be estimated based on the total thrust and total flow:

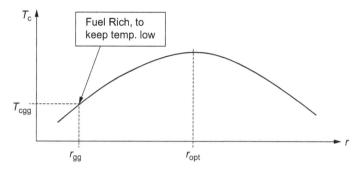

Figure 8.13 Setting r_{gg} to provide an acceptable turbine inlet temperature.

$$I_{speng} = \frac{\text{total thrust}}{\text{total massflow}} = \frac{\dot{m}I_{sp} + \dot{m}_{gg}I_{spgg}}{\dot{m} + \dot{m}_{gg}} \tag{8.14}$$

Equation 8.17 demonstrates the performance loss associated with the open cycle approach since $I_{spgg} \ll I_{sp}$, the engine level result drops a few percent due to the poor turbine exit performance.

The process outlined in this example is a crucial element of the early design process for a bipropellant engine. The decision on engine cycle and overall system topology is fundamental and sets the development path for the propulsion device. As you know, the optimization of chamber pressure is one of the most important system attributes and since that pressure (and the associated injector pressure drop) must be provided by the turbomachinery there is a strong coupling to the overall system operation. While the schematics in Figures 8.8–8.11 provide a simple overview of the engine cycles, they obviously do not convey the large engineering efforts required to develop the plumbing, valving, and mechanical systems required to make a functional device. While the performance of the staged combustion alternative is attractive, this performance gain must be weighed against the additional development cost and time required as compared to simpler cycles, so it is not a foregone conclusion that any one cycle is superior in any given instance.

8.4 LRE Propellant Tanks

All liquid systems require tanks to store propellants prior to injection into the thrust chamber. It is obvious that separate fuel and oxidizer tanks are required for bipropellant systems, while monopropellant systems only require a single tank. As highlighted in Figure 8.6, system redundancy or the desire to control the vehicle c.g. often dictates that multiple tanks be utilized. An example of a case where c.g. travel must be minimized is in the design of kinetic kill vehicles (KKVs) developed for exo-atmospheric interceptors. Figure 8.14 shows a four-tank arrangement to limit c.g. offsets during operation.

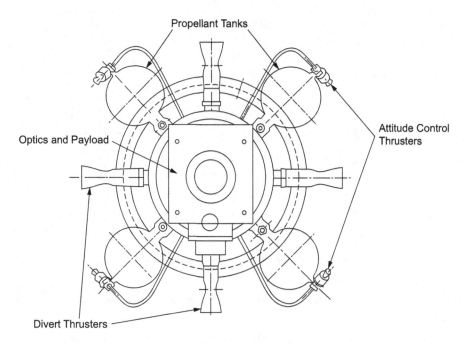

Figure 8.14 Kinetic kill vehicle using four tank arrangement.

In launch vehicle or other airborne vehicle applications, tanks are usually cylindrical in shape in an effort to limit frontal area and thus minimize drag. Tanks may be designed in *tandem*, *concentric*, or *twin* tank arrangements as shown in Figure 8.15. Tandem tanks may employ a common bulkhead design to minimize inert mass as shown in the figure.

For space applications, spherical tanks are desirable since drag is not an issue and a spherical tank is the most efficient way to package propellant. For a given propellant volume, it is easy to show that the wall surface area of a spherical tank is always less than that of a cylindrical tank:

$$\frac{S_{sp}}{S_{cy}} = \frac{1.5^{2/3}}{\frac{1}{2(L/D)^{2/3}} + (L/D)^{1/3}} \tag{8.15}$$

where S_{sp} and S_{ay} are the surface areas of the spherical and cylindrical tanks, respectively, and L/D is the length-to-diameter ratio of the cylindrical tank. This relation proves that the surface area ratio is always less than unity when the tank length/diameter ratio, L/D, is greater than unity, so that a spherical tank is always the most efficient way to house propellant (i.e. tank weight is proportional to surface area). Finally, we should note that in volume limited systems such as OTVs, *toroidal tanks* have been proposed as an efficient means to make use of the available volume. While a clever geometrical approach, there are numerous engineering challenges with toroidal tanks including

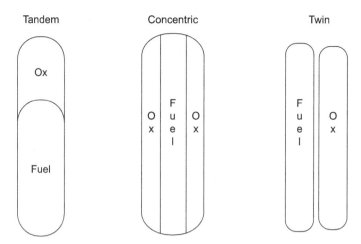

Figure 8.15 Bipropellant tanking arrangements.

fabrication and propellant utilization, and thus far there are no known applications as far as the authors are aware.

The volume requirements for a propellant tank are set primarily by the mass of liquid required and its density. Minor changes in liquid density with temperature must be taken into account in the calculation. In addition, a small gas volume called *ullage* is usually included to account for these volume changes with temperature. Cryogenic tanks must incorporate additional volume to account for *boiloff*, which occurs during the period between tank loading and launch. Finally, a small amount of propellant remains *trapped* at the bottom of the tank and in engine feed lines. Taking all these factors into account, we can estimate the required tank volume as:

$$V_t = V_p + V_u + V_{tr} + V_b \tag{8.16}$$

which includes contributions from propellant, ullage, trapped propellant, and boiloff losses. Generally V_{tr} and V_u are only a few percent of the propellant volume, while V_b can vary considerably depending on propellant type (cryogenic or storable), tank insulation design, tank size, and the time interval between loading and launch. In some applications, pressurant tanks are housed within the main cryogenic propellant tank. This approach permits the pressurant volume to be minimized as it is cooled to low temperature and densified. SpaceX currently embeds composite overwrap pressure vessel (COPV) helium tanks within the liquid oxygen tank on the Falcon 9 vehicle in order to exploit these physics. Helium is used as fuel tank pressurant on the Falcon 9 vehicle so a large volume of gas is required and storage at cryogenic temperature provides a substantial COPV weight savings.

Just like any other pressure vessel, the tank must be designed to withstand the maximum pressure load under all possible operating conditions. In addition, *buckling loads* exerted by structures placed above the tank must also be considered. For our discussion, let us consider the

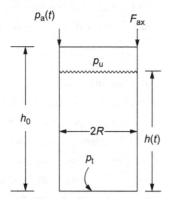

Figure 8.16 Loads on a typical cylindrical tank.

cylindrical tank shown in Figure 8.16. The maximum pressure occurs at the bottom of the tank, where the sum of the tank head and ullage pressure gives:

$$p_t = p_u(t) + \rho g_e \eta(t) h(t) - p_a(t) \tag{8.17}$$

where the vehicle acceleration (η), liquid column height (h), ambient pressure (p_a), and ullage pressure (p_u) are in general all functions of time. Note that the tank head contribution can be significant for high-density or high-acceleration tanks. The maximum pressure is a function not only of tank design and propellant characteristics, but also interacts with the vehicle trajectory through the acceleration, η. In general, the peak pressure will be reached at some time after ignition when the tank level and acceleration give the highest tank head contribution.

If internal pressure loads design the case thickness, for a cylindrical tank we can write:

$$t_c = p_{t(max)} R / \sigma_w \tag{8.18}$$

where σ_w is the maximum allowed working stress level:

$$\sigma_w = \min\left(\frac{F_y}{1.1}; \frac{F_{tu}}{fs}\right) \tag{8.19}$$

where F_y and F_{tu} are material yield and ultimate tensile strengths, and fs is the factor of safety (1.25 for unmanned systems and 1.4 for manned systems).

As mentioned above, we must ensure that a tank with this thickness does not buckle under the imposed load, F_{ax}. The peak axial stress applied to the tank structure is also a function of time since acceleration loads are time dependent:

$$\sigma_{ax} = \max[(p_u - p_a)\pi R^2 - F_{ax}] / (2\pi R t_c) \tag{8.20}$$

where we have assumed that the axial loads are distributed over the entire perimeter of the thin tank wall. In any event, this relation (or an equivalent form) can be used to estimate the peak buckling pressure loads on the tank. If $\sigma_{ax} < 0$ the tank wall is in compression and buckling may occur. An empirical expression used to predict the critical buckling pressure of long tanks is:

$$\sigma_{cr} = -E(9(t_c/R)^{1.6} + 0.16(t_c/L)^{1.3}) \tag{8.21}$$

where E is the Young's modulus for the tank material.

As long as $\sigma_{ax} > \sigma_{cr}$, the tank will not buckle under the imposed loads. Note from Eq. 8.20 that we can reduce the buckling pressure by increasing the tank pressure, which effectively drives the pressure vessel into tension ($\sigma_{ax} > 0$) if the tank pressure is high enough. At this point, buckling is no longer a consideration in the tank design. This technique is called *pressure stabilization*. The most successful example of this design approach is in the liquid tanks used on the Atlas vehicle. The wall material in these tanks is so thin that the tanks are not capable of supporting their own weight under a zero pressure condition (tank pressure equals ambient pressure). A positive tank pressure must be maintained at all times to avoid buckling of these tanks, which are some of the lightest tanks designed for use in a launch vehicle. However, maintenance of a positive pressure at all times can be an engineering challenge, as Atlas engineers would tell you, and there is still debate in the community if this option will ever be incorporated in future designs.

While the above guidelines provide insight into the load balance on a tank, real tank structures are quite complex and require detailed structural analysis, which will likely include finite element analysis techniques accounting for both pressure and external loads such as tank bending or local interfaces for systems. As an example, Figure 8.17 details the cross-section of the external tank used on the Space Shuttle vehicle. This tank employed a tandem arrangement with separate oxidizer and fuel tanks (no common bulkhead). The aft fuel tank contains the lower-density hydrogen and occupies the bulk of the volume, while the upper oxidizer tank contains the higher-density LOX in a much smaller volume but has the bulk of the mass. This large LOX weight located high on the vehicle relative to the vehicle's center of mass necessitates the addition of *slosh baffles* in the tank to mitigate side loads caused by propellant motion in flight. Estimation of slosh and damping benefit provided by the baffles is beyond the scope of this text, but interested readers could consult Abramson (1967) and Yang *et al.* (2012), which are listed in the Further Reading.

Anti-vortex baffles limit the swirling of the propellant as the propellant flows out the bottom of the tank and helps ensure that ullage pressurant gas is not drawn into the feedline as the tank empties (a solution to the vortex formation that we commonly see in draining the bathtub). A LOX feedline takes the propellant from the tank's outlet and carries the propellant outside and along the hydrogen tank, eventually re-entering at the vehicle's aft end and connecting to the engine. Alternatives exist for bringing this propellant line down the middle of the lower tank, particularly in common bulkhead tank configurations. The design of this *downcomer* requires significant attention as any leaks will result the direct mixing of the two propellants in the tank and a large vehicle explosion. For propellants with different freezing temperatures, the downcomer needs to be insulated and in some cases a vacuum-jacketed, two-walled design is employed to minimize any heat transfer and ensure no freezing of either propellant.

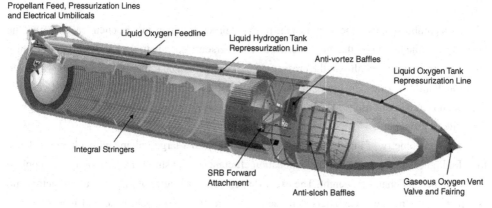

Figure 8.17 Cross-section of Space Shuttle external tank.

Tank pressurization lines are also shown running along the outside of the main structure, which continuously add ullage gas to the tank to fill the additional free volume created as propellant leaves the tank during flight. A vent is included on each tank used during tank loading to minimize the pressure build-up as a significant amount of propellant is boiled during tank chill-in for cryogenic applications and to permit ullage gas to exit if pressure levels build beyond design levels due to heat transfer or excess pressurant flow. Another item not highlighted in the image is the *propellant estimation system*, which provides gauging of the fluid levels in the tank during flight to ensure both propellants are completely used.

Tank walls are typically made from machined plates of aluminum that are rolled and welded together. An *isogrid/orthogrid* structure, comprised of triangular/rectangular webs machined into the tank walls, provides rigidity to transmit compressive thrust loads along tank walls. Machined panels are rolled into the desired cylindrical diameter and friction stir welding is performed to join the ends. Circumferential frames and axial running stringers are often added to provide additional structural strength and stiffness, primarily to increase the buckling strength of the tank. Stiffening may not be required if the compressive loads are limited, such as with the Centaur launch vehicle, whose tank is thin, unstiffened sheet metal. This innovative design maintains the tank in a tension (non-buckling critical) configuration until after the tank is pressurized, at which time the pressure loads keep the tank in tension and therefore immune to buckling. Tank domes can be spun on a mandrel or constructed as welded assemblies. Work continues in the industry on composite tanks; as an example students can access the NASA website dedicated to this technology: https://gameon.nasa .gov/projects-2/archived-projects-2/composite-cryogenic-propellant-tank/. Virgin Galactic has developed impressive capabilities to manufacture liquid oxygen and liquid methane tanks entirely from composite materials.

8.5 LRE THRUST CHAMBERS

The main combustion chamber contains the physical process converting the chemical energy of the propellant to the hot gases that are accelerated at supersonic speeds in the rocket nozzle. Because choked flow occurs at the throat, the pressure in the combustion chamber is directly proportional to the propellant flow rate. The figure of merit for a combustor is the characteristic exhaust velocity, c^*, which is a measure of the stagnation pressure that can be generated in the chamber for a given flow rate of propellant.

The geometry of the chamber is a trade-off between a large volume with a long residence time for higher thermodynamic and combustion efficiencies and small chambers to reduce cooling requirements and chamber weight. The exact shape is often a matter of manufacturing factors and designer experience. For chambers that use pintle injectors, a rounded head end is preferred to better accommodate the radial flow pattern generated by the pintle. Cylindrical chambers combined with a converging section to the throat are common. There is no clear preferred design for the converging section, but well-known engineering approaches for the nozzle throat contour exist which will be discussed in Section 8.7. In low-thrust engines, the combustion chamber and the nozzle may be one piece, but in larger engines the part that is the combustion chamber usually ends downstream of the throat at an expansion ratio that is less the overall nozzle expansion ratio. For the purpose of this discussion, however, the combustion chamber will be considered to end at the entrance to the throat, where the nozzle starts.

After selecting the propellant combination, we would like to understand how this choice influences our chamber design. The chamber should be long enough to permit complete combustion of the propellant at the desired inlet conditions and chamber pressure. We can write the residence time in the chamber as:

$$t_r = l_c/v_c = \rho_c A_c l_c \dot{m} \tag{8.22}$$

where l_c, v_c, ρ_c, and A_c are chamber length, gas velocity and density, and chamber cross-sectional area, respectively. We can write $\dot{m} = p_c A_t/c^*$ and $\rho_c = p_c/RT_c$ to give:

$$t_r = \frac{c^*}{RT_c}(A_c l_c/A_t) \tag{8.23}$$

Note that the fraction on the right-hand side of this relation is a function of the propellant combination alone and is a constant for a given propellant. The term $A_c l_c/A_t$ represents the ratio of the chamber volume to the throat area of the nozzle and is therefore a unit of length. We define this parameter as the *characteristic chamber dimension*, L^* and note that:

$$t_r = \text{const. } L^* \tag{8.24}$$

so that the chamber residence time increases as this dimension increases.

Table 8.1   Characteristic chamber length (L^*) values for selected propellants

Propellant combination	Minimum L^*, m
LOX/RP-1	1.0–1.3
LOX/LH$_2$	0.7–1.0
LOX/GH$_2$	0.5–0.7
Hydrazine family/NTO	0.7–0.9
H$_2$O$_2$/RP-1 (including catalyst)	1.5–1.8

By conducting a series of tests with different chamber lengths (i.e. L^*) values and assessing the combustion efficiency through c^* calculations, we can determine the minimum L^* required to give essentially complete combustion of the propellants. Table 8.1 provides some typical L^* values for a variety of propellant combinations. Heavier hydrocarbon molecules contained in kerosene (RP-1) take longer to combust than lighter fuels like hydrogen which burns very rapidly.

With information of this kind, the designer can provide an estimate of the chamber length required since $l_c = L^* A_t / A_c$. Note that L^* varies with chamber pressure so that this effect should be taken into account in calculations. Also note from Table 8.1 that combinations using hydrogen fuel have the lowest L^* value. This is due to the high evaporation rate of these propellants and the inherent speed of kinetics reactions in the oxygen/hydrogen combustion process. For this reason, cryogenic engines will normally have shorter chambers than a comparable engine using storable propellants.

The chamber *contraction ratio* (chamber cross sectional area to throat area), CR, is another important design parameter and tends to range from about 1.3 to 10. A number of competing criteria must be considered when selecting a contraction ratio. Because reducing CR increases the chamber Mach number, Rayleigh losses become more pronounced in this limit.[1] However, increasing CR increases the chamber diameter and decreases the acoustic frequencies in the chamber, making them more susceptible to excitation. Increasing CR/ chamber diameter also increases chamber weight. These latter considerations become most important in larger engines and they tend to have smaller CR values as a result. The F-1 engine that powered the Saturn V is an example of this as it had CR = 1.3. For mid-sized and smaller engines, the Rayleigh losses tend to be a dominant consideration and as a result these engines tend to have larger CR values.

Other practical considerations can effectively set the contraction ratio. Small thrusters tend to have large CR values as high as 10 as a practical consequence of integration of valves into the injector/chamber body. Staged combustion engines that inject lower density gases can effectively have the injector face area (and hence CR) set by the mass flow and pressure level of the

---

[1]   Rayleigh losses are stagnation pressure losses resulting from heat addition (combustion) at finite Mach number. This loss mechanism has a direct impact on thrust as the stagnation pressure feeding the nozzle determines the thrust production of the engine.

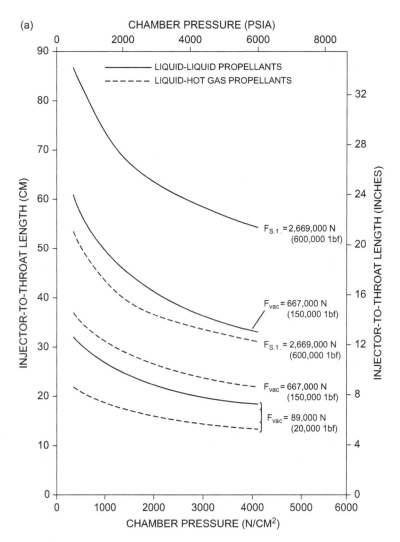

Figure 8.18 Combustor size trends for LOX-Kerosene engines from 1979 NASA/Rocketdyne study. (a) Dimensionless chamber length, L'/D_c, and (b) chamber contraction ratio, CR, are plotted vs. chamber pressure.

propellants. For the peroxide/kerosene propellant combination, where the temperature of the decomposed peroxide (the oxidizer) is high enough to cause autoignition of the kerosene fuel, the contraction ratio tends to be at least 5, but it should be noted that it is also dependent on the specific injector design.

A NASA study conducted by Rocketdyne in 1979 provides some further background on chamber size. Figure 8.18 summarizes the results of sizing studies for oxygen/hydrocarbon combustors, where chamber pressure and thrust level were varied parametrically for a combustor

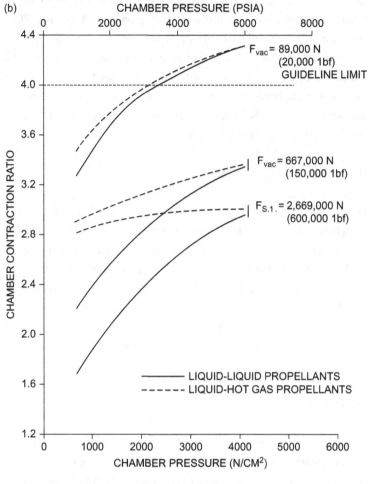

Figure 8.18 (cont.)

that would provide a c^* efficiency of 0.98. Both liquid/liquid and gas/gas injectors were considered. In Figure 8.18(a), the dimensionless chamber length, L'/D_c, is plotted as a function of chamber pressure. Both L' and D_c shrink as p_c increases, but their ratio tends to increase. In Figure 8.18(b), the chamber contraction area ratio, CR, is shown to increase with p_c as throat area decreases more rapidly than chamber cross-sectional area. It should be noted that the mixing and combustion efficiency is highly dependent on the specific injector design, hence the results from these plots or Table 8.1 should only be used as a guide.

It is important to keep the chamber weight as low as possible, so structural analysis is a critical part of chamber and nozzle design. Mechanical and thermal stresses must be considered in the analysis. The chamber life must exceed the number of cycles it is expected to experience in a development test program or the life requirement for a reusable engine.

Because of the inexact nature of the life analysis, safety factors of around 4 are typically used for chamber materials that have a good mechanical properties data base. The life-limiting cause of rocket chambers is usually thermal strain induced low cycle fatigue, which leads to cracks in the chamber. Strains above a few percent are usually too great. Hot gas wall temperatures must be kept below about 1000 K, and care must be taken to minimize thermal gradients. For long-use applications, operating wall temperatures are kept below about one-half the melting temperature of the material to eliminate creep, which is a long-term degradation of material properties that leads to failure.

8.5.1 Preburners and Gas Generators

Preburners and gas generators provide hot gases to drive the turbine stage of stage combustion and gas generator engines, respectively. The combustion gas temperature of these devices is set by the structural requirements of the turbine blades by operating the mixture ratio at a value far from stoichiometric. It is important that the gas composition and temperature at the turbine inlet is as uniform as possible to reduce low cycle fatigue of the turbine blades. Usually a fuel-rich gas is used for the turbine drive gas. In staged combustion LOX/hydrocarbon engines, an oxidizer-rich preburner is used to eliminate the soot that occurs during fuel-rich conditions and fouls the turbine blades. Full-flow staged combustion engines also use oxidizer-rich preburners to drive the oxidizer turbopump.

8.5.2 Ignition Devices

Ignition is the incipient reaction between the propellants that causes an initial heat release sufficient for sustained flame propagation. It is imperative that the ignition delay is very short to prevent the buildup of unreacted propellants in the chamber, which could result in damaging overpressures at ignition or an explosion. The relative timing of propellant entry in the chamber (i.e. fuel-lead or ox-lead) is an important consideration. Ignition devices are not required for hypergolic propellants or when the temperature of one of the propellants exceeds the autoignition temperature of the other propellant.

Two types of ignition devices are generally used – torch ignition and hypergolic "slugs." Combustion wave ignition is a type of torch ignition. There is also current research into using laser ignition for rocket engines. Sometimes a strategically located specialty spark plug is used for an ignition source, especially for propellants that are both gases at startup. Hypergolic ignition is achieved by presenting a momentary flow of a propellant that is hypergolic with one of the propellants, prior to the introduction of the other propellant into the chamber. For instance, triethylaluminum-triethylboron (TEA-TEB) burns spontaneously when contacted with oxygen. In practice, a slug of TEA-TEB is injected simultaneously with oxygen at the start, which results in robust combustion. Kerosene is then injected into the chamber while the TEA-TEB/oxygen combustion is active, igniting the kerosene, and allowing the engine to proceed to main-stage conditions. This method is simple and very reliable, but it requires the repetitive handling of the generally dangerous hypergolic igniter before each run. The effects of the particulate aluminum that is a product of TEA-TEB ignition on downstream parts must be considered. Other hypergolic propellants can be substituted for TEA-TEB, such as inhibited red fuming nitric acid (IRFNA).

Torch ignition is the method of choice where multiple relights are required, or in the preburners of a staged combustion engine where particulate flow into the turbine stage could harm turbine blades and cause blockage in downstream components. In a torch igniter, a separate flow of gases, usually oxygen and hydrogen, are lit in a separate combustion chamber by a spark igniter. The hot gases flow into the chamber where they ignite the main propellants. The torch ignition device is a subsystem and requires its own propellant supply, purge gases, valves, regulators, etc., and thus is very complex and requires development. Quenching of the spark kernel occurs at high pressures and thus ensuring ignition early in the flow process is critical. The flow rate of the torch gas and its location relative to the injector are also important factors, and correlations exist for the torch flow rate and spark energy requirements.

8.6 LRE INJECTORS

The injector design lies at the very heart of the LRE configuration in that its configuration and operating conditions dictate how the propellants mix and burn in the chamber and hence its overall static and dynamic behavior. The injector is responsible for metering, distributing, and atomizing propellants for efficient combustion within the chamber. A successful design must function adequately during engine start, nominal thrust level operation, and any throttling/reduced thrust requirements that stem from a particular application. The design must provide acceptable levels of heating to the injector face and thrust chamber with minimal "hot spots" due to local flame impingement, O/F excursions, or hot gas recirculation. The design needs to utilize an acceptable level of pressure drop for fluid acceleration, atomization, and mixing.

For the novice, it is difficult to appreciate the struggles involved with coming up with a suitable injector design. Because many of the two-phase flow and jet atomization characteristics are not well understood, injector design remains largely an empirical art rather than a science. In many applications, the fluids are injected at supercritical or transcritical conditions, implying that surface tension has vanished and there are no discrete surfaces attributed to droplets. In these cases, the mixing processes are fundamentally different to atomization and more akin to mixing continuous fluids of differing densities. The processes are inherently turbulent and challenge even the most powerful CFD tools in the modern age. The properties of the mixture during the millisecond mixing timeframes are poorly understood or defy accurate measurement so simplified mixing rules are applied as a result. Interaction with the adjacent combustion environment provides an additional feedback mechanism that is poorly understood. These are but a few of the technical challenges one faces when trying to predict what will happen when high flow rocket propellants are introduced into a combustion chamber.

As an example of the tedious exploration of injector configurations, the authors relay a story about the development of the Lunar Module Descent Engine developed by TRW for landing Apollo astronauts on the moon. Due to the landing requirement, the engine had to be throttled over a large range and development of injector concepts went on for a period of several years. The TRW managers became very concerned when presenting concepts that numbered into triple digits (i.e. over 100) so they decided to change the numbering strategy to make this point a bit less obvious to high-level NASA customers. Of course, the end product did its job in famous fashion, but it is a testament to the difficulties in iterating injector concepts as the design space that includes innumerable combinations of configurations and design operating characteristics. This story has been repeated time and time again in engine developments and, as a result, the hard-fought successes gained have been internalized to the point that particular injector configurations are preferred for specific propellant combinations as outlined in the following sections.

8.6.1 Top-level Considerations in Injector Design

Due to the complexity of the complex processes set in motion by the injection process, injector designers have to rely on experience and "engineering judgement" to develop concepts to evaluate experimentally. While the advances in computer power are enabling greater and greater use of CFD simulations to guide the process and to downselect attractive designs, the human element is still a very important part of the design process. Mixing of the two fluids is one of the most challenging issues for the CFD community and the designer. The injector designer is keenly aware that *molecular mixing* is the goal, i.e. to get chemical reaction we need to place a molecule of oxidizer adjacent to a molecule of fuel. One of the prime considerations to the designer is the *volume flow ratio* ($\dot{V}_{ox}/\dot{V}_f$) of the two propellants. Table 8.2 gives volume flow ratios for several bipropellant combinations. Noting that staged combustion applications create gas/liquid or gas/gas main chamber injectors, the volume flows become dependent on chamber pressure via the dependence of gas density on this parameter. One can observe that liquid/liquid and gas/gas systems have volume flow ratios somewhere near unity, but in gas/liquid systems the gas volume flow greatly exceeds the liquid volume flow and dispersion of the small volume of liquid in the gas becomes a challenge.

Individual passages through the injector face are referred to as *injector elements*. The flow through an element is governed by Bernoulli's equation with a discharge coefficient (C_D) to account for mass flow reduction due to viscous and multidimensional flow effects:

$$\dot{m}_e = C_D A \sqrt{2\rho g \Delta p} \tag{8.25}$$

The injector pressure drop is generally set to ensure no interaction between the feed system and chamber and is typically 5–50% of the chamber pressure. Discharge coefficients lie in the range $0.6 < C_D < 0.9$ for most injector elements.

Table 8.2   Oxidizer/fuel volume flow ratios for selected propellant combinations*

Propellant combination	Mixture ratio	$\dot{V}_{ox}/\dot{V}_f$
LOX/kerosene	2.3	1.63
LOX/LH$_2$	6	0.37
LOX/LCH$_4$	3.4	1.25
NTO/MMH	2	1.22
H$_2$O$_2$/kerosene	4.5	2.51
GOX/kerosene	2.3	13.60
GOX/LH$_2$, $p_c = 100$ atm	6	3.10
GOX/LCH$_4$, $p_c = 100$ atm	3.4	10.45
LOX/GH$_2$, $p_c = 100$ atm	6	0.04
LOX/GCH$_4$, $p_c = 100$ atm	3.4	0.23
GOX/GH$_2$, $p_c = 100$ atm	6	0.34
GOX/GCH$_4$, $p_c = 100$ atm	3.4	1.89

* All values at 300 K temperature.

Many liquid rocket injectors exhibit droplet atomization phenomena as a primary feature of the mixing process. Atomization is a rich field with numerous books and papers that describe the process. However, rocket injectors prove to be one of the most difficult challenges due to the density of the sprays and operation at high-pressure conditions. Readers should refer to the Further Reading for materials beyond what we can include here in this introductory text (see Anderson *et al.*, 2006; Ashgriz, 2011; Bayvel and Orzechowski, 1993; Cheng *et al.*, 2003; Doumas and Laster, 1953; Gull and Nurick, 1976; Lefebvre, 1989; Mayer, 1961; Muss *et al.*, 2003; Nurick, 1971; Rodrigues *et al.*, 2015; Ryan *et al.*, 1995; Strakey *et al.*, 2002; Riebling, 1967; Yule and Chinn, 1994).

The *Weber number*, We, is the most important parameter characterizing the atomization process:

$$We = \rho v^2 D/\sigma \tag{8.26}$$

where D is the orifice diameter or other relevant length scale, ρ is the liquid density, v is the injection velocity, and σ is the liquid surface tension. Physically, this parameter represents the ratio of inertial forces attempting to break the drop to surface tension forces that work to keep the drop intact. In realistic rocket conditions, Weber numbers can easily be 10^5–10^6, implying that inertial forces are vastly larger than surface tension. Since the ultimate droplet size obtained will be somewhere near the point where these forces are balanced, this implies that droplets formed can be several orders of magnitude smaller than the orifice/injector dimensions. At rocket flow conditions this implies a staggering number of drops and a challenge beyond any supercomputer or measurement apparatus to practically resolve each one.

Atomization measurements have been conducted using cold flows for most of the injector configurations at low-pressure (i.e. low gas density) conditions, and these data are typically represented as a correlation of *Sauter mean diameter* (SMD), which is the droplet diameter that is reflective of the total surface area present in the spray. Since the drops will vaporize and react based on their surface area, SMD is the most appropriate average drop size to utilize in this analysis. Correlations present SMD as a function of the Reynolds and Weber numbers pertinent to the injection scheme. The correlation for SMD takes the form:

$$\text{SMD} = \text{SMD}(\text{Re}, \text{We}), \quad \text{Re} = \rho v D / \mu, \text{We} = \rho v^2 D / \sigma \qquad (8.27)$$

where ρ, μ, and σ are the liquid density, viscosity, and surface tension respectively, v is the injection velocity, and D is the characteristic orifice dimension (diameter or hydraulic diameter). Note that liquid density is sometimes used in defining We so readers should be aware of this fact. The atomization process is driven by hydrodynamic instabilities that result from gas/liquid or liquid/liquid interactions. Classical instability mechanisms are Kelvin–Helmholtz (instability of high-velocity gas passing over a liquid surface), Rayleigh–Taylor (instability of a dense fluid accelerated by a less dense fluid), and impact waves from liquid/liquid collisions. The atomization process is characterized by larger ligament structures shed from a parent surface that break into large droplets, which can undergo secondary atomization and break into smaller droplets.

Armed with an estimate for the SMD for a candidate injector concept, droplet vaporization rates can be estimated from the classic D^2 *law*:

$$D^2(t) = D_0^2 - kt \qquad (8.28)$$

where D_0 is the initial drop diameter (the SMD value from Eq. 8.27) and k is a constant determined from heat transfer to the droplet and its specific heat and heat of vaporization characteristics. Students unfamiliar with the D^2 law can consult any combustion text for background on this fundamental result (e.g. Turns, 1995). Integrating Eq. 8.28 in time provides the location where complete evaporation has occurred and this point can be compared with the combustor length value to assess if the length is adequate.

Unfortunately, the process outlined in Eqs. 8.26–8.28 is highly idealized and fraught with pitfalls:

- In many high-pressure engines, the liquids are supercritical. This implies that surface tension forces are absent and the mixing process is governed by turbulent interactions of two dense fluids.
- Even when liquids are subcritical and droplets exist, SMD data are often not available, particularly at the high-pressure (i.e. high gas density) conditions of interest.
- It is unclear what environment the droplets are convecting through relative to a D^2 law treatment as they are formed in "cold" regions near the liquid surface and convect into hotter regions in the combustor.

So we see that the injector designer is left with a rather hopeless situation relative to development of an analytic model of the processes. Nevertheless, we do think it is useful for students to think about the underlying physics and try to apply them as guidance for more empirical paths that might be taken.

8.6.2 LRE Injector Types

There are a relatively small number of injector designs employed in current bipropellant LRE injectors. Perhaps the first notion one would have in designing a bipropellant injector is the simple *showerhead* injector concept shown in Figure 8.19. Alternating rows of fuel and oxidizer rings feed axial holes in each ring to form the injector pattern. This element design was abandoned fairly early in the industry as the mixing processes are simply inferior to other concepts. However, it is worth noting that high-precision machining is not required in the showerhead and it is perhaps more forgiving if poor fabrication is the best one has. The basic showerhead configuration has been used for rows of fuel jets arranged at the chamber wall for film cooling (see Chapter 6).

The US began exploring *impinging element* injectors in the early days of rocketry as it was assumed that excellent mixing would result from direct impingement of two liquid jets; effectively the dynamic pressure in the jets can work directly to create atomization in this case. A wide variety of options exist from like or unlike "doublets" comprised of two streams colliding, to triplets or even quadlet elements comprising three and four jets respectively. Typical impingement angles vary from 45° to 60° in many designs. Figure 8.20 provides a schematic of a *like doublet* configuration and an image of the resulting spray development as viewed from a plane perpendicular to the view shown in the schematic. The impact region is characterized by a spray fan that expands radially from the impact point. A large surface area is created in this region to enhance mixing of the two fluids. Instabilities in the thinning film lead to the creation of ligaments that break up into droplets shortly after being shed from the parent sheet.

The *unlike doublet*, comprised of a single fuel jet impacting a single oxidizer jet or *triplet* elements, comprised of two oxidizer jets and a single fuel jet, provides an attractive option for hypergolic propellants as it provides for large interfacial contact between the two fluids. From the volume flow ratios in Table 8.2, we can see that liquid/liquid combinations have volume flow ratios that are near unity and therefore might be good candidates for this style of injector. For NTO/MMH, for example, an unlike doublet configuration looks attractive since both fluids have similar volume flows. A triplet configuration has been employed with LOX/kerosene where two LOX jets impinge on a single fuel jet. Higher volume flow ratios associated with hydrogen peroxide systems tend to configurations with multiple oxidizer jets impinging with a single fuel jet. Unlike configurations are preferred for hypergolic propellants as this promotes rapid ignition. When hypergols are employed, gases will begin to evolve in the contact region and can blow the two streams apart and reduce mixing effectiveness as a result. Impingement velocities need to be high enough to mitigate this unwanted blow-apart effect by essentially reducing the breakup time below that of the gas

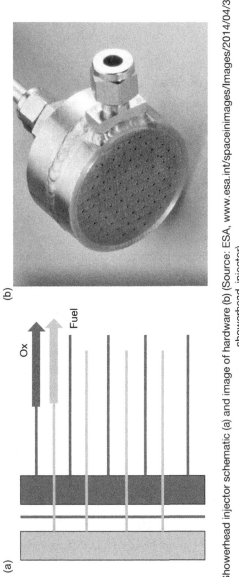

(a)

(b)

Figure 8.19 Showerhead injector schematic (a) and image of hardware (b) (Source: ESA, www.esa.int/spaceinimages/Images/2014/04/3D-printed _showerhead_injector).

Ox

Fuel

(a)   (b)

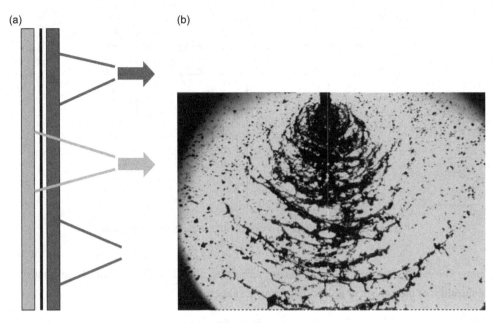

Figure 8.20 (a) Like doublet impinging element injector schematic. (b) High-resolution image of impact region showing spray fan, ligament, and droplet development.

evolution time. Like-on-like configurations may be preferred for non-hypergolic propellants as this permits the mixing/flame region to stand off further from the injector face to mitigate overheating of this surface.

Direct impingement of propellant streams permits for rapid mixing, atomization and ignition. Doublet designs work best when the interacting jets are of nearly the same diameter and high-precision machining is required to ensure that the two streams intersect in the proper plane. NASA SP-8089 provides criteria for triplet and quadlet injectors based on mixing measurements taken at atmospheric conditions. Many of these early experiments performed mainly in the 1960s used heated wax that would solidify upon atomization. Inspection of the resulting particles of solid wax permitted some level of quantitative measurement.

The research community has made strides to improve on these measurements and a couple of references are included in the Further Reading (see Anderon *et al.*, 2006; Ashgritz, 2011). One of the more obvious features in Figure 8.19 is the presence of "impact waves" that create striated ligament structures downstream of the impact point. Atomization models have been developed based on the wavelength of these features.

Figure 8.21 outlines some of the general considerations for design of an unlike doublet injector. The designer is free to choose individual impingement angles, α_f, and α_{ox}. As noted above, it is desirable to have equal injector diameters, so $D_f = D_{ox}$ in this case. In addition, to ensure that the

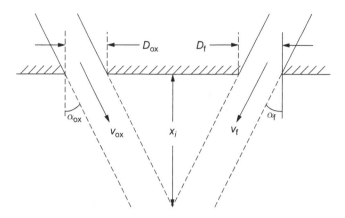

Figure 8.21 Geometry for an unlike doublet impinging element design.

spray fan evolves perpendicular to the injector face, it is desirable to cancel the tangential momentum of the two streams. This requirement suggests:

$$\dot{m}_f v_f \, sin\alpha_f = \dot{m}_{ox} v_{ox} \, sin\alpha_{ox} \tag{8.29}$$

Impingement angles are set to control the impact point location, x_i, as angles that are too steep may lead to splashback onto the injector face and angles that are too shallow can lead to premature aerodynamic distortion or breakup of the liquid jets. As noted previously, typical $\alpha_f + \alpha_{ox}$ values would lie in the 45–60° range. The mixture ratio of the streams is known from thermo-dynamic performance considerations and diameters would be set by manufacturing or cost constraints. The injection velocities are set by atomization characteristics or engine stability (a certain pressure drop is essential to effectively isolate the injector from acoustic disturbances in the chamber). The combination of these criteria permits the designer to set all the variables noted in Figure 8.21.

The *splashplate* injector is an approach that is closely related to the impinging element concept in that it creates similar flowfields downstream of the splashing location. This concept has the advantage that precise machining of the injection orifices is not necessarily required, but the disadvantage of having an additional surface (the splashplate itself) that may be subject to local overheating depending on the installation. Like the impinging element concept, a variety of options exist for impinging individual propellant streams, or combined propellant streams (as in Figure 8.22) onto the splashplate. The US pursued splashplate injectors in early days of LRE development, but as machining capabilities advanced, many developers switched to the impinging element approach. Currently there does not appear to be a large amount of effort dedicated toward this particular injector concept. Design guidelines for splashplates are similar to impinging injectors as the plate can be thought of as the symmetry plane/impingment point of multiple jets.

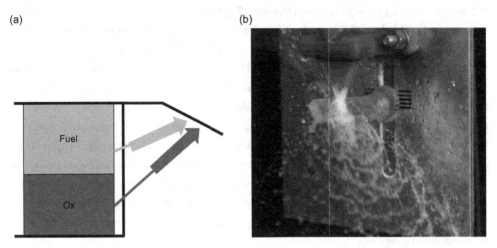

Figure 8.22 (a) Splashplate injector configuration and (b) spray formation.

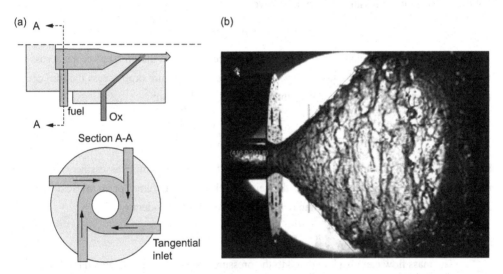

Figure 8.23 (a) Swirl bipropellant injector schematic and (b) spray formation image. Fuel and oxidizer enter
the swirl chambers through several tangential inlets at the entry to each injector.

Like the US, the Soviet Union exploited German talent captured from the Peenemunde group at the end of World War II. While they were aware of the US efforts in impinging element injectors, the manufacturing technology available to them was not as mature and they began to explore alternative injector concepts that did not require as much precision in manufacturing. The result of this exploration was the *swirl bipropellant injector* highlighted in Figure 8.23. Fuel and oxidizer enter respective swirl chambers through several tangential holes such that the momentum of injection is directed into a swirling film. These films interact and exit the injector in a conical sheet that rapidly mixes and atomizes the propellants. Since one fluid tends to reside on the outer part of the sheet, this tends to isolate the injector face from the hottest combustion gases.

The Soviet Union/Russia has used this type of element in numerous applications including both hypergolic and cryogenic propellants. The element can also be used for monopropellants by simply eliminating the oxidizer circuit shown in Figure 8.23. A convergent section can also be included just prior to the exit of the device in order to increase swirl velocity and effect the cone angle of the spray exiting the device. In the bipropellant case shown in Figure 8.23, the large interfacial surface created by the co-swirling streams provides for excellent mixing. The fluid in the swirl chamber is directly interfaced to the chamber by virtue of the fact that a hollow core is formed in the center of the element due to the intense swirl. For this reason, the element can respond to chamber pressure pulsations by forming small waves on the swirling film. It is believed that this behavior can either augment or attenuate combustion instabilities depending on injector and chamber design conditions.

Referencing the cross-section in Figure 8.23, the pressure drop from the manifold to the hollow core chamber pressure condition has two pieces:

$$p_{man} - p_c = p_{man} - p_w + p_w - p_c = \Delta p_{in} - \Delta p_{film} \qquad (8.30)$$

where p_w is the local pressure at the wall, and Δp_{in} and Δp_{film} are the pressure drops across the tangential inlets and across the liquid film, respectively. The pressure drop across the film comes from the hydrostatic pressure formed by the centrifugal acceleration:

$$\Delta p_{film} = p_w - p_c = \rho \frac{w^2}{R_v} h_{film} \qquad (8.31)$$

where w is the swirl velocity in the film (assumed uniform in this simple assessment), R_v is the radius of the vortex chamber, and h_{film} is the height/thickness of the film. For an inviscid flow, the average axial velocity in the swirl chamber is set by mass flow considerations:

$$\dot{m} = \rho v \pi (R_v^2 - (R_v - h_{film})^2) \qquad (8.32)$$

The mass flow can also be related to the pressure drop through the tangential inlets:

$$\dot{m} = C_D A \sqrt{2 \rho g (p_{man} - p_w)} \qquad (8.33)$$

where A is the total cross-sectional area of the inlets. Finally, the spray cone half angle, θ, is set by the respective velocity components in the film:

$$\theta = tan^{-1}(v/w) \qquad (8.34)$$

Assuming the injector geometry, fluid type, and manifold pressure and chamber pressure are set, Eqs. 8.30–8.34 provide five relationships for $\dot{m}$, w, v, p_w, h_{film}, and θ, so it is obvious that the system is underdetermined. Researchers developed another relationship between these quantities based on the principle of maximum flow. This principle states that the injector conditions will adjust

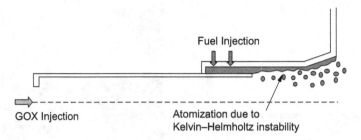

Figure 8.24 The oxidizer-rich staged combustion (ORSC) element developed in the Soviet Union in the 1970s and 1980s.

to pass the maximum flow possible under the imposed constraints but it is also cast in more abstract terms related to the rate information can be translated along the film. The principal has been applied by researchers such as Bayvel and Orzechowski (1993), Bazarov and Yang (1998), and Yule and Chinn (1994), with slightly differing sets of assumptions and differing results. Fundamentally, the difficulties/challenges in the analysis stem from the fact that the tangential channel and the film jointly share the pressure drop applied to the device. All of the models are inviscid and corrections are required to match experimentally measured discharge coefficients. Interested readers are encouraged to investigate the references above as well as the classic atomization text by Lefebvre (1989).

A closely related element developed in the Soviet Union in the 1980s has been termed the *oxidizer-rich staged combustion (ORSC)* element by many modern researchers. Shown in Figure 8.24, this element is comprised of a warm gaseous oxygen axial inner flow that stems from the effluent of the oxidizer-rich (or "ox-rich") preburner that enables the cycle. Fuel is introduced as a swirling film on the outer wall of the passage near its exit. While centrifugal forces tend to throw the fuel outward away from the GOX, the high injection velocity and momentum of the GOX stream tends to play a dominant role and draws the lower momentum fuel inward due to the ejector action and Kelvin–Helmholtz hydrodynamic instabilities that form on the fuel surface.

The ORSC element has been the topic of research in the US for the past decade or so and numerous references are included for interested readers. The sleeve that isolates the oxidizer gas from the fuel injectors plays an important role as initial mixing of the propellants occurs in the tiny base region at the tip of this "post." The bevel that is present just upstream of the injector face provides for deceleration of the gases and for local flameholding in the device. This feature is important as it isolates the sensitive flame anchoring region from acoustic disturbances in the chamber. Chapter 12 provides additional insights relative to the stability characteristics of all the elements presented in this section.

The ORSC element is a variant of a *coaxial injector* element design that is highlighted in Figure 8.25. Coaxial element designs have been used almost exclusively in cryogenic engines employing liquid hydrogen or liquid methane/natural gas fuels. In these applications, fuel has been

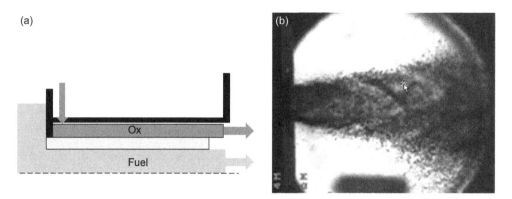

Figure 8.25 Coaxial element design. One or both fluid streams can flow axially to produce a *shear coaxial* design or can be swirled to produce a *swirl coaxial* configuration. The structure separating the two streams is generally referred to as the LOX-post and the post tip is typically submerged a short distance from the chamber as shown in the schematic.

heated in a regenerative cooling jacket and enters the element as a gas or a low-density supercritical fluid. Either of the fluids may be swirled as desired and there is still debate in the community as to the value of the swirl in particular injector configurations as the axial velocity/momentum may still play a dominant role in the atomization process.

In LOX/LH$_2$ engines such as the SSME, liquid oxygen flows down the central channel and is isolated by the annular fuel flow by a *LOX post*. The design of the post is a critical feature and it tends to be slightly submerged as shown in Figure 8.25 in order to isolate the sensitive flame holding region at the post tip from pressure waves in the combustion chamber. The stability of the flameholding region is affected by the hydrogen temperature and often a hydrogen temperature ramping test is done to assess the range of temperatures over which stable flameholding is present. Chapter 12 provides additional discussion on this issue.

The Kelvin–Helmholtz instability mechanism results from waves that form on the oxidizer surface due to the aerodynamic interaction with the high-velocity fuel stream. The crests of the waves represent low-pressure regions and as the waves grow, the pressure continues to drop. Eventually the oxidizer fluid in this region breaks off and mixes with the fuel gases. The high Reynolds number conditions in these injectors create turbulence levels that induce a range of length scales for these physics to play out and an actual gas/liquid interface may not be present as the LOX may be super-critical. The Kelvin–Helmholtz instability is an inviscid mechanism and detailed computations show that viscous effects play a minor role; this fact makes the term "shear" coaxial a bit of a misnomer as it gives the impression that shear is playing a primary role in the atomization and mixing process.

The chief parameter controlling element performance is the momentum ratio, MR:

$$MR = \frac{\rho_g v_g^2}{\rho_l v_l^2} \tag{8.35}$$

Larger momentum ratios support more rapid atomization. If we substitute for velocities in favor of mass flows, Eq. 8.35 can be written:

$$\mathrm{MR} = \frac{\rho_l A_l^2}{r^2 \rho_g A_g^2} \tag{8.36}$$

This form connects the momentum ratio to the geometry of the element and includes the effect of the mixture ratio of the propellant combination. A substantial volume of research exists on coaxial elements due to their usage in such engines as the SSME, RL 10, and Vulcain. One study that investigated the combustion performance of the element concluded that the performance increased when the ratio of residence time, t_r to breakup time, t_b grew large (Sardeshmukh *et al.*, 2017; see also Bedard *et al.*, 2018; Canino and Heister, 2008; Tsohas and Heister, 2011; Umphrey *et al.*, 2017). This ratio was expressed:

$$\frac{t_r}{t_b} = \frac{CL}{h_{\text{film}}} \left[\frac{\mu_l}{\sigma_l} \left(\frac{\rho_g}{\rho_l} \right)^2 \frac{(v_g - v_l)^4}{v_l^3} \right]^{1/3} \tag{8.37}$$

where C is an empirical constant, L is the length of the cup downstream of the LOX post, and μ_l is the liquid viscosity. This form shows that it is really the velocity difference between the gas and liquid that is important, but in practical injector designs, since the gas velocity is more than an order of magnitude higher, the gas dynamic pressure is the principle factor contributing to enhance atomization and mixing relative to the residence time, which is proportional to L.

The ORSC and coaxial designs tend to provide good wall environments on the injector face. In some applications, a porous material called *rigimesh* is employed to provide a low flow of gaseous fuel to isolate the face from the hottest combustion gases. Manufacturing of these types of elements requires great precision as the concentricity of the oxidizer (a.k.a. "ox") tube and the surrounding injector body is very important in order to maintain a consistent oxidizer gap around the annulus. In some applications, designers even use this feature to their advantage for elements adjacent to the wall. By intentionally biasing the ox tube toward the wall, the amount of oxidizer in the near wall region is reduced and thereby reducing combustion gas and wall temperatures.

The *pintle injector* is the last major injector type in use in modern engines. Highlighted in Figure 8.26, this element is comprised of an annular flow of one fluid that is mixed with the other propellant that is injected through a series of radial holes or slots near the pintle tip. The pintle was born in the early days of hypergol testing as coaxial arrangements were being used to evaluate hypergolicity and blow-apart was a frustrating outcome. Engineers at TRW suggested the pintle arrangement and went on to champion that injector concept for a number of space engines developed by the company over the ensuing decades. Perhaps the most famous of the TRW engines is the lunar module descent engine developed to land Apollo astronauts on the moon. The large throttling range for this engine necessitated a translating sleeve that controlled flow areas on both propellant circuits to permit efficient operation at low thrust. While adding complexity, this feature remains attractive today as the potential for a "face shutoff" exists for applications demanding intermittent engine usage.

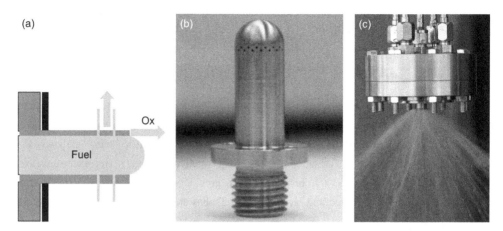

Figure 8.26 Pintle injector. (a) General arrangement, (b) image of typical injector element, and (c) typical spray pattern formed by the device.

As shown in Figure 8.26(b), multiple rows of holes might be included to better cover the perimeter of the injector and improve mixing of the two streams. The atomization performance of these devices is typically attributed to the total momentum ratio, TMR, and the blockage factor, BF:

$$\begin{aligned} \text{TMR} &= \frac{\dot{m}_a v_a}{\dot{m}_r v_r} \\ \text{BF} &= \frac{N D_o}{\pi D_p} \end{aligned} \qquad (8.38)$$

where $()_a$ and $()_r$ refer to axial and radial propellant streams, N is the number of orifices, D_o is the orifice diameter (or slot width), and D_p is the pintle diameter. It is desirable to have TMR values near unity as this produces an initial spray angle near 45° as shown in the image in Figure 8.26. If BF = 1, all of the axial flow perimeter is impacted by radial jets. Practical designs have BF values somewhat less than this limit. Oftentimes two sets of holes/slots of differing diameters are employed with the larger (primary) ring of holes slightly upstream of the smaller (secondary) holes. The complexity of the atomization and mixing processes created by the radial jets impacting the sheet precludes an exact formulation of an optimal TMR/BF combination. The propellant properties (viscosity, density) and volume flow ratios contribute to different solutions depending on the combination under consideration. Another parameter is the skip distance, L_s, the distance the axial stream travels before encountering the first set of radial holes. Typical designs have L_s/D_p ratios near unity.

The pintle is unique in that the other designs in Figures 8.20–8.25 are integrated into a flat face injector plate that fastens to the front end of the combustion chamber. The pintle is similarly installed in the head end of the combustor, which need not necessarily be flat. The spray formed by the pintle is also different in that it is not oriented perpendicular to the pintle axis and a parabolic-shaped flame zone is produced under hotfire conditions. Proponents point to this factor in their claims that engine utilizing pintle injectors are inherently more stable than flat-face injectors that generate a more planar energy release that may be more aligned with the fundamental acoustic wave directions in the combustor.

Pintle injectors have been used extensively on hypergolic propellants as they create a direct impingement of the two fluids, but they have also been employed with LOX hydrocarbon propellant combinations as in the current Merlin engine used by SpaceX on their Falcon launch vehicle. Proponents mention the lower cost of the pintle as there is only a single injection element that lies along the centerline of the chamber. However, to be fair, there is no "free lunch" and large thrust pintle engines have mixing challenges since the size of both the annular gap and the radial holes grow with thrust level and the mixing effectiveness drops as a result.

All of these concepts in Figures 8.20–8.26 share a requirement to keep the two propellants separated upstream of the combustion chamber and as a result a rather intricate manifolding arrangement is typically required. In the end, the L^* data in Table 8.1 presume an excellent propellant mixing. We can't simply plug two fire hoses to the front of the chamber and presume that the fluids will mix in the short times required. We remind readers of the molecular mixing "desirement" – a challenging task indeed for high flow devices such as liquid rocket engines. Because both fluids must be widely distributed across the injector face, an elaborate manifolding design is required for engines using a large number of elements.

For these reasons, it is useful to consider the historical basis for the number of elements on the injector face. A prime consideration for designers is the amount of thrust a single element can produce with high efficiency. There is a good reason for thinking this way, as typical engine developments start with "single element" testing as a lower-cost approach toward optimization of individual injector performance characteristics. *Thrust per element* varies depending on injector type and engine thrust level. In general, smaller engines also have smaller thrust per element as the overall chamber size/length is smaller and a smaller flow per element is required to get excellent performance. Larger engines have larger/longer chambers that support higher thrust/element values. For a given engine thrust requirement, increasing thrust/element decreases the number of injectors, but also tends to decrease the overall efficiency.

NASA SP-8089 has details for a large number of early engines; many modern manufacturers no longer share injector details at this level. Impinging element and splashplate injectors sport thrust/element values roughly in the 100–1100 lbf/element (0.4–5 kN/element) range with the mighty F-1 engine that powered the Saturn V vehicle at the upper end of this range. Coaxial injectors have been developed over a similar range, with the largest known value of 850 lbf/element for the SSME. One might define a thrust/element for a pintle based on the number of radial orifices/ slots. On this basis, prior designs have produced levels roughly in the range 50–1100 lbf/element (0.2–5 KN/element). The ORSC element has displayed the highest thrust/element levels in flight hardware with levels as high as 1700 lbf/element (7.5 KN/element) in Russian engines built for launch vehicle applications.

8.6.3 Injector Manifolds and Throttling Considerations

There are two main topologies that are employed in the manifolding of current large thrust engines. A *ring-type* topology distributes oxidizer and fuel into alternating rings that comprise the injector face. Figure 8.27 shows the ring-type injector arrangement employed in the F-1 engine. Typically one propellant would enter axially into a dome-shaped manifold at the top of the combustion chamber while the other propellant enters radially. Since fuel is used as a regenerative coolant in many designs, it is often convenient to bring this fluid in radially as was done in the F-1. The overall injector face is then comprised of individual rings of fuel injection holes with oxidizer holes being provided in alternate rings. Oxidizer enters alternating rings axially through a series of holes that are clocked properly so as not to interfere with radial fuel spokes feeding successive rings of fuel on the inner portions of the injector. The ring-type manifold arrangement is preferred for impinging element injectors due to the convenience of locating alternating elements of self-impinging fuel and oxidizer jets. It would be highly challenging to locate holes precisely enough to have unlike doublet configurations here as the azimuthal clocking of adjacent rings would require high precision.

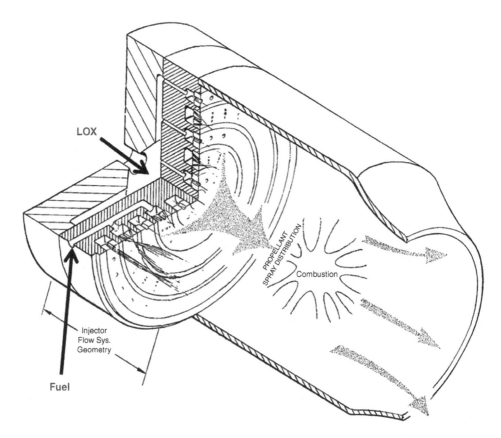

Figure 8.27 F-1 engine injector design and propellant manifold arrangement in a ring-type topology.

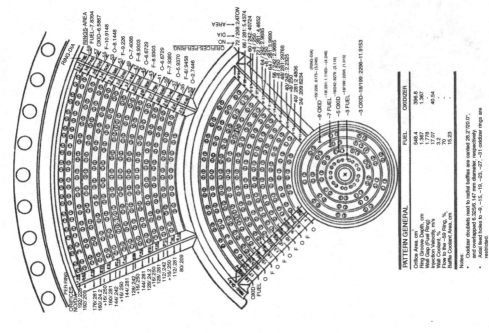

Figure 8.28 Image of F-1 injector face showing baffle design and specific design of alternating rows of oxidizer and fuel orifices.

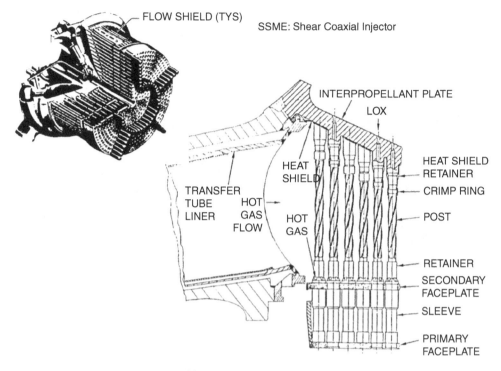

Figure 8.29 SSME injector manifolding arrangement.

Figure 8.28 shows the overall design of the injector face for the F-1. Note the numerous baffles that were introduced to mitigate acoustic waves and the attendant combustion instability that can result when these waves are reinforced by the combustion process.

Coaxial injectors employ an alternate manifolding arrangement that might be termed a *dome and post* topology. Figure 8.29 highlights the overall configuration for the SSME main injector. In this arrangement a LOX dome serves as collector for the main oxidizer flow at the top of the thrust chamber. A series of LOX posts distribute the fluid from a top dome while the gaseous fuel is distributed radially, weaving its way around posts to turn and flow axially into the injector body. The warm fuel gas enters annular slots around the base of each LOX post to form the coaxial injection in the main chamber.

The intricate manifolding required to distribute and meter propellants to numerous individual injection sites serves as an additional challenge for the injector designer. Readers are encouraged to revisit NASA SP-8089 relative to guidelines for these critical interfaces; a leak between propellant circuits can lead to catastrophic results and numerous hardware failures have resulted from breaches of this nature. In an attempt to obtain excellent mixing, the designer is tempted to premix portions or all of the fluids upstream of the chamber. Experience shows that this approach will inevitably end in disaster as it is difficult to understand all the dynamics of this interaction during start and shutdown transients where potential exists for a flame to be introduced with catastrophic consequences.

If we know the total flow rate, the number of injector elements (for fuel or oxidizer) can readily be computed from the element flow rate:

$$N = \dot{m}/\dot{m}_e \qquad (8.39)$$

Then the chamber diameter D_c would simply be set by the element density (i.e. the number of elements per unit area), which could be tolerated based on structural considerations. If we let N_D represent this number of elements per unit area then:

$$A_c = (N_{ox} + N_{fuel})/N_D \qquad (8.40)$$

which permits a rapid sizing of the injector face area required to distribute propellants. The individual element areas can also be related to each other if we specify the mixture ratio for a certain group of elements:

$$r = \frac{[C_D A \sqrt{\rho \Delta p}]_{ox}}{[C_D A \sqrt{\rho \Delta p}]_f} \qquad (8.41)$$

The injector pressure drop is a critical design variable for the injector designer. For many designs, this parameter varies as 5–20% of chamber pressure. While low-pressure-drop injectors reduce peak pressures for turbomachinery to provide, they tend to "soften" the injection system and make it more susceptible to coupling with chamber combustion leading to instabilities. Raising Δp tends to "stiffen" the system and decouple the injection process from the dynamic response in the combustion chamber. Since increased Δp increases the injection velocity, it also tends to improve both the atomization and mixing processes as a result. Liquid/liquid injectors such as impinging element types typically demand higher Δp in the 20% range while gas/liquid coaxial systems can employ smaller pressure drops as low as 5% of p_c.

Throttling conditions provide challenges to the injector designer since the lower flow rates associated with part-power operation inherently lead to a reduction in injection velocity, a softening of the injector response, and a reduction in atomization and mixing effectiveness. One can maintain larger injection velocities if the orifice area can change with throttle setting but this approach demands a variable-area injector. This technique was applied in the Lunar Module Descent Engine developed by TRW for the Apollo lunar lander as mentioned previously. Using this approach, deep throttling to 20% of maximum thrust was achieved.

8.7 LRE Combustor/Injector Design and Analysis

The chamber/injector design process begins with the thrust requirements, propellant and injector topology selections, and resulting mixture ratio for the system, all defined by the *cycle balance*. The cycle balance typically starts with assumed values for design parameters like pressure drop and efficiency that are based on previous experience with similar combustors. The overall flow rate can be computed from the thrust requirement and estimate of delivered I_{sp}:

$$\dot{m} = F/I_{sp} = F/(I_{spth}\eta_{ere}) \tag{8.42}$$

where the theoretical I_{sp} is computed from CEA with an estimate of the nozzle expansion ratio one might utilize in the application. The energy release efficiency, η_{ere}, reflects combustion/mixing losses and nozzle losses:

$$\eta_{ere} = \eta_{c*}\eta_{cf} \tag{8.43}$$

Typical $c*$ efficiencies range from 95% for a well-designed LOX/RP engine to 99% for a well-designed LOX/H$_2$ engine. Nozzle divergence, boundary layer, and kinetics losses are reflected in η_{cf}, which is 97–99% for large nozzles, but can drop into the low 90s for smaller engines.

Given the flow rate estimate, the chamber throat area is then:

$$A_t = \dot{m}\eta_{c*}/(gp_c) \tag{8.44}$$

The chamber cross-sectional area is set by the contraction area ratio, CR:

$$A_c = \text{CR } A_t \tag{8.45}$$

Historically, contraction ratios have varied from 1.3 to 10 depending on engine thrust level. For example, the high-thrust F-1 engine had CR = 1.3, while SSME had CR = 3. Larger CR values increase chamber diameter and weight and tend to reduce acoustic frequencies (which may place more combustion instability modes in play), but reduce chamber gas velocities and Rayleigh losses. Smaller CR values provide for lighter chambers, but higher Rayleigh losses associated with the increased chamber Mach number. Another important consideration is having sufficient chamber face area to accommodate all the injection orifices. Small thrusters tend to have a very high CR, simply due to packaging and valve interface considerations. Once the chamber cross-section area is set, a chamber length required for acceptable efficiency can be determined from $L*$ or L' data presented previously, or from component tests.

The designer must choose the thrust level per injector element as a prime consideration for the injector design. This design parameter is often based on heritage design, and will affect both combustion efficiency and the thermal environment at the injector face, Given this parameter, Eqs. 8.32–8.35 provide a mechanism to compute the element flow rate and orifice sizes for a stipulated injector pressure drop, Δp.

Using this rather simple approach, a basic injector and chamber configuration can be created to support a more detailed analysis. Inherent in the design process is the requirement to pick parameters such as CR, thrust per element, $L*$ (or L'), and η_{ere}, whose suitability needs to be confirmed with a more detailed analysis or test data.

Detailed analyses can take on a range of complexity depending on the maturity of the design and corporate philosophy. As mentioned in Chapter 6, US industry uses the Two-Dimensional Kinetics (TDK) code as an analytic tool to attempt to predict LRE performance with multidimensional flow effects and detailed boundary layer treatments for frictional losses and heat transfer. The TDK code, and similar tools that exist in Europe and Asia, provide the working engineer with a capability to estimate η_{c*} and η_{cf} from first principles and input from empirical mixing efficiencies given a chamber and injector design. This code features integral method boundary layer development with inviscid axisymmetric base flow solutions and method of characteristics capabilities for nozzle design. Modules also exist for estimating mixing and kinetics losses depending on injection configuration and chemistry of combustion.

While modern CFD techniques continually are advancing for unsteady, multidimensional analysis, simulation of reacting two-phase flows is still a very immature capability. For this reason, most of the detailed CFD analyses conducted assume a gas/gas behavior or model liquid phases as supercritical with continuous variation in properties and negligible surface tension forces. Arguably, the most useful CFD contribution in the current computational environment is an axisymmetric, unsteady reacting flow computation that might parallel a single element injector experiment. With evolving CFD capabilities over the last 10–15 years, we have learned that the highly turbulent, high Reynolds number conditions pertinent to rocket combustors tend to create highly unsteady flows with vortex shedding and other hydrodynamic instabilities contributing to the vigorous mixing and intense combustion processes. Figure 8.30 shows a recent research result for a single ORSC injector element.

While a result like that depicted in Figure 8.30 is very useful, it is still difficult to implement in a design environment that could contain hundreds of injector elements, and simpler 1-D or empirical approaches still have merit in navigating a large design space.

Mixing processes are difficult to characterize, even with multidimensional CFD treatments. An approximate approach based on a streamtube analysis for a given injector element, would bias the local mixture ratio to account for mixing efficiency, E_m:

$$r_{\text{ox-rich}} = r_{\text{des}}/E_m, \quad r_{\text{fuel-rich}} = r_{\text{des}}E_m \tag{8.46}$$

where r_{des} is the design/target mixture ratio for the element in question, and $r_{\text{ox-rich}}$ and $r_{\text{fuel-rich}}$ are the mixture ratios for elements that are oxidizer and fuel rich, respectively. Typical E_m values lie in the 0.7–0.9 range.

Correlations for single element mixing efficiencies exist for a number of injector types, or they may be measured during the development phase. After measuring the single-element mixing efficiency, a second correlation takes account of the overall injector pattern, for example, the number of elements, to arrive at an overall E_m for the oxidizer and fuel. The CEA tool is then used to compute a mass flow weighted average c^* for the chamber reflecting fuel-rich, ox-rich, and wall elements:

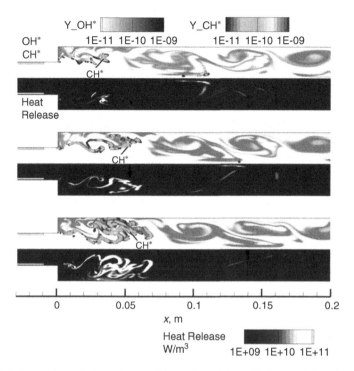

Figure 8.30 Modern high resolution axisymmetric CFD result for a single ORSC-type injector element showing intermediate species CH* and OH* as marker of the flame zone (upper images) and the overall heat release profiles (lower images) at different instances in time (Sardeshmukh *et al.* 2017).

$$c* = x_{\text{core,fuel-rich}} c*(r_{\text{fuel-rich}}) + x_{\text{core,ox-rich}} c*(r_{\text{ox-rich}}) + x_{\text{wall}} c * (r_{\text{wall}}) \qquad (8.47)$$

where the *x* values represent the fraction of the flow in the element type in question. Equation 8.48 also allows for the use of so-called barrier elements near the wall that operate at a lower O/F ratio to provide a relatively cooler barrier of combustion products near the wall. The r_{wall} value is selected to keep gas temperatures within acceptable limits for the wall material to be utilized. Fuel-film cooling may also be used along with, or instead of, barrier elements, to provide an acceptable thermal environment. The effect of fuel-film cooling on performance is also analyzed using empirical methods. A low limit on the performance contribution from fuel used in film cooling is to assume it is totally unreacted, and that it exists at its saturation temperature corresponding to the chamber operating pressure. In fact, some of the fuel film will be entrained into the core flow. For more background here, see Chapter 6.

8.8 LRE Unsteady Systems Analysis Using Lumped Parameter Methods

One criticism we might level against existing propulsion texts is that there is insufficient treatment of transient operation of propulsion devices. Students learn how to do quasi-steady

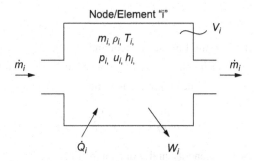

Figure 8.31 Control volume for lumped parameter analysis of LRE system. This volume is for the most general case – in practice only a subset of the interactions shown here would be present in a given element.

analyses that provide a top-level approach for system analysis and design, but don't necessarily develop an appreciation for the large amount of efforts required to assess system operability in the "real world." Substantial engineering time is dedicated to the assessment of safe and reliable approaches to starting and throttling engines – a topic that is of tremendous importance and a source of great consternation during the tedious engine development process. One must gain appreciation for the tremendous power that is unleashed during the ignition/startup process and the repercussions of not ramping that power up in a smooth and effective fashion. In this section, we assess modern analysis techniques currently employed to aid in the development of these processes.

In Chapters 7 and 11 we discuss the lumped parameter approaches that are employed in SRM and HRE analyses – and the LRE community has their own analogue as well. Recall that the term "lumped parameter" implies that we are willing to neglect spatial variations in a quantity such as pressure or temperature to simplify analyses and characterize a given component or region of the engine with a single value that can vary in time. These techniques are powerful in that they reduce the analysis to a coupled set of ordinary differential equations, rather than having to resort to the full multidimensional Navier–Stokes and energy equations that would require complex meshes to resolve the labyrinth-like flowpath fluids can take inside an engine system.

In the US, rocket engine analyses are conducted with the ROCket Engine Transient Simulation (ROCETS) code. Both codes are arranged to solve a system of ODEs resulting from specification of an engine system as a series of "nodes"; each of which are characterized by a unique time-varying pressure, temperature, and flow rate. For a given node, heat transfer, mass transfer (vaporization or condensation), work interaction (in pump or turbine), and energy addition from combustion might be present, so in the most general sense the control volume would be represented by Figure 8.31.

As in the case of SRM lumped parameter ballistics in Chapter 7, we allow for mass storage in the element, so in this case conservation of mass can be written:

$$\frac{\mathrm{d}m_i}{\mathrm{d}t} = \dot{m}_i - \dot{m}_{i-1} \tag{8.48}$$

where m_i is the current mass of fluid in the element and $\dot{m}_i$ and $\dot{m}_{i-1}$ are the instantaneous mass flows exiting and entering the element, respectively. Since the density of the fluid in the element is simply its mass divided by its volume (V_i), we can eliminate m_i from Eq. 8.39 and write:

$$\frac{\mathrm{d}\rho_i}{\mathrm{d}t} = \frac{\dot{m}_i - \dot{m}_{i-1}}{V_i} \tag{8.49}$$

In general, there can be energy and work interactions on the element as highlighted in Figure 8.34. The rate of heat addition/loss, $\dot{Q}_i$ can be used to model combustion-related heat addition or heat transfer into/out of the element based on local heat transfer rates. The rate of work interaction, $\dot{W}_i$, includes p-dV work, but also can be used to model turbopumps or turbines that input/extract energy from the flow. On this basis, conservation of energy (the First Law of Thermodynamics) can be written:

$$\frac{\mathrm{d}U_i}{\mathrm{d}t} = \frac{\mathrm{d}(m_i \cdot u_i)}{\mathrm{d}t} = \dot{E}_i - \dot{E}_{i-1} + \dot{Q}_i + \dot{W}_i \tag{8.50}$$

where U_i is the internal energy in the element and u_i is the specific internal energy. The rate of fluid energy leaving the element is $\dot{E}_i$ and depends on enthalpy of the gases entering:

$$\dot{E}_i = \dot{m}_i h_i \tag{8.51}$$

with an analogous result for the energy entering the element, $\dot{E}_{i-1}$. The work interaction can be divided into p-dV work and shaft work attributed to a turbopump ($\dot{W}_{sh_i}$):

$$\dot{W}_i = p_i \frac{\mathrm{d}V_i}{\mathrm{d}t} + \dot{W}_{sh_i} \tag{8.52}$$

In general, the element volume does not change with time ($\mathrm{d}V_i/\mathrm{d}t = 0$), and the p-dV throttling of the fluid is tracked in the enthalpy change measured in the $\dot{E}_i$ terms. However, if a flexible element, such as a POGO accumulator is present, then the $\mathrm{d}V_i/\mathrm{d}t$ interaction becomes important. Differentiating the left-hand side of Eq. 8.48 and combining with Eqs. 8.51 and 8.52 gives:

$$\frac{\mathrm{d}u_i}{\mathrm{d}t} = \frac{1}{m_i} \left(\dot{m}_i \cdot h_i - \dot{m}_{i-1} \cdot h_{i-1} - u_i \cdot \frac{\mathrm{d}m_i}{\mathrm{d}t} + \dot{Q}_i - p_i \cdot \frac{\mathrm{d}V_i}{\mathrm{d}t} + \dot{W}_{sh_i} \right) \tag{8.53}$$

Substituting in the result from continuity equation in 8.49 and the definition of gas density gives the final form of the energy balance:

$$\frac{\mathrm{d}u_i}{\mathrm{d}t} = \frac{1}{\rho_i V_i} \left(\dot{m}_i \cdot h_i - \dot{m}_{i-1} \cdot h_{i-1} - u_i(\dot{m}_i - \dot{m}_{i-1}) + \dot{Q}_i - p_i \cdot \frac{\mathrm{d}V_i}{\mathrm{d}t} + \dot{W}_{sh_i} \right) \tag{8.54}$$

To close the system, we need some constitutive relations for the other parameters that appear in the control volume in Figure 8.31. Basically this requirement is met by recognizing that a thermodynamic state can be stipulated with any two known thermodynamic variables, so we may write:

$$p_i = p(\rho_{i,}, u_i), \quad T_i = T(\rho_{i,}, u_i), \quad h_i = h(\rho_{i,}, u_i) = u_i + p_i/\rho_i \tag{8.55}$$

For example, if we were to assume calorically perfect gases, these relationships become:

$$p_i = \rho_{i,}RT_i, \quad T_i = T_{i-1} + (u_i - u_{i-1})/c_v, \quad h_i = u_i + p_i/\rho_i \tag{8.56}$$

In many, if not most, applications the fluids in question cannot be treated as ideal gases and more elaborate property routines need to be created. However, the functionality of these tables is still captured in Eq. 8.46. Equations 8.39, 8.45 and 8.47 provide five relationships for the thermo-dynamic variables p, ρ, T, u, and h. A sixth relationship closes the system by relating the mass flow to these thermodynamic parameters, i.e.

$$\dot{m}_i = \dot{m}_i(p_i, \rho_i) \tag{8.57}$$

For example, for a flow through a propellant line or an injector orifice, this relationship would involve the discharge coefficient:

$$\dot{m}_i = C_D A \sqrt{2g\rho\Delta p} = C_D A \sqrt{2g\rho_i(p_{i-1} - p_i)} \tag{8.58}$$

where A is the orifice/line cross-sectional area and C_D is the discharge coefficient. Analogous expressions can be developed for other engine components (i.e. other "nodes" in the network). A combustor node is particularly interesting and can require substantial additional development. Propellants enter the combustor at individual rates and affect the instantaneous and overall stoichiometry; the ignition device firing time is of course set by the user. Flame spreading occurs upon firing of the igniter and pressure builds in the chamber while gases are also escaping through the throat. Hypergolic ignition has a similar sequence, but of course we don't get to stipulate when the ignition event occurs as it is dictated by the hypergolicity of the propellants.

Turbomachinery also provides unique modeling challenges. As noted in Chapter 9, the rotor speed, N, varies in time as the turbopump spools up to full power:

$$\frac{d^2N}{dt^2} = \frac{\tau_{turb} - \tau_{pump}}{I} \tag{8.59}$$

where I is the moment of inertia of the rotating spool, and τ_{turb} and τ_{pump} are the torque output/demand of the turbine/pump, respectively. Ultimately these quantities depend on $\dot{m}_i, p_i, \rho_i$, etc. from nodes on either side of the pump/turbine, so Eq. 8.59 becomes highly coupled to the properties upstream/downstream of the device. Managing the way the turbopump spools up is every bit as critical as the ignition/combustion transient and of course the two are coupled. Hopefully this gives an appreciation of the level of engineering effort required to understand all these transient phenomena in a real system. In a real engine design environment, engineers spend a lot of time with these types of analyses to sort out engine transient behavior during startup, throttling excur-sions, and shutdown transients.

The overall mathematical solution of this highly coupled set of ODEs is a bit beyond what we can afford to include in this introductory text, but a general approach is outlined here. Ultimately, the overall system will be specified by a discrete number of nodes, n, and a set of six equations will be stipulated for $p, \rho, T, u, \dot{m}$, and h at each node. Coefficients that multiply each of these unknowns will be determined and any known parameters comprising the "right-hand side" of the equation set will be collected with the result eventually assembled into classic linear algebra problem $AX = B$. Here, A is a $6n \times 6n$ vector of coefficients that multiply the unknown system variables in column vector X and B is a column vector comprising known quantities on the right-hand side of the individual equations. Mathematically, the problem is a nonlinear coupled set of ODEs and solution techniques like Newton's method (Newton–Raphson method) or trapezoidal integration are employed as elucidated in Appendix A for single equation "systems."

8.9 A Note on Additive Manufacturing

At the time of writing (Summer 2018), a revolution is underway relative to our ability to make metal parts with highly detailed features. These processes are called direct laser metal sintering (DMLS) by many in the industry and a number of vendors are providing machines that can sinter/ melt metal powders to "additively" manufacture parts from metals such as stainless steel, Inconel, aluminum, and even copper alloys. While there is still substantial debate about the microstructural characteristics of the parts relative to usage as a primary structure, for parts that are not carrying primary loads like beams or pressure vessels there is a current/near term path toward application.

Feature sizes less than 500 microns (0.5 mm) are available from numerous vendors and the ability to make parts with these minute features exists at a growing rate. Part diameters are currently limited to 10–12 inches (25–30 cm) or so with many of the existing machines, but the field is evolving rapidly to capabilities to make even larger parts. It is difficult to think of an application that benefits more from these developments than a rocket injector and it is no coincidence that all major manufacturers are working to exploit this nascent capability. Both NASA and the German Space Agency (DLR) have manufactured entire injector chamber assemblies from stainless steel alloys that contain dozens of individual injectors. Figure 8.32 gives readers a good idea of the capabilities that are being exercised at present in the form of a full-scale, additively manufactured, Vinci engine injector.

Copper and its alloys are of great interest in LRE chambers due to the high thermal conductivity characteristics, but it is this feature that makes the material difficult to sinter (i.e. laser energy rapidly departs the melt zone). However, strides are being made here at NASA and in both US and European firms. Mesh-type structures, like the Rigimesh material that is used on some injector faces, can also be printed directly using the DMLS technology. In the US GE has already qualified DMLS injectors for use in the LEAP gas turbine engine and firms like Aerojet Rocketdyne, SpaceX and Blue Origin have advanced capabilities due to the strong interest in improving injector manufacturing capabilities and greatly reducing costs. Ducts and lines can also

Figure 8.32 Ariane Vinci engine injector additively manufactured from a stainless steel alloy as a single part. The coaxial injector features 123 individual elements with hundreds of radial ports for gaseous hydrogen injection into individual elements.

be printed and the complex 3-D transfer ducts can be printed based on "as measured" dimensions as the engine is being built up to mitigate post-manufacture rework that has often been required. It is rumored that over 60% of the mass of the Merlin engine used in the SpaceX Falcon 9 vehicle is being built using this technology.

Like any technology, DMLS has its limitations. Surface finish is much poorer than with traditional "substractive" manufacturing approaches and may require rework depending on the demands for a given application. Here, the local finish is ultimately dictated by particle size of the powder employed in the DMLS process. Typical powder mass median diameters are of the order of 30–50 microns, so the region near the edge of a melted/sintered surface might display roughness features at this scale. Since the parts are built layer-by-layer, overhang regions (say, at the top of a round hole) are more challenging to build and often require support material (material that has not been sintered, but occupies the space one wants to retain as a void region) to fabricate. In ornate passages such as injectors, removal of support material becomes a challenge.

Retaining tight tolerances at the micron level is still a challenge in that ultimately DMLS tolerances are tied to powder size used in the process (as well as ability to control position and quality of the beam). For those who desire the tiniest of holes, the "as built" dimensions might vary substantially from those specified. Micro-porosity, or local voids within the melted material itself, can also be present, and this presents a challenge if one is trying to realize the full strength capability of the material. For this reason, some parts may need to be slightly thicker to account for uncertainty in strength. This normally isn't a large limitation for injectors, but other structural members such as lines/ducts may be altered as a result.

FURTHER READING

Abramson, H. N. (1967) "The Dynamic Behavior of Liquids in Moving Containers," NASA SP-106.

Anderson, W. E., Ryan, III, H. M., and Santoro, R. J. (2006) "Impact Wave-Based Model of Impinging Jet Atomization," *Atomization and Sprays*, 16(7): 791–805.

Ashgriz, N. (ed.) (2011) *Handbook of Atomization and Sprays*. Springer.

Bazarov, V. G. and Yang, V. (1998) "Liquid-Propellant Rocket Engine Injector Dynamics," *Journal of Propulsion and Power*, 14(5): 797–806.

Bayvel, L. and Orzechowski, Z. (1993) *Liquid Atomization*. Taylor & Francis.

Bedard, M. J., Austin, B. J., and Anderson, W. E. (2018) "Detailed Measurement of ORSC Main Chamber Injector Dynamics in a Model Rocket Combustor," in *2018 AIAA Aerospace Sciences Meeting*. AIAA, p. 1186.

Canino, J. and Heister, S. D. (2008) "Contributions of Orifice Hydrodynamic Instabilities to Primary Atomization," *Atomization and Sprays*, 19(1): 91–102.

Cheng, G. C., Davis, R. R., Johnson, C. W., Muss, J. A., Greisen, D. A., and Cohn, R. K. (2003) "Development of GOX-Kerosene Swirl Coaxial Injector Technology," in *39th AIAA Joint Propulsion Conference*. AIAA.

Doumas, M. and Laster, R. (1953) "Liquid-Film Properties for Centrifugal Spray Nozzles," *Chemical Engineering Progress*, 49(10).

Gill, G. S. and Nurick, W. H. (1976) "Liquid Rocket Engine Injectors," NASA SP-8089, Cleveland, OH.

Lefebvre, A. (1989) *Atomization and Sprays*. Hemisphere Publishing Corp,.

Mayer, E. (1961) "Theory of Liquid Atomization in High Velocity Gas Streams," ARS Journal, 31: 1783–1785.

Muss, J. A., Johnson, C. W., Cheng, G. C., and Cohn, R. K. (2003) "Numerical Cold Flow and Combustion Characterization of Swirl Coaxial Injectors," in *41st AIAA Aerospace Sciences Meeting*. AIAA.

Nurick, W. H. (1971) "Analysis of Sprays from Rocket Engine Injectors," *Journal of Spacecraft*, 8 (7): 796–798.

Rodrigues, N. S., Kulkarni, V., Gao, J., Chen, J., and Sojka, P. E. (2015) "An Experimental and Theoretical Investigation of Spray Characteristics of Impinging Jets in Impact Wave Regime," *Experimental Fluids*, 56(3) 1–13.

Ryan, H. M., Anderson, W. E., Pal, S., and Santoro, R. J. (1995) "Atomization Characteristics of Impinging Liquid Jets," *Journal of Propulsion Power*, 11(1): 135–145.

Strakey, P. A., Talley, D. G., Tseng, L. K., and Miner, K. I. (2002) "The Effects Of LOX Post Biasing On SSME Injector Wall Compatibility," *Journal of Propulsion and Power*, 18(2).

Riebling, R. W. (1967) "Criteria for Optimum Propellant Mixing in Impinging-Jet Injection Elements," *Journal of Spacecraft and Rockets*, 4(6): 817–819.

Sardeshmukh, S., Bedard, M., and Anderson, W. (2017) "The use of OH* and CH* as Heat Release Markers in Combustion Dynamics," *International Journal of Spray and Combustion Dynamics*, 9(4): 409–423.

Sutton, G. P. and Biblarz, O. (2016) *Rocket Propulsion Elements*, 9th edn. Wiley.

Sutton, G. P. (2006) *History of Liquid Propellant Rocket Engines*. AIAA.

Tsohas, J. and Heister, S. D. (2011) "Numerical Simulations of Liquid Rocket Coaxial Injector Hydrodynamics," *AIAA Journal of Propulsion & Power*, 27: 793–810.

Turns, S. R. (1996) *An Introduction to Combustion*. Vol. 499. McGraw-Hill.

Umphrey, C., Harvazinski, M. E., Schumaker, S. A., and Sankaran, V. (2017) "Large-Eddy Simulation of Single-Element Gas-Centered Swirl-Coaxial Injectors for Combustion Stability Prediction," in *53rd AIAA/SAE/ASEE Joint Propulsion Conference*. AIAA, p. 4689.

Yang, H. Q., Purandare, R., Peugeot, J. and West, J. (2012) "Prediction of Liquid Slosh Damping Using a High Resolution CFD Tool," in *48th AIAA Joint Propulsion Conference*. AIAA.

Yule, A. J. and Chinn, J. J. (1994) "Swirl Atomizer Flow: Classical Inviscid Theory Revisited," presented at: *The 6th International Conference on Liquid Atomization and Spray Systems*, ICLASS-94, Rouen, France, July, pp. 334–341.

HOMEWORK PROBLEMS

8.1 In this problem, consider the design of an OTV using a single LOX/LH$_2$ engine capable of placing a 10,000 lb payload into a geosynchronous orbit from an initial circular parking orbit 100 Nm above the Earth's surface.

(a) Determine ΔV values for both burns of the engine required to utilize a Hohmann transfer.

(b) Given $p_c = 400$ psi, $\epsilon = 50$, $\eta_o = 0.90$, $F = 15,000$ lb, determine the optimal mixture ratio, c^*, and I_{sp} of the engine to be used.

(c) Determine the useful propellant load required to obtain the total ΔV in part (a) assuming $\lambda = 0.8$.

(d) To allow for boiloff, we must carry 5% more oxidizer and fuel than is necessary to complete the mission. The tanks must accommodate this boiloff volume as well as a 5% volume reserve for ullage. Calculate the volume of fuel and oxidizer tanks including boiloff and ullage volumes.

(e) Calculate tank dimensions assuming a tandem arrangement with flat-ends (circular cylinders) and that the L/D of the fuel tank is 1.25.

(f) Determine the burning times for each of the two burns in the Hohmann transfer. Determine nozzle throat and exit diameters. You are given the following data:

$$L^* = 45 \text{ in}, \ R_i = 0.6, \ R_{\omega td} = 0.4, \ A_c/A_t = 3.0$$

$$\theta_i = 60°, \ \theta_d = 17°, \text{ conical nozzle}$$

Calculate all chamber and nozzle dimensions (including chamber length). Sketch the chamber/nozzle design noting all dimensions.

(g) Assume the payload is a cylinder 10 ft diameter and 6 ft long. Allow 0.5 ft between tanks and payload and 1.5 ft between tankage and engine for interstage/pumps/etc. Make a scale drawing of the entire system including the engine/nozzle assembly.

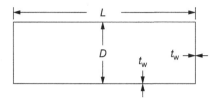

Figure 8.33 Diagram for Problem 8.2.

(h) Plot vehicle velocity and acceleration (in g) for each of the two burns assuming constant thrust.

8.2 One can prove that a spherical tank can enclose a given volume, V, with a minimum surface area, S. However, if we are sizing a tank, we must also consider the thickness of material if we are concerned about weight. Is it possible that a long, skinny tank could weigh less than that of a sphere? Consider the simple cylindrical tank design in Figure 8.33 as a basis of comparison. Presume that tank material properties (ρ, F_{tu}, f_s) and design pressure, p_t, are known. Derive an expression for the weight of this cylindrical tank (M_t) for a fixed internal volume, V, and configuration L/D assuming that the end caps are the same thickness as the cylinder material, t_w. Find the ratio M_t/M_{ts}, where M_{ts} is the mass of an equivalent spherical tank; this ratio should only be a function of L/D. Are longer tanks lighter than spheres?

8.3 In this problem, consider the design of an OTV using a single LOX/LH$_2$ engine capable of placing a 10,000 lb payload into a geosynchronous orbit from an initial circular parking orbit 100 Nm above the Earth's surface.

(a) Determine ΔV values for both burns of the engine required to utilize a Hohmann transfer.

(b) Given $p_c = 650$ psi, $\epsilon = 75$, $\eta_o = 0.95$, $F = 20,000$ lb, determine the optimal mixture ratio, c^*, and I_{sp} to maximize the density impulse.

(c) Determine the useful propellant load required to obtain the total ΔV in part (a) assuming $\lambda = 0.8$.

(d) To allow for boiloff, we must carry 5% more oxidizer and fuel than is necessary to complete the mission. The tanks must accommodate this boiloff volume as well as a 5% volume reserve for ullage. Calculate the volume of fuel and oxidizer tanks including boiloff and ullage volumes. You may assume that all boiloff occurs prior to the first burn of the engine.

(e) Calculate tank dimensions assuming a tandem arrangement with flat-ends (circular cylinders) and that the L/D of the fuel tank is 1.6.

(f) Determine the burning times for each of the two burns in the Hohmann transfer. Determine nozzle throat and exit diameters. You are given the following data:

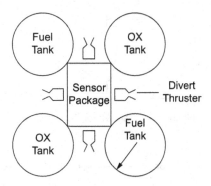

Figure 8.34 Diagram for Problem 8.4.

$$L^* = 45 \text{ in}, R_i = 0.6, R_{\text{ctd}} = 0.4, A_c/A_t = 4.0$$

$$\theta_i = 60°, \theta_d = 17°, \text{ conical nozzle}$$

Calculate all chamber and nozzle dimensions (including chamber length). Sketch the chamber/nozzle design noting all dimensions.

(g) Assume the payload is a cylinder 8 ft diameter and 6 ft long. Allow 0.5 ft between tanks and payload and 1.5 ft between tankage and engine for interstage/pumps/etc. Make a scale drawing of the entire system including the engine/nozzle assembly.

(h) Plot vehicle velocity and acceleration (in g) for each of the two burns assuming constant thrust.

8.4 Consider the Kinetic Kill Vehicle (KKV) concept shown in Figure 8.34. Designers like to use the N_2O_4/MMH propellant combination in this application because it permits the use of equal volumes for both fuel and oxidizer tanks. Suppose you are given the following information:

Payload (sensor) mass = 10 kg (includes tank support structure)
$I_{sp} = 300$ s, required divert $\Delta v = 1$ km/s, $\lambda = 0.5$

where λ accounts for all inert components in all four tanks/thrusters and associated plumbing. With this information, determine:

(a) the propellant mixture ratio;
(b) the total propellant mass required;
(c) the radius of the spherical tanks used in the KKV; and
(d) the tank wall thickness assuming a design burst pressure of 5 MPa and titanium construction.

8.5 Suppose you decide to go to work for a major propulsion contractor after graduation. As your first task, your boss asks you to develop a correlation for the cost of liquid rocket engine turbopumps. He/she has compiled a database of existing turbopumps which includes virtually

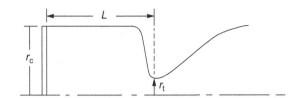

Figure 8.35 Diagram for Problem 8.6.

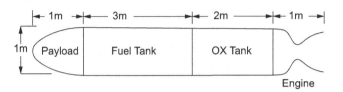

Figure 8.36 Diagram for Problem 8.7.

everything there is to know about these devices, such as: size, weight, flow rate, pressure rise, materials of construction, date of manufacture, manufacturer, parts count, size/shape/thickness of all parts, number of stages, the name of the operator who made the device, etc. Discuss the parameters you would choose to use in your correlation, with a justification as to why these parameters should influence the cost of the device.

8.6 Consider the liquid rocket engine shown in Figure 8.35. The engine will employ LOX and Ethanol propellants which deliver a c^* of 5900 ft/s at $\gamma = 1.2$. We desire a vacuum thrust level of 30,000 lbf and a nozzle expansion ratio of 50 with a chamber pressure of 1200 psi. To permit adequate combustion we wish to keep chamber Mach numbers at or below 0.1 with an engine L^* value of 40 inches. Using this information, determine:

   (a) the engine vacuum I_{sp};
   (b) the Nozzle throat radius, r_t;
   (c) the chamber radius, r_c; and
   (d) the chamber length, L.

8.7 Consider the bipropellant vehicle being developed for a ground-based interceptor shown in Figure 8.36. The vehicle has a liftoff mass of 20,000 kg and a constant-thrust engine which provides an initial acceleration of $3g$ to the rocket. The vehicle has a burnout weight of 5000 kg, 1000 kg of which is payload. The designed would like to employ an aluminum fuel tank ($E = 6900$ MPa, $F_{tu} = 565$ MPa, $v = 0.3$) with a skin thickness of 2 mm to make the required weight budget. Assuming the walls of this tank must not only contain the ullage pressure stresses, but must also support axial loads from the payload, determine a satisfactory ullage pressure for use in this application. You may assume that a safety factor of 1.25 is adequate for determining a working tensile or compressive stresses.

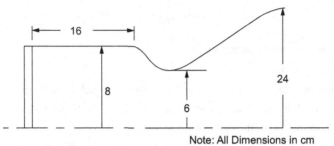

Note: All Dimensions in cm

Figure 8.37 Diagram for Problem 8.10.

8.8 Consider a cylindrical propellant tank of height h_o filled completely to the top (no ullage) with a liquid propellant of density ρ. Let p_u represent the gauge pressure in the ullage at the top of the tank and assume that this pressure is kept fixed during firing using an autogenous pressurization system. The vehicle acceleration (in g) can be represented as:

$$\eta = 1 + kt;\ k = \text{const};\ 0 \leq t \leq t_b$$

Assume the propellant mass flow is constant over the burn duration.

(a) Derive an expression for the tank level, h, as a function of time in terms of h_0, t, and t_b.

(b) Derive expressions for the maximum tank pressure and the time at which this pressure occurs in terms of p_u, k, t_b, ρ, and h_o.

·8.9 Historically, some liquid propellant launch vehicles have utilized a helium pressurant tank which is actually immersed inside the LOX tank. Gases from the pressurant tank are subsequently run through a heat exchanger, then used for tank pressurization.

(a) Why did the engineers decide to put a pressurant vessel inside a cryogenic liquid propellant tank?

(b) From an operational standpoint, is this a good idea? What issues might be of concern in this application?

8.10 Consider the liquid rocket engine using LOX and RP-1 ($CH_{1.97}$) as propellants shown in Figure 8.37. The following measurements were recorded during a 30 second constant thrust firing:

Vacuum thrust = 56.9 kN

Chamber pressure = 3 MPa

Oxidizer flow rate = 16.6 kg/s

Fuel flow rate = 8.3 kg/s

In addition, by running the NASA code for one-dimensional equilibrium performance, we determined:

$$I_{spv} = 311 \text{ s (vacuum)}, c^* = 1700 \text{ m/s}$$

Using this information, determine:

(a) The mixture ratio.

(b) The stoichiometric mixture ratio. Is this mixture fuel rich or fuel lean?

(c) The engine delivered characteristic velocity and the c^* efficiency, η_{c^*}.

(d) The engine thrust coefficient efficiency, η_{cf}.

(e) If your η_{c^*} value is less than 98%, suggest a design modification that should improve performance to an acceptable level.

8.11 Consider the design of an orbital transfer vehicle using LOX/LH$_2$ propellants.

(a) Determine the ΔV required to boost the payload from a 100 Nm circular orbit to geosynchronous orbit.

(b) Given the following:

$$p_c = 500 \text{ psi}, \ \epsilon = 60, \ \eta_o = 0.95$$

determine the optimal mixture ratio, I_{sp}, and c^* of the liquid rocket engine propelling the vehicle.

(c) Given that the payload weight is 8000 lb, calculate the total weight of propellant required to achieve the ΔV value in part (a). Assume a propellant mass fraction of 0.85.

(d) To allow for boiloff, we must carry 5% more oxidizer and fuel than is necessary to complete the mission. In addition, the tanks must accommodate this boiloff volume as well as a 3% volume reserve for ullage. Calculate:

  (i) the total propellant required as in part (c); and

  (ii) the weight and volumes of fuel and oxidizer.

(e) Calculate tank lengths and diameter for cylindrical tanks arranged in a tandem fashion. Assume the L/D of the *fuel* tank is two.

(f) Given $L^* = 45$ in and $t_b = 100$ s, calculate the nozzle throat and exit radii and the flow rates of LH$_2$ and LOX through the chamber. Determine the burning times required for each of the burns in the Hohmann transfer.

(g) Given:

$$R_i = 0.5, \ R_{\omega tu} = 1.0, \ R_{\omega td} = 0.6, \ A_c/A_t = 2.5$$

$$\theta_i = 45°, \ \theta_d = 15°, \text{ conical nozzle}$$

calculate the length of the chamber and draw the nozzle contour for this rocket engine.

(h) Assume the payload is a cylinder with 8 ft diameter and 6 ft length. Draw the payload, tankage, and chamber/nozzle of the rocket to scale. Allow 1 ft between tankage and payload for interstage and 2 ft between tankage and chamber for pumps. What is the gross weight of the OTV?

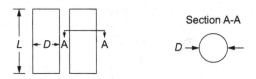

Figure 8.38 Diagram for Problem 8.12.

(i) Plot the vehicle velocity, mass, and acceleration (in g) as a function of time for each of the two burns assuming constant thrust. The payload has an acceleration limitation of $8g$; will throttling be required? If so, how much (in %)?

8.12 In this problem, you will develop a preliminary design of a rocket pack for human use in hovering and short duration flights. We will assume that a monopropellant called hydrazine (N_2H_4) will be used in the system. This propellant has a density near that of water ($\rho = 62.6$ lb/ft^3) and a sea level I_{sp} of 180 s at 300 psi average chamber pressure and a characteristic velocity of 3000 ft/s. Assume a propellant mass fraction of 0.65 for this rather small system and that our Rocketeer weighs 170 lb.

(a) Determine the ΔV required for our Rocketeer to be able to hover for one minute. [Hint: It's *not* zero!]

(b) Determine the amount of propellant required to accomplish the mission in part (a). If you can't find the ΔV from part (a), assume a value of 1500 ft/s in your calculations.

(c) Determine the inert weight of the propulsion system.

(d) Assume the system contains twin cylindrical tanks with $L/D = 3$ as shown in Figure 8.38. Determine L and D for the tanks.

8.13 In this problem, consider the design of a tandem tankage system capable of holding 1,500,000 lb of the cryogenic propellant combination LOX/H$_2$ at an overall mixture ratio of 5.5. Assume the common bulkhead spherical dome design shown in Figure 8.39.

In addition, assume ullage, boiloff, and trapped propellant volumes to represent 5, 1, and 6% usable propellant volume for each tank, respectively.

(a) Determine the volume requirements for each of the tanks.

(b) Assuming we desire $L_f/D = 4$, determine the lengths L_f, L_o, and D as shown in Figure 8.39.

(c) Assuming that nothing else is stacked on top of the tank, calculate the ullage pressure required in the fuel tank at liftoff to ensure pressure stabilization. You may neglect oxidizer tank structural mass in this calculation.

8.14 (a) An ideal cryogenic liquid rocket engine utilizing LOX/LH$_2$ is pictured in Figure 8.40. The engine is operating at a mixture ratio of 5.5, which implies a characteristic velocity of 2410 m/s, a molecular weight of 12.5 kg/kg-mol, a flame temperature of 3570 K, and $\gamma = 1.2$. If the *total* mass flow into the engine is 500 kg/s determine the chamber pressure and vacuum thrust of the engine.

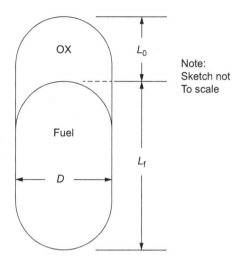

Figure 8.39 Diagram for Problem 8.13.

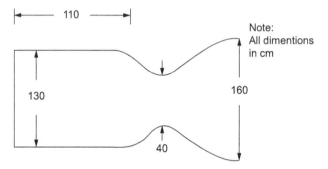

Figure 8.40 Diagram for Problem 8.14.

(b) Ten seconds prior to engine shutdown, the fuel turbopump fails catastrophically. Presuming that the fuel flow into the chamber instantaneously drops to zero, determine the rate of change of chamber pressure with time. You may neglect changes of chamber temperature with time and assume a cylindrical chamber of diameter and length shown in the Figure 8.40.

(c) How fast does the thrust of the engine change with time?

8.15 Some authors have noted that there are structural costs (larger nozzles, heavier chambers, etc.) associated with increased I_{sp} of high-performance space systems. Suppose the final mass of such a system can be written $aM + bI_{sp}$, with a and b representing constants that are positive in sign; where aM includes the payload and all inert components not affected by I_{sp}, and bI_{sp} is the mass of structural components that are affected by increasing I_{sp} levels. For a system

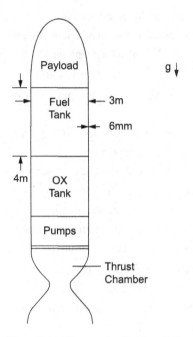

Figure 8.41 Diagram for Problem 8.16.

like this, increased I_{sp} has the tendency of increasing performance (through increased energy) while decreasing performance (through increased inert weight).

(a) Prove that for a given propellant mass, M_p, that an optimal I_{sp} exists for a system of this type.

(b) Find the optimal I_{sp} if $M_p/Ma = 1$ and $b = Ma$/unit time.

8.16 The bipropellant launch vehicle shown in Figure 8.41 is to be used to accelerate a 30,000 kg payload (containing an upper-stage propulsion system). The vehicle has a gross liftoff weight (mass) of 400,000 kg and the propulsion system has an overall mass fraction of 0.80. The liquid rocket engine provides a constant thrust of 4.5 mN, while the pressurization system maintains a constant ullage pressure of one atmosphere (absolute pressure). Assuming the fuel is kerosene ($\rho = 800$ kg/m³), and that the engine burns out in vacuum conditions, determine if the tank thickness of 6 mm is adequate to avoid buckling of the tank wall at burnout of the rocket. The tank is fabricated from aluminum which has a Young's modulus of 6900 MPa and a Poisson's ratio of 0.3.

8.17 A fixed volume of propellant (V) is to be stored in a cylindrical tank at a given design burst pressure (P_b). The designer has two choices in tanks, a long, skinny tank with $L/D = 5$, or a short stubby tank with $L/D = 1$. Further, assume that both tanks are manufactured from the same material with ultimate tensile strength F_{tu}.

Derive an expression for the structural weight ratio of the long tank to the short tank (M_{t5}/M_{t1}) under these assumptions. You may assume that the ends of the tank are the same thickness as the wall of the cylinder. Which tank will be lighter?

8.18 Consider the design of an orbital transfer vehicle using liquid oxygen/liquid hydrogen.

  (a) Determine the ΔV required to boost the payload from a 100 Nm circular orbit to geosynchronous orbit.

  (b) Given $p_c = 1000$ psi, $\epsilon = 40$, $\eta_o = 0.95$, determine the optimal mixture ratio, I_{sp}, and c^* of the liquid rocket engine propelling the vehicle.

  (c) Given that the payload weight is 5,000 lb, calculate the total weight of propellant required to achieve the ΔV value in part (a). Assume a propellant mass fraction of 0.85.

  (d) Those of you using cryogenic propellants must include a 5% weight margin to account for boiloff losses during the mission. In addition, the tanks must accommodate this boiloff volume as well as a 3% volume reserve for ullage. Calculate:

   (i) the total propellant required as in part (c); and

   (ii) the weight and volumes of fuel and oxidizer.

  (e) Calculate tank lengths and diameter for cylindrical tanks arranged in a tandem fashion. Assume the L/D of the *fuel* tank is two.

  (f) Assume $L^* = 45$ in unless you have hydrogen as fuel. In this case, assume $L^* = 20$ in. Also, assume the total $t_b = 150$ s. Calculate the nozzle throat and exit radii and the flow rates of fuel and oxidizer through the chamber. Determine the individual burning times required for each of the burns in the Hohmann transfer.

  (g) Given:

$$R_i = 0.5, R_{\omega tu} = 1.0, A_c/A_t = 2.5$$

$$\theta_i = 45°, \theta_d = 18°, \text{ conical nozzle}$$

calculate the length of the chamber and draw the nozzle contour for this rocket engine.

  (h) Assume the payload is a cylinder with 6 ft diameter and 4 ft length. Draw the payload, tankage, and chamber/nozzle of the rocket to scale. Allow 1 ft between tankage and payload for interstage and 2 ft between tankage and chamber for pumps. What is the gross weight of the OTV?

  (i) Plot the vehicle velocity, mass, and acceleration (in g) as a function of time for each of the two burns, assuming constant thrust. The payload has an acceleration limitation of $8g$. Will throttling be required? If so, how much (in %)?

8.19 Consider the LRE-propelled, space-based, rocket shown in Figure 8.42. The engine operates with a constant thrust of 200,000 N and I_{sp} of 350 s for a duration of 25 s. The vehicle mass at ignition of the engine is 2500 kg, while the payload mass (which includes all structure forward of the fuel tank) is 500 kg. The tank is fabricated from aluminum, which has

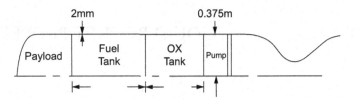

Figure 8.42 Diagram for Problem 8.19.

a Young's modulus of 6900 MPa and a Poisson's ratio of 0.3. Using this information, determine the minimum ullage pressure in the fuel tank to prevent buckling of the tank wall.

8.20 The first-stage engine uses an ablative chamber (a phenolic resin impregnated with silica fibers) for cooling. The mixture ratio is biased to keep the near wall hot gas temperature sufficiently low so the ablative can work. The upper limit of temperature is usually taken as about 3200 °F – above that the ablation rate is quite high and the throat goes away really fast. A striated injector is being designed. For what near-wall O/F should the injector be designed? Assume a stagnation recovery factor as described in Chapter 6 of 0.92. Using a two streamtube analysis, calculate the performance loss associated with providing this fuel-rich gas near the wall. Tests indicate that, at these conditions, the throat recession rate is 0.001 in/s. After a 120 s burn, what is the thrust level?

8.21 The second-stage engine uses a regeneratively cooled engine. Hydrogen is the coolant, and the combustion chamber wall is made of a copper alloy. Material specialists say that the wall temperature must be kept below 1100 °F to eliminate creep. What heat flux must be absorbed by the regenerative coolant at (a) the face and (b) the throat? Again assume that the stagnation recovery factor is 0.92.

8.22 A ground-launched interceptor utilizes a cylindrical fuel tank 30 inches in diameter and 100 inches long. The tank has a ullage pressure of 30 psi and a maximum internal pressure of 120 psig. The vehicle has a maximum acceleration of 20g and the fuel tank supports 1000 lb in upper-stage propulsion, interstage, and fairings. The tank is to be built from aluminum 6061 alloy, which has the following properties:

Ultimate tensile strength = 42,000 psi
Yield strength (tension) = 36,000 psi
Modulus = 10×10^6 psi
Density = 0.098 lb/in^3

(a) Determine the minimum acceptable wall thickness for this tank.
(b) Assuming constant wall thickness throughout, estimate the weight of the tank skin.

CHAPTER 9   LIQUID ROCKET PROPELLANTS

In this chapter, we provide an in-depth review of liquid rocket propellants, their properties, and benefits for specific applications. The chapter begins with an overview of the major classes of liquid propellants and provides a historical perspective to help explain why the industry uses specific propellants and retired others. Following an introduction and definitions of fuels and oxidizers in Sections 9.1 and 9.2, Section 9.3 focuses on the desirable properties of liquid rocket propellants. The next five sections look at key liquid propellant types. Section 9.4 covers monopropellants, starting with hydrazine since it is, to date, the most commonly used mono-propellant. Green monopropellants, or "non-toxic" monopropellants, follow the hydrazine discussion as interest is growing all over the world for easier to handle yet high-performance options. Section 9.5 continues from Section 9.2 on the theme of storable propellants but switches the discussion to bi-propellants and hypergolic propellants: two classes of propellants for upper-stages and space applications. Section 9.6 provides an overview of gelled propellants. These propellants fit in this chapter as they are designed to behave as liquids under shear conditions but, typically, remain solid-like at rest. Section 9.7 focuses on a critical class of liquid propellants: the cryogens. A significant emphasis is placed on liquid oxygen, liquid hydrogen, and liquid methane as they are the most common options in aerospace systems. Section 9.8 offers a few conclusions on liquid rocket propellants.

Key chapters in this text related to liquid propellants include Chapter 5, which describes performance analysis/thermochemistry aspects, Chapter 8 focuses on liquid rocket engines, and Chapter 10 is dedicated to turbomachinery. In addition, Chapter 6 provides discussion of regenerative cooling systems and Chapter 12 looks at feed system and combustion dynamics.

9.1 Introduction: Classification of Liquid Propellants and Historical Perspective

Any discussion on liquid rocket propellants should include a reference to *Ignition*! by John D. Clark (1972). *Ignition*! is a must read for anyone interested in learning about rocket propellants and, to a large extent, what not to do with rocket propellants. Many quotes could be pulled from this fascinating book but, and to set the stage in this definition section of the chapter, the following quote should get the reader's attention:

The aniline is almost as bad [as Red Fuming Nitric Acid], but a bit more subtle in its actions. If a man is splashed generously with it, and it isn't removed immediately, he usually turns purple and then blue and is likely to die of cyanosis in a matter of minutes. (Clark, 1972)

Implicitly, this quote conveys several of the key properties of rocket propellants. In addition to the obvious toxicity of the referenced propellants, the extreme energy contained within rocket propellants originates from chemical bounds and the exchange of electrons. This chapter provides an engineering oriented review of the main features that make a fuel be a fuel and an oxidizer be an oxidizer. In addition, the properties of the most common liquid rocket propellants are reviewed and compared to each other. Readers who are interested in chemistry and chemical kinetics should refer to specialized textbooks.

In general terms, propellants can be compared using three qualities, energetic, kinetic, and utilization (Barrière *et al.*, 1960):

- The energetic quality relates to the specific impulse of the propellant combination. This thermodynamic property indicates how effectively each pound of propellant is used in imparting thrust to a vehicle.
- The kinetic quality is quantified by the ignition properties of the propellant combination. The kinetic quality gives information about the amount of time necessary for the energy to be released. It is a very important property because it affects the conditions of ignition and the combustion stability.
- The utilization quality involves several metrics that describe the behavior of the propellant apart from its combustion. All the propellant properties that come into play in the feeding and cooling systems of a rocket combustor are included. A non-exhaustive list of these properties includes density, toxicity, melting and boiling points, and vapor pressure. The utilization quality also takes into consideration such problems as supply, maintenance, and storage, as well as cost.

These qualities are particularly important to consider in the early stages of the design of a new liquid propulsion system; whether it is for an endo-atmospheric application including all rocket testing facilities, the first stages of large launch vehicle systems and many missiles, or for an "out of the atmosphere" (exo-atmospheric) application, including the second and third stages of launch vehicles, satellites, space probes, etc.

Key considerations to take into account when selecting a liquid propellant, or any propellant for that matter, include the desired performance of the propulsion system (provided generally in terms of specific impulse), but also a host of properties including the density of the propellant combination, which is particularly important for tank sizing and center of gravity predictions, the storability of the propellants under the expected storage and operating conditions of the propulsion system, as well as cost, availability, manufacturing processes, material compatibility, and toxicity.

Other considerations that relate more closely to the detailed design of the mission expected for the selected liquid propellants include the expected lifetime of the mission, from seconds in a missile or expandable rocket stage, to years and decades in the case of the deep space probe such as the twin Voyager probes, which are still in operation 40 years into a journey around the outer solar system and beyond. For all these missions, possible liquid propellants range from simple monopropellants such as hydrazine to storable combinations like nitrogen tetroxide and mono-methylhydrazine, better-performing (and cheaper) combinations like liquid oxygen and kerosene, to the highest performance combinations of liquid oxygen and liquid hydrogen.

Many propellant combinations have been proposed and tested over the last 75+ years of liquid rocket experimentations. Not all these combinations, as high performing as they may be, can (or should) in practice be used for on-board rocket systems; no matter the end application. For example, a liquid fluorine and liquid hydrogen "enhanced" with beryllium particles combination has been proposed (Sarner, 1966). Its specific impulse of 480 seconds at p_c = 1000 psia, sea-level expansion is certainly highly enticing, but the toxicity and handling "nightmares" associated with fluorine and beryllium make that combination utterly impractical.

Current trends in propellants research includes a transition from LOX/kerosene to LOX/ methane for boost engines in particular. In the storable propellants, with a somewhat loose definition of "storable" as being liquid under "normal Earth ambient conditions," key advances have been made and on-going research aims at the replacement of nitrogen tetroxide/monomethyl hydrazine. Similarly, in monopropellant research, significant efforts have been put into replacing hydrazine systems with much less toxic and higher performance options.

The trend towards lower toxicity and non-carcinogenic propellants aims primarily at enhancing personnel safety. This lower toxicity requirement also includes a need for easier manufacturing and disposal procedures which may be related to an increase in environmental responsibility but certainly should lead to a reduction in costs and enable faster turn-around. On the other hand, the liquid propellant systems currently in use have an amazingly high success rate and many propellants have remained highly attractive choices because of relatively good performance characteristics and highly desirable exhaust properties. For example, as toxic as hydrazine is, it is an excellent non-cryogenic fuel, readily storable over a wide range of orbits around the Earth, with good and well-understood material compatibility characteristics, and simple, carbon-free exhaust products. While less toxic alternatives to hydrazine exist and new ones are being developed and qualified for space operation, it remains an excellent fuel for many applications.

To close this introduction on propellant toxicity, the reader will note in the text and other articles that the term "non-toxic" is typically used in a relative manner compared to hydrazine – mostly from a handling standpoint. Some people use the terms "less-toxic" or "green." For all purposes, we are talking about rocket propellants so it is reasonable to assume that the user will exercise standard safety precautions in handling them. When a propellant can be handled in a safe manner without a respirator or is not a carcinogen upon (brief) exposure to vapors, it is either "less-toxic" or "non-toxic"; the distinction here is academic and left to the reader.

9.2 WHAT IS A FUEL? AND WHAT IS AN OXIDIZER?

What makes a fuel a fuel, and an oxidizer an oxidizer? It will be implicit for all readers that oxygen, for example, is a great oxidizer but the question here is why? What makes oxygen a "great oxidizer"? Conversely, all of us are familiar with hydrocarbon fuels, from the gasoline we fill (most) cars with to the "Jet-A" or jet fuel A we run airplanes with and the "Rocket Propellant 1" or RP-1 we put in rockets, but what makes these fuels, and countless others, fuels?

The short answer to these critical questions is: electron transfer. It is from the ability of a molecule to donate or accept electrons that its fundamental nature as a fuel or an oxidizer originates. Fuels donate electrons and oxidizers accept electrons. A stoichiometric mixture is one in which there are just as many electron donors as there are electron recipients. For example, one molecule of methane with eight reducing electrons mixes stoichiometrically with two molecules of oxygen with four oxidizing electrons each. Such fundamental consideration of the ability of a molecule to donate or accept electrons is important to understand as it provides the basis for our current and future selection of any chemical propellant combination, no matter the end application.

To expand on the roles of electron transfer in the combustion of (rocket) propellants, we need two fundamental chemistry notions: oxidization states and electronegativity. Oxidation states will help us see that electron transfer relates to how electrons are distributed in an atom or molecule relative to its monatomic, uncharged, elemental state. From there, in molecules, how electrons are distributed is based on the electronegativities of the component atoms.

In molecules such as nitrogen or oxygen, which contain only one type of atom, neither atom exerts greater preference for the shared electrons. When a molecule contains more than one kind of atom, the oxidation states are determined by the respective electronegativities of the elements. Typically, the oxidation state of the molecule corresponds to the number of bonds that the atom can form (valency), and the charge is often determined by which side of the periodic table it is on. Thus, hydrogen atoms (left-hand side of periodic table) typically have a +1 oxidation state, while halogens (right-hand side of periodic table) have a –1 oxidation state. Oxygen and nitrogen typically have oxidation states of –2 and –3, respectively, and carbon can be either +4, +2 or 4. The sum of the oxidation states of the atoms in a molecule must equal the total charge on the molecule, whether it is charged or uncharged. For example, in Figure 9.1, the uncharged molecule methane (CH_4) follows the rule with the one carbon atom in the –4 oxidation state, and the four hydrogen atoms in the +1 oxidation state, resulting in a net neutral molecule ($-4 + 1 + 1 + 1 + 1 = 0$). Similarly, the charged ammonium cation (NH_4^+), follows the rule, with the one nitrogen atom in the –3 oxidation state and four hydrogen atoms in the +1 oxidation state ($-3 + 1 + 1 + 1 + 1 = +1$). In hydrogen peroxide (HOOH), which is uncharged, the hydrogens are in the +1 oxidation state, and each oxygen is in the –1 oxidation state as opposed to the –2 oxidation state, bringing the net charge to zero.

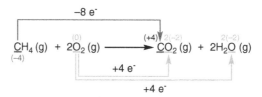

Figure 9.1 Electron transfer in methane combustion with oxygen.

Changes in oxidation state can occur during chemical reactions. The change in oxidation state can be related to the number of electrons associated with each atom. In the combustion of gaseous methane, carbon is oxidized from the −4 oxidation state to the +4 oxidation state, thus losing a total of 8 electrons. Similarly, 4 atoms of oxygen go from an oxidation state of 0 to +2, resulting in a net gain of 8 electrons. The hydrogen atoms do not change in oxidation state; this situation occurs often.

In this example, the methane is causing the oxygen to gain electrons, or be reduced.[1] So methane is called a reducing agent. The diatomic oxygen is causing the carbon to be oxidized (lose electrons), and so is called an oxidizing agent, or simply oxidizer.

When a strong oxidizer comes into contact with a strong reducing agent, both chemicals have strong impetus to spontaneously react. In most cases, the reaction releases a large amount of energy in the form of heat and light. The heat of this reaction is highly dependent on the starting chemicals and possible end products, but typical combustion reactions result in the formation of stronger bonds, corresponding to a net exothermic reaction.

Only a few elements in the periodic table have sufficiently large electronegativities to be oxidizers. Fluorine attracts shared electrons to the greatest extent, followed by oxygen, then nitrogen, and finally chlorine. However, nitrogen, despite its high electronegativity, is not known as an oxidizer. Nitrogen forms a triple bond, which is one of the strongest chemical bonds known ($\Delta H_{dissociation} = 944.7$ kJ mol^{-1}). We can therefore limit useful oxidizers for rocket applications to three compounds, known as the prime oxidizers: oxygen, fluorine, and chlorine. No rocket system currently uses fluorine due to its toxicity and handling difficulties. Chlorine is also hard to handle, but has attractive properties, particularly in solid propellants. Bromine and iodine are also good oxidizers but neither have seen any substantial use in rocketry given the highly desirable properties of oxygen and oxygen containing molecules.

Additional oxidizing molecules can be created by combining these prime oxidizers with other elements. The central atom should be electronegative and have a low atomic weight. Electropositive elements create strong bonds with large energy losses, and heavy elements increase

---

[1] The process of gaining electrons is called reduction, because early chemists thought that oxidation involved gaining oxygen atoms, and so the oxidizer must lose oxygen atoms, or be reduced. Later, scientists realized that oxidation involves transfer of electrons, but, the terminology remains: gain of electrons is reduction, and loss of electrons, whether due to oxygen or not, is oxidation.

the molecular weight of the combustion products (which decreases the I_{sp}, see Chapter 5). For example, nitric acid (HNO_3) has three oxygen atoms bonded to a nitrogen and a hydrogen atom. The nitrogen atom provides a convenient "anchor" for the oxygen atoms, but the oxidation process requires first breaking the bonds with the hydrogen and nitrogen atoms, reducing the energy released by the reaction. Hydrogen peroxide (H_2O_2) also oxidizes less well than pure oxygen because energy is released in the formation of O–H bonds.

Almost anything can be used as a fuel, but not all fuels make sense, particularly for rocket propulsion applications. Most useful fuels contain carbon, hydrogen, or light metals. Reducing agents which when oxidized yield high energy and low molecular weight gaseous products are good candidates for rocket propulsion.

Fuels are either inorganic or organic. Inorganic fuels include hydrogen compounds such as hydrogen (H_2) itself, ammonia (NH_3), hydrazine (N_2H_4), and though rarely used anymore: hydrides (e.g. LiH) and amides of low molecular weight metals such as lithium amide ($LiNH_2$). Organic fuels include saturated and unsaturated hydrocarbons, alcohols, amines, ethers, and aromatic compounds.

Liquid propellants can be divided into cryogenic propellants (or cryogens, e.g. liquid oxygen, hydrogen, or methane), and storable propellants, typically defined as those that are liquid at standard Earth conditions (e.g. kerosene fuels). Cryogens typically have higher specific impulses for two, somewhat interlinked, reasons. First, oxygen and fluorine are the strongest oxidizing agents (compounds with high electronegativity). Both are low-boiling gases. Storable oxidizers rely on additional elements, added to oxygen (or less commonly nowadays fluorine) to create liquids with boiling points that are near ambient temperatures. As we discussed above, adding atoms, which by nature have lower electronegativity than the prime oxidizers oxygen and fluorine, inhibits the compound's ability to accept electrons and therefore reduces the energy of the oxidation reaction. Second, common cryogens have simple molecular structures which results in low molecular weight exhaust products, and hence higher specific impulses. For example, the maximum shifting equilibrium specific impulse of the combination liquid oxygen and liquid hydrogen is about 400 s at a chamber pressure of 1000 psia and sea level expansion. In contrast, the maximum specific impulse for a common storable combination of nitrogen tetroxide and monomethylhydrazine is nearly 100 s lower.

For convenience, the properties of the most common liquid propellants in contemporary rocket systems are provided in Appendix A.2.

9.3 Desirable Properties in Liquid Propellants

Prior to reviewing the most common liquid propellants in use in contemporary rocket systems, it is of interest to list the properties one should strive for or should check for in designing or selecting a propellant combination for liquid rocket engines (for any chemical rocket engine for that matter, liquid, hybrid, and solid).

The list below is adapted from the excellent, albeit in this first quarter of the twenty-first century a bit dated, discussion on the combustion of liquid propellants by Altman and Penner (1960). Section L11 of Altman's and Penner's chapter concludes with eight key properties provided here in the same order but all critical important to the safe and reliable operation of a liquid rocket engine.

The eight key, or fundamental, properties for propellant selection relate to:

1. the heat of formation of the propellants;
2. the low molecular weights of the exhaust products;
3. the density of the propellants;
4. the state of the propellants at standard temperature and pressure;
5. the cooling the liquid rocket engine;
6. the stability of the propellants;
7. the toxicity, fire, and explosion hazards of the propellants; and
8. the duty cycle of liquid propellant engines.

First, high performance is linked to the *heat of formation of the propellants*, or reactants. A key aspect to remember with heats of formation of reactants is that small values are, usually, better. Referring back to Chapter 5, the reader will recognize that maximizing the heat release from a chemical reaction (or a highly negative heat of reaction for an exothermic reaction) means establishing a large heat of formation difference between the products and the reactants of the reaction. In turns, this implies the need for highly negative heat of formation values for the products of the reactants and moderate or small negative or positive heat of formation values for the reactants (or the propellants). To complement this first key property of liquid propellants, while most propellants will have a small negative heat of formation, a high negative heat of formation would mean a highly stable and hard to dissociate propellant while a high positive heat of formation would imply stability issues (as in some Class B monopropellants, see Section 9.4).

Second, the reaction products should have *low molecular weights* and large negative heat of formation values. As discussed earlier, the propellant combination of liquid oxygen and liquid hydrogen is a good example of a combination with small heat of formation values for the reactants (since both are, by convention, zero), high heat of formation values, and relatively low molecular weight for the products (water).

Third, the *densities of the liquid propellants* should be as high as possible to maximize volumetric efficiencies. An excellent illustration of the gains provided by high-density propellant is the relatively recent demonstration and use of "densified" or super-cooled kerosene by SpaceX in an effort to load more kerosene in the first stage of its Falcon 9, thus enabling longer burn times and the return of the stage back to a landing platform (or pad). While high densities are always desirable, propellant combinations for which densities and mixture ratios combine to provide nearly identical fuel and oxidizer tank volumes are of strong interest in many applications because they facilitate center of gravity management. This consideration is especially important in upper-stage propulsion systems, satellites, and high-performance missile interceptors.

Fourth, despite excellent performance originating from its excellent ability to accept electrons, cryogenic oxygen suffers from high handling costs and logistics surrounding the management of any cryogenic system. Similarly, while liquid methane offers many advantages over liquid kerosene, its handling is greatly complicated by its cryogenic nature, which in some cases, precludes its use for long-term applications. Generally speaking, propellants that are in a *natural liquid state around Earth ambient conditions*, often listed from –40 °C to +60 °C for aerospace systems, are desirable. That said, cryogenic propellants are excellent candidates in launch vehicles and, with improvements in storage conditions in space, one can expect to see an increase in their use in space vehicles in coming years.

Fifth, and as noted in Chapter 5, *cooling a liquid rocket engine* is extremely challenging with heat fluxes and combustion temperatures in most bipropellants well above 3000 K. It is therefore clear that any propellant used for cooling should have specific properties compatible with these conditions. The two most important propellant properties towards a good liquid coolant are a high specific heat, or the ability to store heat before it begins to get hot, and a high heat of vaporization, or the ability to store a lot of heat before vaporizing. Additional desirable properties include:

- a high boiling point to maintain a liquid state over a wide range of temperatures;
- a low freezing point to avoid freezing (particularly in deep space systems);
- a low viscosity enabling a steady and low-pressure-drop system;
- a high thermal conductivity to avoid large thermal gradients within the cooling fluid itself; and
- low corrosiveness to preserve hardware.

Sixth, the *stability of the propellants* must be well defined and understood. The stability of a propellant includes its behavior under normal storage conditions as would be expected in "standard" storage containers and the high-temperature stability as would be encountered in regenerative cooling passages in injector passages and manifolds under steady-state operation. Storage stability takes into account chemical processes that lead to the decomposition and deterioration of the propellant, which in turn can lead to self-pressurization and reaction with the storage vessel itself. In contrast, standard boiloff or vaporization of the propellant under certain ambient conditions relate to the storage life of the propellant but not its stability. Hydrogen peroxide is an interesting exception as a liquid, and storable, propellant that exhibits excellent stability under the right storage conditions but also a finite rate of homogeneous decomposition, which depend not only on ambient storage conditions but also on catalytic interactions with its storage containers. In general though, it is important to determine stability as a function of temperature, container size (which affects surface area to volume ratio) and material, seal and wetted component materials, ullage volume (gas to liquid volume ratio), and wetted surface conditions.

Seventh, the *fire and explosion hazards, sensitivity, and the physiological toxicity* of rocket propellants are critical considerations for any system. Entire volumes are dedicated to each of these topics but, for the sake of conciseness, an emphasis will be placed in this chapter on "exposure guidelines," which are available from a variety of agencies around the world. In the US, the most widely used (and stringent) measure of human tolerance to toxins is the threshold limit value (TLV) of the American Conference of Industrial Hygienists (ACGIH). TLVs are defined as the "airborne concentrations of substances and represent conditions under which it is believed that nearly all workers may be repeatedly exposed day after day without adverse effects."[2] TLV values are used as guidelines for airborne contaminants and are acknowledged by federal regulations such as OSHA, the Occupational Safety and Health Administration. Like the TLV, OSHA defines the permissible exposure limits (PEL) of workers, and both TLVs and PELs are typically reported as time-weighted average values (e.g. TLV-TWA) expressed in parts-per-million or ppm. This is an average concentration, assuming a normal 8-hour workday and a 40-hour workweek. PELs differ from other exposure limits in that they are regulations, not recommended guidelines. The National Institute of Safety and Health (NIOSH) also defines similar exposure guidelines along with "Immediately Dangerous to Life and Health" (IDLH) guidelines that are particularly well suited for many applications in the rocket industry. Whether or not a propellant is considered "toxic," these guidelines and regulations must be understood and defined prior to any use beyond a well-controlled laboratory environment.

The eighth and last key property defined by Altman and Penner (1960) relates to the *duty cycle of liquid propellant engines*. More broadly speaking than their original target of hypergolic propellants and their inherent ignition delays, and particularly with recent advances in reusable launch vehicles, the meaning of duty cycle should be understood in this context as how often an engine may be ignited and operated. Clearly, repeatable and stable operation is the objective for any rocket engine and this involves both highly repeatable ignition schemes (hypergolic or not) and the control and, if needed, mitigation of any propellant residue that may build up in injector passages or on the walls of the chamber.

Many other considerations come into play in the selection of liquid propellants. These include topics as wide ranging as the possible erosion of the hardware from potentially (likely) corrosive exhaust gases, a host of physical properties that affect the transfer of the propellants into storage tanks and subsequently in the combustion chamber, as well as critically important plume signature properties.

All these properties affect the cost of the propellants and of the propulsion system in general.

---

[2] See: https://www.acgih.org/tlv-bei-guidelines/tlv-chemical-substances-introduction [accessed March 2018].

9.4 MONOPROPELLANTS

A monopropellant is a fluid capable of undergoing exothermic reaction to yield gaseous products at a rate and with sufficient heat generation to sustain the decomposition process. Or, put another way, "a monopropellant is a substance which does not require the addition of another ingredient (fuel or oxidizer) to bring about release of its thermochemical energy" (Warren, 1958).

A monopropellant may be a single substance, with hydrazine as the most common example, or a mixture of compounds like the US Air Force monopropellant AF-M315E developed to alleviate the toxicity concerns of hydrazine while providing higher density and specific impulse.

Key advantages of monopropellants compared to other liquid propellants center on their simplicity of implementation. With only one fluid to load and connect to a thruster, minimal constraints on the precise propellant injection locations within a reaction chamber, and, relatively, low reaction temperatures (with hydrazine, for example, under 1500 K), monopropellants have been used primarily for attitude control devices with the propulsive energy generated, typically, through catalytic decomposition. This relative simplicity comes at the cost of lower performance than bipropellants and solid propellants. Further, the inherent reactivity of monopropellants makes any operation potential dangerous with risks of self- or accidently triggered decompositions than can lead to the possibility of detonations.

Like any other propellant, a fluid suitable for use as a monopropellant must satisfy three qualities, or requirement types. First, as mentioned, the thermodynamics of the fluid decomposition process must such be that "heat and mass transport from the reaction zone occur at a rate sufficient to maintain the decomposition processes" (Altman *et al.*, 1960: 83) Second, the kinetics quality of the monopropellant should be such that the fluid is stable, i.e. does not decompose "appreciably," at standard storage conditions and over extended durations (of months to years) but readily and reproducibly decomposes upon reaction with a catalyst or a heat source. The utilization quality refers to all handling aspects of the fluid (including transport requirements) and the conditions under which the monopropellant can be safely operated without significant changes in its decomposition rate.

To date, only four monopropellants have been used in spacecraft: hydrazine, hydrogen peroxide, propyl nitrate, and LMP-103S. Several others have been considered, however, and at least one, AF-M315E, is scheduled for flight with the Green Propellant Infusion Mission (GPIM) on the second flight of the SpaceX Falcon Heavy rocket.

Out of the four monopropellants used in space flight systems, hydrazine is by far the most common and will therefore be reviewed first. The second most common monopropellant is hydrogen peroxide with significant use up to the 1970s, current use in the Soyuz spacecraft, and a regular interest within the aerospace community for various missions. LMP-103S is a Swedish-developed monopropellant based on ammonium dinitramine (ADN). It has similar properties to the hydroxylammonium nitrate (HAN)-based AF-M315E so they will be reviewed together. Finally, propyl nitrate, or more precisely n-propyl nitrate, is a liquid monopropellant over a large temperature range but is significantly more shock sensitive than other flight-proven options. The same less-common monopropellants may be found by the reader in older textbooks or specialized literature in which fluids as reactive as nitromethane and nitroglycerin are mentioned for use in rockets.

One final but important aspect of monopropellants is the three classes used to distinguish between them. The three classes, A, B, and C, have the merit of providing clear distinctions between different "types" of monopropellant and will help the description of monopropellant currently available or in development.

- Class A includes single molecule compounds with both a fuel and an oxidizer included in the molecular structure. As can be expected, these compounds tend to exhibit stability issues under certain conditions or simply, like nitroglycerin as an example, are explosives. Others, like hydrogen peroxide, are purely the definition of a fuel and an oxidizer included in a molecular structure with two hydrogen atoms each connected an oxygen atom (H_2O_2).
- Class B encompasses single molecules which structures make them either fuels or oxidizers. These molecules behave as monopropellants when, upon activation with a catalyst or a heat source, the chemical structure of the atoms in the molecule becomes unstable. Hydrazine is the most common Class B monopropellant.
- Class C monopropellants are those synthetic mixture of two or more compounds of fuel and oxidizer. Unlike Class A monopropellants in which a single molecule is involved, constituents of a Class C monopropellant are not spontaneously reactive with one another but react sustainably once activated via ignition has occurred. AF-M315E and LMP-103S are recently developed Class C monopropellants.
- The densities, freezing points, boiling points, and heats of formation of representative and current monopropellants are provided in the Appendix.

9.4.1 Hydrazine (N_2H_4)

Hydrazine is a derivative of ammonia with the chemical formula N_2H_4 and the chemical structure shown in Table 9.1.

Liquid hydrazine has been, by far, the most commonly used monopropellant in space propulsion since the 1950s. It is both a monopropellant and a bipropellant fuel used in early rocket system with oxidizers such as hydrogen peroxide (with which it is self-igniting, or hypergolic). The applications for hydrazine include altitude control thrusters, insertion stages, and gas generators.

All the monopropellant engines that use hydrazine require a catalyst which decomposes it into ammonia, nitrogen gas, and hydrogen gas. While several catalyst materials have been suggested and tested with hydrazine, a key aspect, valid for any monopropellant system, is its

Table 9.1   Hydrazine chemical structure

2D representation	3D representation	Brief description
H_2N—NH_2		Hydrazine can be thought of as an ammonia derivative which contains a nitrogen–nitrogen linkage. As with any case where functional groups are mixed, it would be highly dangerous to classify hydrazine as an amine. The addition of a second nitrogen to the molecule increases both its hydrogen bond donating and accepting capability, elevating its melting and boiling point to a range more usable for propellants.

ability to start readily and operate steadily from low temperatures while sustaining the stresses induced by the exothermic decomposition process. The definition of "low temperature" is naturally quite subjective, but the high-altitudes thrusters typical of monopropellant systems certainly mean below Earth ambient temperatures. Most tested materials did not qualify as catalysts for hydrazine until the development of Shell 405, a solid catalyst composed of 30 wt.% iridium deposited on aluminum oxide (Alumina, Al_2O_3). Alumina is a highly crystalline support with interlinked micro and macro-pores (Schmidt and Walter, 2001), providing a large surface area for the iridium and a highly efficient catalyst material.

The hydrazine monopropellant engines themselves are quite simple and similar to their hydrogen peroxide counterpart with a pressurization and feed system upstream of the catalyst bed and measures to preheat the catalyst bed to ensure rapid and reliable ignition of the hydrazine. This preheating is not mandatory as Shell 405 can decompose room temperature hydrazine (operating temperatures as low as 275 K have been reported) without preheating but, in practice, consistent operation in ensured with preheating. Consistent operation means no appreciable mechanical degradation of the catalyst with bed temperatures as high around 1373 K, all with time delays below 10 ms. Shell 405 is now the most common catalyst in commercial hydrazine monopropellant thrusters with Aerojet Rocketdyne and Moog Isp as major suppliers in the US. Other catalysts such as carbon nanofibers (Vieira *et al.*, 2002) and molybdenum nitride on alumina (Chen *et al.*, 2002) have been proposed, and recent advances in additive manufacturing, including with high metal point metals such as iridium, will undoubtedly provide new avenues to develop more compact or less temperature sensitive catalyst beds.

Decomposition

The decomposition of hydrazine can be described with two consecutive reactions outlined in Eqs. 9.1 and 9.2:

$$N_2H_4 \rightarrow \frac{4}{3}NH_3 + \frac{1}{3}N_2 \tag{9.1}$$

$$NH_3 \rightarrow \frac{1}{2}N_2 + \frac{3}{2}H_2 \tag{9.2}$$

The first reaction is a low-temperature reaction, typically below 400 K, and relates to the dissociation of the N_2H_4 molecules into ammonia (NH_3) and hydrogen (H_2). It is an exothermic reaction generating ~112 kJ/mol of hydrazine. The second reaction describes the dissociation of ammonia into its fundamental constituents of nitrogen and hydrogen. It is an endothermic reaction requiring about 46 kJ/mol of ammonia. The dissociation of ammonia lowers the decomposition of temperature of the overall reaction (which negatively affects performance) but also decreases the molecular weight of the exhaust products (which positively affects performance). The net effect of these two reactions hydrazine can be described by the generalized reaction:

$$3N_2H_4 \rightarrow (1-x)NH_3 + (1+2x)N_2 + 6xH_2 \tag{9.3}$$

where x is a reaction parameter between 0 and 1 and reflective of the amount of ammonia dissociation in the exhaust. Naturally, this dissociation amount is directly related to the catalyst structure and its type. The size and operating pressure in the catalyst bed, and the residence time of the hydrazine within it, also affects the proportion of the NH_3 that is dissociated in the reaction in Eq. 9.2.

Ultimately, both the characteristic velocity (c^*) and the specific impulse of the catalyst bed are affected by the amount of ammonia decomposition, with a maximum c^* value near 1350 m/s with about 30% of the ammonia decomposed and a maximum theoretical vacuum I_{sp} near 260 s with no ammonia decomposition. In practice, the ammonia dissociation is limited from 30% to 80% by limiting the catalyst bed thickness (with a typical value around 55% depending upon on the design).

Overall, the exothermicity of the hydrazine decomposition reaction leads to a demonstrated vacuum specific impulse (I_{sp}) exceeding 220 seconds.

Properties

Hydrazine is colorless and hygroscopic liquid with a distinct, ammonia-like odor and a liquid range surprisingly close to that of water. The reported freezing point typically varies from 1 to 2 °C depending upon the grade (or amount of impurities). Similarly, the boiling point of hydrazine is ~112 °C, close to that of water at standard temperature and pressure. Hydrazine is a highly polar solvent, which is miscible with other polar solvents such as water (Haws and Harden, 1965).

There are a surprisingly large number of industrial uses for hydrazine, from pharmaceuticals to polymers foams where hydrazine is used as foaming agent and power plant applications in which

Table 9.2   Composition of monopropellant and high-purity grade hydrazine per military specification MIL-PRF-26536G (July 2017)

Composition in wt.% unless noted	Grade	
	Monopropellant	High purity
Hydrazine, min	98.5	99
Water, max	1.0	1.0
Particulate, max (in mg/L)	1.0	1.0
Ammonia		0.3
Aniline, max	0.5	0.003
Carbon dioxide, max	0.003	0.003
Chloride, max	0.0005	0.0005
Iron, max	0.0004	0.0004
Nonvolatile residue, max	0.005	0.001
Other volatile carbonaceous material (UDMH, MMH, and alcohol), max	0.02	0.005

Source: www.hydrazine.com

hydrazine serves as an oxygen scavenger. All these applications use either hydrazine hydrate (typically a 64 wt.% solution of hydrazine in water) or commercial grades of hydrazine with water concentrations and impurities that are not compatible with rocket grade catalysts. The monopropellant grade hydrazine defined by the military specification MIL-PRF-26536G (2017) contains less than 1 wt.% water, less than 1 wt.% particulates, and less than 0.5 wt.% aniline ($C_2H_5NH_2$). The full specification is provided in Table 9.2, adapted from www.hydrazine.com, along with the high-purity grade of hydrazine defined in military specification MIL-P-26536D (1987) used in most contemporary catalyst beds.

Safety

The surprisingly widespread use of hydrazine (at various grades) in industry comes from the fact that it is a powerful reducing agent. This property must be carefully weighted with the high toxicity of hydrazine and the precautions one must take to handle it. As with many monopropellants (and rocket propellants in general), improper storage and handling of hydrazine can quickly lead to exposure, fire, explosion, and material incompatibility.

Hydrazine is highly reactive with oxygen in the air, oxidizing agents, organic matter (which includes humans!), and oxides of metals such as iron, copper, lead, manganese, and molybdenum. This reactivity implies a necessary and absolute high level of cleanliness with any system wetted by hydrazine (any monopropellant or, really, propellant in general). While it is not susceptible to decomposition through normal impact or friction (hydrodynamic shear), hydrazine must be stored

Table 9.3   Hydrogen peroxide chemical structure

2D representation	3D representation	Brief description
HO⌒OH		Clear liquid

in a DOT-approved shipping container, kept tightly closed, and vented carefully when opening. Any metal container of hydrazine should be connected to a proper ground to avoid issues with static electricity discharges. In general, any propellant container should be protected from electrical sparks, open flames, and heat sources. The reactivity of hydrazine with oxidizing agents means that it should be stored under a nitrogen blanket.

Hydrazine and its fuel derivatives monomethylhydrazine (MMH) and unsymmetrical dimethylhydrazine (UDMH) pose very similar risks to humans. As strong reducing agents, all are carcinogens (destructive to living tissues). Consequently, both the American Conference of Industrial Hygienists (ACGIH) and the Occupational Safety and Health Administration (OSHA) define very low exposure limits of 0.01 ppm for the TLV-TWA from ACGIH and 1.0 ppm for the PEL-TWA from OSHA. A compilation of the available exposure guidelines for hydrazine, MMH, and nitrogen tetroxide (NTO, a common hypergolic oxidizer with MMH) is provided in the Appendix.

These hazards require extreme propellant ground handing precautions that drive up operational costs. The expensive handling and storage cost is also due to infrastructure requirements associated with hydrazine transport, storage, servicing procedures, and cleanup of accidental releases (Whitmore *et al.*, 2013).

9.4.2 Hydrogen peroxide (H_2O_2)

Hydrogen peroxide is the simplest "peroxide," a compound with an oxygen–oxygen single bond, with the chemical formula H_2O_2 and the chemical structure shown in Table 9.3.

In July 1818, Louis-Jacques Thenard reported to the Paris Academy of Sciences a method of preparation of what he first believed to be "oxygenated acids." Later that year, Thenard realized that he had discovered a new compound: oxidized water (or hydrogen peroxide). In all Thenard tested more than 130 substances, including metals, oxides, salts, acids, and bases. For some of these Thenard gave a complete description of the chemical action accompanying the catalytic decomposition of hydrogen peroxide. While this work constitutes one of the earliest recognitions of catalysis, its mechanism remained obscure to Thenard. He could not "conceive how an organ, without being absorbed or changed, can continually act on a liquid to transform it to new products" (Schumb, 1955). Without knowing it, he had given the definition of a catalyst. Another major accomplishment by Thenard was the preparation of essentially anhydrous hydrogen peroxide. This achievement is particularly impressive in that up until 1936 nobody made hydrogen peroxide concentrations higher than 30% (at least not commercially). Thenard's work was the most comprehensive study of hydrogen peroxide made in the nineteenth century.

Hydrogen peroxide has been extensively used as a rocket propellant. In fact, it has the longest history of any liquid propellant, which deserves at least a short review here as it provides insights into propellant selection strategies.

Brief History

The first propulsive application of hydrogen peroxide can be traced back to 1937 in the Heinkel He 52 aircraft designed by Helmuth Walter. The 80% hydrogen peroxide (T-stoff) was decomposed by aqueous solutions of calcium permanganate (Z-stoff) to propel assisted take-off units (ATOs). A few years later, a similar technology was used for the V-1 ATOs and to drive the turbopump on the V-2. Another major program involving hydrogen peroxide was the Messerschmitt Me-163 Komet. Ready for testing in 1938, the first model of the Me-163 was propelled by a "cold" motor. This motor also operated on the decomposition of T-stoff by Z-stoff.

After World War II, the production of concentrated hydrogen peroxide moved to the US, UK, and Soviet Union. The first major improvement over liquid/liquid decomposition techniques was the development of heterogeneous catalysts capable of decomposing rocket grade hydrogen peroxide (RGHP). RGHP is characterized by a high mass ratio of hydrogen peroxide to water (usually greater than 85%) and by less than 1.0 mg/l of phosphorus, tin, or sodium ions. RGHP is also referred to as high test peroxide (HTP).

In the US, the first practical heterogeneous catalyst enabled 90% hydrogen peroxide to be used as a monopropellant in the attitude control thrusters on both the X-15 and the Mercury spacecraft. The US Redstone, Jupiter, and Viking all used hydrogen peroxide to drive the turbo-pumps. The UK also pursued hydrogen peroxide for applications in aircraft ATOs, submarine, and torpedo systems. The most significant use of hydrogen peroxide for propulsion purposes was the Black Knight program carried out in the UK in the 1950s. Hydrogen peroxide was pre-decomposed in a silver screen catalyst bed and the resultant hot gases were burned with kerosene in four pump-fed regeneratively cooled gas generator cycle engines. The success of the Black Knight vehicle, launched 22 times between 1952 and 1968, ultimately led to the development of a more powerful rocket named the Black Arrow. On its fourth attempt the Black Arrow successfully placed into orbit a British satellite in October 1971. However, a few months earlier, financial and political concerns had led Britain's minister for aerospace to announce the cancellation of the program.

To this day, use of hydrogen peroxide as a monopropellant in space vehicles continues in Russia for the control systems in the Progress and Soyuz spacecraft (Iarochenko and Dedic, 2001). In both the US and UK, propulsion devices using hydrogen peroxide were largely abandoned after the 1970s in favor of higher performance monopropellants and bipropellants systems.

Interest in hydrogen peroxide as a low-toxicity and low vapor pressure propellant resurfaced in the US in the 1990s and 2000s with industrial and government investments into catalyst bed technologies and hypergolic fuels. Programs such as the X-37, ISTAR, an upper-stage bipropellant engine, the Airborne Laser (ABL) program, and a liquid target vehicle for the Missile Defense Agency are examples of these investments. Several of these programs are well documented in professional publications and achieved several important milestones with RGHP. However, by the mid 2000s, hardware failures with RGHP led NASA to significantly reduce its, and subsequently the industry's, investments in RGHP propulsion systems. Around the same timeframe of the 1990s and 2000s, private entities such Beal Aerospace and as Blue Origin demonstrated the potential of RGHP systems. Beal Aerospace developed and demonstrated several 90 wt.% RGHP/RP-1 engines, including a 800,000 lbf thrust engine. Blue Origin demonstrated its BE-1 and BE-2 engines, respectively a monopropellant and a RGHP/Kerosene engine. Interest in hydrogen peroxide, both as a monopropellant and a bipropellant, continues to this day in the US with a variety of development programs.

In Europe, interest in hydrogen peroxide resurfaced in the late 2000s with the "Green Advanced Space Propulsion" program, the first large European-wide, comprehensive effort to investigate green propellants. The GRASP consortium consisted of industry, universities, and research institutes and incorporated 12 entities from seven European countries (Austria, France, Germany, Italy, Sweden, Poland, and the UK). GRASP was run from 2008 to 2011. Starting from a list of about 100 green propellant candidates, 27 were selected for further evaluation as they showed significant reduction in toxicity, good performance potential, good storage and handling qualities, and in some cases a large reduction in procurement cost compared to their more toxic counterparts. Eleven of those 27 candidates were scheduled to be experimentally investigated and hydrogen peroxide, along with ammonium dinitramide (ADN)-based blends (e.g. FLP-106, described below) were identified as the two most promising European options for a short-term replacement of hydrazine seem. Additional details on the GRASP program are available at www.grasp-fp7.eu/grasp/.

Decomposition

All hydrogen peroxide solutions, including RGHP, decompose naturally at a very low rate of nominally 0.5–2% a year. The decomposition rate of hydrogen peroxide increases by a factor of 2.3 for each 10 °C rise in temperature, in accordance with Arrhenius' Law. The rate of decomposition of high-purity 90% hydrogen peroxide is listed at various temperatures in Table 9.4 (Paushkin, 1962).

Table 9.4   Rate of decomposition of 90% hydrogen peroxide at
various temperatures

Temperature [°C]	Approximate rate of decomposition
30	1% per annum
66	1% per week
100	2% per day
140	Rapid decomposition

If the container is not equipped with a pressure relief mechanism, the exothermic decomposition process will self-accelerate and very large volumes of gases will be generated. The volume of gas generated by the adiabatic decomposition of a 90% solution of hydrogen peroxide is approximately 5000 times the volume of the original solution. Consequently, even small volumes of hydrogen peroxide can result in an explosive rupture of equipment if pressure relief is not provided.

RGHP can be decomposed reliably into superheated oxygen gas and water vapor by a variety of metals, oxides, and salts. The most common catalyst in RGHP monopropellant systems is silver, typically arranged in screens (either pure or coated on stainless steel or another alloy). As for any catalyst bed, preheating can accelerate decomposition and provide smoother start transients. However, any preheating is a tax on the power systems of a spacecraft and should be therefore be minimized, no matter for which monopropellant it is. With RGHP, even cold RGHP, catalyst beds can and have been designed that do not require preheating.

The governing reaction for the decomposition process involving 100% hydrogen peroxide is given by

$$2H_2O_2(l) \rightarrow 2H_2O(l) + O_2(g) + 689.3 \text{ cal/g of } 100\% \ H_2O_2 \qquad (9.4)$$

The reaction features first-order chemical kinetics, with the rate of reaction being strongly dependent on temperature, hydrogen peroxide purity and concentration, and surface activity, and less dependent on surface roughness and pH. The non-catalyzed reaction is quite slow and unsuitable for a thruster application, but several catalysts are known to dramatically increase the reaction speed. One of the chief advantages of hydrogen peroxide as a propellant is the ease with which it can be catalytically decomposed into oxygen and steam. Furthermore, the reactivity of hydrogen peroxide is a great asset in propulsive applications since it tends to be hypergolic with compounds less hazardous than hydrazine fuels (Blevins et al., 2004).

The temperature of the decomposition gases and, therefore, the amount of heat generated is a function of the hydrogen peroxide concentration (Nimmerfroh et al., 1999). Hydrogen peroxide concentrations below 67 wt.% do not contain sufficient energy to sustain their self-decomposition, or in other words: they do not contain enough decomposition energy to boil off water in the solution.

The hydrogen peroxide concentrations of interest in this study are for the most part above 90%. The adiabatic decomposition temperatures associated with these concentrations can be accurately estimated using a linear regression equation. The regression coefficient of Eq. 9.5 is 0.99:

$$T_{ad}(K) = 25.339(H_2O_2 \text{ wt.\%}) - 1266.7 \qquad (9.5)$$

Equation 9.5 was obtained by correlating the adiabatic decomposition temperatures obtained from the *Hydrogen Peroxide Physical Properties Data Book* (Buffalo Electro-Chemical Company, 1955).

Properties

Hydrogen peroxide is a relatively simple material which has most of the desirable properties for liquid propellant propulsion systems. Solutions of hydrogen peroxide are clear, colorless, water-like in appearance, and can be mixed with water in any proportion. At high concentration, hydrogen peroxide has a slight acidic odor. When compared with traditional hypergolic oxidizers such as nitric acid and nitrogen tetroxide, hydrogen peroxide solutions have similar density, lower vapor pressure and reduced corrosiveness. Properties of typical hypergolic oxidizers are shown in the Appendix.

The density of hydrogen peroxide increases with concentration. For concentrations greater than 90%, the density of hydrogen peroxide is equivalent to that of nitric acid and nitrogen tetroxide. The high density of hydrogen peroxide is an advantage in rocket systems in which volume is limited. Furthermore, high-density propellants require smaller and lighter tanks, which in turn may help reduce drag during flight through the atmosphere.

Another desirable property of hydrogen peroxide is its high specific heat. As shown in the Appendix, the specific heat of hydrogen peroxide is comparable to that of water, which is a very good coolant.

The relatively high freezing point of hydrogen peroxide is a disadvantage in some cold weather applications. However, it should be noted that, unlike water, hydrogen peroxide contracts upon freezing. This means that the storage container should not burst or suffer major structural damage upon freezing of hydrogen peroxide. Such a feature is highly desirable in long-term storage.

Pure hydrogen peroxide is markedly insensitive to detonation (Warren, 1958). However, mixtures with soluble organics can detonate, and vapors containing more than 26 mole percent hydrogen peroxide are detonable. Vapors of this concentration are formed above 90% hydrogen peroxide only when the liquid is heated above 230 °F.

Hydrogen peroxide is produced in commercial quantities for use by several industries with the major user being the paper pulp processing industry. Other applications include industrial and potable water treatment, environmental contamination cleanup, mining, electronics, and metals processing. There are five major producers of hydrogen peroxide in North America: Chemprox, a joint venture between Air Liquide and Elf Atochem; Degussa; Eka Nobel; Peroxychem (a company derived from the FMC Corporation); and Solvay Interox. The use of hydrogen peroxide for propulsive applications is orders of magnitude lower than the commercial hydrogen peroxide markets and, in 2018, Peroxychem is the only manufacturer of RGHP in the US. In Europe, the industrial company Evonik produces concentrated hydrogen peroxide.

RGHP used in the US satisfies the military specification MIL-P-16005E issued in 1968. Despite the fact that this specification was cancelled in 1988, it is still considered by some hydrogen peroxide manufacturers to be the only one that has demonstrated efficacy.

Safety

Hydrogen peroxide solutions are non-toxic. However, suitable materials for construction of hydrogen peroxide systems must be carefully chosen and only those materials with a high degree of compatibility should be used for long-term storage. Potential health effects and proper personal protective equipment required for handling hydrogen peroxide are described in Material and Safety Data Sheets (MSDS) provided by manufacturers. In short, hydrogen peroxide solutions and vapors are skin corrosive agents. Skin burn and respiratory tract inflammation can occur when exposed to high concentrations. A readily available source of running water, adequate protective equipment, and proper safety training of personnel greatly mitigate the effect of exposure to high hydrogen peroxide concentrations. In fact, when handled properly, hydrogen peroxide is a very safe material and most operators prefer to handle hydrogen peroxide rather than other rocket grade oxidizers.

Handling and storage considerations generally require non-toxic and non-corrosive propellants with low vapor pressure, low freezing point, high shock stability, and high ignition energy. The low vapor pressure of hydrogen peroxide is an advantage because it reduces the likelihood of propellant vaporizing in the feed system of a rocket engine. As shown in the Appendix, the vapor pressure of 100% hydrogen peroxide is approximately 500 times lower than that of NTO and approximately 30 times lower than that of 99% nitric acid.

As for any fluid, the exposure guidelines and regulations for hydrogen peroxide must be reviewed with respect to the concentration of the fluid to be used. This is particularly important for hydrogen peroxide since most, if not all, rocket applications use concentrations above the 67 wt.% mark. This is an important point when one wants to make a situation "safe", dilution of hydrogen peroxide below 67 wt.% provides some measure of safety. However, concentrations even as low as 15% can be hazardous. Previously at Purdue University, a 27% mixture was accidentally stored in a sealed vessel and underwent decomposition/reaction causing the pressure in the vessel to rise to the rupture point. Lower-concentration mixtures can also cause burns and other exposure symptoms, so caution is advised even for these less hazardous concentrations.

Both OSHA and NIOSH define time weighted average exposure limits of 1 ppm. The ACGIH recommends a threshold limit value (TLV) of 1 ppm for hydrogen peroxide. While

Figure 9.2 Purdue University graduate student ready to load RGHP.

surprisingly low, given than it is the same limit given for hydrazine, the hydrogen peroxide TLV is somewhat arbitrarily set at 10% of the limit of irritation. For hydrogen peroxide, it is not directly related to toxicity. A probably much more useful limit for rocket applications is the immediately dangerous to life or health (IDLH) value of 75 ppm defined by NIOSH for hydrogen peroxide. This value is 3.75 times higher than the NIOSH-defined IDLH for nitrogen tetroxide. Combined with a factor of 500 in vapor pressures of 100% hydrogen peroxide vs. nitrogen tetroxide, the relative ease of handling the two liquids becomes obvious.

However, it must be very clear to the reader that hydrogen peroxide is a strong oxidizer and a monopropellant. As with any monopropellant, one should always be concerned about spontaneous decomposition. Exposure to hydrogen peroxide includes symptoms of irritation of the eyes, nose, and throat, corneal ulcer, skin redness, and bleaching hair. The use of proper personal protective equipment is critical and includes clothing to prevent skin contact, face shields, and gloves.

Finally, to conclude on this brief safety overview: with hydrogen peroxide, water is your strongest ally. Ensure that a ready source is at your testing or launch site.

9.4.3 Ionic Liquid-based Monopropellants

In general terms, ionic liquids (ILs) are salts in which the ions are poorly coordinated. This property makes ionic liquids melt at or below 100 °C or, even more conveniently for many applications, at room temperature. Along with low to almost non-existent vapor pressure, high thermal stability, and the ability to dissolve a wide range of chemical species, "ILs have brought a green revolution to chemistry and chemical engineering in the past dozen years" (Zhang and Shreeve, 2014).

While highly promising, the relative novelty of ILs and the virtually infinite number of building blocks to form new ILs complicates the generalization of their potential application to rocket systems.[3] However, over the past 20 to 25 years, several organizations around the world have made significant progress in developing rocket-grade and low-toxicity monopropellants taking advantage of the properties of ILs. The key objective of this work is the development of candidates to replace hydrazine as a primary propellant for space propulsions systems.

Particularly, the US Department of Defense (DoD) and the Swedish Space Corporation (SSC) subsidiary Ecological Advanced Propulsion Systems (ECAPS) have developed IL propellant options based on ammonium salts, and, respectively, on hydroxylammonium nitrate (HAN) and ammonium dinitramide (ADN). Both HAN and ADN are outlined in Table 9.5. These propellants, called AF-M315E for the HAN-based option (Figure 9.3(a)) and, for the ADN-based options LMP-103S and FLP-106 (Figure 9.3(b)), contain an IL as an ionic oxidizer associated with a fuel as a reducing agent. Water is added as a solvent, a stabilizer, and to tame the combustion temperature in catalyst beds and thrusters. Fuel components are added in the blend to increase propellant performance through higher energy release and higher flame temperature. The common fuel components are methanol, glycine, or ethanol. Overall, AF-M315E, LMP-103S, and FLP-106 can offer up to a 50% improvement in available density-specific impulse over hydrazine. One of the main advantages of these two salts is that they have a very low vapor pressure at room temperature and thus do not give off any toxic vapors characteristic of hydrazine-based propellants.

The undeniable toxicity benefits of AF-M315E and LMP-103S over hydrazine should be counterbalanced by the potential hazards associated with the crystalline solids of HAN and ADN,

---

[3] Scientists estimate there are as many as 10^{18} possible binary ILs ("Ionic Liquids to Replace Hydrazine. Customized ionic liquids offer better safety and convenience than traditional propellants." Dryden Flight Research Center, Edwards, CA. NASA Tech Briefs, March 2011).

Table 9.5   HAN and ADN chemical structures

2D representation	3D representation	Brief chemical description
		Hydroxylammonium nitrate (HAN) is an ionic complex comprised of the hydroxylammonium cation and nitrate anion. HAN is the neutralized acid–base pair of hydroxylamine ($HONH_2$) and nitric acid (HNO_3). Ammonium dinitramide (ADN) is an example of a (semi-) stable nitrogen anion paired with an ammonium cation, forming a salt. The nitramide anion is a form of substituted amide (NH_2^-) anion, where the two hydrogens are replaced with nitro (NO_2) groups.

Figure 9.3 (a) AF-M315E, an HAN-based monopropellant. (b) FLP-106, an ADN-based monopropellant.

two highly energetic salts with both reducing and oxidizing components. This means that in the pure form both are unstable and potentially explosive. In propellant applications, however, both HAN and ADN are used in concentrated aqueous solutions to limit the explosion potential. In addition, the water content in both AF-M315E and LMP-103S complicates their ignition processes and, at least to date, has meant that significant preheating on the catalyst beds is required for those to operate reliably. Cold-starts with AF-M315E and LMP-103S are not yet possible, thus presenting significant challenges to the power systems on-board smaller spacecraft.

9.4.4 AF-M315E: Hydroxyl Ammonium Nitrate (NH_3OHNO_3, HAN)-based Monopropellant

As a Class C monopropellant, AF-M315E is a synthetic ionic liquid mixture of the salt hydroxyl-ammonium nitrate (HAN), water, and a hygroscopic fuel (Hawkins *et al.*, 2010). The basic propellant formulation was developed by the US Air Force Research Laboratory (AFRL) in 1998 as part of AFLR's Integrated High Payoff Rocket Technology Program, but its origins go back to the US Army's investigation into HAN for use in liquid gun propellants. A patent awarded to the US AFRL (patent under a secrecy order) covers the exact formulation of AF-M315E.

While the formulation of AF-M315E is not public, published liquid gun propellant research provides a good basis to understand the properties of typical HAN-based liquid propellants (Jankovsky, 1996). These propellant blends consist of HAN as the oxidizing agent, water as the solvent and buffering agent, and tri-ethanol ammonium nitrate (TEAN, $(OHCH_2CH_2)_3NHNO_3$) or diethyl-hydroxyl-ammonium nitrate (DEHAN, $(CH_3CH_2)_2NHOHNO_3$) as the fuel components. While the water serves as a diluent in the HAN-based liquid propellant in solutions, it is not too surprising that it affects performance. More water leads to a lower exhaust temperature but also a lower specific impulse (Jankovsky, 1996).

AF-M315E is typically ignited and decomposed through the use of a preheated catalyst generating an adiabatic flame temperature close to 1800 °C (Tsay *et al.*, 2013). Not many catalyst materials can survive such combustion temperatures as many commonly used catalyst materials would sublimate or sinter well below 1800 °C. Several technologies do exist, however, with, for example, the Green Propellant Infusion Mission (GPIM), where Aerojet Rocketdyne's catalyst bed design makes use of a patent-pending LCH-240 (iridium on hafnium oxide) catalyst. Similarly, Busek Co. Inc. is developing a family of space thrusters ranging from 0.1 to 22 N for AF-M315E using their patented monolithic catalyst. Examples of AF-M3155E thrusters are shown in Figure 9.4.

AF-M315E exhibits significant advantages over hydrazine, from a toxicity standpoint as mentioned earlier but also from a performance standpoint. With a density of 1.47 g/cm^3 at standard temperature and pressure, it is nearly 50% denser than hydrazine. Combined with an expected vacuum specific impulse of 250 s (22 N Aerojet Thruster, p_c up to 400 psia, and an expansion ratio

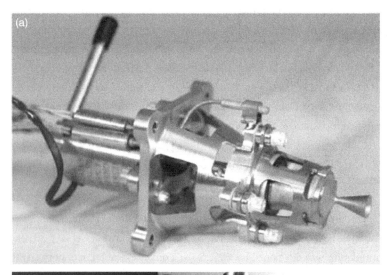

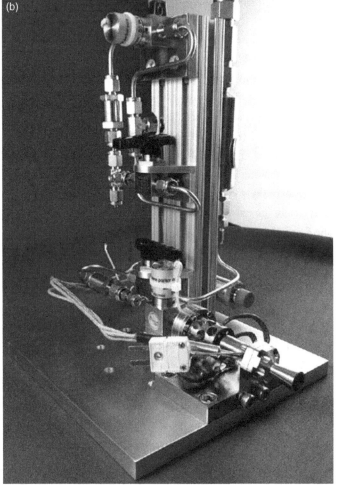

Figure 9.4 AF-M315E Thrusters: (a) Aerojet Rocketdyne 1 N thruster, and (b) Busek Co. Inc. 0.5 N thruster.

of 50), the density specific impulse of AF-M315E is ~366 s-g/cm^3, a value comparable to traditional storable bipropellants for low ΔV missions.

In terms of handling, AF-M315E is a mild acid that demonstrates long-term compatibility with only a limited set of metals, including Ti-6Al-4V, but not any iron-bearing material. A series of broad characterization tests concluded that AF-M315E has a significantly reduced sensitivity to adiabatic compression. For transportation purposes, AF-M315E is typically shipped in 19 L composite pails and has received a hazard classification of 1.4C, allowing it to be transported on cargo aircraft (Hawkins *et al.*, 2010; Sackheim and Masse, 2014).

Personal protective equipment typical for use with AF-M315E includes gloves, eye protection, and splash suits or coveralls (Hawkins *et al.*, 2010). Additional properties of AF-M315E are listed in Appendix A.2.

Overall, the development and existing qualification work with AF-M315E makes it a high-TRL "green monopropellant" with established handling guidelines but, as with any monopropellant, rigorous procedures must be followed at dedicated facilities.

9.4.5 LMP-103S and FLP-106: Ammonium Dinitramide (H$_4$N$_4$O$_4$, ADN)-based Monopropellants

LMP-103S is a Class C monopropellant and is an ionic aqueous solution of 63% ammonium dinitramide (ADN), 18.4% methanol, 14% water, and 4.6% ammonia per mass developed by the Swedish company ECAPS and the Swedish Defense Research Agency (FOI) beginning in 1997. LMP-103S is quite unique in contemporary rocket propellants in that it is one of very few new propellants demonstrated in space operation since the introduction of monomethylhydrazine in the late 1960s. The propellant and thrusters developed specifically for LMP-103S were flight demonstrated aboard the PRISMA Mango satellite, launched in 2010. The PRISMA Mango satellite mission was a Swedish-led technology mission to demonstrate formation flying and rendezvous technologies.

The LMP-103S mixture combines the oxidizing properties of ADN and the reducing properties of methanol. As with AF-M315E, water acts as a stabilizer and moderates the catalyst bed temperature. Ammonia is another stabilizer in LMP-103S but also serves to balance the pH level and dissolve the ADN crystals into the solution. Similarly, as with AF-M315E, preheating of the catalyst bed is required to achieve the adiabatic flame temperature of LMP-103S of ~1600 °C (Neff *et al.*, 2009).

In July 2016, ECAPS and Orbital ATK signed a cooperation agreement to continue the development of propulsion systems for LMP-103S. The companies also successful tested a 22 N HPGP thruster with support from NASA Marshall Spaceflight Center and NASA Glenn Research Center.

FLP-106 is another, albeit much less developed than AF-M315E, ADN-based monopropellant developed by the Swedish Defense Research Agency. It consists of ADN (64.6%), water

(23.9%), and a "low volatile hydrocarbon fuel" (11.5%) referred to as "F-6." FLP-106 reportedly provides an eight-second specific impulse improvement over LMP-103S and a 13% increase in density-specific impulse.

Although FLP-106 is not as developed as LMP-103S, the GRASP review team identified it as one of two viable hydrazine replacements.

Publications on FLP-106 discuss its physical characteristics, its theoretical performance along with ignition data obtained in laboratory experiments. Thruster qualification data is not readily available.

From a safety standpoint, FLP-106 and LMP-103S have similar oral toxicity but FLP-106 has a lower vapor pressure. As with LMP-103S, FLP-106 is reportedly not shock sensitive (Wingborg, 2011).

One final word of caution to conclude this section: with monopropellants, one should always be concerned about spontaneous decomposition.

9.5 STORABLE AND HYPERGOLIC PROPELLANTS

As their name indicates, the design or the selection of a storable propellant leads to a readily useable propulsion system. In effect, the ubiquitous gasoline in our cars and trucks is a perfect example of a storable propellant. This ready-to-use characteristic comes from a range of requirements that may themselves vary as some propulsion systems rely on storable but not hypergolic propellants such as hydrogen peroxide and kerosene.[4] In turn, other propulsion systems rely on the extremely fast ignition delay between storable and hypergolic propellants such as nitrogen tetroxide and monomethylhydrazine. In general, however, and unlike cryogenic propellants which require extensive conditioning measures, storable propellants are liquids over a range of temperatures characteristic ambient Earth conditions. Relatively significant temperature ranges have been defined, however, particularly for military applications with –50 to +70 °C as lower and upper bounds for liquid propellants. One drawback of storable propellants is that, with the exception of kerosene-based fuels, they are often toxic, reactive, corrosive, or generally difficult to handle.

Other requirements for storable propellants include the ability to be stored for extended periods of multiple years, with no change in properties or off-gassing, in perfectly sealed containers. This implies a high level of compatibility with standard aerospace materials, and, typically, low vapor pressures.

In practice, no propellant, storable or not, is perfect but it is clear that storability typically implies a loss in performance in return for greater usability.

---

[4] Whether hydrogen peroxide is a "true" storable propellant is subject to debate since it should not be stored in a perfectly sealed container; a feature typically attributed to "storable" propellants.

9.5.1 Storable Hydrocarbon Propellants

The evolution in the use of kerosene fuels $((CH_2)_n)$ for rocket propulsion since the first experiments with liquid fuels by Robert H. Goddard provides an interesting perspective on the reasons for the widespread use of storable hydrocarbon propellants in contemporary rockets.

Around the early to mid 1920s, when Goddard worked on his liquid-fueled rocket, hydrocarbon fuels such as ether and gasoline were common. Goddard's rocket itself burnt gasoline with liquid oxygen. Following World War II and the use of alcohols in the V2 rockets and several derivatives of the V2, the Navajo and Jupiter rockets reignited (no pun intended) the use of hydrocarbon fuels in the US. JP-4, or Jet-Propellant 4, was the kerosene of choice for these late 1940s and 1950s programs that demonstrated the critical performance improvements of kerosene fuels over alcohols. From there, several generations of kerosene fuels ensued, mainly with more refined versions leading to the definition of the RP, or Rocket Propellant class, of kerosene fuels. Today, RP-1, or Rocket Propellant 1 is the most common hydrocarbon fuel in rocket systems with liquid methane gaining significant interest in the past few years. Both RP-1 and liquid methane are typically combusted with liquid oxygen.

Fundamentally, a hydrocarbon is a combination of hydrogen and carbon to yield a large number of compounds. Hydrocarbons come from crude oil, and hence exhibit a wide range of properties – between hydrocarbons and within each compound itself. They may contain both saturated and unsaturated compounds, aromatics, and other complex structures. The degree of saturation of a hydrocarbon refers, simply, to how many *pi* bonds (double bonds or triple bonds) and/or rings are in a molecule. Benzene and acetylene, shown in Table 9.6, are convenient representations of degree of saturation of hydrocarbon molecules.

In general, any number of hydrocarbons are suitable for use as fuels and the simpler structures, for example methane (CH_4), should give higher performance since no energy is lost in extra carbon–carbon bonds (Sarner, 1966).

The refinement of crude oil into rocket fuels follows acid washing, sulfur dioxide extraction, distillation, and cracking processes in which the less desirable constituents are "cracked" then removed by distillation. These processes are also used to generate ultra-low-sulfur content versions of RP-1 or RP-2, for example. Another variant of liquid storable hydrocarbon fuels used in some rocket systems, particularly experimental rockets and combustors, is the so-called "JP" series of fuels, or Jet Propellants, specifically developed for jet engines.

The general chemical composition of RP-1 is 41% paraffins, 56% naphthenes, and 3% aromatics. The low concentration of aromatics and olefins is advantageous as it limits the possibility for coke residues in cooling passages and injectors. RP-1, like most liquid storable hydrocarbon fuels, has a low freezing point of -34 °C and a relatively high density of ~0.8 g/cm^3 at room condition. Additional properties for RP-1, JP-8, and other hydrocarbon fuels are provided in Appendix A. It should be noted here that these fuels are not pure compounds, and therefore exact properties are not exact.

Table 9.6 Benzene and acetylene chemical structure

2D representation	3D representation	Brief chemical description
		Benzene is the most commonly used example of an aromatic system. It satisfies the four Hückel rules for aromaticity. Benzene is commonly found as a combustion product, and the benzene substructure is suspected of being a key partner in coke/soot formation.
		Benzene has four degrees of unsaturation (three double bonds and one ring).
		Acetylene has two degrees of unsaturation (one triple bond = two *pi* bonds).

Hydrocarbon fuels can be toxic by inhalation, ingestion, or prolonged skin contact. Concentrations above 1000 ppm produce symptoms similar to intoxication. Standard protective equipment for storable hydrocarbon fuels depend on the application and the quantity but should include gloves and safety googles. As with the gasoline in our cars and trucks, extreme caution should be exercised in the event of a large spill due to the fire hazards associated with the flammable vapors emanating from such spills.

Corrosion and material compatibility issues with storable hydrocarbon fuels are well documented and relate mostly to the presence of mercaptans (sulfur-containing compounds) and naphthenic acids (known for their role in corrosion damage to oil refinery equipment via a phenomenon called naphthenic acid corrosion (NAC)) as minor components of the fuel's compounds.

9.5.2 Hypergolic Propellants: A Brief History

Hypergolicity was discovered independently in Russia in 1933, Germany in 1936, and the US in 1940 (Sutton, 1949). The etymology of the word "hypergol" comes from the Greek language: *hyper-*, meaning extreme and *ergon-*, meaning work; literally it means "extreme work." The term was used by German scientists to characterize the reaction between a fuel and an oxidizer that would ignite spontaneously when brought in contact with each other.

In practice, ignition is not instantaneous but occurs after some period of mixing between the fuel and oxidizer. The ignition delay of a given propellant combination is typically defined as the observed difference between the time at which the fuel and the oxidizer come into contact and the time at which they ignite. Since the detection of a luminous flash can be a significant challenge in fielded systems, ignition delay is also sometimes defined by a fixed temperature or pressure increase. Ignition is a transient process, where the development of a flame represents only the final step following the complex interactions between many chemical and physical phenomena. In most liquid bipropellant systems, the main practical challenges are to shorten and reproduce consistent ignition delays. The integrity of rocket engines at start-up relies on this repeatability. Prediction of the change in ignition delay with variations in environmental conditions is also critical since, like any other chemical reactions, ignition processes can be significantly influenced by changes in ambient conditions.

The early applications of hypergolic fuels are linked to the work of German scientists just prior to World War II and to the production of highly concentrated hydrogen peroxide. Officially designated as "109-509," the motor powering the Messerschmitt Me-163B was the first bipropellant rocket motor designed to propel an aircraft. It burned concentrated hydrogen peroxide and a fuel mixture called C-stoff. The fuel consisted of 30% hydrazine hydrate ($N_2H_4 \cdot H_2O$), 57% methyl alcohol, and water. By itself, hydrazine hydrate fuel is spontaneously combustible with hydrogen peroxide with no ignition device required. However, as an energy source, it is greatly inferior to alcohol, hence only the minimum amount was put into the mixture to make the combination self-igniting. The Me-163B first appeared in combat over Germany in 1944.

Between the end of World War II and the development of NTO/MMH in the 1970s, attempts were made at finding hypergolic fuel and oxidizer combinations. Many of these oxidizers were toxic and highly corrosive, even on aerospace-grade materials. In most cases, it was found that the handling problems and safety risks outweighed the performance benefits (Clark, 1972). The most common oxidizers with reduced toxicity and corrosiveness were some form of nitric acid (HNO_3) and hydrogen peroxide (H_2O_2). However, both of these oxidizers require special care for storage and handling. Several types of nitric acid mixtures can be used as oxidizers. White fuming nitric acid (WFNA) is concentrated nitric acid containing approximately 2% water and 0.5% nitrogen dioxide (NO_2) (Wright, 1977). Red fuming nitric acid (RFNA) consists of concentrated nitric acid which contains about 14% dissolved nitrogen dioxide (NO_2) and 1.5 to 2.5% water. Inhibited red fuming nitric acid (IRFNA) is RFNA with the addition of 0.7% hydrogen fluoride to inhibit corrosion. Another combination, used by the Germans during World War II, consists of concentrated nitric acid with a small percentage of concentrated sulfuric acid (H_2SO_4) (Sutton, 1949). The most promising hypergolic fuels for use with FNAs reported before the 1970s were furfuryl alcohol ($C_5H_6O_2$) and aniline (C_6H_7N) (Morrell, 1957).

Hydrazine derivatives quickly became the fuels of choice after the 1970s. Today, the most commonly used hypergolic propellants in spacecraft are monomethyl hydrazine (MMH) and nitrogen tetroxide (NTO). The NTO/MMH propellant combination is used, for example, in the Ariane 5 upper-stage Aestus engine which is used in the final phase of the mission to propel the payload to the point of orbit injection. This propellant combination was also used in the orbital maneuvering subsystems (OMS) and the reaction control systems (RCS) of the Space Shuttle orbiter for orbital insertion, major orbital maneuvers, and attitude control. It is now used by SpaceX and others for the same OMS and RCS. An important milestone in the history of hypergolic propellants was their use in the Lunar Excursion Module (LEM) descent engine with a mixture of Aerozine-50 and NTO. Aerozine-50 is a trade name for a 50 wt.% mixture hydrazine (N_2H_4) and 50 wt.% UDMH.

When compared with traditional hypergolic oxidizers such as nitric acid and nitrogen tetroxide, hydrogen peroxide solutions have similar density, lower vapor pressure, and reduced corrosiveness. Properties of typical hypergolic oxidizers are shown in the Appendix.

Despite their widespread use, good performance, and dual mono- and bipropellant capability, hydrazine-based fuels suffer from the major drawbacks of being highly toxic, suspected carcinogens, and requiring the use of specialized personal protective equipment depending on exposure risk (AIAA, 1999; Schmidt and Walter, 2001).

Nitrogen tetroxide-based oxidizers suffer from similar toxicity concerns, are corrosive, and have high vapor pressures, thus rendering their transport and vapor containment very challenging.

The properties of MMH and NTO are summarized in Table 9.7 (along with the properties in the Appendix). Included in Table 9.7 are each propellant's Short-Term Emergency Guidance Levels (SPEGLs) averaged over one hour, as used by NASA for predictions of mean hazard distance for release of the maximum liquid propellant load (Davis et al., 1999).

9.5.3 Nitric Acid

Nitric acid (HNO_3), with the chemical structure shown in Table 9.8, is a relative low vapor pressure oxidizer with a rich history in rocket propulsion. Key physical properties of nitric acid are listed in the Appendix.

Concentrated nitric acid, with at least 90% nitric acid and mixed oxides of nitrogen (MON), received a lot of attention in the 1950s through the 1970s when nitrogen tetroxide replaced most

Table 9.7   Characteristics of monomethylhydrazine and nitrogen tetroxide

Component	Formula	Density, g/cc	Vapor pressure, torr (at 20 °C)	Boiling point, °C	SPEGLs, ppm
Monomethylhydrazine	CH_3HNNH_2	0.86	36	87	0.12
Nitrogen tetroxide	N_2O_4	1.44	720	21.1	1

Table 9.8   Nitric acid chemical structure

2D representation	3D representation	Brief chemical description
		Nitric acid is a strong mineral acid composed of a nitrogen atom in the +5 oxidation state. It is both a strong acid and a strong oxidant. Nitric acid readily donates a proton (H^+) to compounds bearing lone pairs of electrons available for bonding, such as hydrazine derivatives and other amines.

uses of nitric acid as a rocket oxidizer. In fact, in the 1960s, nitric acid was the most widely used of the nitrogen compounds. Concentrated nitric acid served as an oxidizer in bipropellant systems and, in the manufacture of many of the ingredients in solid propellants.

Pure or anhydrous nitric acid is rare but, for rocket application, a common grade is referred to as white fuming nitric acid (WFNA). WFNA releases characteristic pale, almost invisible fumes upon venting and contains not over 2 wt.% water and impurities. Red fuming nitric acid (RFNA) consists of nitric acid with between 5 and 20 wt.% dissolved NO_2. The color of the acid varies from orange to dark red, depending upon the amount of dissolved NO_2 with characteristic brownish-red fumes. A small addition of fluorine ion (less than 1% of HF) causes a fluoride layer to form on the wall, and greatly reduces the corrosion with many metals. The addition of fluorine to RFNA generates inhibited red fuming nitric acid (IRFNA).

Nitric acid is quite easily storable with, at a concentration of 99 wt.%, a low freezing point of approximately –42 C and a vapor pressure (44 mmHg) in between that of rocket grade hydrogen peroxide (1.36 mmHg) and nitrogen tetroxide (720 mmHg). The properties of high concentration nitric acid are compared with those of hydrogen peroxide and nitrogen tetroxide in Table 9.9. All three have similar densities.

Nitric acid exhibits good stability characteristics upon mechanical shock or impact, but easily reacts with organic materials presenting a potential fire hazard if not properly handled or stored. Most aluminum alloys and 316 stainless steel are compatible for the storage of nitric acid; however, any other kind of ferrous or nonferrous metals should be avoided or carefully checked for compatibility.

The largest market for nitric acid is in the production of fertilizers, ammonium nitrate (AN) and calcium ammonium nitrate (CAN). In 2013, this accounted for 80% of the total world consumption of nitric acid.

Table 9.9   Hypergolic oxidizer properties

Oxidizer	Density, g/cm^3 (at 20 °C)	Vapor pressure, mmHg (at 20 °C)	Boiling/ freezing points, °C	References
RGHP (100%)	1.4422	1.36	150/–10	FMC (1955), CPIA/M4 (1996), Aerojet (1961)
Nitric acid (99%)	1.51	42	83/–42	https://pubchem.ncbi.nlm.nih .gov/compound/nitric_acid
NTO	1.44	720	21/–11	Aerojet (1961)

9.5.4 Nitrogen Tetroxide and Mixed Oxides of Nitrogen

Dinitrogen tetroxide, or nitrogen tetroxide (NTO), is the storable oxidizer of choice in most contemporary spacecraft. NTO is most often combined with nitric oxide (NO) to produce a MON-type acid. The molecular representations and brief chemical descriptions of NTO and NO are listed in Table 9.10. The properties of hydrazine, MMH, and UDMH are also listed in the Appendix.

The corrosion challenges experienced in the 1950s and 1960s with concentrated nitric acid solutions and the higher performance inherent to nitrogen tetroxide resulted in its implementation in many spacecraft systems. Liquid NTO displays a reddish-brown color, a property that arises from the quantity of NO_2 presents in the solution.

A key part of the dissociation of NTO into NO_2 molecules arises from the weak N–N bond. Due to the large electron density associated with the two N–O double bonds on each nitrogen, the N–N bond length is longer than would typically be seen in other molecules bearing that linkage. As a result of this longer bond length, the bond strength decreases significantly.

Like nitric acid, the density of NTO is high (1.43 g/cm^3) but other properties are more challenging since the freezing point of NTO is quite high (–11.2 °C) and its boiling point is low (21.15 °C). The freezing point of NTO can be depressed by using nitric oxide (NO), which also contributes to reducing corrosion stresses. The addition of NO in NTO produces a MON-type acid. Of these mixed oxides of nitrogen, MON-1 and MON-3 are most often used. MON-10 and MON-25 are also used or considered for specific missions with more stringent operational requirements.

The toxicity of NTO and MON in general calls for strict regulations surrounding transport and handling. Extensive safety information related to NTO and MON can be found in the American Institute of Aeronautics and Astronautics standard AIAA SP-086, "Special Project: Fire, Explosion, Compatibility, and Safety Hazards of Nitrogen Tetroxide" (AIAA, 2001).

Table 9.10   NTO and NO chemical structure

2D representation	3D representation	Brief chemical description
		NTO is an oxide of nitrogen with strong oxidizing capability. It exists in the +4 oxidation state. NTO readily dissociates into NO_2 radicals, again in the +4 oxidation state, which readily abstract hydrogen atoms (specifically the hydrogen nucleus AND one electron) to propagate radical mechanisms. The abstraction of a hydrogen atom results in an exotherm and, in addition to the radical formed from the molecule the hydrogen was abstracted from, forms a molecule of nitrous acid or HONO, which can react further.
		Nitric oxide is added to NTO to produce MON with freezing points at low as -55 °C with 25 wt.% NO in NTO (thus producing MON-25).

Performance

The vacuum specific impulse of NTO with a variety of fuels are shown in Figure 9.5. The optimal mixture ratio with MMH combines conveniently with the density of both liquids to yield identical size tanks and consumption rates on-board spacecraft, thus facilitating center of gravity management.

9.5.6 MMH and UDMH

Perhaps the most widely used hypergolic fuels today are monomethylhydrazine (MMH) and unsymmetrical dimethylhydrazine (UDMH). Since both are hydrazine derivatives, their molecular representations and brief chemical descriptions are provided along with that of hydrazine in Table 9.11. The properties of hydrazine, MMH, and UDMH are also listed in the Appendix.

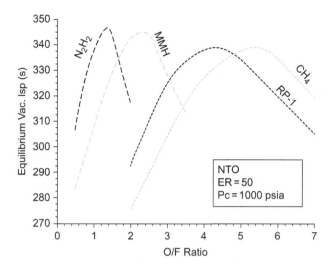

Figure 9.5 Equilibrium vacuum specific impulse of nitrogen tetroxide with common fuels.

Table 9.11   Hydrazine, MMH, and UDMH chemical structures

2D representation	3D representation	Brief chemical description
		Hydrazine can be thought of as an ammonia derivative which contains a nitrogen–nitrogen linkage. As with any case where functional groups are mixed, it would be highly dangerous to classify hydrazine as an amine. Compared to ammonia, the addition of a second nitrogen to the molecule increases both its hydrogen bond donating and accepting capability, elevating its melting and boiling point to a range more usable for propellants.
		Monomethylhydrazine, or methylhydrazine, is the first carbon-substituted homologue of hydrazine. The added methyl group (CH_3-) both adds molecular weight to the molecule and increases the nucleophilicity of the attached nitrogen.

(continued)

Table 9.11   (*cont.*)

2D representation	3D representation	Brief chemical description
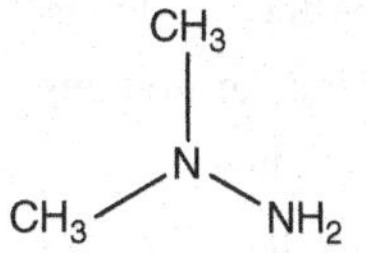 CH₃—N(CH₃)—NH₂ structure	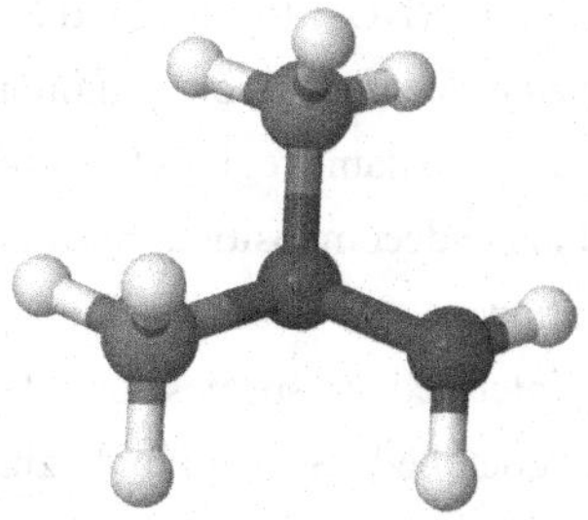 3D molecular model	Unsymmetrical dimethylhydrazine, or *N,N*-dimethylhydrazine or 1,1--dimethylhydrazine, is the second carbon-substituted homologue of hydrazine, with the second carbon substituent placed on the same nitrogen as the first (as opposed to 1,2-dimethylhydrazine with one methyl substituent on each nitro-gen). The additional methyl substi-tuent further increases the nucleophilicity of the internal nitrogen, but both increases the stearic bulk around that nitrogen's lone pairs and removes its hydrogen bond donation capability. This, in turn, lowers both the melting and boiling point of UDMH relative to MMH.

MMH (CH_6N_2) is a clear, colorless hygroscopic liquid. It is a mildly alkaline base and a very strong reducing agent. This fuel has an odor similar to ammonia or organic amines and is miscible with water, hydrazine, hydrazine derivatives, amines, and lower weight monohydric alcohols. Unlike hydrazine with its high freezing point near 2 °C, MMH does not freeze until −52 °C. This feature alone, along with its performance with NTO, makes MMH the most common storable, hypergolic bipropellant fuel for satellites and upper stages. MMH can be used as a regenerative coolant in bipropellant engines.

UDMH ($C_2H_8N_2$) is very similar to MMH but possesses a slightly higher molecular weight and lower melting and boiling points. It is used mostly in Russian engines.

Reaction Pathways

The intimate contact between these agents and a strong oxidizer can lead to an immediate and energetic reaction. The oxidation–reduction reactions between the fuel and the oxidizer ultimately lead to ignition and combustion. However, the vigorous oxidation–reduction reaction is so power-ful that the liquid-phase reactants can, under certain conditions, be forced apart by the rapid production of gaseous products (Schmidt and Walter, 2001). The separation of the propellants leads to incomplete combustion.

Sawyer (1966) proposed a two-step reaction process between NTO and hydrazine that comprises the fast reduction of nitrogen dioxide (NO_2) to nitric oxide (NO) and then a slower reduction of NO with simultaneous hydrazine decomposition. Additionally, hydrazine droplets burning in NTO atmospheres are shown to have two flame regions (Lawver, 1966). An inner yellow flame at the droplet surface consists of hydrazine decomposition. Surrounding the yellow flame is a reddish-orange flame constituting oxidation of the decomposition products. Many detailed studies have been performed investigating intermediate species formation and reaction mechanisms for hydrazines with various oxidizers. A detailed review of hydrazine decomposition kinetics and combustion is beyond the scope of this chapter. The interested reader is directed to Sawyer (1966) for additional information on reaction pathways, intermediate species formation, and possible effects on rocket engine performance.

Performance

When compared to other hypergolic propellants, the MMH and NTO combination yields vacuum specific impulse values over 340 seconds (see Figure 9.6) along with ignition delay times of typically around 3 ms. These performance characteristics in combination with operational reliability have become the standard in the aerospace market today.

9.5.7 Green (or Less Toxic) Hypergolic Propellants

The most commonly used hypergolic propellants in spacecraft are monomethyl hydrazine (MMH) and nitrogen tetroxide (NTO). They are high-density storable propellants that react, typically,

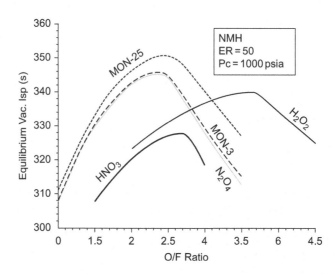

Figure 9.6 Equilibrium vacuum specific impulse of monomethylhydrazine with common storable oxidizers.

within 3 milliseconds of contact. Despite their high performance, hydrazine-based fuels suffer from a major drawback as they are carcinogens and require the use of full body protection suits.

Several alternatives to NTO/MMH have been proposed over the years. Alternatives include ionic liquid hypergolic fuels, boron-based fuels and a variety of fuel classes defined to overcome most of the toxicity issues encountered with MMH, and as much as possible with NTO although oxidizer choices are limited.

Note that the term "non-toxic" is used herein in a relative manner compared to hydrazine – some people use the term "less-toxic" – for all matter of purposes, we are in fact talking about rocket propellants, it is therefore reasonable to assume that one would not try to drink such propellant and would exercise standard safety precautions in handling. When a propellant can be handled in a safe manner without a respirator or is not known to be a carcinogen upon brief exposure to vapors, it is either "less-toxic" or "non-toxic"; the distinction here is academic and left to the reader.

A variety of factors have resulted in a renewed interest in propellants with reduced toxicity. These factors include heightened sensitivity to cost, environmental concerns, and most importantly the personnel hazards associated with toxic propellants. Further driving the move toward other propellants is the realization that absolute performance in terms of specific impulse is not, for many applications, the most appropriate metric for selecting a propulsion system. For example, as a monopropellant, 90% hydrogen peroxide has a specific impulse approximately 25% below that of a hydrazine monopropellant thruster in similar conditions. However, because of its high density, 90% hydrogen peroxide has a density specific impulse approximately 6% higher than hydrazine, making it potentially more advantageous for volume-constrained systems. Density specific impulse is defined herein as the product of the monopropellant density with its specific impulse.

Several institutions around the world have made significant progress towards the development of hypergolic rocket propellants with reduced toxicity and greater performance than traditional hydrazine-based compounds (Pourpoint and Anderson, 2007). In many studies, the oxidizer of choice was RGHP although recent work on ionic liquid fuels and boron-based fuels focus on hypergolicity with WFNA, RFNA, or NTO and its derivatives.

While many studies deserve to be reported here, the reader is instead encouraged to consult scientific journal publications for specific studies. For the sake of conciseness, the following discussion focuses on green fuels hypergolic with RGHP, as such a combination can be viewed a truly "green" propellant combination.

Melof and Grubelich (2001) recognized that the challenge is to find a low-toxicity fuel that gives rapid hypergolic ignition with RGHP, and exhibits good performance, is safe to handle, and is storable

Promising results were obtained with non-toxic hypergolic miscible fuels (NHMF), a class of fuels developed and patented by the Naval Air Warfare Center-Weapons Division (NAWC-WD) (Rusek et al., 1999). The formulation of NHMF is the result of earlier research involving a traditional jet fuel, JP-10, in which a manganese containing compound was dissolved. While the manganese catalyzed JP-10 initiated the decomposition of RGHP, reliable hypergolic ignition could not be achieved. The difficulties associated with the hypergolic ignition of that fuel were attributed to the non-miscibility of JP-10 and RGHP. The non-polarity of JP-10 and the polarity of RGHP prevented intimate mixing and reaction of the two compounds. Furthermore, theoretical energy calculations of the combination of the fuels and RGHP resulted in an optimum oxidizer to fuel ratio (O/F) in the range 5:1 to 7:1. Such high O/F meant that the propellant combination could not be used in traditional injection systems developed for NTO and hydrazine-based fuels.

A significant breakthrough was reached by finding a transition metal compound that was soluble in a lower molecular weight polar solvent that could also serve as the fuel (Siede, 2003). The result was the development of Block 0, a mixture of 78 wt.% methanol and 22 wt.% manganese (II) acetate tetrahydrate (MAT). Block 0 is a heterogeneous NHMF tested with RGHP by several institutions (Pourpoint and Anderson, 2007). Calculated vacuum specific impulse values of the combination RGHP/Block 0 are about 10% lower than with NTO/MMH. A major drawback of Block 0 is that it does not age well, especially at elevated temperature or in the presence of water and oxygen. Attempts at correcting the aging problem with various additives were unsuccessful (Purcell *et al.*, 2002). However, while poor aging characteristics can be unacceptable in space launch applications, the database of experimental ignition delay values that were obtained with Block 0 remains an extremely valuable tool in any comparison study of hypergolic fuels.

In 2003, the US Navy was granted a new patent on "Reduced Toxicity Hypergolic Bipropellant Fuels" (Diede, 2003). Two fuel formulations are described in the patent application. The first fuel, referred to as HF-2B is a solution of tri(ethylene glycol) dimethyl ether or triglyme with dissolved sodium borohydride. This hydride was chosen within a group of potential reducing agents. It was chosen partly because most hydrides are known to react instantly and violently with RGHP and partly because it led to a fuel mixture that should ignite reliably with hydrogen peroxide concentrations as low as 70%. Diede reports successful test firings of the fuel in a rocket engine with ignition delays on the order of 1 to 3 milliseconds with 96% RGHP. Another interesting result reported in the patent application is that the fuel ignited within 2 to 3 milliseconds in a pressure environment equivalent to an altitude of 110,000 ft. It should be noted that, while the triglyme compound has a low overall toxicity rating, it is listed as a chemical substance that can cause malformations of an embryo or fetus. In this case, inhalation hazards are mitigated by the very low vapor pressure of the compound. The second fuel, HF-25J, also used sodium borohydride as the reducing agent and diethylenetriamine as the solvent. Ignition delays of the order of 2–3 ms are mentioned when this fuel is reacted with RGHP. In this case, higher specific impulse is a trade off to the increased toxicity of diethylenetriamine compared to triglyme. For both fuels, Diede emphasizes the role played by the miscibility of the reactants.

9.5.8 Hypergolic Propellant Additives

Numerous chemical additives have been added to hypergolic fuels and oxidizers. Additives serve many purposes, including determining reaction mechanisms, altering reaction rates, and altering the ignition delay. The following discussion is limited to chemical additives reported to change the hypergolic ignition delay or influence the ignition limits.

Weiss *et al.* (1964) studied the change in ignition delay of hydrazine/NTO with 33 distinct additives tested as thermal moderators, free radical traps, free radical sources, and surface active agents. The average ignition delay ranged from 1.2 to 4.0 ms, compared to 3.0 ms for neat hydrazine. Twenty-two of the additives shortened the ignition delay of neat hydrazine. The two surface active agents, or wetting agents, used also shortened the ignition delay. Bernard (1955), using furfuryl alcohol and nitric acid, found that wetting agents decreased the interfacial tension between the bipropellants, leading to increased mixing and shortened ignition delays.

The effect of adding hydrazine nitrate, a known intermediate species in hydrazine/RFNA reactions and hydrazine/NTO reactions, has a different effect on hydrazine/NTO than on UDMH/NTO. Weiss *et al.* (1964) observed the addition of hydrazine nitrate to hydrazine decreased the ignition delay of hydrazine/NTO from 3 ms to 1.6 ms in a swirl chamber. Corbett *et al.* (1964) measured an increase from 1 ms to 4.7 ms when adding hydrazine nitrate to UDMH in an unlike doublet injector using UDMH/NTO. The addition of the intermediate species hydrazine nitrate to its reactants promotes ignition, but its addition to UDMH/NTO, where it is not an intermediate species of the reactants, inhibits ignition.

Corbett *et al.* (1964) studied the change in the minimum ignition pressure of MMH/NTO when adding furfuryl alcohol, phenylether, methylbutynol, ethlyether, and benzene at 10% by weight to MMH. The MMH/additive mixture and NTO were vaporized and introduced into a vacuum chamber until the minimum ignition pressure was reached. Furfuryl alcohol decreased the minimum ignition pressure from ~6 mmHg to ~4.5 mmHg and was the only additive to influence the minimum ignition pressure.

The most common method to alter the ignition delay is to place the additive into the fuel, but some results are available when an additive was mixed with the oxidizer. As discussed earlier, small amounts of NO are usually placed into NTO to suppress stress corrosion cracking or lower the freezing point of NTO, creating the MON family of oxidizers (Schmidt and Walter, 2001). The ignition delay of hydrazine fuels and NTO increases with increasing NO content. Corbett *et al.* (1964) added Compound R ($CF(NF_2)_3$) to NTO at 2 and 3 wt.% weight. Compound R was chosen for tests because of its miscibility with NTO, its similar vapor pressure with NTO, and its high fluorine content. The addition of Compound R raised the minimum ignition pressure and contaminated the test apparatus.

Hundreds of additives have been used to alter the ignition delay of hydrazine fuels and NTO. Although many additives can decrease the ignition delay, no additive (besides NO) has been used commonly in hypergolic propellants.

9.6 GELLED PROPELLANTS

Gelled propellants have received regular attention since the early days of rocketry. A gelled propellant must be shear thinning to be useful in a rocket application. This property allows gelled propellants to be stored as a pseudo-solid but flowed and injected in the same way as neat propellants (Pein, 2005). In theory at least, this shear thinning property means that gelled propellants benefit from the high-performance, wide-range thrust control and reignitability of liquid propellants with the storage behavior of solid propellants. The ability to store a propellant, a hypergolic propellant in particular, without worry for spillage is certainly very appealing. Gelled propellants also provide the potential to add insoluble metallic particles such as aluminum, magnesium, or boron in the gel composite with minimal particle settling to increase energy density of the propellant and engine thrust (Dennis *et al.*, 2011).

In order to fully characterize the properties of a gelled propellant, it is necessary to conduct rheological experiments to determine the viscosity (η) and yield stress (τ_y). Viscometry experiments subject the propellants to a defined shear rate ($\dot{\gamma}$) and measure the resistive force response to determine η. Yield stress experiments are performed by increasing shear stress until the stress strain relationship becomes nonlinear.

The shear thinning properties of gelled propellants can be evaluated with a variety of models, with the most common one, the power law model, shown in Eq. 9.6, with a consistency index (K) and power law exponent (n). This model allows for the prediction of η and provides some insight into the repeatability of gelled propellant mixes:

$$\eta(\dot{\gamma}) = K\dot{\gamma}^{n-1} \tag{9.6}$$

Common gelling agents include Klucel® HF Pharm grade HPC (typically referred to as HPC) and Aerosil® R974 hydrophobic fumed silica (referred to as Aerosil). The hydrophobic coating on Aerosil® R974 allows for gel mixing without pre-drying the fumed silica. RFNA has been gelled with Cabosil© M5 fumed silica, but the hydrophilic surface allowed significant water absorption. Aerosil® R974 reduces the required inert gelling agent concentration by 1 wt.% allowing the gelled RFNA propellant to contain more reactive material. Typical Aerosil gelled propellants are shown in Figure 9.7.

The ignition process is suspected to be heavily influenced by the rheological properties of the propellants after injection. The silica–silica bonds form a network inside the liquid propellant. This network is made up of hydrogen bonding between the silica particles. When shear is applied, this network begins to break down allowing the gel viscosity to decrease. After some time at rest, the hydrogen bonds begin to reform and the gel viscosity increases again.

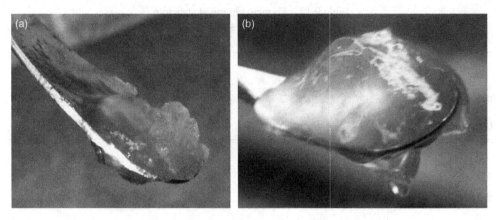

Figure 9.7 MMH/6 wt.% Aerosil gel (a) and RFNA/3 wt.% Aerosil gel (b) (Heister *et al.*, 2014).

The thixotropic property of silica gels prevents them from instantly reforming a full network after shear is removed. Therefore, it is suspected that the propellants that are impinging are still at reduced viscosities similar to neat propellants. The particulate gel mechanism is in contrast to the entanglement mechanism that allows HPC to form a gel. HPC forms a network of entangled polymers which impart elastic properties to the final gel. HPC gels do not possess thixotropic properties and are suspected to return to the pre-shear viscosity very quickly. Ultimately, the mechanisms of gel formation and breakup upon shear have a major impact on the usefulness of a gelled propellant.

9.7 Cryogenic Propellants

Cryogenic propellants are liquefied gases at extremely low temperatures. Temperatures as low as –250 °C are indeed necessary to liquefy hydrogen. These low temperatures also mean that the simplest reducing compounds (fuels) and oxidizing molecules can be used as propellant without the need for carrier atoms. For example, oxygen can be pumped into an oxidizer manifold with the benefits of the liquid density at temperatures without the need for a nitrogen atom. While cryogenic propellants, also referred to as cryogens, are typically the most energetic types of propellants, their extremely low temperatures mean that they are also more difficult to store for any length of time. Vaporization losses require provisions for venting the propellant tank.

Today liquid oxygen is the most common cryogenic oxidizer. It is used in many launch vehicles around the world. Liquid hydrogen and more recently liquid methane are the most commonly used liquid cryogenic fuels.

Table 9.12   Oxygen chemical structure

2D representation	3D representation	Brief chemical description
(depiction of O_2 is debated) (depiction of O_2 is debated)		Two oxygen atoms each share two electrons to form two covalent bonds and make an oxygen molecule (O_2).

9.7.1 Liquid Oxygen

The oxygen molecule is quite simple with two oxygen atoms connected by a double bond, with the chemical formula O_2 and the chemical structure shown in Table 9.12. Key physical properties of liquid oxygen are listed in the Appendix.

Nowadays, liquid oxygen is the most common oxidizer in large launch vehicles where it is reacted with a kerosene fuel, liquid methane, or liquid hydrogen. Historically, liquid oxygen has been used with alcohols, with the famous example of the German V-2 rocket, and starting around 1953 with "gasoline fractions" then later kerosene fuels for performance reasons. Today, the SpaceX Merlin engine is one example among many of a liquid oxygen/kerosene engine and the Blue Origin BE-4 engine is a good example of a liquid oxygen/liquid methane engine. This popularity is easily explained by the relatively high density of liquid oxygen and its high performance. While nitric acid was an early choice of oxidizer, the energy content of pure oxygen can only be exceeded by fluorine, with obvious handling and toxicity issues. Liquid ozone was also considered as a potential oxidizer, but highly questionable stability characteristics have eliminated it from any serious contention.

From the discovery of oxygen in 1774 by Priestley to its subsequent first liquefaction in 1883 by two Polish professors Zygmunt Wróblewski and Karol Olszewski, many scientists have attempted to explain or demonstrate the properties of oxygen, but it was Lavoisier in 1787 who demonstrated that oxygen was a constituent of the atmosphere. He also showed that it is that constituent that gives air its property of supporting combustion.

While a number of production methods exist, liquid oxygen is most commonly produced by fractional distillation of liquid air. For rocket applications, it is produced according to military specification MIL-PRF-25508F. It is inexpensive, non-toxic, and non-flammable. However, while oxygen will not react spontaneously with organic materials at ambient pressures, mixtures of liquid oxygen and organic materials can be detonated when confined. Therefore all containers, lines, and pumping equipment for liquid oxygen must be free of organic matter.

Oxygen is the second largest component of the atmosphere, comprising 20.8% by volume. Liquid oxygen is a pale blue liquid that is non-toxic and non-corrosive but, like any cryogen, will cause severe burns upon prolonged contact with the skin. A main factor in the selection of materials used with liquid oxygen (oxygen in general) is its reactivity with combustible materials. Cryogenic temperatures also bring in issues with embrittlement and decreased ductility. Suitable materials and safety guidelines for oxygen are available in several references, including the National Fire Protection Association (NFPA) standards 55 and 99. Overall, since liquid oxygen is highly reactive with most organic materials, it is critical that all equipment in contact with it be kept rigorously clean and free of contaminants. Goggles and lined leather gloves should be worn when dealing with an open system. As with any cryogens, adequate ventilation is critical in areas where liquid oxygen is in use.

Performance

Today, liquid oxygen is *the* oxidizer of choice for launch vehicles and has had this status of "preferred oxidizer" since the early days of rocketry. There is an entire chapter dedicated to cryogenic oxidizers in *Ignition!* (Clark, 1972). From Clark's discussion and countless references, liquid oxygen has been used as an oxidizer with a large variety of fuels, from hydrazine, to alcohols in the sounding rocket Viking and several experimental vehicles of the early 1950s, as well as the Redstone missile. More recently, liquid oxygen has been used with kerosene fuels (which replaced alcohols in the 1960s), hydrogen, and liquid methane.

Most contemporary launch vehicles use liquid oxygen with either liquid hydrogen, liquid methane, or kerosene (with liquid methane gaining significant traction). The LOX/kerosene combination is used on many US and Russian launchers, for example. The chamber pressure of LOX/kerosene engines can be limited by coking in the cooling channels and combustion instabilities; however, advances in staged combustion cycles, LOX-rich gas generators, and special kerosene blends to reduce coking have made LOX/kerosene engines highly competitive. Significant and current development work on LOX/liquid methane engines is outlined below.

As shown in Figure 9.8, the combination of liquid oxygen (LOX, liquid O_2) and liquid hydrogen (LH$_2$) provides quite significantly higher specific impulse than with any other contemporary (and practical) fuel. The relatively flat I_{spvac} curve characteristic of LOX/LH$_2$ is another significant advantage over other fuel options since it provides flexibility in engine operation. The temperatures involved with liquid hydrogen, however, make it a difficult propellant combination to work with. In fact, the first flight of a LOX/LH$_2$ engine, the RL10 developed by Pratt and Whitney, only occurred in 1963, more than 20 years after the V2 rocket . The J-2 upper-stage engine is another 1960s LOX/LH$_2$ engine developed by Rocketdyne in support of the Apollo program. It first flew in 1966. The LOX/LH$_2$ propellant combination was used for several decades on the Space Shuttle main engine, the RS-25, and is still used today on various upper-stage engines and is planned for use on the first stage of the Space Launch System (SLS) rocket. Additional discussion on the performance and rocket systems using LOX/LH$_2$ is provided along with the description of liquid hydrogen.

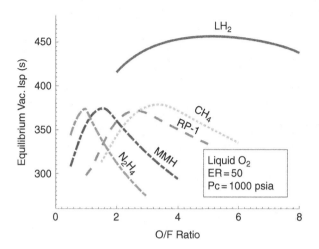

Figure 9.8 Equilibrium vacuum specific impulse of liquid oxygen with common fuels.

9.7.2 Liquid Hydrogen

The hydrogen molecule is quite simple with two hydrogen atoms connected by a single bond, with the chemical formula H_2 and the chemical structure shown in Table 9.13.

Hydrogen is an amazing compound. It can be produced by multiple sources, including several that are inherently renewable, and has the highest specific energy of any currently used aerospace fuel at 142 MJ/kg, or ~2.5 times that of methane, ~3 times that of kerosene, and ~4.5 times that of an early fuel: methanol. This performance advantage of hydrogen comes from its unique ability to produce extremely low molecular weight products of combustion. Despite this unquestionable performance benefit, two key properties run against hydrogen as the ideal rocket fuel:

1. its low density (0.07 g/cm^3, compared to 0.66 g/cm^3 for liquid methane, 0.78 g/cm^3 for kerosene, and 0.79 g/cm^3 for methanol)
2. its low boiling point of -252.9 °C making it a "hard cryogen" compared to the "soft cryogen" liquid methane (boiling point -161.5 °C) and the storable liquids RP-1 (boiling point 78 °C) and methanol (boiling point 64.7 °C). The low density of hydrogen results in an energy density about four times lower by volume than that of RP-1, for example.

In rocket applications, liquid hydrogen is reacted with liquid oxygen to form water with a stoichiometric heat of reaction of 286 kJ/mol or 572 kJ total:

$$2H_{2(g)} + O_{2(g)} \rightarrow 2H_2O_{(l)} + 572 \text{ kJ} \tag{9.7}$$

The combustion of pure hydrogen and pure oxygen results in flames that emit ultra-violet light. The flame is therefore invisible to the eye but often results in ice crystals downstream of the

Table 9.13   Hydrogen chemical structure

2D representation	3D representation	Brief chemical description
H——H		Diatomic hydrogen is the simplest possible molecule. It acts as a good reducing agent, given its propensity to fulfill its duet valence shell requirement. Hydrogen is ubiquitous in nature and bonds with pretty much anything, albeit usually by giving up one electron to be in the +1 oxidation state for covalently bonded molecules, although there are several exceptions, e.g. many metal hydrides.

exhaust plume and a white looking trail. Hydrogen is also hypergolic with fluorine and chlorine, a convenient property often considered in the early days of space exploration but vastly overshadowed, despite also having high specific impulse values, by the toxic nature of both oxidizers.

In terms of specific impulse and thanks to the high combustion temperature and low exhaust molecular weight, no other (practical) combination comes close to hydrogen/oxygen. It should be noted, however, that although the maximum specific impulse for the O_2/H_2 propellant combination is obtained at a mixture ratio of 4.7, engines using these propellants usually operate at mixture ratios between 5.5 and 6 because of the difference in the densities of the oxidizer and fuel.

The inherent performance of liquid hydrogen, typically associated with liquid oxygen, has made it the fuel of choice in many launch vehicles, across several space agencies and launch providers around the world. It is used in both booster stages and upper stages with a variety of engine cycles, with the systems listed in Table 9.14 as examples.

The *Guide to Safety of Hydrogen and Hydrogen Systems* published by the American Institute of Aeronautics and Astronautics is a great resource for the design and operation of any hydrogen system (AIAA, 2017).

Properties

Liquid hydrogen is a colorless and non-toxic liquid but, combined with its low boiling, hydrogen gas is naturally an extremely flammable material. In fact, the flammable limits of gaseous hydrogen in air are 4.0 to 75 volume percent with an autoignition temperature near 500 °C.

A key property of hydrogen relates to its storage limitations. The low boiling point of hydrogen means that large amounts of energy can be spent for small gains in storage efficiency and liquid hydrogen is only cost effective for highly specific uses (rocket propulsion being a highly

Table 9.14   Liquid hydrogen and liquid oxygen powered engines around the world

Vehicle (country/region)	Engine	Max. sea level thrust, lbf
Space Shuttle (USA)	SSME, RS-25	418,000
Delta IV (USA)	RS-68	660,000–705,000
Ariane (Europe)	Vulcain	220,000
H-II Series (Japan)	LE-7/A	189,600
Energia (Soviet Union, Russia)	RD-0120	343,000
Atlas, Titan, Saturn I, Delta IV (USA)	RL 10	25,000
Ariane (Europe)	HM7B	13,980
H-I, H-II (Japan)	LE-5	Up to 30,000
Mk-2 (India)	CE7.5	16,500

specific and relevant one). It also means that a liquid hydrogen supply lasts only on the order of weeks regardless of usage with relatively high hazard for layperson end-user operation.

Insulation is key to preserving the liquid state of hydrogen and to protecting exterior elements. The main insulation technologies used with aerospace grade liquid hydrogen systems include foams (as one the Space Shuttle main engine propellant tanks), active and passive vacuum jackets, and a large variety of heat-rejection mechanisms.

For transfer, it is generally easier to use a pressurization system rather than a pumping system due to the low density and high vapor pressure of liquid hydrogen. For pressurization, only gaseous helium or gaseous hydrogen can be used as pressurants due to their freezing points.

Alternatives to liquid hydrogen include compressed gas, slush, metal hydrides, carbon nanotubes, ammonia, and amine borane complexes, although none of these options are used nearly as widely as liquid hydrogen in rocket systems based on density and heat of reaction considerations.

Finally, as with any hydrogen system, hydrogen embrittlement must be accounted for in liquid hydrogen material selection and inspection.

9.7.3 Liquid Methane

Methane is the simplest possible alkane (hydrocarbons or organic compounds made up of only carbon–carbon single bonds). The chemical formula of methane is CH_4 with the chemical structure shown in Table 9.15. Additional properties for liquid methane are shown in the Appendix. An excellent reference on liquefied natural gas is the NFPA standard number 59A. This standard, focused on the production, storage, and handling of liquefied natural gas (LNG), is highly relevant for liquid methane since it is the major component of LNG.

In its cryogenic liquid form, methane is an attractive fuel for rocket applications since, in addition to being non-toxic and inexpensive to produce, its density (0.422 g/cm^3) is nearly six times that of liquid hydrogen (0.07 g/cm^3). This density improvement comes with the added benefit that

Table 9.15 Methane chemical structure

2D representation	3D representation	Brief chemical description
		Methane is the simplest possible hydrocarbon subunit. All four bonds between the central carbon atom and surrounding hydrogen atoms are equivalent, resulting in a tetrahedral structure. Due to the net neutral dipole moment, methane is nonpolar and thus has very little intermolecular interaction with other molecules, resulting in very low melting and boiling points.

the temperature of liquid methane is much warmer (-162 °C) than that of liquid hydrogen (-253 °C) meaning less insulation and cooling requirements. The relative ease of use of liquid methane compared to liquid hydrogen is amplified further by the fact that it is stored as a liquid at a temperature similar to liquid oxygen (-183 °C).

Compared to other liquid hydrocarbon fuels, methane allows for a fuel-rich gas generator without soot formation, very high cooling efficiency, and its gaseous state immediately downstream of rocket injectors lowers the risk of combustion instabilities.

Safety requirements for liquid methane relate naturally primarily to its cryogenic nature. In addition, methane forms explosive compositions with air in concentrations of 5–14%. The maximum allowable concentration for methane is over 20,000 ppm. As with any cryogen, its expansion from a liquid to a gas state is accompanied by a significant volume increase, which, in addition to explosion risks, could lead to a lowering of the oxygen content in the air to below a tolerable level.

The advantages of liquid methane make it a desirable fuel for a variety of rocket applications, ranging from launch vehicles to space reaction control and maneuvering engines, as well as descent and ascent engines for planetary missions. The atmosphere of Mars could also provide a means of producing methane in-situ.

The last 20 or so years, however, have seen a significant rise in the interest for liquid methane with ORBITEC (Orbital Technologies Corporation; since 2014 part of the Sierra Nevada Corporation) and XCOR Aerospace developing liquid oxygen/liquid methane engines in the mid to late 2000s. Since the early 2010s, progress towards liquid oxygen/liquid methane engines has been very significant with, among others, SpaceX developing the 1,000,000 lbf class Raptor engine and Blue Origin developing the 550,000 lbf BE-4 engine.

Similarly to methane, other light hydrocarbons such as cyclopropane and propene offer attractive properties with reported specific impulse values greater than either kerosene or methane and densities higher than methane. Significant additional development work is required, however, before these light hydrocarbons will be able to compete with the level of maturity liquid methane systems are rapidly gaining.

9.8 FINAL CONSIDERATIONS

Since the dawn of space exploration, several catastrophic system failures have led to the loss of lives, hardware, and missions, as well as environmental damage on land, at sea, and in space. Recently, the public results of the investigation into the loss of a SpaceX Falcon 9 rocket on September 1, 2016 indicate that the explosion was caused by a breach in the cryogenic helium system of the vehicle's upper oxygen tank. Along with many others before it, this latest incident emphasizes the importance of propellant characterization research under relevant conditions.

These failures stem from a huge number of potential factors, each associated with the inherently limited characteristics of materials, as aerospace engineers around the world push boundaries to provide more cost-effective solutions. Clearly, on-going development work into soft-cryogen liquid methane or subcooled (or densified) kerosene, for example, will lead to new discoveries in materials behavior at low temperatures and, over time, the fundamental properties of each propellant considered for aerospace propulsion systems will help elucidate the observed behaviors.

Similarly, the development of "green" storable propellants, from the monopropellants AF-M315E and LMP-103S to green hypergolic propellants to replace monomethylhydrazine, has already shown great benefits with additional mission capabilities at reduced costs. On-going research aims at solidifying these benefits across a wider range of platforms, whether for commercial use, governmental space agencies, or military systems.

It is interesting as we conclude this chapter on liquid propellants to consider, briefly, where propellants (whether well-established or in development) may be going next. As humankind looks to the stars and dreams of a Mars colony or Moon bases appear to gain traction with private investors and public support, new challenges will arise and new research will be needed.

For example, while the Earth's magnetosphere shields us from significant exposure to radiation from the sun and from space, beyond this shield both high-energy protons and charged particles can damage electrical systems, biological systems, and any spacecraft component through which the highest energy radiation (the ionizing radiation) can pass. As noted in *Space Faring: The Radiation Challenge* (Rask *et al.*, 2008), the "amount, or dose, of space radiation is typically low, but the effects are cumulative." It will therefore be critical to understand and quantify the effects high-energy radiation can have on the whole spacecraft, including its propellants, pressure vessels, and valves. Existing research on the effects of radiation on liquid propellants is quite limited but a 1970 report to the NASA JPL by Best *et al.* (1970) provides an interesting insight into the potential of FLOX (a fluorine–oxygen mixture used in the 1950s and 1960s), oxygen, difluoride, nitrogen tetroxide, diborane, hydrazine, and propane. The study reports that "The most striking

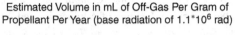

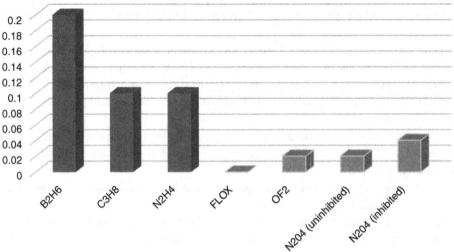

Figure 9.9 Effects of radiation on liquid propellants (Source: adapted from Best *et al.*, 1970).

result is the obviously much larger sensitivity to radiation induced decomposition of the fuels, as a group, compared to the oxidizers." More specifically, and depending upon the isotope source, a common storable fuel like hydrazine is one to two orders of magnitude more sensitive than nitrogen tetroxide on a relative gas volume formed per unit of propellant per unit dose. And, while the investigated oxidizers appear much less susceptible to radiation than the fuels, the presence of 1 to 3 wt.% of nitryl fluoride to inhibit the nitrogen tetroxide and react with any water which might intrude into the propellant leads to a factor of about two in the decomposition gas volume generated upon irradiation compared to the uninhibited version. These results are represented in Figure 9.9.

Best *et al.* conclude that the vapor is more susceptible than the liquid to decomposition and that there are "some interactions with the walls of the vessel." Finally, in a particularly troublesome but critical conclusion for long-term space missions, they clearly state: "it appears that the pressure would continue to build up over the propellant linearly with time over mission times of 1 to 10 years. The rate per unit time will depend on the size, configuration, and shielding of the radio-isotope source" (Best *et al.*, 1970).

FURTHER READING

Aerojet-General Corporation (1961) *Performance and Properties of Liquid Propellants*, June.
AIAA (1999) *Special Project: Fire, Explosion, Compatibility, and Safety Hazards of Hypergols –
Hydrazine*, AIAA SP-084. American Institute of Aeronautics and Astronautics (AIAA).

AIAA (2001) *Special Project: Fire, Explosion, Compatibility, and Safety Hazards of Nitrogen Tetroxide*, AIAA SP-086. American Institute of Aeronautics and Astronautics (AIAA).

AIAA (2017) *Guide to Safety of Hydrogen and Hydrogen Systems*, ANSI/AIAA G-095A. American Institute of Aeronautics and Astronautics (AIAA). https://arc.aiaa.org/doi/abs/10 .2514/4.105197.001 [accessed March 2018].

Altman, D. and Penner, S. S. (1960) "Chapter 3: Combustion of Selected Bipropellant Systems – Section L: Combustion of Liquid Propellants," in *Liquid Propellant Rockets*, Altman, D., Carter, J. M., Penner, S. S., and Summerfield, M. (eds.). Princeton Aeronautical, pp. 101–105

Altman, D., Carter, J. M., Penner, S. S., and Summerfield, M. (1960) *Liquid Propellant Rockets*. Princeton Aeronautical.

American Conference of Industrial Hygienists (ACGIH) (2018) Introduction – Threshold Limit Value (TLV). See: https://www.acgih.org/tlv-bei-guidelines/tlv-chemical-substances-introduction [accessed March 2018].

Barrère, M., Jaumotte, A., Fraeijs De Veubeke, B., and Vandenkerckhove , J. (1960) *Rocket Propulsion*. Elsevier Publishing Company.

Bernard, M. L. J. (1955) "Surface Properties of Liquid Bipropellants and Their Effects on the Kinetics of Ignition," in *Fifth Symposium (International) on Combustion, Combustion in Engines and Combustion Kinetics*, University of Pittsburgh. Reinhart Publishing Corporation, p. 7.

Best, R. E., Kircher, J. F., and Rollins, J. E. (1970) "Radiation Effects on Liquid Propellants". NASA Report.

Blevins, J. A., Gostowski, R., and Chianese, S. (2004) "An Experimental Investigation of Hypergolic Ignition Delay of Hydrogen Peroxide With Fuel Mixtures," in *42nd AIAA Aerospace Sciences Meeting and Exhibit*, Aerospace Sciences Meetings. AIAA-2004-1335.

Buffalo Electro-Chemical Company, Research & Development Department (1955) *Hydrogen Peroxide Physical Properties Data Book*. Bulletin No. 67. Becco Chemical Division.

Chen, Xiaowei, *et al.* (2002) "Catalytic Decomposition of Hydrazine Over Supported Molybdenum Nitride Catalysts in a Monopropellant Thruster," Catalysis Letters, 79: 21–25.

Clarck, J. D. (1972) *Ignition! An Informal History of Liquid Rocket Propellants*. Rutgers University Press.

Corbett, A., Seamans, T., Dawson, B., and Cheetham, C. (1964) "Hypergolic Ignition At Reduced Pressures," Technical Report No. AFRPL-TR-64–175. www.dtic.mil/dtic/tr/fulltext/u2/610144 .pdf [accessed March 2018].

CPIA/M4 (1996) *Liquid Propellant Manual – Hydrogen Peroxide*. Chemical Propulsion Information Agency.

Davis, D. D., Mast, D. J., and Baker, D. L. (1999) "Recent Developments in Chemically Reactive Sensors for Propellants." Chemical Propulsion Information Agency.

Dennis, J., Fineman, C., Yoon, C., Santos, P., Pourpoint, T., Son, S., Heister, S. and Campanella, O. (2011) "Characterization of Gelling Systems for Development of Hypergolic Gels," in *European Conference for Aerospace Sciences (EUCASS)*, Saint Petersburg, Russia, July 4–8.

Diede, A. (2003) "Reduced Toxicity Hypergolic Bipropellant Fuels," US Patent 6695938.

Hawkins, T., Brand, A., Mckay, M., and Tinnirello, M. (2010) *Reduced Toxicity, High Performance Monopropellant at the U.S. Air Force Research Laboratory.* Report N.P. Air Force Research Laboratory.

Haws, J. L. and Harden, D. G. (1965) "Thermodynamic Properties of Hydrazine," Journal of Spacecraft and Rockets, 2(6): 972–974.

Heister, S. *et al.* (2014) *Final Report: Spray and Combustion of Gelled Hypergolic Propellants,* Contract #, W911NF-08-L-0171, www.dtic.mil/dtic/tr/fulltext/u2/a623637.pdf [accessed March 2018].

Iarochenko, N. and Dedic, V. (2001) *Hydrogen Peroxide as Monopropellant Catalysts and Catalyst Beds Experience from More than Thirty Years of Exploitation.* European Space Agency Special Publication No. 484. European Space Agency.

Jankovsky, R. S. (1996) "HAN-Based Monopropellant Assessment for Spacecraft," in *32nd AIAA/ASME/SAE/ASEE Joint Propulsion Conference and Exhibit.* AIAA 1996-2863.

Lawver, B. R. (1966) "Some Observations on Combustion of N_2H_4 Droplets," *AIAA Journal,* 4: 659–662.

Melof, B. M. and Grubelich, M. C. (2001) "Investigation of Hypergolic Fuels with Hydrogen Peroxide," in *37th Joint Propulsion Conference and Exhibit,* AIAA 2001-3837.

Morrell, G. (1957) "Summary of Naca Research on Ignition Lag of Self-Igniting Fuel – Nitric Acid Propellants," *NACA Research Memorandum.*

Neff, K., King, P., Anflo, K., and Mollerberg, R. (2009) "High Performance Green Propellant for Satellite Applications," in *45th AIAA/ASME/SAE/ASEE Joint Propulsion Conference & Exhibit.* AIAA 2009-4878.

Nimmerfroh, N., Feigenbaum, H. and Walzer, E. (1999) "Propulse™ Hydrogen Peroxide: History, Manufacture, Quality, and Toxicity," in *2nd International Hydrogen Peroxide Propulsion Conference.*

Paushkin, Y. M. (1962) *The Chemical Composition and Properties of Fuels for Jet Propulsion.* Pergamon Press.

Pein, R. (2005) "Gel Propellants and Propulsion," in *5th International High Energy Materials Conference and Exhibit Drdl,* Hyderabad, India, November.

Pourpoint, T. L. and Anderson, W. E. (2007) "Hypergolic Reaction Mechanisms of Catalytically Promoted Fuels With Rocket Grade Hydrogen Peroxide," Combustion Science and Technology, 179: 2107–2133.

Purcell, N., Diede, A., and Minthorn, M. (2002) "Test Results of New Reduced-Toxicity Hypergols for Use With Hydrogen Peroxide Oxidizer," in *5th International Hydrogen Peroxide Propulsion Conference.*

Rask, J., Vercoutere, W., Navarro, B. J., and Krause, A. (2008) *Space Faring: The Radiation Challenge – Middle School Educator Guide.* NASA. www.nasa.gov/pdf/284277main_radiation_ms.pdf [accessed March 2018].

Rusek, J. J., Anderson, N., Lormand, B. M., and Purcell, N. L. (1999) "Non-Toxic Hypergolic Miscible Bipropellant," US Patent 5932837.

Sackheim, R. and Masse, R. (2014) "Green Propulsion Advancement: Challenging the Maturity of Monopropellant Hydrazine," *Journal of Propulsion and Power* 30(2): 265–276.

Sarner, S. F. (1966) *Propellant Chemistry.* Reinhold.

Sawyer, R. F. (1966) "The Homogeneous Gas Phase Kinetics of Reactions in the Hydrazine-Nitrogen Tetroxide Propellant System". Tech. report, http://www.dtic.mil/docs/citations/AD0634277.

Schmidt, E. and Walter, E. (2001) *Hydrazine and its Derivatives: Preparation, Properties, Applications*. Wiley-Interscience.

Schumb, W., Satterfield, C. N., and Wentworth, R. L. (1955) *Hydrogen Peroxide*. Reinhold Publishing Corporation.

Sutton, G. P. (1949) *Rocket Propulsion Elements – An Introduction to the Engineering of Rockets*: John Wiley & Sons, Inc.

Sutton, G. P. (2006) *History of Liquid Propellant Rocket Engines*. AIAA.

Tsay, M., Lafko, D., Zwahlen, J., and Costa, W. (2013) "Development of Busek 0.5 N Green Monopropellant Thruster," in *27th Annual AIAA/USU Conference on Small Satellites*. AIAA.

Vieira, R., Pham-Huu, C., Keller, N., and Ledoux, M. J. (2002) "New Carbon Nanofiber/Graphite Felt Composite for use as a Catalyst Support for Hydrazine Catalytic Decomposition," *Chemical Communications*, 9: 954–955.

Warren, F. A. (1958) *Rocket Propellants*. Reinhold Publishing Corporation.

Weiss, H. G., Johnson, B., Fisher, H. D., and Gerstein, M. (1964) "Modification of the Hydrazine-Nitrogen Tetroxide Ignition Delay," *AIAA Journal*, 2: 2222–2223.

Whitmore, S. A., Merkley, D. P., and Eilers, S. D. (2013) "Hydrocarbon-Seeded Ignition System for Small Spacecraft Thrusters Using Ionic Liquid Propellants," in *27th Annual AIAA/USU Conference on Small Satellites*, SSC13-VII-6. AIAA.

Wingborg, N. (2011) "ADN Propellant Development," in *4th European Conference for Aerospace Sciences (EUCASS)*, Saint Petersburg, Russia, July 4–8.

Wright, A. C. (1977) *USAF Propellant Handbooks. Nitric Acid/Nitrogen Tetroxide Oxidizers, Vol. 2*. Martin Marietta Aerospace.

Zhang, Q. and Shreeve, J. M. (2014) "Energetic Ionic Liquids as Explosives and Propellant Fuels: A New Journey of Ionic Liquid Chemistry," *Chemical Review*, 114 (20): 10527–10574.

CHAPTER 10   ROCKET TURBOMACHINERY
FUNDAMENTALS

This chapter provides an introduction to turbomachinery used in chemical rocket propulsion devices and is applicable to both LRE and HRE systems that utilize liquid (storable or cryogenic) propellants. Section 10.1 provides an introduction and general description of elements of a turbopump. Section 10.2 provides background on pump design fundamentals while Sections 10.3 and 10.4 provide more detail on inducer and impeller design. Thrust balance is briefly discussed in Section 10.5 and overall operating envelope and CFD analysis results are discussed in Section 10.6. Turbine design is covered in Section 10.7 and bearings, seals, and shaft design in Section 10.8. Sections 10.9 and 10.10 briefly discusses rotordynamic issues and trends toward additive manufacturing.

10.1 INTRODUCTION: ELEMENTS OF ROCKET TURBOPUMPS AND HISTORICAL PERSPECTIVE

The term *turbopump* is used to describe a pump and a turbine that are integrated into a single assembly. Rocket engine turbopumps permit the delivery of propellants at high flow rates and pressures and support the use of lightweight, low-pressure tanks for applications with large volumes of propellant. Development of turbopumps is a very real engineering challenge since most applications demand minimal pump inlet pressures, large pressure rises, and high flow rates in combination with a desire for low weight. Materials selection is critical as fluids being pumped may be extremely flammable, highly corrosive, cryogenic, and susceptible to catalytic decomposition. The devices must endure the loads and vibration associated with a flight environment and in general have to operate over a wide range when considering startup and full-thrust limits. For cryogenic propellants, temperatures within the turbopump can change by over 1600 °F (850 °C) in a matter of inches/centimeters. These issues place the turbomachinery development challenge on par with that of the thrust chamber in many instances. Modern turbopumps operate at speeds as high as 100,000 RPM and transmit shaft power as high as 100,000 horsepower (750 MW). The SSME (now RS-25) turbopumps process volumetric flows high enough to empty an average swimming pool in 25 seconds while sporting peak discharge pressures in excess of 8000 psi.

While there are many applications for pumping liquids to high pressures, the rocket application is unique because of the fluids being pumped and desire for very high pressure rise and minimum weight. The "industry" that has developed these unique devices really amounts to a small community of dedicated designers and analysts at a handful of companies around the globe. The evolution of the design process has validated many "rules of thumb" or empirical approaches that were developed in the early days of liquid rocket engines in the 1950s and 1960s. In the early 1970s, NASA set out to document the elements of rocket engine design with a Special Publication (SP) series that is still quite valuable today. Several of the NASA SPs are listed in the Further Reading at the end of this chapter and a complete list is attached in the Appendix. In particular, the seminal work by D. Huzel and D. Huang (1992), documented as NASA SP-125 and later published by AIAA as Vol. 147 in the Progress in Astronautics and Aeronautics series has stood the test of time and remains a primer for those interested in LRE design and, in particular, the design of rocket turbopumps. While this chapter is intended to give students a first glimpse into the elements of turbopump design, it in no way replaces these valuable references that share much more detail on the design process.

10.1.1 Major Elements of a Rocket Turbopump

Figure 10.1 shows elements of a simplified turbopump. The *housing* is the major stationary structure that contains the rotating assembly and has provisions for routing fluids to or from the device. The rotating section of the turbopump is referred to as a *rotor*, an image of a rotor assembly is provided in Figure 10.2. Liquid propellant enters the pump inlet in the axial direction, and is directed radially by the blades within the pump by a rotating part referred to as an *impeller*. The pump must be designed to achieve the desired pressure rise such that the propellants are at sufficient pressure to be injected into the combustion chamber with its desired chamber pressure. Any pressure drops due to frictional losses in lines or valves must be accounted for in setting the desired pump outlet pressure. In regeneratively cooled engines, there may be substantial pressure losses in the regenerative cooling system as discussed in Chapter 6. The design shown in Figure 10.1 has two pump stages as the liquid that has been pushed outward by the first stage is then directed back toward the shaft for a second stage of pressure rise. For many rocket applications, the desired pressure rise is larger than can be accommodated with a single stage. While the J-2 hydrogen pump is a notable exception, modern pumps typically require three or less stages to achieve the desired exit pressure.

On the right-hand side of Figure 10.1, high-pressure warm/hot turbine drive gases enter the turbine manifold and are accelerated radially inward through a row of stationary blades, referred to as *turbine nozzles*, onto a row of rotating blades, referred to as *rotor blades* or *turbine buckets*, on the rotor. The design in Figure 10.1 is called a *radial turbine*, and can be employed in lower-pressure engines that have a limited amount of power to be supplied to the pump. For higher-power applications, an *axial turbine* may be preferred as it is straightforward to integrate multiple turbine stages as required to meet the power demands of the pump(s). Figure 10.2 shows a typical arrangement of two stages of axial turbine rotors on the right-hand end of the shaft. The function of the turbine is to provide the torque necessary to spin the pumps to their desired speed in order to achieve the required pressure rise. As described in Chapter 8, there are a variety of options for

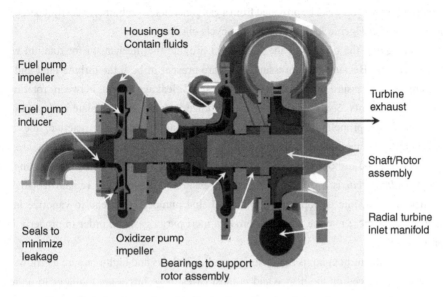

Figure 10.1 Cross-section of typical turbopump showing major components.

Figure 10.2 Photo of rotating section of SSME fuel turbopump. Pump impellers are made of titanium, the turbine disk is a high-temperature steel alloy, Waspaloy and turbine blades are MAR-M-246-Hf high-temperature steel.

turbine drive gases as they can be provided from a gas generator, a preburner, a catalyst bed, or from expanded/heated cryogenic fuel in an expander cycle engine.

Bearings on the outside of the rotor shaft provide the mechanism for rotation without excessive friction. Because high-pressure fluids are present in both the turbine inlet and pump exit, there is a pressure gradient that drives small leakage flows between rotating and stationary components. *Seals* are installed to minimize leakage and to isolate hot/warm turbine gases from liquid propellants. Because rocket propellants are highly reactive, these seals provide a critical function of isolating turbine drive gases from the liquids that are being pumped. Since some turbopumps may have fuel and oxidizer being pumped on the same shaft, an *interpropellant seal* is a critical element that ensures the two fluids remain isolated from each other. In cryogenic turbopumps, the fluids being pumped may tend to vaporize in these small spaces and there may be a need to provide inert purge gases in order to properly isolate the propellants.

Because the turbopump is redirecting large liquid flows and high-pressure gas flows, axial thrust forces are placed on the device and in most cases these forces are too large to be reacted against the bearings themselves. In this case, a *thrust piston*, or other thrust balancing mechanisms are required. See Section 10.5 for additional discussion on this issue.

10.1.2 Turbopump Design Challenges

Figure 10.2 is an image of the rotor of the SSME fuel turbopump. The large forces on the impellers, combined with a desire to minimize weight led designers to choose titanium for this portion of the turbopump. The high-temperature environment and high mechanical forces in the turbine dictated the use of high-temperature steel alloys in the turbine section. The propellants being utilized have a large impact on materials selection in turbopumps. Strong oxidizers such as liquid oxygen or nitric acid demand materials with oxidizer compatibility. Use of incompatible materials can be disastrous as strong oxidizers are happy to see metal parts as fuel. While less reactive than oxidizers, material considerations are also very important for fuels. In particular, hydrogen can be a challenging fluid. The tiny size of the molecule means it can permeate grain boundaries in metals and cause hydrogen embrittlement, a problem that was encountered in the Space Shuttle program since the fuel-rich preburner was essentially producing hot hydrogen gas that was used to drive turbines. While a materials treatise is beyond the scope of this text, the student needs to realize that materials selection and qualification for the environments present is a major issue for the turbopump designer.

Turbopumps also present significant thermostructural challenges, particularly when cryogenic propellants are being utilized. Figure 10.3 shows a cross-section of the SSME high-pressure fuel turbopump showing the three pump stages and two turbine stages illuminated in the photo in Figure 10.2. Note the proximity of the cryogenic hydrogen at the exit of the pump to the hot turbine inlet gas manifold to get an appreciation for the large temperature gradients that can be present.

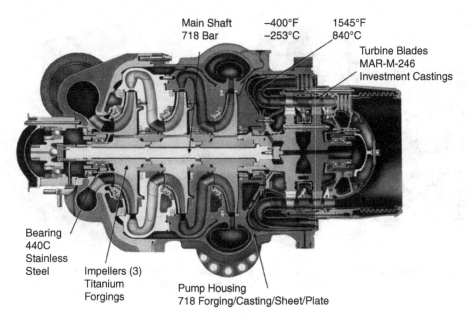

Figure 10.3 Cross-section of SSME three-stage high-pressure fuel turbopump showing materials of construction and flow conditions.

Since thermal stresses depend on the temperature gradient, they can represent very significant loads on the metal parts that have to react to these stresses.

The low pressures at the pump inlet can provide challenges to the pump designer since local pressures on the moving blades may be lower than the fluid vapor pressure. In this instance, cavitation can result and influence performance of the pump. This problem is exacerbated in situations where high vapor pressure fluids (such as cryogenic propellants) are being utilized. To mitigate this problem, an *inducer* is often placed on the inlet of many pumps. Figure 10.4 shows a cross-section of the SSME low-pressure oxidizer turbopump with inducer at the inlet as well as a separate photo of a typical inducer (an inducer is also present at the first pump stage inlet in Figure 10.1). The inducer is similar to a screw in that the angle of the blades is very low relative to the azimuthal/axial plane.

There are a number of different arrangements available to the turbopump designer based on the fact that a single turbine can drive one or more pumps. A particular configuration may also be driven by packaging constraints given spaces available within the engine/vehicle. In Section 10.2 we will see that pump design/speed is influenced by volumetric flow requirements and since oxidizer and fuel have differing densities and flow rates there is always a desire to operate pumps at different speeds. Similarly, the turbine drive will have its own speed (that in general would be different than either of the optimal pump speeds) for maximum efficiency. For this reason, geared systems were considered, particularly in early designs, in an effort to operate each component near an optimal speed to maximize efficiency. Figure 10.5 shows an arrangement of a gear-

(a)   (b)

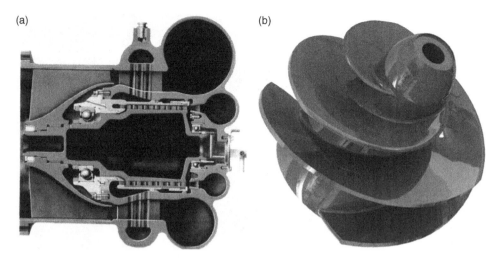

Figure 10.4 SSME low-pressure oxidizer pump with inducer (a) that mitigates cavitation on inlet. (b)
Rendering of a two-blade inducer.

driven turbopump that has a single turbine (lower-right portion of image) driving two back-to-back
pumps (top of image). This approach has several shortcomings since the gearbox assembly requires
a separate lubrication system, creates some frictional losses, and adds weight to the device and
these shortcomings must be weight against the benefits being near optimal efficiency (and hence
reduced size) of the rotating components. In general, the rocket industry has moved away from
geared configurations in modern designs.

Figure 10.6 shows some topology alternatives for turbopumps that use a *direct drive*
configuration with no gearbox. Modern pump technology affords operation at non-optimal
speeds with improved inducer designs, as discussed in Section 10.2. In direct drive configurations, the operating speed results from a compromise of efficiency of all components on the drive
shaft. Numerous alternatives are available as shown in Figure 10.6. Placing the pumps back-to-
back helps to cancel some of the thrust loads and packaging issues may favor placing the turbine
on the end of the stack or in the middle between the two pumps. Having separate turbines for each
pump permits each shaft to operate at a different speed and improves overall efficiency. Here the
two turbines could be driven from a common gas source either in a parallel or series
configuration.

Recently some firms are investigating the potential for electrically driven pumps for smaller
engines. As battery technology progresses, there is potential that this option could feed some niche
markets where the total volume of propellant to be delivered is not terribly large. The elimination of
the turbine and in most cases the combustion device that goes with the turbine can be a major
consideration, especially for smaller systems. The high torque available from electric motors
provides for rapid start and speed control.

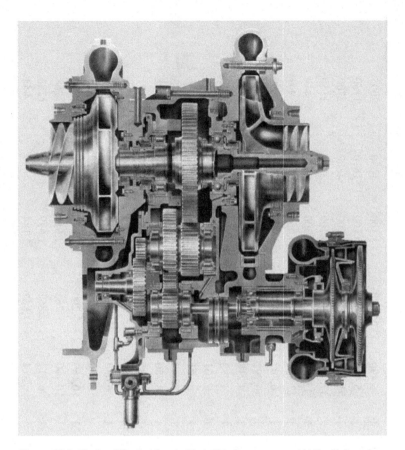

Figure 10.5 Photo of Rocketdyne's Mark III turbopump used in the H-1 engine.

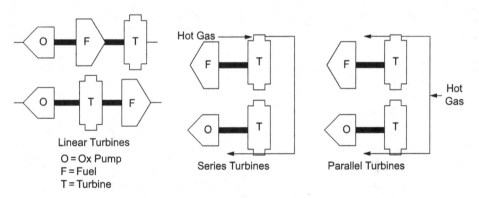

Figure 10.6 Direct drive, geared drive, and dual shaft options for turbopump configurations.

Table 10.1   Design characteristics of turbopumps (n = number of pump stages)

Pump	Fluid density, lbm/ft³	n	Speed, RPM	Flow rate, GPM	Head per stg, ft	delta p per stg, psi	Specific speed, Ns
Saturn V booster – F1 engine: oxygen	71.4	1	5,490	25,080	3,100	1530	2095
Saturn V booster – F1 engine – fuel (RP-1)	50.5	1	5,490	15,640	5,100	1810	1130
Space Shuttle main engine: oxygen	71.4	1	31,100	7,240	9,700	4800	2700
Space Shuttle main engine: fuel	50.5	3	37,400	16,300	66,700	2280	1150
RD-170 – oxygen	71.4	1	13850	25,008	17,300	8534	1448
RD-170 – fuel	50.5	1	13,850	14,485	20,000	7100	991
Rocketdyne Mark 3 fuel (RP-1)	50.5	1	6,792	2,188	2,923	1025	799
Rocketdyne Mark 3 LOX	71.4	1	6,792	3,148	2,067	1025	1243
A-7 Redstone (oxygen)	71.4	1	4,718	1,290	636	356	1338
A-7 Redstone (alcohol)	56.6	1	4,718	1,190	1,139	464	830
MB-3 Thor (oxygen)	71.4	1	6,101	2,870	1,651	862	1262
MB-3 Thor (Rf-1?)	51.2	1	6,101	1,700	2,117	913	806
LR87-AJ-3 Titan 1 (oxygen)	71.4	1	7,949	2,600	1,510	798	1673
LR87-AJ-3 Titan 1 (RP-1)	50.5	1	8,780	1,630	2,881	1034	901
H-1 Saturn 1B (oxygen)	71.4	1	8,985	1,100	1,613	819	1171
H-1 Saturn 1B (RP-1)	50.5	1	25,207	659	1,024	1091	3575
MA-5 Atlas (oxygen)	71.4	1	6,680	1,410	1,851	980	889
MA-5 Atlas (fuel)	50.5	1	6,650	2,130	2,719	1020	815
YLR81-BA-11 Agena (oxygen)	71.4	1	10,160	1,200	1,879	982	1233
YLR81-BA-11 Agena (RP-1)	50.5	1	10,160	765	2,616	996	768
YLR87-AJ-7 Gemini-Titan 1st stage (Oxygen)	71.4	1	6,314	2,862	1,679	877	1288
YLR87-AJ-7 Gemini-Titan 1st stage (Fuel)	50.5	1	6,314	1,867	2,184	839	854
Rocketdyne Mark 3 fuel (RP-1)	50.5	1	6,792	2,188	2,923	1025	799
Rocketdyne Mark 3 LOX	71.4	1	6,792	3,148	2,067	1025	1243

10.1.3 Turbopump Historical Data and Design Requirements

While not all engine manufacturers provide data on turbopump design characteristics, information is available for a good number of designs and some of these data are summarized in Table 10.1. We thank Mr. Bill Murray of Ursa Major Technologies for sharing this information. As you can see from the table, modern turbopump manufacturers are less inclined to share details of this nature, but the historical information in the table still provides background on performance delivered to date. The community still recognizes the Space Shuttle Main Engine and RD-170 turbopumps as the most powerful developed to date and the SSME fuel pump is the only multistage design in the historical database.

The discharge pressure and flow rate are two of the most important requirements for the turbopump designer. The pump discharge pressure is determined from an overall engine system analysis since the desired chamber pressure, injector pressure drop, and pressure losses from the pump to the injector all factor into this value. Since fuel is typically used as coolant in regeneratively cooled engines, the fuel pump discharge pressure is generally higher than the oxidizer pump. Cryogenic propellants (particularly LH_2) can go through very significant density changes in the cooling jacket and as a result have very large pressure drops in this section of the flowpath. Expander cycle engines rely on the enthalpy input to the fuel as a power source for turbine drive gases, so there is generally a larger pump discharge pressure required to accomplish this need. The pressure demands have some influence with chamber pressure due to the details of fluid property changes with pressure. In general, a systems analysis is required to ascertain the desired pump discharge pressure, but Table 10.1 provides a rough estimate for students wanting to start a design process.

10.2 PUMP DESIGN FUNDAMENTALS

Figure 10.7 provides a schematic of the cross-section and frontal view of a typical pump. The inlet region of the pump is typically referred to as the *suction side*, while the exit portion is called the *discharge side* or "pressure" side of the pump. The inducer is the screw-shaped section at the inlet that provides the initial pressure rise for fluids coming directly from propellant tanks. High-pressure pumps that accept effluent from a low-pressure pump, such as the SSME high-pressure fuel turbopump in Figure 10.3, do not require inducers. The impeller is comprised of a number of blades and is responsible for providing additional flow turning and pressure rise in the device. As shown in Section A-A in the schematic, these blades are typically swept backward relative to the direction of rotation.

The blades can be *shrouded* to provide a flowpath with rotating walls on both sides, or *unshrouded* to provide a flowpath that has a rotating wall only on the lower side. The rotor in Figure 10.2 is an example of a shrouded design, while the schematic in Figure 10.7 shows an unshrouded design. Shrouded impellers provide higher efficiency at the expense of higher weight.

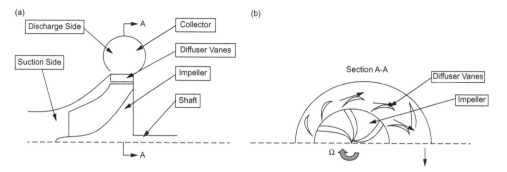

Figure 10.7 Schematic of a typical pump stage showing major components (a) and overall configuration (b).

The increased weight creates higher loads for bearings and shaft to react, but the fact that the blade tips are supported by the shroud greatly reduces the potential for blade vibrations due to aeroelasticity effects. The presence of a shroud also reduces rubbing risks when compared to an unshrouded design that must maintain tight clearance control throughout the meridional distance.

Fluid exiting the impeller enters the *collector* or *volute* whose function is to decelerate the fluid and route it for use in the next stage of the pump or for use within the engine. A row of *diffuser vanes* are sometimes included to aid in the turning process between the impeller exit and collector entrance as shown in Figure 10.7. If we consider the very last pump stage, then the flow in the collector is typically directed to a pipe at a specific azimuthal location such that the diffuser vanes would turn the impeller exit flow to the circumferential direction as shown in Section A-A in Figure 10.7. If we are considering an intermediate pump stage then some axial velocity would be retained in order to move the flow toward the next stage of the pump.

The azimuthal velocity of the blades ($\vec{\text{U}}$) will depend on the pump or rotor speed N and the radial distance (r) to the point in question:

$$\vec{\text{U}} = U\hat{e}_\theta = \frac{2\pi N}{60} r \tag{10.1}$$

where $\hat{e}_\theta$ is the unit normal vector in the azimuthal direction. Note that since U is always in the azimuthal direction, we generally just talk about the scalar form. Also, note that N is generally specified in revolutions per minute (RPM), it is converted to radians/second with the $2\pi(60)$ factor that appears in the denominator in Eq. 10.1.

Because we have rotating parts, we could define fluid velocities in the *absolute*, or *relative*, frame of reference. The absolute (sometimes called "laboratory") frame of reference is that of a stationary observer and the velocity in this scale is defined as $\vec{c}$. The relative frame refers to the velocity that would be observed if one were rotating with the blade itself. We define the relative velocity as $\vec{w}$. The absolute and relative velocities differ by the azimuthal velocity $\vec{\text{U}}$:

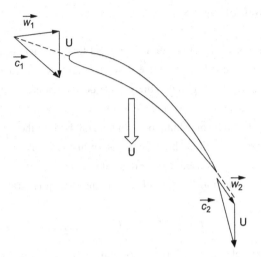

Figure 10.8 Velocity triangles on a rotating blade row. At optimal design, $\vec{w}$ should be aligned with the moving blade as shown in the triangle at the exit of the blade.

$$\vec{c} = \vec{U} + \vec{w} \tag{10.2}$$

This simple relationship is very powerful in turbomachinery analysis as it allows us to relate the three velocities at any radial location within the flowpath. Using Eq. 10.2, velocity triangles can be constructed for inlet and outlet of a moving blade row as shown in Figure 10.8. The relative velocity vector is closely related to the shape of the blades since an observer riding on the blade would see the flow inclined to the angle of the blade. In fact, for an "on design" condition where optimal performance is being obtained, we would typically assume that *flow angles equal blade angles*. This relationship is shown at the blade exit in Figure 10.8; there is no angle of attack on the blades and the angle of the velocity vector $\vec{w}$ must be aligned with the blade. At off-design conditions, this requirement will no longer be valid and angles of attack can develop on the rotating blades as shown in the inlet velocity triangle in Figure 10.8. This situation results when the pump is being operated at off-design conditions. As with any airfoil, the flow over the blades could separate and in a turbopump this condition would lead to a drastic drop in mass flow through the pump.

Finally, we note that both the $\vec{c}$ and $\vec{w}$ vectors can be decomposed into contributions in axial and azimuthal directions, respectively. If we let $\hat{e}_z$ and $\hat{e}_\theta$ represent unit vectors in the axial and circumferential directions respectively, we can always write:

$$\vec{c} = c_z \hat{e}_z + c_\theta \hat{e}_\theta \text{ and } \vec{w} = w_z \hat{e}_z + w_\theta \hat{e}_\theta \tag{10.3}$$

at any location in the flowpath. At any point in the flow, we can also define a *meridional velocity*, c_m (actually the meridional speed), which is the velocity of the flow aligned with the local centerline of the flowpath. As the flow is turned radially outward, the direction of c_m is always aligned with the local centerline of the duct.

10.2.1 Pump Inlet Requirements

The pump inlet pressure is a critical design requirement that is generally provided as a result of vehicle-level trade studies aimed to minimize system weight or maximize payload for a given vehicle architecture. Figure 10.9 highlights the vehicle design features that contribute to the inlet pressure p_1. The tank ullage pressure, p_u, and the hydrostatic pressure imposed by the weight of the column of fluid of height z both contribute to p_1. In general, both of these pressures can vary with time as the height of the liquid, the vehicle acceleration history, tank pressurization system, and structural loads on the tank all contribute to the desired p_u history and the hydrostatic contribution.

Using Bernoulli's equation, we can relate p_1 to the other pressures:

$$p_u + \rho \eta g Z - \Delta p_f = p_1 + \frac{\rho v_1^2}{2} \tag{10.4}$$

where η is the vehicle acceleration in g ($\eta = \eta(t)$) and Δp_f is any pressure drop due to frictional losses. The inlet velocity v_1 in the dynamic pressure term in Eq. 10.4 depends on the desired mass flow and the inlet area (inlet diameter) from continuity:

$$v_1 = 4\dot{m}/(\pi d_1^2 \rho) = 4Q/(\pi d_1^2)$$

where $Q = \dot{m}/\rho$ is the volumetric flow delivered by the pump. We typically work with Q rather than $\dot{m}$ in pump analysis because pumps are inherently volumetric flow devices. Volumetric flow is

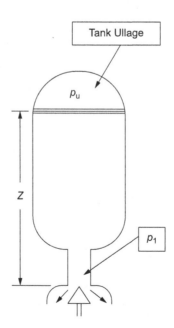

Figure 10.9 Pump placement at bottom of propellant tank. Tank ullage and hydrostatic pressure (influenced by vehicle acceleration) all contribute to the pump inlet pressure, p_1.

typically specified in gallons per minute (GPM) in the US system or liters per minute (LPM or l/min) in the SI system. Since the fluids being pumped in rocket applications can generally be assumed to be incompressible, it is straightforward to compute mass flow from volume flow and density. Using the result for v_1, the inlet pressure can be expressed:

$$p_1 = p_u + \rho \eta g Z - \Delta p_f - \frac{8\rho Q^2}{\pi^2 d_1^4} \qquad (10.5)$$

Equation 10.5 emphasizes the fact that the inlet pressure can be driven quite low depending on the inlet diameter and volume flow in the system and there is potential for the pressure to be near the vapor pressure in the liquids being pumped. For cryogenic fluids, they are typically loaded at temperatures near their ambient boiling point so the vapor pressure is near 1 atmosphere. Storable propellants have lower vapor pressures of course. If p_1 is near p_v, the locally higher velocities near rotating blades create local pressures below the vapor pressure and cavitation of the fluid can result. Gross cavitation would be disastrous since the pump would begin to pump large amounts of vapor (which is vastly lower density than the liquid), and would quickly overspeed as the turbine continues to put out power for pushing liquids in the pump.

Typically the engineer responsible for the pump will set a requirement for the minimum pump inlet pressure that the tank must supply. That requirement will force the vehicle design engineer to figure out a tank size and tank pressurization scheme that will optimize the tank for weight over the flight of the rocket. To achieve this purpose, the ullage pressure is varied during flight to compensate for the constantly changing fluid level and acceleration rate. The "negotiation" between vehicle/tank designers and turbomachinery folks can get very interesting relative to this fundamental parameter.

As we will see in the Section 10.2.2, pump designers tend to focus on the *head rise* provided by the device. Physically, the static head (H) is the height of a column of fluid that gives the hydrostatic pressure p, i.e. $H = p/(\rho g)$. Using this definition, the total head available at the inlet of the pump is then:

$$H_u + \eta Z - \Delta H_f = H_{avail} \qquad (10.6)$$

A parameter called the *net positive suction head*, *NPSH*, measures the margin (in feet or meters) available to avert cavitation at the inlet to the pump:

$$\text{NPSH} = H_{avail} - P_v/(\rho g) = H_u + \eta Z - \Delta H_f - H_v \qquad (10.7)$$

As the community currently has the desire to use high-speed direct drive turbopumps, NPSH is always much greater than zero. The value of NPSH depends on the fluid being pumped and the engine architecture.

Using the definition of static head, Eq. 10.6 can also be written in terms of pressure, with the resulting definition for the *net positive suction pressure*, *NPSP*:

$$\text{NPSP} = P_u + \rho g \eta Z - \Delta P_f - P_v = P_{01} - P_v \qquad (10.8)$$

Finally, the suction specific speed is another important performance measure of a pump. If one conducts dimensional analysis, a "scaled" speed results from considerations of varying volumetric flow and pressure/head rise. This scaled speed is called the suction specific speed, N_{ss}, and is defined as:

$$N_{ss} = N\sqrt{Q}/\mathrm{NPSH}_c^{0.75} \tag{10.9}$$

where NPSH_c is the critical NPSH value at the worst-case condition. The units of N_{ss} are intimately tied to the US system $(\mathrm{RPM(GPM)}^{0.5}/(\mathrm{ft})^{0.75})$ and pump designers frequently omit the actual units when specifying values. Using this nomenclature, $10{,}000 < N_{ss} < 100{,}000$ for rocket pumps.

The selection of pump speed is probably the most critical design decision in the pump development process and the ultimate choice results from a negotiation of many different considerations. The overall architecture plays an obvious role, for example, if a direct drive arrangement is selected then the pump and the turbine share the same speed. Higher speeds are good for keeping the impeller diameter small and for improving the efficiency of the turbines, but are bad for suction performance. Structural limits tend to be the speed limiter on a liquid hydrogen pump while the limiting factor is more often the suction performance of the inducer for denser propellants that have lower Q demands.

Blade stresses, impeller and turbine disk stresses, bearing speed limits and rotordynamics are all additional considerations for this fundamental decision. Last but certainly not least is the effect that pump speed has on pump efficiency. Figure 10.10 provides efficiency

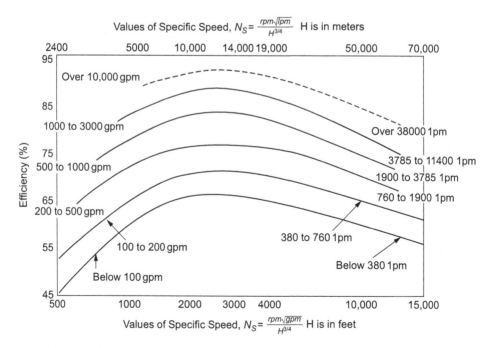

Figure 10.10 Typical pump efficiency as a function of specific speed for a range of volumetric flow rates.

trends as a function of specific speed and volumetric flow rate from historical data and theoretical considerations. Oftentimes, the designer is forced to compromise a bit on efficiency in order to meet many of the other considerations as outlined above. The use of a boost pump can alleviate NPSH issues and permit higher operating speeds; this option was employed on the SSME.

10.2.2   Pump Pressure Rise and Power Demand

Fundamentally, pressure rise is created in a pump by imparting swirl to fluid elements and increasing their kinetic energy. This kinetic energy, reflected as dynamic pressure, can be converted to static pressure by decelerating/diffusing the flow in the volute. The *Euler momentum equation* relates the torque imposed on the fluid to the change in tangential momentum:

$$\tau = \dot{m}[(rc_\theta)_2 - (rc_\theta)_1] \tag{10.10}$$

so that changes in the azimuthal component of the absolute velocity, and the radial distance the fluid lies from the shaft center determine the total torque. In a pure axial pump, r = constant, so the torque on the fluid depends on the change in swirl velocity imparted to the fluid (i.e. Δc_θ). The power ($\dot{W}$) required to provide this torque can be written:

$$\dot{W} = \tau N = \dot{m}[(rc_\theta)_2 - (rc_\theta)_1]N \tag{10.11}$$

where N is converted into units of rad/s. Recall that in Chapter 8 we related the pump power to the pressure rise. Assuming incompressible flow, one may write:

$$\dot{W} = \dot{m}\Delta p\eta_p/\rho \tag{10.12}$$

where η_p is the pump efficiency. This parameter accounts for frictional losses, leakage, and multidimensional flow effects (flow separation, vortical flows) within the flowpath. Equating Eqs. 10.11 and 10.12 relates the pressure rise to changes in velocity and pump geometry:

$$\Delta p = \rho[(rc_\theta)_2 - (rc_\theta)_1]N \tag{10.13}$$

Equation 10.13 shows that there are a variety of ways to achieve the desired pressure rise as it is a function of pump speed, total radial motion of the fluid element, and the change in tangential velocity imparted to the element.

The head coefficient, Ψ, is a dimensionless measure of the head rise (or pressure rise using the definition of head) that is provided by the inducer or impeller:

$$\psi = g\Delta H/U_t^2 \tag{10.14}$$

where ΔH is the physical headrise in feet or meters and U_t is the tip speed of the blades at the rated power condition. Head coefficients can vary from values as low as 0.1 for an inducer to 0.5 for a modern high performance impeller. The flow coefficient, ϕ, is another important parameter characterizing pump performance:

$$\phi = c_m / U_t \qquad (10.15)$$

where c_m is the meridional velocity as discussed previously. Physically, ϕ represents the c_m value that can be accepted for a given tip speed and we can see from Eq. 10.13 that this ultimately relates to the pressure rise that can be attained in the component in question. Inlet flow coefficients typically range from 0.07 to 0.3 and discharge flow coefficients range from 0.01 to 0.15 on typical rocket turbopumps.

10.3 Inducer Design

Armed with knowledge about minimum acceptable inlet pressures and NPSH values, the pump designer can consider the need for an inducer. In modern systems inducers are used in nearly all cases as they tend to minimize required tank ullage pressures, and hence minimize tank and system weight. A notable exception here would be a very high-pressure engine system that may employ both low-pressure and high-pressure turbopumps such as the SSME designs shown in Figures 10.3 and 10.4. As noted previously, an inducer is not required on a high-pressure pump. The actual delivered Ψ value depends on the pressure rise desired to meet the N_{ss} requirement for the impeller. Generally, an inducer is required if $N_{ss} > 10{,}000$.

The NPSH$_c$ value results from investigating pump inlet pressure and NPSH values delivered per Eqs. 10.7 and 10.8 over all operating conditions for a given application. In general, a safety factor (sf) is also applied in order to ensure that there is some margin for uncertainty so that NPSH$_c$ = NPSH$_{min}$/sf. Typical safety factor values vary from 1.2 to 2.0. Lower values would apply for shorter life missions as cavitation erosion on the blade tips can result from local pressures dropping below vapor pressure. Higher values of sf ensure that local pressures remain above vapor pressures such that cavitation erosion is eliminated for long life operation.

Figure 10.12 provides a schematic of the cross-section of an inducer and a local velocity triangle as viewed from observer viewing radially inward from outside of the device. The meridional velocity at the inlet is set from the required inlet pressure to meet NPSH$_c$ considerations:

$$\frac{\rho c_m^2}{2} = p_{in} - p_a = \rho g \text{NPSH}_c / c_f \qquad (10.16)$$

where c_f is an experience-based correction factor developed based on the performance of real rocket inducers (Huzel and Huang, 1992). Experience has shown that liquid hydrogen inducers follow the theoretical result that Bernoulli would give with cf = 1. However, liquid oxygen inducer designs tend to demand a lower c_m value consistent with c_f =2. For non-cryogenic liquids such as RP-1, a c_f value of 3 is recommended.

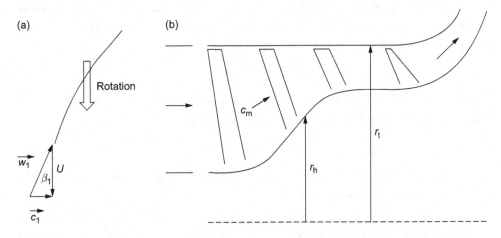

Figure 10.11 (a) Velocity triangle at inlet to inducer (radially inward view of the inducer blade as viewed from above) and (b) cross-sectional view of inducer showing hub radius variation and blade cant. Blades are canted forward for structural reasons. The leading edge of the inducer blade is shaped azimuthally (smoothly cut back from maximum extent r_t as viewed azimuthally) in order to reduce structural loads. At the inlet, the absolute velocity $\bar{c}$ is aligned in axial direction (no swirl). Typical inducer inlet angle (β_1) is 6–11°.

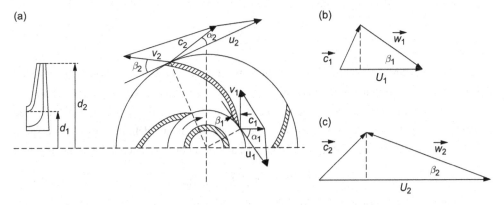

Figure 10.12 (a) Cross-sectional and front views of a shrouded impeller and inlet and exit velocity triangles. (b) The inlet velocity triangle (station 1) is in the axial/azimuthal plane while (c) the exit triangle (station 2) is in the radial/azimuthal plane.

From mass conservation considerations, c_m must also be related to the prescribed volume flow and the annular area available (A):

$$Q = c_m A = \pi r_t^2 c_m (1 - (r_h/r_t)^2) \tag{10.17}$$

where r_h and r_t are the local hub and tip radii as shown in Figure 10.11. In the inlet, a rule of thumb for the radius ratio is $r_h/r_{tt} = 0.3$. Given this information and the desired volume flow allows one to solve for c_m, and r_t using Eqs. 10.16 and 10.17.

Inducer designers typically design to a tip flow coefficient value of 0.1. Using this value and the resulting c_m from solution of Eqs. 10.16 and 10.17 permits one to immediately compute the inducer blade tip speed:

$$U_t = c_m/\phi = 10c_m \tag{10.18}$$

Note that setting tip and meridianal speeds also amounts to specifying β, i.e. from the velocity triangle shown in Figure 10.11 (and the definition of ϕ) we have:

$$\beta_t = \tan^{-1}(U_t/c_m) = \tan^{-1}\phi \approx 6^o \tag{10.19}$$

Actual β_t values are typically a bit larger to ensure that there is a slight positive angle of attack on the blade. A typical value is roughly 10–15°.

The radial variation in blade profile is determined to maintain a balanced radial pressure gradient. For a pure swirling flow, the pressure varies radially as a function of r^2 due to hydrostatic forces resulting from the swirling velocity component. If the pressure distribution varies from this value, an imbalance will result and create secondary flows within the passage. Along the inlet, we can use this requirement to determine the profile (twist) of the blade:

$$r \tan \beta = \text{constant} = r_t \tan \beta_t \tag{10.20}$$

Knowing the tip speed and tip radius immediately allows one to compute the speed of the impeller:

$$N = \frac{60U_t}{2\pi r_t} \tag{10.21}$$

As the inducer is designed to provide sufficient suction head to mitigate cavitation in the impeller, this requirement fundamentally sets the head/pressure rise in the inducer. The impeller designer sets the minimum acceptable suction specific speed for that device, $N_{ss,impeller}$. Using the speed computed in Eq. 10.19 and the desired Q value for the pump, the definition of specific speed (Eq. 10.9) gives the required NPSH (call it NPSH_I) at the end of the inducer/inlet to the impeller:

$$\text{NPSH}_I^{0.75} = N\sqrt{Q}/N_{ss,impeller} \tag{10.22}$$

Computing NPSH_I from Eq. 10.22 permits us to compute the required head rise for the inducer itself:

$$\Delta H = \text{NPSH}_I - \text{NPSH}_c \tag{10.23}$$

and the resulting required pressure rise within the inducer is simply $\Delta p = \rho g \Delta H$. This head/pressure rise stipulates the required velocity components at the exit of the inducer. Using Eq. 10.13 we have:

$$(rc_\theta)_e - (rc_\theta)_i = \frac{g\Delta H}{N\eta_I} \tag{10.24}$$

Here, η_1 is the inducer efficiency and subscripts "e" and "i" denote exit and inlet of the inducer, respectively. We define η_1 as the ratio of theoretical (left-hand side of Eq. 10.24) and actual head rise characteristics of the inducer. If we assume there is no swirl of the flow entering the inducer ($c_{\theta i}=0$) and that r_t is constant at the value we computed at the inlet, then Eq. 10.24 gives $c_{\theta e}$:

$$c_{\theta e} = \frac{\Delta H}{N \eta_1 r_t} = \frac{\Delta H}{\eta_1 U_t} \tag{10.25}$$

Use of continuity at the exit (Eq. 10.17) and the exit velocity triangle resulting from the $c_{\theta e}$ result from Eq. 10.25 and the prior computed U_t permits one to compute the exit blade angle at the tip of the blade and the hub radius at the exit location. To generate additional pressure, additional swirl must be imparted to the fluid per Eqs. 10.11–10.13 and the blade angle β must increase as one advances along the inducer surface.

The axial contour of the blade itself is developed to create a smooth, monotonic pressure rise along the length of the inducer. Since the local c_m value is related to the local pressure, this quantity can be obtained from the applicable form of Eq. 10.16. Knowing the local c_m value permits one to compute the local hub radius from Eq. 10.17.

10.4 IMPELLER DESIGN

Figure 10.12 provides an image of a centrifugal impeller with velocity triangles at the inlet and exit (stations 1 and 2 respectively) planes. As mentioned in Section 10.2.3, the impeller design begins with an assessment of the suitable suction specific speed. If an inducer is placed upstream of the impeller, then N_{ss} can be larger due to the fact that the inducer has raised pressures well above cavitation levels.

We presume that Q, N, and ΔH have all been specified as a result of overall pump requirements or conditions fed from an inducer that lies upstream of the impeller. If an inducer is present, the meridional velocity (principally in the axial direction) feeding the impeller would be assumed to be equal to the value exiting the inducer. If no inducer is present then this parameter would be set by NPSH considerations and a treatment similar to what is described in Eq. 10.16 would be required. Here we will neglect any leakage flows and assume that flow angles are equal to blade angles (no flow slip) in order to simplify the analysis. This latter assumption is somewhat tenuous and readers interested in refining the analysis are referred to Huzel and Huang (1992).

As with the inducer, a flow coefficient is generally specified for the inlet at the tip (or "eye") of the inlet. Values for impellers tend to be larger than for inducers, but values ranging $0.08 < \phi < 0.4$ have been used in prior designs. Given ϕ and the meridional velocity, we can compute U_{1t} from the definition of the flow coefficient:

$$U_{1t} = c_{m1t}/\phi \tag{10.26}$$

where c_{m1t} is the corresponding meridional tip velocity exiting the inducer or the required value to meet NPSH constraints if no inducer is present. Given the U_{1t} value, the corresponding speed at the hub is then:

$$U_{1h} = U_{1t}(r_h/r_t) \tag{10.27}$$

If an inducer is present, we can also assume that azimuthal velocities at its exit will be equivalent to inlet conditions on the impeller. If no inducer is present, one would generally assume that $c_\theta = 0$, which is consistent with the no-swirl assumption. Given the appropriate c_θ value, the impeller inlet angle is then computed from the inlet velocity triangle:

$$\beta_{1t} = \tan^{-1}(c_{m1t}/(U_{1t} - c_{\theta 1t})) \tag{10.28}$$

with an analogous result at the hub of the impeller inlet blades.

At the impeller exit, the blade angle β_2 represents one important consideration and values of 17–40° have been employed for backward-leaning blades typically used in rocket impellers. Acceleration of the meridional flow will typically give c_{m2} values in the range of 100–150% of c_{m1} values computed in the inlet. If we regard these two quantities as known, then the other impeller exit conditions r_2, U_2, and $c_{\theta 2}$ can be determined from the impeller head rise requirement, the exit velocity triangle, and the pump rotation rate via:

$$(rc_\theta)_2 - \overline{(rc_\theta)_1} = \frac{g\Delta H_{imp}}{N\eta_{imp}} \tag{10.29}$$

$$\beta_2 = \tan^{-1}(c_{m2}/(U_2 - c_{\theta 2})) \tag{10.30}$$

$$r_2 = \frac{60 U_2}{2\pi N} \tag{10.31}$$

where ΔH_{imp} and η_{imp} are the required impeller head rise and efficiency, respectively. Typical impeller efficiencies lie in the range 70–90%. The quantity $\overline{(rc_\theta)_1}$ is the average value across the impeller inlet plane. Other considerations for a good design include the following:

- Impeller eye to tip radius ratio (r_{1t}/r_2) should be less than 0.7 for a good design.
- Tip speed (U_2) limits. Good designs will lie within allowable tip speeds for the fluid being pumped and the materials of fabrication. Titanium (Ti-5–2.5) provides highest tip speeds of 1800–2000 ft/s for hydrogen and about 1200 ft/s for methane fuels but is incompatible with liquid oxygen. Aluminum is also reserved for fuels and tip speeds of 600 ft/s and 1200 ft/s are attainable with methane and hydrogen fuels respectively. Inconel (IN 718) can be used with any of these three fluids covering speeds of 760–900 ft/s for LOX, 1400 ft/s for hydrogen and 1300–1500 ft/s for methane.
- Exit flow coefficient values typically lying in the range 0.05–0.3 have been used on successful designs.

The numerous criteria provide for a challenging design exercise. Assumed values for c_{m2} and β_2 can be adjusted to better match the other desired attributes. If the head rise is too large, a single-stage answer may not be available and multiple impeller stages may be required. Once a suitable design is developed, the width of the impeller at its exit, b, may be computed from volume flow considerations:

$$b = Q/(2\pi r_2 c_{m2}) \tag{10.32}$$

10.5 THRUST BALANCE

The axial momentum of the large liquid flow entering the pump is directed outward at some angle by the impeller and as a result a net thrust force is communicated to the shaft. For example, if a radial impeller is employed as in the analysis in the prior section in reference to Figure 10.12, all of the axial momentum entering the pump is effectively pushing the impeller aft. Figure 10.13 shows this situation in more detail and emphasizes the fact that the pressure in the cavity between the rotating impeller and the stationary wall also enters the momentum balance. In this case, the net thrust force is then:

$$F_t = p_1 \pi r_1^2 + \dot{m} c_1 - p_{cav} \pi r_{cav}^2 \tag{10.33}$$

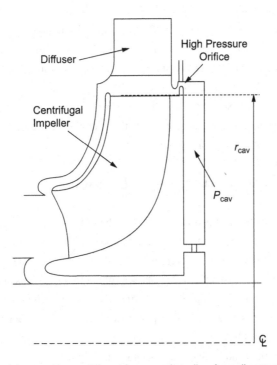

Figure 10.13 Balance piston used to react thrust forces on impeller. A small amount of high-pressure exit flow is metered to the back side of the impeller to create the thrust balance.

where c_1 is the velocity of liquid entering the pump, p_{cav} is the average pressure in the cavity and r_{cav} is the radial distance to the end of the cavity. Here we have neglected the small cross-sectional area of the shaft. The flow in the cavity is swirled due to viscous interactions and in general the cavity azimuthal velocity will vary with r, and hence p_{cav} will vary as r^2. The bearings that serve as interface between the stationary housing and the rotor have some capability to react axial loads, but in rocket turbopumps the loads are typically too large to react by the bearings alone. In these cases, a *balance piston* is employed as shown schematically in Figure 10.13. The general idea of the balance piston is that a controlled flow is allowed to enter the region behind the impeller and by metering this flow with an orifice we can control p_{cav} and the size of the cavity r_{cav} so as to keep F_t in Eq. 10.33 to levels acceptable for the bearings to absorb. The pressure in the cavity is always higher than the inlet pressure, and a sump and return path is included such that this fluid is effectively recirculated back into the inlet. Note that the designer has to evaluate the performance of the balance piston over the entire operating range of the pump in order to ensure that loads are within acceptable limits at all operating points.

In general, thrust balancing is much more complex than the simple situation described in Figure 10.13 and turbopump designers have to expend major efforts in the design process to account for all thrust forces in the assembly and find ways to control them or react them appropriately in bearings and other portions of the structure. For example, in cases where a turbine is mounted in-line with the pump, a global axial force balance is required and one must consider thrust forces from both components. In general, the turbine thrust force can be manipulated to counteract the force on the pump. Huzel and Huang (1992) have a nice discussion of this approach as applied to SSME high-pressure fuel turbopump. The impeller forces can be manipulated by shrouding the front side of the blades or introducing labyrinth seals in convenient locations. Radial vanes can be placed on the backside of the impeller and vane height, vane/wall gap, and number of vanes can be varied to create the desired balance. Using these more advanced notions permits designers to eliminate balance pistons on modern designs as these devices increase rotor length and have deleterious effect on shaft rotordynamics and weight.

10.6 Pump Operating Envelope and CFD Analysis

The overall efficiency of the pump is a reflection of the efficiency of the respective inducer and impeller components, but also reflects losses due to frictional processes not accounted for within either the inducer or impeller stages. Larger pumps (higher Q values) tend to have larger Reynolds numbers and lower viscous losses. Size also plays a role in terms of seals and leakage as a given clearance that may be attainable becomes a larger fraction of the overall flowpath as the dimensions of the pump shrink. For these reasons, larger pumps tend to be higher efficiency, as was shown in Figure 10.10 earlier in this chapter.

While Figure 10.10 shows that there is an optimal specific speed to maximize efficiency, one can also think of a given set of hardware producing the highest efficiency at a given volumetric

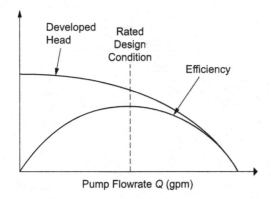

Figure 10.14 Typical pump characteristics over operational envelope. Pump efficiency is a maximum at rated design condition while the pressure rise is largest at low flow conditions.

flow. Pump designers would typically want this optimal performance to exist at the rated design condition as shown in Figure 10.14. This figure also shows how pump headrise and power demand change with volume flow, or throttle setting. At off-design conditions away from the rated flow, angles of attack will develop on pump blades and flow separation regions can appear. These irreversible processes lead to a reduction in efficiency. While efficiency can be rather low at low flow conditions, the head rise actually increases as a larger amount of work is done on a smaller amount of fluid. At flows above the rated condition, head output begins to fall dramatically as well as pump efficiency. Pump power demands actually peak in this region as power is the product of volume flow, pressure rise and efficiency as noted in Eq. 10.12.

The modern pump design process begins with many of the fundamental calculations described in this section. A quasi-one-dimensional code is typically written to generate the "mean line" contour of blades in both inducers and impellers and this meanline contour, supplemented with hub and tip blade angles at critical locations, provides for an initial shaping of the blade contour. While frictional effects are largely ignored in this process, the results are surprisingly good when these blade shapes are run in multidimensional CFD tools. Figure 10.15 shows a typical result of a fully 3-D calculation for an inducer and a volute. Because the flow rates and Reynolds numbers are so high in rocket turbopumps, viscous losses tend to be second order effects when considering repercussions on the base flowfields.

While CFD analysis may not have great influence on the basic blade shapes, it is providing useful corrections to the simple meanline theory. In addition, it is very useful to assessing performance at off-design conditions where separated flows become a significant feature. Finally, CFD analyses are very helpful in estimating secondary and leakage flows; the understanding of which being essential for axial thrust balance predictions as discussed in Section 10.5.

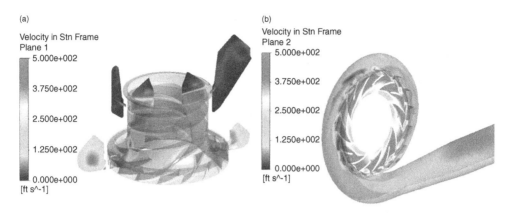

Figure 10.15 Modern multidimensional CFD computations of inducer/impeller (a) and diffuser/volute (b) velocity fields. Both results depict velocities in stationary (lab) frame in ft/s.

10.7 TURBINE FUNDAMENTALS

Figures 10.1–10.3 provide images of typical turbine designs in rocket turbopumps. The turbine typically includes one or more stages in order to develop the power required to drive the pumps. The turbine stage begins with a stationary row of blades called *stators*, or more commonly turbine *nozzles*. The function of this stage is to accelerate gases and turn them to the proper angle to align the velocity with the leading edge of blades on the moving blade row. The moving blades are termed *rotors*. Figure 10.17 provides nomenclature and velocity components for a typical turbine stage. In contrast to pumps, the change in tangential velocity is opposite the direction of rotation noted by the blade/rotor speed U_T. A parameter that characterizes turbine design is the degree of reaction, R that relates the enthalpy drop in the rotor to the overall enthalpy drop in the stage:

$$R = \frac{h_2 - h_3}{h_1 - h_3} \qquad (10.34)$$

If all of the enthalpy drop occurs in the turbine nozzles, $R = 0$ and this type of turbine is called an *impulse turbine*. Since the enthalpy change is directly related to the pressure change ($h = e + p/\rho$), an impulse turbine will have a large pressure drop in the turbine nozzles and constant pressure in the rotor stage. Impulse turbines have historically been employed in gas generator, tapoff, or expander bleed cycles due to the desire to get the maximum amount of power from the precious turbine gas.

If $R \neq 0$, the device is referred to as a *reaction turbine*. Reaction turbines tend to be utilized in staged combustion engines that derive power from large amount of flow from preburners as they provide the highest efficiency. The gas velocities leaving the turbine nozzle are supersonic and can be much higher than Mach. However, since the rotor is moving away from these gases the relative velocity impacting the blade is subsonic such that no shocks are

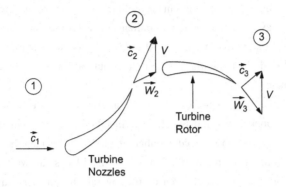

Figure 10.16 Velocity triangles and nomenclature for flow through an axial turbine stage.

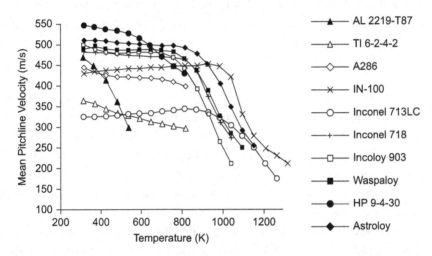

Figure 10.17 Allowable mean pitchline velocity (U_T) as a function of operating temperature for a variety of different turbine materials (Humble *et al.*, 1995).

generated. The velocity triangle at station 2 in Figure 10.16 emphasizes this fact. The nozzle design can be an axisymmetric de Laval type nozzle or a rectangular flow path vaned style. The vaned style has been made with the nozzle throat well within the nozzle flow path with two expansion ramps and some designs the throat is closer to the exit and more representing a single expansion ramp nozzle.

The power developed by a turbine stage can be determined from the Euler momentum equation as we used in developing pump power requirement. From the velocity triangles in Figure 10.17:

$$P = \tau N = \dot{m} U_T (c_{\theta 2} - c_{\theta 3}) \tag{10.35}$$

so as in the case of the compressor, changes in tangential velocity determine power (or work) output of the stage. Since we have favorable pressure gradients in the turbine, large amounts of turning are possible and we can have situations where $c_{\theta 2} > 0$ and $c_{\theta 3} < 0$ (as shown in Figure 10.16). If a multi-stage design is employed, then one must compute the total power output by adding results from Eq. 10.35 for each stage.

Partial admission nozzles are used in some applications. In this case, a portion of the annular flow area is blocked off and nozzles are installed over the remaining open portion. The obvious benefit here is the reduced number of blades to be manufactured and potential weight savings resulting from removal of some of the blades, as well as in reduction of manifold volume and weight. In this case, the overall torque created in the device is distributed over a smaller number of rotor blades and their design must reflect the proportionally higher loads that result.

In Eq. 10.35, U_T is the rotational velocity. In the case of a turbine, this parameter is called the *mean pitchline velocity*, defined as the circumferential velocity at the center of the blade passage:

$$U_T = N\bar{r} \tag{10.36}$$

where $\bar{r}$ is the radial height at the center of the blade passage. The mechanical loads on the turbine blades scale with this velocity, so its selection is critical and dependent on the strength of the material used in the blades. Figure 10.17 provides allowable U_T values as a function of temperature for numerous different turbine materials. Typical turbine inlet temperatures are in the 1200–1400 °F (920–1030 K) range with some applications as high as 1600 °F /1140 K. Figure 10.17 shows several materials hold strength in this range to support mean pitchline velocities over 400 m/s (1300 ft/s). It is worth mentioning here that in direct drive configurations the turbine and the pump share the same shaft speed so the designer must allow for this fact in setting the mean radius of the turbine passage to achieve the desired pitchline speed. While in most cases, the speed limit is set by turbine performance, one must evaluate capability on the impeller side of the rotor as well. Impeller tip speeds comparable to turbine pitchline velocities implies that both parts would be of comparable size. Figure 10.2 demonstrates this fact for the SSME hardware.

The power developed from a given turbine stage can also be developed from thermo-dynamic considerations. If we have an adiabatic system, the ideal work done is related to the change in stagnation enthalpy across the stage and since power is work (per unit mass) times flow rate, we can write:

$$p = \dot{m}\eta_t c_p(T_{t1} - T_{t3}) \tag{10.37}$$

where η_t is the overall efficiency of the stage. Physically, η_t is the ratio of the power delivered by the turbine and the theoretical power obtained from isentropic expansion of the gases through an equivalent pressure ratio. For an isentropic expansion, conservation of energy provides an ideal velocity called the *spouting velocity, c_o,* by turbine designers:

$$c_o = \sqrt{2c_p T_{t1}(1 - (p_{t3}/p_{t1})^{(\gamma-1)/\gamma})} \tag{10.38}$$

It is useful to correlate the turbine efficiency with the U_T/c_o velocity ratio as shown in Figure 10.18. Deviations from perfect efficiency are attributed to frictional processes such as boundary layers within fluids, secondary flows, friction and leakage through sealing surfaces, and thermal losses due to heat transfer. For a single-stage turbine, ideal performance is achieved for U_T/c_o in the 40–50% range. For a multistage turbine, the reader should recognize that c_o is really a fictitious value since only a fraction of the overall pressure drop would occur in one stage, i.e. the pressure ratio in Eq. 10.37 would then reflect the overall inlet and outlet stagnation values for the device. In this case, optimal performance is achieved at lower U_T/c_o ratios as indicated in Figure 10.18.

10.7.1 CFD Analysis in Turbines

As in the case of turbopumps, modern analysis of rocket turbines is carried out with multidimensional CFD and finite element packages. Figure 10.19 shows a typical result for a turbine cascade (an unwrapped row of blades showing the x–θ plane). For turbines, the CFD results provide detailed pressure and temperature loading on the blades for use in thermostructural analyses. Regions of flow separation or secondary flows can be identified.

Thermostructural analysis of the blades is a major consideration since these elements are subjected to high mechanical and thermal loads. The right image in Figure 10.19 shows a typical result for a turbine rotor. Mechanical bending loads induced by the pressure distribution on the blades is counteracted to some degree by centrifugal forces due to rotation. Thermal stresses can be important; particularly at startup when hot gases impinge on cold blades. Finally, the

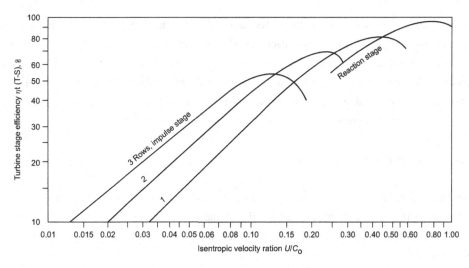

Figure 10.18 Turbine efficiency as a function of mean pitchline/nozzle spouting velocity ratio for 1–4 stage turbines (Source: NASA SP-8110).

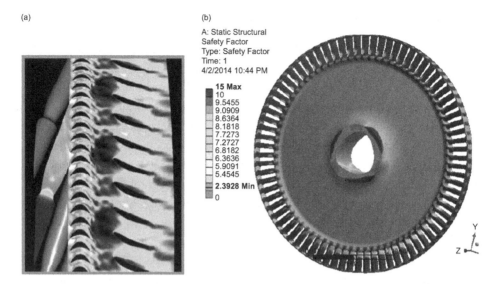

Figure 10.19 (a) Mach number distribution from multidimensional CFD analysis showing high values exiting turbine nozzles on the left and low values downstream of turbine buckets in the middle. Expansion wave structure is evident in the purple region downstream of the nozzle throats. The dark structures on the right are turning vanes to remove residual swirl in order to provide an axial flow out of the device. (b) Thermostructural analysis of turbine rotor showing safety factor at maximum speed condition.

multidimensional analyses provide opportunities to assess component efficiencies from first principles and improve on any estimates employed earlier in the design process.

10.8 SHAFTS, BEARINGS, AND SEALS

The mechanical components supporting operation of the rotor assembly see unique environments when compared with most industrial machines. Highly corrosive and potentially reactive liquids are being moved through the device over temperatures that can range from hard cryogenic conditions to extreme heat in turbine sections. These unique requirements generally imply that specialized designs and materials must be considered such that design of these mechanical components is much more involved than picking items out of a catalogue.

10.8.1 Shaft for Rotor Assembly

The shaft must be large enough diameter to communicate the torque provided by the turbine (Eq. 10.35) to the pumps (Eq. 10.11). At steady state conditions, these two torques are equivalent and the value input by the turbine is communicated along the shaft to the pump(s). The stress induced in a rod under torsion (σ_t) is:

$$\sigma_t = 16\tau / (\pi d_s^3) \qquad (10.39)$$

where d_s is the shaft diameter. This result can be used to estimate the minimum acceptable shaft diameter assuming σ_t is a worst-case allowable stress reflective of all operating conditions the shaft might be exposed to. When one considers the power levels being communicated along the shaft, it is easy to understand that structural properties are crucial. Since the shaft may be exposed to propellants or hot turbine gases, material considerations must also reflect this fact. In addition the shaft must provide stiffness to react bending loads imposed by the weight of assemblies that are cantilevered from location of bearings.

10.8.2 Turbopump Bearings

As mentioned in the introduction to this section, rocket turbopump bearings can see some very unique environments given the liquids that are being utilized and the reactivity of these liquids. In contrast to conventional bearing usage where a lubrication system reduces frictional losses and provides bearing cooling during operation, the reactivity of the propellants with typical lubricants negates this option. For this reason, the propellants themselves may serve as lubricants or the bearings can be designed to function with no lubricant at all. This implies that bearings must be fabricated from materials that are compatible with the liquids being pumped. The bearing system can be exposed to highly transient thermal and mechanical loading during transients such as engine start and shutdown.

The bearing system has three major functions:

1. Provide radial control/positioning of the rotor assembly. The bearings must be of sufficient stiffness to prevent major rubs of moving blades against the outer casing and must maintain tight clearances to reduce leakage flows.
2. Axial control of the rotor assembly. The bearing system must react any axial (thrust) loads over all transient operation that represents off-design conditions for any axial thrust balancing system employed. Any residual thrust loads during nominal operation must also be supported by the bearing system.
3. Control of rotordynamics. The stiffness of the bearings is a prime restoring force for vibrations of the rotor system (see Section 10.9) and also serves as a damping mechanism to limit amplitudes of vibrations.

The two major classes of bearings that are typically used in rocket turbopumps are *rolling element* and *fluid film* or *hydrostatic* bearings. Rolling element bearings are the most common type of bearing design employed in today's devices. Figure 10.20 shows typical arrangement of rolling bearings using balls or cylindrical rollers as the main element that permits rotor rotation relative to the casing. The rolling elements are supported by *raceways* (or races) that act to contain the spherical or cylindrical rotating elements. Rolling element bearings provide high stiffness for reacting radial loads and controlling vibration and they also have some capacity to react axial thrust loads. Cylindrical bearings have higher radial load capacity due to the fact that a greater surface is in contact with the rotating and stationary elements being supported. For example, ball

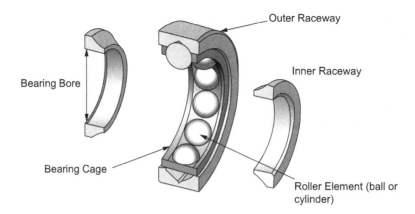

Figure 10.20 Elements of a typical bearing assembly.

Figure 10.21 Conrad-type roller bearing (left) and cylindrical roller bearing (right).

bearings have radial stiffness in the range $0.2–1.5 \times 10^6$/ft while cylindrical bearing stiffness can approach 2 million/ft. Figure 10.22 provides images of cylindrical roller bearings and Conrad-type ball bearings. The angular raceway in the Conrad-type bearing supports both axial and radial loads.

A major consideration for any rolling element bearing is a speed limit that is imposed by a parameter that is related to the product of the bearing inner race bore diameter (D) and the rotor speed (N). The DN limit is fundamental to the type of bearing chosen and is typically specified in units of millimeters times RPM. While cylindrical bearings sport a greater load capacity, they are more limited in speed than ball type designs. A typical DN limit for cylindrical bearings is 1.6 million while ball bearings are capable up to 2.1 million (the bearing community typically omits the mm-RPM units associated with DN).

While the bearing designs in Figure 10.20 have seen reliable usage in numerous rocket turbopumps, the reliability (life limits) and speed (DN) limits of these devices has led rocket

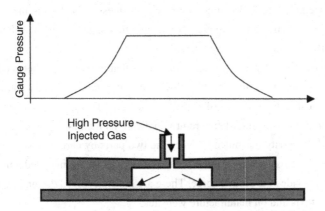

Figure 10.22 Schematic of hydrostatic bearing design. These systems are also called balance pistons in some applications. The locally higher pressure beneath the bearing surface acts to support the loads transmitted from the rotor assembly.

turbopump designers to explore the use of hydrostatic bearings in more recent designs. In the hydrostatic bearing, the shaft loads are supported by the fluids themselves and no physical bearing system is present. Figure 10.22 provides a schematic of a design that permits for fluid flow into the raceway in order to support rotor loads.

Advantages of hydrostatic bearings are their potential for long life and the fact that they have no speed limit. These systems also display desirable stiffness/damping characteristics for rotordynamic considerations. The lifetime advantage is particularly important in the current industry where firms such as SpaceX and Blue Origin are developing/operating reusable rocket first stages. Inspection and replacement of bearings is a major consideration and motivation for developing hydrostatic solutions. However, the main downside of implementing hydrostatic lies in the fact that the rotor assembly rotation must begin or end with almost no fluid present, i.e. one can't use the fluid as a bearing if the fluid hasn't arrived or has just been depleted. For these systems, the rotor must be designed to physically rub against the casing during startup and shutdown and rub-tolerant materials must be employed as a result.

10.8.3 Turbopump Seals

When a novice stares at a complex cross-section such as that in Figure 10.1 it may not be easy to understand the environments that a seal may encounter. If you think about the turbopump in an integral sense, we have high-pressure liquids from pump discharges and high-pressure gases at turbine rotor inlets that are all contained within an integral outer shell we call the *housing*. In between these two high-pressure sources we place a rotating element so there is always a path for the high-pressure fluids to meet. Given the fluids that we are working with, a meeting of this type will generally be disastrous and lead to unintended combustion within the device and eventual "unplanned disassembly," as folks often say in our business. Add to this the complications of working with highly reactive and potentially cryogenic fluids, and the novice should begin to get a picture of the life of a turbopump seal designer.

There are two main classes of seals: *static seals* that mitigate flow between two stationary components and *dynamic seals* that mitigate flow between a stationary and a rotating component. O-rings are the most common type of static seal device. The devices are typically made of elastomeric materials, but for high-temperature applications metallic options exist. The O-ring's function is to seal the cavity machined between the two parts by careful design of the size of the cavity (relative to the cross-sectional diameter of the O-ring) and in intended amount of squeeze/compression that provides the sealing force. There are highly evolved design practices for O-ring seals in numerous texts and on manufacture websites.

The term dynamic seal is actually a misnomer, because all dynamic "seals" leak. In the best of cases, the leakage rates might be so low that they are not measureable, but frequently measureable leakage rates must be managed as a part of the design process. However, turbopump designers have exploited these leakage flows to significantly improve the damping and stiffness characteristics of the rotor. Figure 10.23 provides schematics of some of the more common seal types used in rocket turbopumps. Face contact seals are created by intimate rubbing of the sealing surface. While this type of seal provides minimal leakage, heat generated by the rubbing of the two materials limits the application. Both operating pressure and local rotation speed contribute to this frictional heating and the seals are limited to pressure × velocity (pV) values in the range 10,000–50,000 psi ft/s as a result. There are also lifetime limitations for these seals as material is mechanically removed due to the frictional action during operation.

Labyrinth-type seals are commonly used for longer life/higher reliability applications and where pV values exceed the capabilities of face seals. The downside of using this type of seal system is the relatively higher leakage rates. Labyrinth seals create a tortuous flow path for leaking fluid by employing a number of teeth that meet the sealing surface and rely on viscous forces within the flow around these teeth to minimize the leakage flow. Seating surfaces do rub against walls, but may erode back under operation such that leakage rates increase during the break-in period. Labyrinth seals are commonly used to seal impeller surfaces and special materials (Kel-F) are employed for oxygen compatibility when LOX is the working fluid.

Since dynamic seals are inevitably providing some leakage flow, interpropellant seals (IPS) are typically purged with high-pressure inert gas (helium or nitrogen) in order to ensure that the two fluids never meet inside the device. Figure 10.24 shows the basic arrangement of the IPS. This seal is incredibly important since leakage could mean mixing of the two propellants and disastrous consequences.

These IPS would be required at the interpropellant seal between the two fluids is always a critical failure mechanism considered in system reliability. Helium is typically used as purge gas as its low boiling point assures it will remain a gas even when cryogenic propellants are used. Figure 10.25 shows purge gas flow in a turbopump designed by Purdue students for Dr. Pourpoint's

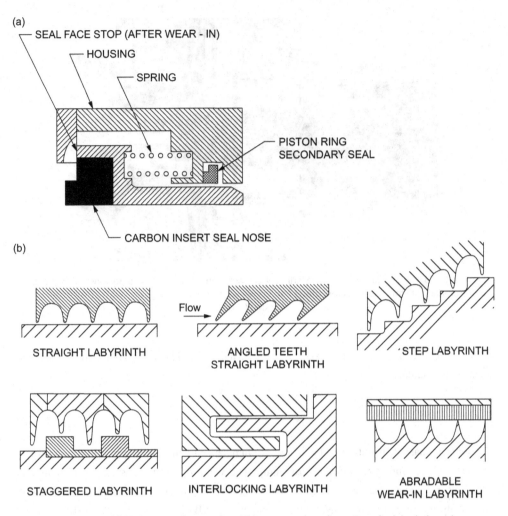

Figure 10.23 (a) Turbopump face seal and (b) a range of configurations for labyrinth seals
(Source: NASA SP-8121).

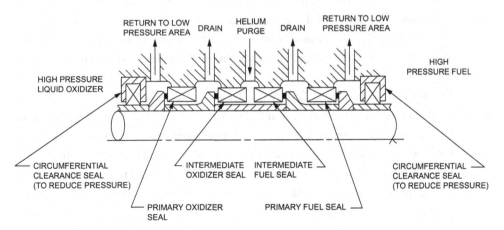

Figure 10.24 General arrangement for an interpropellant seal (IPS) with helium purge
(Source: NASA SP-8121).

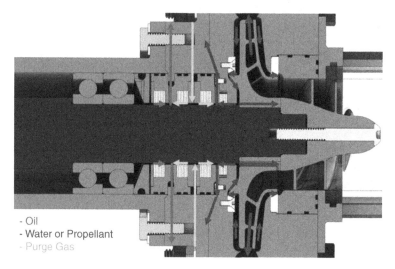

- Oil
- Water or Propellant
- Purge Gas

Figure 10.25 Purging of an interpropellant seal on demonstration turbopump designed and built by Purdue students. In this case, oil lubricated bearings were utilized since water and high-pressure air are the working fluids.

design/build/test class at Purdue. Purge gas is routed into the seal gland through a small hole at one point in the annulus.

10.9 ROTORDYNAMICS

Asymmetric mass distribution or loads that are unbalanced azimuthally provide driving forces that can induce vibrations into the turbopump. The vibration problem is treated in numerous texts and is a standard exercise in most undergraduate engineering curricula, so the details of such analyses will not be repeated here (for example, see Balje, 1981; Huzel and Huang, 1992; Japikse et al., 1997). In this section, we simply provide perspectives that are unique or important for the rocket turbopump application. While vehicle loads can also enter the analysis, experience has shown that these lower-frequency forcing functions don't normally impact the rotordynamics. Aeroelasticity effects relative to forced vibration of individual blades could also be a consideration but risks here are greatly reduced when shrouded designs are employed.

The high rotation rates and relatively low mass of rocket turbopumps leads to a situation where the natural damping of the device is insufficient to critically damp the system so at resonance conditions large vibrations can result. These vibrations could limit seal life or even lead to mechanical failure. As a general design practice, the designer wants to ensure that critical speeds (where vibration amplitudes are largest) are far removed from the operating range of the device. It is desirable to design the rotor such that the operational speeds all lie below the first critical speed of

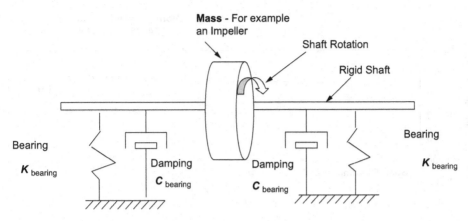

Figure 10.26 Turbopump rotordynamics idealization of rotor/bearing assembly as mass/spring damper system.

the system. Given the uncertainties in identifying critical speeds analytically, a margin (typically 5–20% depending on application) is included to account for uncertainties.

Figure 10.26 shows an idealized turbopump configuration with the rotor of mass m supported by two bearings with stiffness k and damping coefficient, c. The deflection of the rotor can be mapped out for various forcing frequencies in order to assess the behavior. Recall the classic mass/spring/damper differential equation:

$$F \sin(\omega t) = m\ddot{x} + c\dot{x} + kx \qquad (10.40)$$

The textbook result for this single mass system gives the natural (in this case critical) frequency of oscillation as

$$\omega_{\text{crit}} = \sqrt{k/m} \qquad (10.41)$$

While this result is applicable for a single mass/spring/damper system, the conclusion that increasing stiffness is the prime mechanism for moving the critical frequency above the operating frequency (i.e. shaft speed N) range. In the complex situation of a rotor with numerous blade rows and bearing support locations, the shaft can be broken into a number of separate masses (m_i, for $i = 1$ to n) with individual stiffness and damping values at each location such that shaft deformations (x_i) can be determined for a number of different modes of vibration. Mass elements that are overhung in axial extent outside of the bearing locations can be important considerations that point to the importance of the bearing location in this problem.

Existing commercial codes exist for this type of analysis. Figure 10.27 presents results for a turbopump designed by Purdue students using the XLRotor software. Critical speeds for various bending modes of the shaft are plotted as a function of rotor speed demonstrating that all critical modes lie above the operating line (with sufficient margin) all the way to redline

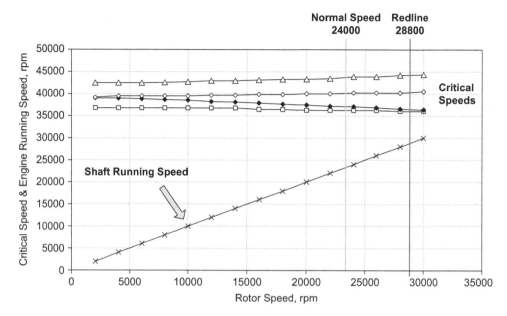

Figure 10.27 Typical Campbell diagram result of rotordynamic analysis showing critical speeds derived from various vibration modes and rotor speeds.

operating speed. This result is typically referred to as a Campbell diagram for those working in the industry.

10.10 A NOTE ON ADDITIVE MANUFACTURING

In Chapter 8 (Section 8.10) we briefly discussed current trends toward "printed" injectors and/or chamber assemblies using laser sintering/melting manufacturing techniques. The turbomachinery world is also changing due to the evolving capabilities, but the applicability of "AM" is a complex discussion. In general, AM does not offer as many advantages for turbopump rotor components as it does for other areas of the engine, the manufacturing precision and strength requirements of the rotor may be one of the last areas where the technique is adopted in industry. The technique is attractive for turbopump manifolds and volutes because it avoids the need for long-lead castings. In general, there is a strong desire to eliminate castings to the greatest extent possible due to the long development and delivery times for these parts. As the diameter capability of AM devices increases, more and more parts can exploit this capability to reduce fabrication times.

Figure 10.28 highlights two components recently printed from Inco 718 for a Purdue student design project. Note the surface finish that is indicative of the roughness resulting from the AM process, i.e. at the surface, there is some memory of the particle sizes as the surface particles are only partially melted by the beam. Unfortunately, tolerance and surface roughness issues still make additive manufacturing problematic for many rotor components although there are

Figure 10.28 Additively manufactured inducer (a) and turbine rotor (b) for turbopump developed by Purdue students. Parts were printed from Inco 718 and the diameter of these parts is 4.6 and 5.7 inches, respectively. Note the surface finish that is characteristic of the additive manufacturing process.

a number of firms printing entire turbopumps for smaller engines in development. There is still healthy debate in the industry on the true impediment that the surface roughness provides. After all, rocket turbopumps are high flow, high Reynolds number devices and viscous losses due to skin friction tend to be minor, so the additional loss due to a rougher surface is not as big a penalty as one might first expect. As with any manufacturing method you will just work around your limitations. For example, one can design in levels of material non-uniformity as well as have extra material for balancing. At the end of the day, it trades out with the weight savings you get with greater design creativity – for example, with the housings, if you're printing those as well. Additional viscous loss in the pump can be traded against additional gas generator/preburner performance to achieve the same flow and head rise characteristics.

Most rotor components don't require significant design compromises due to any existing manufacturing limitations. That being said, one exception might be the impeller where the ease of building in a shroud with additive manufacturing would offer performance and structural benefits. At low flow rates and high pressures, shroudless designs have serious performance shortfalls due to tip losses, so it is really best to use a shrouded impeller design. Unfortunately, at this size, conventional machining of a shrouded design is neigh impossible, so that may be one reason (among many) why smaller high-pressure engine designs have rarely been developed. At higher flow rates the manufacturing penalties are still present but the performance impact of a shroudless design is less severe, so many designs in this range make use of shroudless impellers. At the very highest flow rates the impellers become large enough that even a shrouded design can be built using conventional five-axis machining, so that is where we again start to see shrouded designs employed. At least that may be one reason the shroud/shroudless trade studies have gone the direction they have on existing engines.

As a final note, it is worth mentioning that parts such as those shown in Figure 10.28 are not simply used "out of the box." Particularly where strength is critical, hot isostatic pressing (HIP'ing as it is called in the industry) improves structural characteristics. A heat treatment cycle is then employed to remove residual stresses from the HIP'ed parts and then post-process machining to meet areas with tight tolerances. During balancing of the rotor, additional material may need to be removed.

FURTHER READING

AK Steel (2014) *15–5 PH Stainless Steel* [Product Data Sheet]. Retrieved from: www.aksteel.com /pdf/markets_products/stainless/precipitation/15–5_ph_data_sheet.pdf.

ASM International (1990) *Properties and Selection: Irons, Steels, and High Performance Alloys, Vol. 1*, 10th edn. ASM International.

Balje, O. E. (1981) *Turbomachines: A Guide to Design, Selection and Theory*. John Wiley and Sons.

CF Turbo (2014) *CF Turbo: Software Package*. CF Turbo Software and Engineering, Dresden, Germany.

Humble, R. W., Henry, G. N., and Larson, W. J. (1995) *Space Propulsion Analysis and Design*. McGraw Hill.

Huzel, D. and Huang, D. (1992) *Design of Liquid Propellant Rocket Engines*. AIAA.

Japikse, D., Marscher, W. D., and Furst, R. B. (1997) *Centrifugal Pump Design and Performance*. NREC, Chapter 6.

Stoffel Polygon Systems (2014) *Polygon Interface Design Guidelines* [White paper]. Stoffel Polygon Systems.

Sutton, G. P. (2006) *History of Liquid Propellant Rocket Engines*, AIAA. Chapters 3, 4.

XLRotor (2014) *XLRotor, Software Package*. Rotating Machinery Analysis, Inc., North Carolina.

Young, W. C. and Budynas, R. G. (2002), *Roark's Formulas for Stress and Strain*, 7th edn. McGraw-Hill, Chapter 16.

CHAPTER 11   HYBRID ROCKET ENGINES

This chapter provides an introduction to hybrid rocket propulsion. Section 11.1 provides an introduction to the technology with a bit of a historical perspective. Section 11.2 provides a brief overview of unique aspects of HRE combustion with a more detailed evaluation in Section 11.5. Sections 11.3 and 11.4 discuss ballistic analysis using lumped parameters and ballistic element methods, respectively. Some unique aspects of HRE propellants are covered in Section 11.6 and in Section 11.7 we discuss HRE design processes and alternatives.

11.1 INTRODUCTION: GENERAL ARRANGEMENT AND HISTORY

The term "hybrid" is used for this type of chemical rocket propulsion system because it includes aspects of both solid rocket motor and liquid rocket engine devices. The community has referred to these systems as both hybrid rocket engines (HRE) and hybrid rocket motors (HRM) over the years, with the former definition seeming to be more commonly used today. Presumably, we would use the term "engine" to apply when the device contains turbomachinery and the word "motor" to apply when turbomachinery is absent. The HRE is comprised of one liquid propellant (generally oxidizer) and one solid propellant (generally fuel). Figure 11.1 shows the general arrangement and major features of the device. Liquid oxidizer is injected into the forward section of the solid fuel section and combustion occurs near the periphery of fuel "ports" formed in the fuel section.

As in solid rocket systems, the solid fuel absorbs heat from combustion and vaporizes/ pyrolyzes to provide gaseous fuel to feed the combustion reactions. The liquid oxidizer is also vaporized by heat from the combustion process. Generally, a separate ignition device is required to initiate the process, although hypergolic propellant combinations have been employed in some systems. Section 11.6 provides a more complete description of propellant combinations typically used in these devices. Oftentimes, an aft mixing region is included to mix oxidizer that might be issuing through the middle of the fuel port with the relatively fuel-rich mixture at the periphery of the port.

Employing one liquid and one solid propellant has several advantages. If the liquid oxidizer tank is empty (for transport and handling), the system is non-hazardous as the fuel typically does not contain any oxidizer and is therefore need not be treated as a hazardous material. For example, one of the most popular fuels, HTPB, is basically the same rubber that is used in car tires, so if the system with a fuel like this is employed it is completely safe (relative to inadvertent combustion)

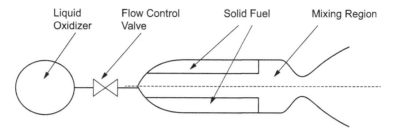

Figure 11.1 Major elements of hybrid rocket engines (HREs).

until the oxidizer is loaded in its tank. HRE proponents will point out that this safety feature also translates to reduced operations costs as special handling procedures are not required when the oxidizer tank is empty (and presumably the ignition system is not installed).

In contrast to solid rocket propellant grains, cracks, voids, and debonds within the fuel grain are not nearly as catastrophic as combustion in an HRE is inherently oxidizer limited. The designer is afforded substantial freedom in arriving at the fuel configuration; Section 11.7 highlights some unique alternatives that have been suggested in the past. If one utilizes a control valve, throttling capability can be realized by varying oxidizer flow. Because only one fluid is employed, HREs are inherently simpler than LREs and the entire logistics train of one fluid is eliminated. However, if one considers a pump-fed HRE, this advantage may evaporate as a separate combustion source (gas generator) would be required to power the pump. Section 11.7 discusses potential cycles and operational configurations for large HREs.

The main disadvantages of the HRE stem from the combustion process itself. Fuel regression rates tend to be low (roughly 1/10th that of a solid propellant), so a large surface area with numerous fuel ports is required to generate high thrust. The large number of ports leads to volumetric and inert weight inefficiencies when compared to SRM or LRE cousins. Because the fuel flow is not controlled, HREs are subject to mixture ratio shifts during operation and therefore realize performance advantages losses when the oxidizer/fuel (O/F) ratio deviates from the optimum value. This performance loss is not present in SRM or LRE systems where the propellant formulation, or individual fuel and oxidizer flows, can be tightly controlled. Overall the I_{sp} of HRE systems lies somewhere between that of the lower-performing SRMs and the higher-performing LREs.

Fuel utilization is another challenge as the system needs to be very carefully designed to ensure that all the fuel is combusted prior to oxidizer flow termination. While many universities and hobbyists have explored small, single-port devices, larger multi-port HREs have been more difficult to develop due to combustion instabilities that stem from the large energy required to vaporize the high incoming oxidizer flow.

While the first HRE systems were developed in the 1950s, there have been surprisingly few operational HREs. In the 1960s, the US-developed Sandpiper, High Altitude Supersonic Target

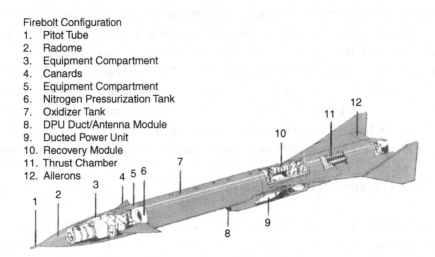

Firebolt Configuration
1. Pitot Tube
2. Radome
3. Equipment Compartment
4. Canards
5. Equipment Compartment
6. Nitrogen Pressurization Tank
7. Oxidizer Tank
8. DPU Duct/Antenna Module
9. Ducted Power Unit
10. Recovery Module
11. Thrust Chamber
12. Ailerons

Figure 11.2 Firebolt High Altitude Supersonic Target employing hybrid propulsion system using IRFNA and PMMA propellants. The propulsion system was developed by United Technologies Chemical Systems Division and had 10:1 throttling with 400 s burn duration.

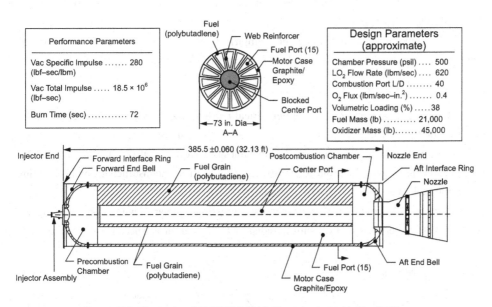

Figure 11.3 Elements of AMROC 250 klbf booster tested in the 1990s.

(HAST), and Thunderbolt systems were used as high-speed targets in the 1960s. An image of HAST is shown in Figure 11.2. This HRE system employed inhibited red-fuming nitric acid (IRFNA) as oxidizer and poly methyl-methacrylate/polybutadiene (PMM/PB) fuel. AMROC (American Rocket Company) developed a large multi-port HRE in the 1990s using liquid oxygen (LOX) and PB propellants. This motor, shown in Figure 11.3, demonstrated 250,000 lbf thrust level for potential application for launch vehicle boost propulsion. Today, the prominent example is the system developed by Virgin Galactic to power the Spaceship I/II vehicles. A nice film of the motor operating in flight can be found on the SpaceFlight Insider website (see www.youtube.com/watch? v=yVETvuGsv-w). Perhaps this is the ideal application of HRE technology as the safety advantages are greatest in this human-rated system and the overall mission requirements are not terribly onerous relative to packaging constraints or Δv. With this one current exception, one might conclude that the HRE has most frequently served as a "bridesmaid" role in that it can't match the performance of an LRE propellant combination nor the packaging efficiency of an SRM system.

11.2 HRE COMBUSTION FUNDAMENTALS

Figure 11.4 provides a close-up view near the fuel surface within a typical HRE fuel port. Pure oxidizer enters the port and begins to mix and combust near the fuel surface. Below the flame zone, a mixture of fuel and product gases exist, while above the flame zone a mixture of oxidizer and product gases is present. The flame height is set by a competition between energy from the flame vaporizing the fuel, the turbulent boundary layer development along the fuel port, and the blowing of the fuel gases departing the surface. Section 11.5 provides a more complete description of the theory of hybrid fuel regression.

The community has determined both empirically and analytically that, for a given propellant combination, the fuel regression rate, r, is primarily influenced by the port massflux, G:

$$r = aG^n \tag{11.1}$$

where $G = \dot{m}/A_\mathrm{p} = \rho v$ and a and n are empirical constants. As designers, we are motivated to operate with the very highest G values possible since that will minimize the port area (and hence fuel section volume) for a given desired propellant flow. However, if G is too large, we risk the potential of "blowing the flame out" due to excessive velocities (and insufficient combustion times) within the fuel port.

Figure 11.5 provides a view of how the regression rate varies with G on a log–log scale. At low G values (very low port velocities), radiation to the fuel surface becomes the principle mode

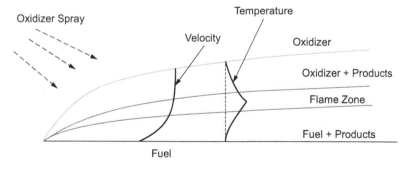

Figure 11.4 View of the flowfield and combustion near the burning fuel surface.

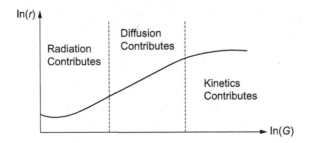

Figure 11.5 Typical behavior of fuel regression rate (r) with port massflux (G), showing radiation-dominated, convection/diffusion dominated, and kinetics dominated regimes.

of heat transfer and the regression rate becomes less dependent on G under these conditions. There is a fairly wide range of G values where nearly linear behavior is attained (note that this implies r is proportional to G^n on a log–log plot). This is the region where convection is the chief energy transport mechanism to the fuel, and the region where practical devices operate over the majority of the burn duration. At the very high G values coincident with the highest port velocities and shortest residence times, combustion chemistry becomes a limiting factor. Operation in this regime is risky as the potential to extinguish the flame exists.

11.3   HRE LUMPED PARAMETER BALLISTICS

In Chapter 6, we discussed the process for computing the chamber pressure and thrust histories for solid rocket motors under the lumped parameter assumption. In this section we address the analogue process for an HRE. The lumped parameter assumption presumes that one can character-ize the entire combustion chamber with a single average pressure that varies only with time. This assumption necessarily neglects and pressure drop down the length of a fuel port and within any aft-mixing section that may be present upstream of the nozzle. Additionally, since we are lumping all variations along the fuel port into a total flow emanating from that port, we must presume that the regression rate is constant along the length of said port. While this assumption may seem quite restrictive given the discussion in the prior section associated with Figure 11.4, but experience shows that under many circumstances axial variation in fuel regression is small. We will discuss this at more length in Section 11.4.

The fuel section geometry (number of ports, port shape, port length, motor internal diameter) must be known to conduct a ballistics calculation. Functionally, this geometry is distilled down to two pieces of information, the port cross-sectional area, A_p, and the port perimeter, Per. As in our lumped parameter discussion on SRM ballistics in Chapter 7, we can define a web distance, w, that can be used to describe how A_p and Per change with time, i.e. $A_p = A_p(w)$, Per $=$ Per(w). Since the port length typically changes little over time (i.e. little combustion on forward or aft-faces of the fuel grain), the burning surface area at any instant in time can be written:

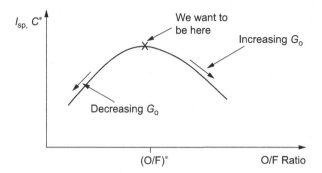

Figure 11.6 Variation of I_{sp} and c^* with the instantaneous oxidizer/fuel mass flow ratio (O/F).

$$A_b = N \times \text{Per} \times L \tag{11.2}$$

where N is the number of ports and L is the overall length of the fuel grain.

Additionally, the most important wrinkle that arises for a lumped parameter analysis of an HRE is the tendency for mixture ratio to shift during motor operation. Since the port geometry is changing with time, even if the oxidizer flow is constant, there will generally be some changes in fuel flow over the duration of the firing – for this reason, the characteristic velocity c^* can no longer be regarded as constant. Figure 11.6 emphasizes this fact. As with any chemical propulsion device, there is an optimum mixture ratio, or oxidizer/fuel mass flow ratio, O/F, which optimizes performance of either I_{sp} or c^*. As the O/F shifts from this desired value (let's call it (O/F)*), we have a departure from the maximum performance. Keeping this in mind, we will always need to know c^* (O/F) to do any HRE ballistics calculation.

Finally, we need to have some information about the fuel regression rate itself in order to compute instantaneous fuel flow and overall performance. While Eq. 11.1 provides an empirical relation for fuel flow, it is recursive in nature because the fuel flow itself must be computed in order to get the local G value in the port. Also, since fuel is continually added along the length of the port, $G = G(x)$, which is inconsistent with our lumped parameter assumption. For these reasons, many HRE experts tend to fit the regression rate using just the oxidizer massflux:

$$r = aG_{ox}^n \tag{11.3}$$

where a and n are empirical constants as before. This form greatly simplifies the analysis since $G_{ox} = \dot{m}_{ox}/A_p$, where A_p is the instantaneous port area that can be expressed as a function of the current web distance. Using this form of the regression rate, the instantaneous fuel flow rate can be expressed:

$$\dot{m}_f = r\rho_f A_b = a(\dot{m}_{ox}/A_p(w))^n \rho_f N \text{Per}(w)L \tag{11.4}$$

Armed with these relationships, a simple lumped parameter HRE ballistics algorithm is shown in Table 11.1. The user can choose to step in increments of web distance or time; the regression rate connects these two variables, since by definition $r = dw/dt$. As with any numerical method, one must choose a time-step (or web distance step) such that results become insensitive to this parameter. Formally, the algorithm show employs a forward Euler time integration scheme that is only accurate to first order, i.e. to Δt^1 or Δw^1. Experience shows that this accuracy is adequate if one employs at least 100 time-steps, but since this type of code is very fast, one can get arbitrary accuracy with little computational expense. As shown in the algorithm, the instantaneous chamber pressure is determined from the mass flow balance at the throat:

$$p_c = (\dot{m}_{\text{ox}} + \dot{m}_f)c * /(gA_t) \tag{11.5}$$

which basically stems from the definition of $c*$ itself. Once again we remind the reader that $c*$ is not constant here and must be computed based on the instantaneous fuel and oxidizer flows. While $c*$ is a function of pressure, it tends to be a weak function so if one generates the $c*$ vs. O/F function at the desired average pressure accuracy should be sufficient given the level of the of the other assumptions in the model. Given the instantaneous chamber pressure (and a thrust coefficient for the nozzle being used), thrust and I_{sp} can then be computed using the methodology outlined in Chapter 4.

11.3.1 Example: Single-Port Hybrid Motor/Engine

If the fuel section geometry is simple enough, the methodology outlined in Table 11.1 can be used to develop analytic expressions for the port geometry, fuel flow, and chamber pressure as functions of time. From a practical perspective, this configuration (as shown in Figure 11.7) is frequently used by academe for basic combustion studies or for small flight vehicles. For these reasons, it is an ideal example to investigate.

   If we neglect any fuel regression on the end faces of the cylindrical fuel grain, the port and burn surface areas can be written in terms of the time-dependent port radius $R(t)$:

$$A_p = \pi R^2; A_b = 2\pi RL \tag{E1}$$

If we assume the regression rate is correlated with oxidizer massflux, we may write:

$$r = aG_{\text{ox}}^n = a\left(\frac{\dot{m}_{\text{ox}}}{\pi R^2}\right)^n \tag{E2}$$

so if $\dot{m}_{\text{ox}} = $ constant $r \propto R^{-2n}$. Since R grows with time, the regression rate will decrease as the port opens up with fuel consumption. This is true of all hybrids, not just this example. At any instant in time, fuel flow rate can then be written:

$$\dot{m}_f = r\rho_f A_b = a\rho_f\left(\frac{\dot{m}_{\text{ox}}}{\pi R^2}\right)^n 2\pi RL = 2a\pi^{(1-n)}P_f L\dot{m}_{\text{ox}}{}^n R^{(1-2n)} \tag{E3}$$

Table 11.1   Psuedo-language algorithm for lumped parameter hybrid ballistics assuming steady operation, negligible pressure drop down the fuel grain, and constant regression along length of grain

Algorithm steps	Notes
Given $\dot{m}_{ox} = \dot{m}_{ox}(t)$, Per = Per$(w)$, $A_p = A_p(w)$	Inputs
Given $c^* = c^*(O/F)$, Regression parameters a, n	More inputs
Given N, A_t, ρ_f, L, Δt or Δw, C_f	Input thrust coefficient (could be time varying if vehicle is flying in atmosphere)
Then	
$t = 0$; w = 0; $A_p = A_{p0}$; Per $=$ Per$_0$	Initial conditions
$A_{b0} =$ Per$_0$ LN, $G_{ox} = m_{ox}(t)/A_p$	Initial conditions
$r = aG_{ox}^n$	Initial regression rate
$\dot{m}_f = r\rho_f A_b$	Initial fuel flow
$O/F = \dot{m}_{ox}/\dot{m}_f$	Initial O/F
$c^* = c^*(O/F)$	Initial c^*
$p_c = (\dot{m}_{ox} + \dot{m}_f)c * /(gA_t)$	Initial p_c, alternately, one could set the desired p_c and use this equation to compute the required A_t
For $i = 1 \ldots 100$	For loop assuming 100 steps – can use arbitrary size here to give desired accuracy
$w_i = w_{i-l} + r_{i-l}\Delta t$	Updating web distance assuming time-step is prescribed – alternately can update time assuming web increment is prescribed
$A_{pi} = A_p(w)$	
$G_{oxi} = \dot{m}_{ox}(t_i)/A_{pi}$	
$r_i = aG_{oxi}^n$	
Per$_i$ = Per(w_i)	
$A_{bi} = N$ Per$_i L$	
$\dot{m}_{fi} = r_i\rho_f A_{bi}$	
$(O/F)_i = \dot{m}_{oxi}/\dot{m}_{fi}$	
$c^*_i = c^*((O/F)_i)$	
$p_{ci} = (\dot{m}_{oxi} + \dot{m}_{fi})c*_i/(gA_t)$	
$F_i = C_f p_{ci} A_t$	
$I_{spi} = C_f c^*_i/g$	
End For	End for loop and plot arrays of p_c, F, I_{sp}, etc.

so that $\dot{m}_f \propto R^{1-2n}$. To find $\dot{m}_f$ at any instant, we need $R = R(t)$. Using the definition of r, we note:

$$r = \frac{dR}{dt} = a\left(\frac{\dot{m}_{ox}}{\pi}\right)^n R^{-2n} \tag{E4}$$

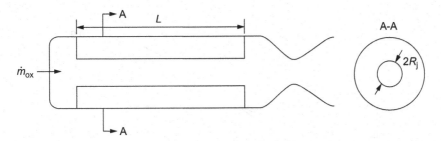

Figure 11.7 Schematic of single-port hybrid motor with port radius $R = R(t)$.

or if we separate variables, we can write:

$$R^{2n}dR = a\left(\frac{\dot{m}_{ox}}{\pi}\right)^n dt \tag{E5}$$

Equation E5 can be integrated analytically if at $t = 0$ we assume $R(0) = R_i$, to give:

$$R(t) = \left[a(2n+1)\left(\frac{\dot{m}_{ox}}{\pi}\right)^n t + R_i^{2n+1}\right]^{\frac{1}{2n+1}} \tag{E6}$$

Substituting this result into Eq. E3 gives the fuel flow history:

$$\dot{m}_f(t) = 2a\pi^{1-n}\rho_f L \dot{m}_{ox}^n \left[a(2n+1)\left(\frac{\dot{m}_{ox}}{\pi}\right)^n t + R_i^{2n+1}\right]^{\frac{1-2n}{1+2n}} \tag{E7}$$

The resulting O/F history can be obtained by dividing this result into the oxidizer flow that is assumed to be constant in this example. Here, the key parameter is the regression rate exponent, n, as it determines how fuel flow changes with time and how the O/F shifts with time. We can see from Eq. E7 that $\dot{m}_f$ is constant only when $n = 0.5$ (we will show later that $n = 0.8$ from theory). For $n > 1/2$, $\dot{m}_f$ will **decrease** with time. Therefore, the O/F will **increase** with time. Figure 11.8 depicts this situation for a given port geometry, oxidizer flow, and fuel characteristics for differing values of n. As n grows above 0.5, the O/F shift during motor operation will tend to increase.

Since we want the average O/F to be optimal, we start out on the fuel-rich side and end up on the fuel-lean side. Figure 11.9 depicts this situation in terms of the I_{sp} history of the HRE system. While we have introduced this notion in terms of the simple single-port hybrid, the trend in Figure 11.10 is a general one characteristic of larger multi-port configurations as well.

11.3.2 Fuel Section Design Effects on HRE Performance

Design of the fuel section also affects the size of mixture ratio shifts. Vonderwell *et al.* (1995) investigated this effect for a multi-port configuration for both LOX/HTPB and hydrogen peroxide (HP)/HTPB propellant combinations. The wagon-wheel fuel port design and overall performance

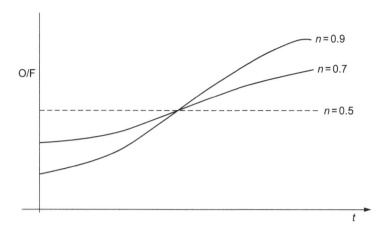

Figure 11.8 O/F histories for different regression rate exponent (n) values for a single-port hybrid motor with port radius $R = R(t)$. Multiport HREs display similar behavior.

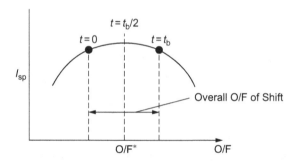

Figure 11.9 I_{sp} variation with O/F assuming constant oxidizer flow. We need to ignite the HRE at fuel rich condition (O/F < O/F*) in order to reach peak performance near the middle of the firing. Generally HREs lose 1–2% in average I_{sp} (as compared to the peak value) due to this O/F shift.

of the two propellant combinations are summarized in Figure 11.10. Because LOX/HTPB tends to optimize at lower O/F, a larger fuel section is required as compared to HP/HTPB.

Figure 11.11 shows the overall I_{sp} shift for both propellant combinations for wagon-wheel fuel grains with differing number of ports. As we increase the number of fuel ports, their change in geometry during the firing becomes more important and we get larger shifts in O/F and I_{sp} as a result. Complicating this issue is the overall behavior of I_{sp} vs. O/F as shown in Figure 11.11.

Performance changes are also affected by the initial oxidizer flux level – as we increase this parameter we increase fuel regression rate and the subsequent changes in port geometry attributed to such. Figure 11.12 shows the behavior for both propellant combinations for two different initial flux levels of 0.4 and 0.8 lbm/in²-s. The higher flux level allows us to make a smaller fuel section (since port area is reduced for the required mass flow), but larger I_{sp} excursions will result.

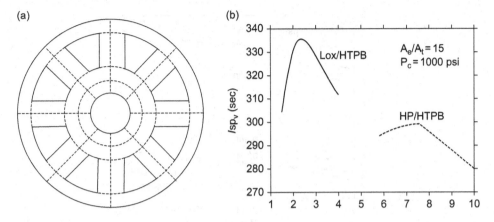

Figure 11.10 (a) Wagon-wheel fuel port design and (b) I_{sp} vs. O/F for LOX/HTPB and HP/HTPB propellant combinations.

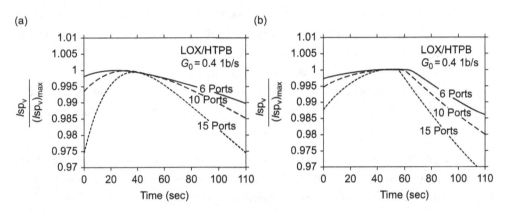

Figure 11.11 I_{sp} shifts due to changes in number of fuel ports. Larger number of smaller ports creates larger changes in fuel section geometry with time and larger shifts in performance.

11.3.3 Throttling Effects on HRE Performance

At this point you may have thought about ways to counteract this loss in performance by varying the oxidizer flow in time as well. In fact, if we had a throttleable HRE and we had intimate knowledge of the fuel flow history, we could, in principle, vary the oxidizer flow such that O/F = O/F* at all times and we get peak performance during the entire duration of the firing. However, there are three practical difficulties with this approach:

1. Extra infrastructure is required to permit variable oxidizer flow. A control valve and associated control system will be necessary and these items add both weight and complexity to the device.

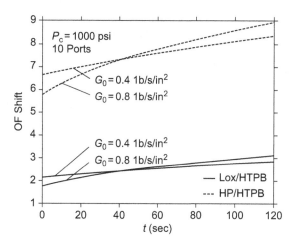

Figure 11.12 Shifts in O/F for LOX/HTPB and HP/HTPB propellant combinations for differing initial oxidizer massflux. Larger massflux leads to higher regression rates and larger changes in port geometry over time.

2. We really do not know the fuel flow at each instant in time. The modeling of the fuel regression is somewhat approximate and empirical coefficients are typically selected to give the integrated total fuel consumed. At a given instant in time, the fuel flow could be slightly greater or less than the value we predict.

3. Most missions have a thrust requirement that is "non-negotiable." Changing the oxidizer flow directly changes the thrust and therefore we may not have latitude in a given mission to make this type of excursion.

Despite these challenges, throttleable configurations certainly have been considered and their merits may overcome the difficulties highlighted above. For the booster study highlighted in Figures 11.10–11.12, the relative mass flow/throttle setting history is shown in Figure 11.13. Boosters on launch vehicles typically need a thrust history like this since the vehicle is getting lighter toward the end of the burn, less thrust is needed to maintain the desired acceleration.

Figure 11.14 shows the dramatic effect this throttling history has on the I_{sp} shifts for both LOX/HTPB and HP/HTPB propellant combinations. The reduction in oxidizer flow late in the firing helps to balance with similar reductions in fuel flow (per Eq. E6) resulting in lower overall O/F and I_{sp} shifts. In some respects, the launch vehicle booster application is ideal for HREs for this reason.

11.4 HRE Ballistic Element Analysis

In Chapter 7, we discussed the ballistic element treatment for SRMs – this section highlights an equivalent approach for HREs. By using ballistic elements, we can discover variations along the fuel port and allow for more complex regression rate models. Probably the most important outcome from an analysis like this (relative to the lumped parameter approach) is the pressure drop along the

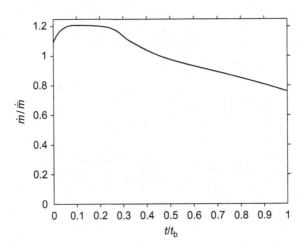

Figure 11.13 Throttle history for an HRE booster – less thrust is needed toward the end of the burn as the vehicle mass is dropping with propellant consumption.

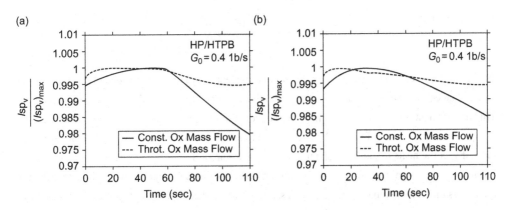

Figure 11.14 I_{sp} shifts due to throttling for an HRE booster mission. The commanded reduction in oxidizer flow late in the mission drastically reduces I_{sp} shifts since the fuel flow is also dropping with time.

length of the port. Both heat and mass addition along the port lead to losses in stagnation pressure and as this quantity is a prime performance measure, it becomes important to estimate these losses for real systems. The approach outlined here is documented in AIAA Progress in Astronautics and Aeronautics, Vol. 218 (Heister and Wernimont, 2007), so readers interested in details should consult this source.

Figure 11.15 outlines the nomenclature for a single-port configuration – extension to multiple ports is straightforward although readers should recognize that the method is limited to considering equivalent variations in properties along all ports. For this reason, one only needs to analyze a single port in order to elucidate the time-dependent behavior of interest. If an HRE utilizes multiple ports

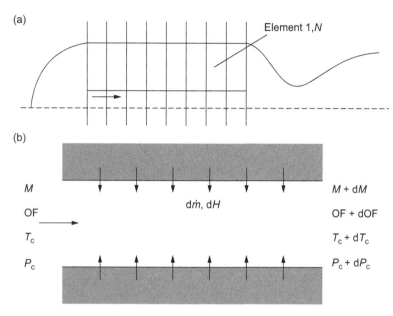

Figure 11.15 Ballistic element analysis. (a) A given fuel port is divided axially into N elements. (b) Property changes across a given element.

that are fed non-equivalent flows or have non-equivalent geometry, a more elaborate treatment is required since all ports will share the same overall static pressure drop under quasi-steady operation. This requirement would dictate that flow be passed from one port to another in some instances.

As emphasized in Figure 11.15, we now have to account for variation of c^*, Mach number, ratio of specific heats (γ), and gas specific heat c_p as a function of O/F. The specific heat, c_p, is also a function of temperature so a strong coupling exists within the overall energy balance. As the method is one dimensional, mixing is necessarily neglected such that average flow properties are computed at each axial location. In this case, we must account for temperature, Mach number (velocity), and pressure variations along the length of the port; conservation of mass, momentum, and energy for this situation provide the three necessary relationships:

$$\frac{dM}{M} = F_1 \left(\frac{dH}{c_p \overline{T}} + \eta \frac{d\dot{m}}{\dot{m}} - \frac{d\mathfrak{M}}{\mathfrak{M}} \right) - \frac{1}{2} \frac{d\gamma}{\gamma} \tag{11.6}$$

$$\frac{dP}{P} = F_2 \left(\frac{d\mathfrak{M}}{\mathfrak{M}} - \frac{dH}{H} - 2\eta \frac{d\dot{m}}{\dot{m}} \right) \tag{11.7}$$

$$\frac{dH}{c_p \overline{T}} = \frac{d(c_p T_c) - d(c_p \overline{T})}{c_p \overline{T}} \tag{11.8}$$

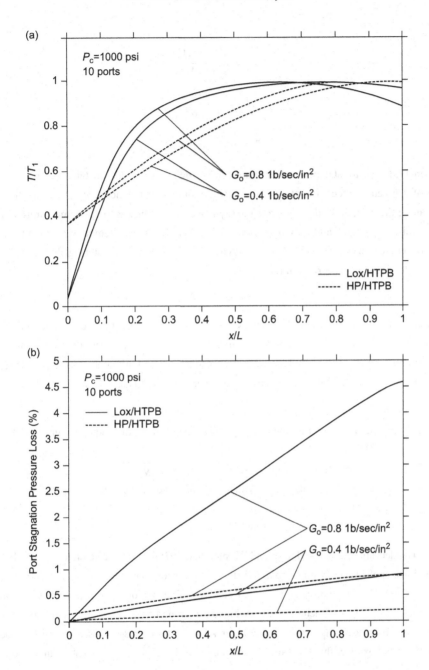

Figure 11.16 Typical results from ballistic element ballistics model for the HRE booster case discussed in Section 10.3. Local port temperature reaches a maximum at stoichiometric conditions and decreases slightly toward the aft-end of the port under fuel-rich conditions. Port stagnation pressure losses are strongly dependent on oxidizer massflux as increases in this parameter increase dynamic pressure and Mach numbers in the port.

where:

$$F_1 = \frac{1}{2}\frac{1 + \gamma M^2}{1 - M^2}$$

$$F_2 = \frac{\gamma M^2}{1 - M^2}$$

$$\eta = 1 + \frac{\gamma - 1}{2}M^2$$

Equation 11.6 is just a statement of continuity ($\dot{m} = \rho v A$), assuming constant port area over the length of a given element. Equation (11.7) is the momentum balance neglecting any frictional forces along the port, and Eq. 11.8 is the energy balance reflecting enthalpy/temperature changes due to combustion energy addition. In this expression, T_c is the "flame" temperature computed from the local combustion of fuel in the element with oxidizer. The mass flow entering element "i" can be written in terms of local regression rate:

$$d\dot{m} = r_i\rho_f A_{bi} \tag{11.9}$$

Equations 11.6–11.9 provide a well-posed system for M, P, T, $\dot{m}$ as a function of length down the port. An iterative process is required at each time level because the pressure drop down the port is unknown and hence the initial starting pressure is also unknown. Ultimately, the head end pressure must be iterated until the flow rates balance. The overall process is as follows:

1. Guess a head-end chamber pressure, p_{c1}.
2. Integrate down the fuel port to get oxidizer/fuel ratio at the end of the grain $(O/F)_N$ and $c^* = c^*(O/F)_N$.
3. Compute mass flow entering nozzle: $\dot{m} = \dot{m}_{ox} + \dot{m}_{fN}$.
4. The mass flow must also agree with the stagnation pressure that is computed at the end of the fuel port, i.e. $\dot{m} = g p_{cN} A_t / c^*$.
5. Iterate on p_{c1} until the mass flows balance.

Interested readers are referred to Heister and Wernimont (2007) for more detail on the approach. Since this iteration is applied at each time-step, clearly the best guess for p_{c1} is the value obtained from the prior time level. Overall, the iteration converges rapidly and typical stagnation pressure losses at the beginning of the firing, when the port Mach numbers are highest, is around 5%. Figure 11.16 summarizes results from the HRE booster example discussed in Figures 11.12–11.14. Pressure losses are strongly dependent on the initial oxidizer flux level since higher fluxes lead to higher Mach numbers in the port.

11.5 HRE Combustion Theory

The theory for HRE combustion has its roots in the work of Marxman and colleagues in the 1960s (see Marxman and Gilbert, 1962). As outlined in Section 11.1, there are three regimes of

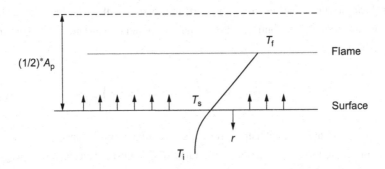

Figure 11.17 Temperature field near the surface of regressing fuel port.

combustion: (1) radiation dominated combustion at low G values; (2) diffusion/convection controlled combustion at moderate G values; and (3) kinetics limited combustion at high G values. As mentioned previously, most practical applications live in the diffusion dominated regime where convective heat transfer to the fuel surface balances with the energy required to vaporize the fuel itself. We will review the theoretical underpinnings of this situation here. *Space Propulsion Analysis and Design* (Humble *et al.*, 1995) listed at the end of the chapter has discussion on the radiation dominated region for those who have interest. Some fuels like paraffin/wax have an additional mechanism for fuel regression via a melt layer hydrodynamic instability. Readers interested in analytic treatment of these physics are referred to references at the end of the chapter.

For the analysis, we employ the following assumptions:

- neglect radiation – all heat transfer from convection;
- assume fast kinetics (reaction time ≪ flow time);
- use average properties (simplifies algebra but retains the salient physics); and
- steady-state conditions.

Under these assumptions, a local view near the fuel surface is shown in Figure 11.17. The bulk fuel temperature is assumed to be T_i and there is some subsurface heating as energy from the flame zone is conducted down into the fuel. As most common fuels are good insulators, this energy transport plays only a minor role in establishing the regression rate; however, it is crucial as it helps to bring the fuel to vaporization/sublimation conditions at the interface. Since the flame is typically anchored close to the surface, we can assume a linear temperature gradient from the surface (at temperature T_s) to the flame zone (at temperature T_f). Fuel gases emanating from the surface have a blowing effect that reduces heat transfer and insulates the surface from the hottest combustion gases.

The oxidizer massflux in the fuel port is as before ($G_{ox} = \dot{m}_{ox}/A_p$) and the fuel massflux is related to the regression rate we seek to compute:

$$G_f = \dot{m}_f/A_w = \rho_f r \tag{11.10}$$

At a given location along the length of the port, the total massflux can be computed from the oxidizer flow and the fuel flow contributed up to that location. For a distance x from the entry to the fuel port, we have:

$$G = \frac{\dot{m}_f + \dot{m}_0}{A_p} = G_0 + \frac{\dot{m}_f}{A_p} = G_0 + 4 \cdot \rho_f \cdot \int_0^x \frac{r \cdot dx}{D_h(x)} \qquad (11.11)$$

The fuel regression rate is established from the overall energy balance at the fuel surface. The energy contributed from convection to the surface $(\dot{q}_s)$ must balance against energy required to vaporize fuel plus any energy conducted into the fuel subsurface:

$$G_f \cdot h_v + G_f \cdot c_s \cdot (T_s - T_i) = \dot{q}_s \qquad (11.12)$$

where c_s is the specific heat of the fuel and h_v is the energy to liquefy (or sublime) the fuel, and heat to the surface temperature T_s. We can think of an "effective" heat of vaporization by lumping in the subsurface energy to simplify the energy balance to:

$$G_f \cdot h_{veff} = \rho_f r = \dot{q}_s \qquad (11.13)$$

where we have used Eq. 11.10 and note that $h_{veff} = h_v + C_s \cdot (T_s - T_i)$.

We can write the convective heat transfer to the surface in terms of the Stanton number (St):

$$\dot{q}_s = St \cdot G \cdot \Delta h_T \qquad (11.14)$$

where Δh_T is the enthalpy difference between gas and solid: $\Delta h_T = c_p^*(T_f - T_s)$. As we discussed in Chapter 6, we can employ the Reynolds analogy between frictional processes and heat transfer processes. This approach provides a connection between the Stanton number and the friction coefficient, C_f:

$$St = \frac{C_f}{2} \qquad (11.15)$$

Combining Eqs. 11.14 and 11.15, we get:

$$r = \frac{G}{2 \cdot \rho_f} \cdot C_f \cdot \frac{\Delta h_T}{h_{veff}} \qquad (11.16)$$

Remember that $r = r(x)$, $C_f = C_f(x)$ and $G = G(x)$. For a flat plate with a turbulent boundary layer and no blowing, the friction coefficient may be expressed as:

$$\frac{C_{f0}}{2} \simeq 0.03 \cdot Re_x^{-0.2} = 0.03 \cdot G^{-0.2} \left(\frac{\mu_g}{x}\right)^{0.2} \qquad (11.17)$$

where C_{fo} is the friction on a non-vaporizing wall (i.e. no blowing). The problem is that we have blowing at the wall, which tends to reduce C_f, and it also makes $\Delta h_T/h_{veff}$ difficult to calculate. Physically, if $\Delta h_T/h_{veff}$ grows, the blowing at the wall becomes more significant and the wall friction and heat transfer should drop. Using test data, *Marxman* developed the empirical relation to account for these physics:

$$\frac{C_{\mathrm{f}}}{C_{\mathrm{f0}}} \cong 1.2 \cdot \left(\frac{\Delta h_{\mathrm{T}}}{h_{\mathrm{veff}}}\right)^{-0.77} \tag{11.18}$$

Combining Eqs. 11.16–11.18 gives:

$$r = \frac{0.036}{\rho_{\mathrm{f}}} \cdot G^{0.8} \cdot \left(\frac{\mu_{\mathrm{g}}}{x}\right)^{0.2} \cdot \left(\frac{\Delta h_{\mathrm{T}}}{h_{\mathrm{veff}}}\right)^{0.23} \tag{11.19}$$

Equation 11.19 is the classic result for HRE combustion in the diffusion-dominated regime where most engines live. The theoretical exponent on G is 0.8; as we have seen in Chapters 6–8, this exponent appears for all processes dominated by convective heat transfer. Note that massflux and x effects are in the opposite direction; i.e. r tends to grow with G as we move down the port, but tends to reduce with increased x. This is a fortuitous since these counteracting effects tend to show that in many applications the regression is quite uniform axially and that the forward end of the fuel port is totally consumed at about the same time the aft portion of the port vanishes. There is a practical limitation of using this result at the entry to the fuel port, where $x \rightarrow 0$ as the result is singular at this location. Physically, the expression is really not valid at the entry to the port as boundary layers are just beginning to develop there and Eq. 11.18 is not valid.

While it is nice to have theoretical results such as Eq. 11.19, experience shows that they tend to fall short when a propellant combination that differs from that used by Marxman is being considered. For a given fuel the parameters ρ_{f} and $\Delta h_{\mathrm{T}}/h_{\mathrm{veff}}$ are fixed and if we consider we can lump μg, Δh_{f}, and h_{veff} into a constant a, we get:

$$r = aG^n/x^{0.2} \tag{11.20}$$

Remember, in general, this is an **implicit** expression since:

$$G = \frac{\dot{m}_{\mathrm{f}} + \dot{m}_0}{A_{\mathrm{p}}} = \frac{\left[\dot{m}_{\mathrm{o}} + \displaystyle\int_0^x r \cdot \rho_{\mathrm{f}} \cdot \mathrm{d}A_{\mathrm{b}}\right]}{A_{\mathrm{p}}} \tag{11.21}$$

While the form in Eqs. 11.20–11.21 provides for mechanism to give axial variation in regression, the result is singular near the port entry as discussed previously. Most users are happy to consider an average regression rate along the fuel port, which might suggest a forms like:

$$r = aG^n \text{ or } r = aG_{\mathrm{ox}}^n \tag{11.22}$$

where the constants a and n are derived from experimental data. While the form using the complete massflux G is more correct, as this quantity varies axially and introduces an implicit form, most folks resort to the simplest version using G_{ox} as we used in examples in Section 11.3.

In Section 11.6, we provide examples of regression rate data for several different propellant combinations.

As you may have inferred from this empirically based discussion, our knowledge of the details of HRE combustion is still rather poor. Unfortunately, there has been little interest from basic research sponsors because of the dearth of HRE systems used by governments and military organizations around the world. Detailed mixing processes, behavior of fuel pyrolysis, two-phase processes associated with the use of liquid oxidizers, and detailed chemistry of the combustion reactions have gone largely unstudied. Given all the complex physics in this application, it is remarkable that in many instances the fuel regression remains fairly constant over the length of the port. While the massflux increases continually along the port length, oxidizer consumption also increases so at the higher flux levels at the end of the port, there is less oxidizer available for combustion. The building of boundary layers along the length of the port also serve to offset the increasing fluxes as we highlighted in the simple analysis in this section. Hopefully, future HRE development programs will help us shed light on these interesting physics.

11.6 HRE Propellants

Chapter 9 provides information on liquid propellants and therefore provides valuable information for HRE systems as well and we refer readers to this chapter for information on liquid properties and handling issues. Chapter 7 provides information on solid propellant binders that are often used in HRE fuels. Figure 11.18 shows the performance of many popular or previously used HRE propellant combinations. Gaseous oxygen is a popular choice for laboratory studies, but liquid oxygen is a more realistic oxidizer for flight applications due to the high density. Nitrous oxide, N_2O, currently used in the SpaceShip 2 motors developed by Virgin Galactic, is another oxidizer currently receiving attention. Inhibited red fuming nitric acid (IRFNA) and its cousin white-fuming nitric acid (WFNA) have received attention for some military systems due to their storability and overall energy density. However, the toxicity and corrosivity of these oxidizers does create challenges. Hydrogen peroxide (HP), typically used in concentrations between 90 and 98% (remaining constituent is water) has been used in several systems and has been the oxidizer of choice for many of the prior works at Purdue University. Both N_2O and HP tend to optimize performance at high mixture ratios, which is an advantage for these systems as the size of the fuel section is reduced as a result.

There is tremendous flexibility in choices for fuel grains as any solid hydrocarbon material can provide the necessary hydrogen and carbon molecules to create high flame temperatures. Rubber-type materials like HTPB and its cousins CTPB (carboxyl-terminated polybutadiene) and PBAN (polybutadiene acrylonitrile) have been popular choices for many applications. However, more plastic-like materials such as DCPD (dicyclopolybutadiene, a strong plastic used in boat hulls), polyethylene (milk jug material) and polystyrene (infused with air to make styrofoam cups) have also been used. As mentioned in Section 11.5, parafin-based fuels sport enhanced regression

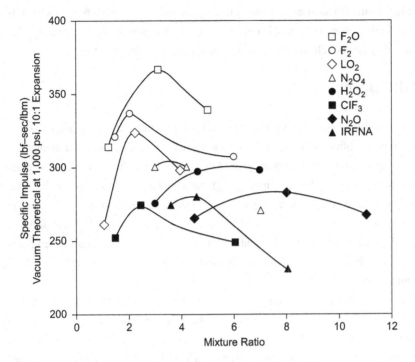

Figure 11.18 Performance of several different oxidizers as a function of O/F ratio assuming HTPB fuel, 1000 psi chamber pressure, and expansion to sea-level pressure.

rates due to the melt layer instability that permits physical shedding of droplets into the oxidizing gas/combustion product mixture. Any of these fuels can be loaded with metal powder (aluminum being the most common choice), although the performance enhancement in terms of I_{sp} or c^* is rather modest in most cases. Fuels can also be loaded with catalytic material in order to create oxidizer decomposition and combustion without the need for a separate ignition system. This option is particularly attractive for systems that demand restart capabilities.

Both HP and N_2O are monopropellants as well, i.e. they can be decomposed into hot gases as their decomposition liberates heat. In particular, HP can be decomposed into steam and oxygen when placed in proximity to a catalyst material. Potassium permanganate, sodium borohydride, and manganese dioxide are some of the few known catalysts for HP. Unfortunately, N_2O catalysts have been harder to identify and this oxidizer does not decompose as readily as HP, but there is ongoing work in this area. Numerous materials are catalytic with WFNA/IRFNA, so this option is also available when these oxidizers are employed.

Some researchers have exploited the monopropellant characteristics of HP to make "staged" systems whereby some fraction of the HP flows through a catalyst bed and the hot gases autoignite the fuel grain downstream. In these systems, the remaining HP can be injected in liquid form and will thermally decompose in the hot combustion products from the staged reaction thereby minimizing the size of the catalyst bed required for the system. Alternately, catalytic fuel grains can be manufactured by mixing catalyst into the main fuel constituent(s). These staged and catalytic systems have a distinct advantage in that they do not require a separate ignition system and they can be relit for missions demanding multiple thrust pulses.

11.7 HRE Design

Figure 11.19 highlights the overall design process for an HRE system. The main system requirements must be specified before we can size the system. Typically, these requirements might be specified in terms of an ideal velocity gain (Δv) or an average thrust and burning duration (t_b). The payload may have acceleration limits (g_{max}) that would place a limitation on the instantaneous thrust/weight ratio of the vehicle. For booster systems, the initial thrust (F_i) is an important consideration as the vehicle is heaviest at liftoff and thrust must exceed weight at this time. For systems flying in the atmosphere, the dynamic pressure of the air rushing over the vehicle sets all the aerodynamic loads. For this reason, a constraint is typically applied to limit this value (q_{max}) in many applications.

The designer must also stipulate the propellant combination and propellant densities, a regression law (e.g. $r = aG_{ox}{}^n$) and the thermochemical performance (I_{sp} and c^* vs. O/F) must be defined at an average chamber pressure (p_c) in order to determine the instantaneous performance of the engine/motor. While p_c can vary with time due to changes in flow rate and c^*, I_{sp} and c^* tend to be weak functions of this parameter and an average value is generally acceptable for preliminary design. However, many groups have integrated the NASA CEA tool (described in Chapter 5) into a time-stepping algorithm for motor performance and this would permit one to get the instantaneous I_{sp} and c^* values at arbitrary chamber pressure conditions.

The final set of user inputs involves the fuel section topology. The traditional fuel topology employs a number of ports integrated into a high L/D cylindrical chamber. The wagon-wheel design shown in Figure 11.10 is but one example; four-port and seven-port circular cluster have also been employed in prior works. Multiport configurations such as these must reconcile the presence of fuel slivers that are formed as ports begin to intersect with fuel depletion. In larger systems, physical structure may be required between the ports to support the very thin fuel slivers that form near the end of the burn. These *web stiffeners* and typically made from insulating phenolic materials such as those described in Chapter 7 on SRMs. In smaller motors, it may be possible to ignore the tiny slivers and permit them to be ejected through the motor throat in the latter parts of the firing. At Purdue, we fired a four-port polyethylene fuel grain that routinely ejected the tiny slivers between ports – there was no deleterious effect on motor performance other than the fact that the sliver material basically acts as an inert and does not contribute to motor impulse.

Non-conventional fuel section topologies are also available to the designer. For example, a "pancake" motor configuration can be developed wherein the oxidizer flows between parallel cylindrical plates of fuel as illustrated in Figure 11.20. Here oxidizer flow could be directed radially inward toward a centrally located nozzle (Figure 11.20a) or radially outward toward a collection of

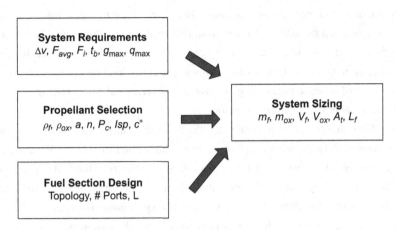

Figure 11.19 HRE design process.

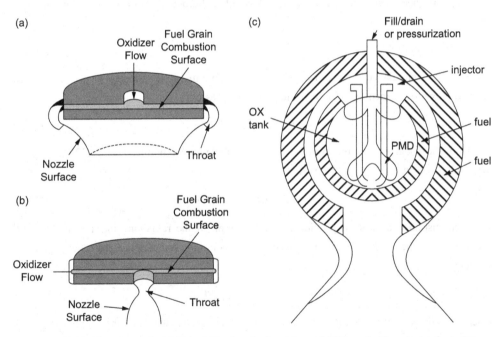

Figure 11.20 Non-conventional fuel section topologies. (a) and (b) "Pancake" topologies featuring outward-flowing or inward-flowing oxidizer. (c) Concentric spherical topology with oxidizer tank embedded in spherical fuel section.

nozzles or an aerospike annular nozzle (Figure 11.20b). This configuration may be desirable for severely length-limited systems. While the flux level changes continually as the gases move radially, some limited investigations (for the outward-flowing topology) indicate that regression was fairly uniform (see Caravella *et al.*, 1998). The concentric spherical topology (Figure 11.20c) is another option that has received some attention. Here a spherical oxidizer tank would lie in the center of the system and the wall of the tank would be lined with fuel (presumably with a layer of insulation directly adjacent to the tank to prevent overheating). A second concentric fuel layer would lie adjacent to an outer chamber that serves as pressure vessel for combustion. This system has the advantage of improved packaging (relative to a conventional design) and minimum weight pressure vessels that employ spherical topology.

As a final consideration in the design, the issue of oxidizer vaporization is one that is important in larger HRE systems, particularly those employing liquid oxidizers. With large oxidizer flows comes a potential to quench a flame that develops at the entry to the fuel ports. Consider an energy balance at the entry to the fuel port with the heat of combustion being contributed by fuel reaction with some small portion of the oxidizer flow. If we presume that this combustion occurs at locally stoichiometric conditions, the energy contributed from the combustion can be expressed:

$$\dot{E}_{in} = (\dot{m}_f + \dot{m}_{oxb})h_c \qquad (11.23)$$

where h_c is the heat of combustion and $\dot{m}_{oxb}$ is the oxidizer burned at a stoichiometric mixture ratio O/F^*:

$$\dot{m}_{oxb} = \dot{m}_f (O/F)^* \qquad (11.24)$$

If we assume a circular fuel port of radius R_p with a simple fuel regression law based on a desired oxidizer flux of G_{ox}, then the fuel flow rate may be expressed:

$$\dot{m}_f = aG_{ox}^n 2\pi R_p x \qquad (11.25)$$

where x is the distance measured from the entry to the fuel port. Combining Eqs. 11.22–11.24 gives the combustion energy input as a function of x:

$$\dot{E}_{in} = aG_{ox}^n 2\pi R_p x(1 + (O/F)^*)h_c \qquad (11.26)$$

This combustion energy must provide for vaporization of the remaining oxidizer in the port. Assuming that the oxidizer enters the engine at its saturation temperature, this energy may be written:

$$\dot{E}_v = (\dot{m}_{ox} - \dot{m}_{oxb})h_v \qquad (11.27)$$

where h_v is the heat of vaporization of the oxidizer. For oxidizers that enter the engine subcooled, the heat of vaporization must also account for the enthalpy required to heat the oxidizer to its boiling point. For a desired oxidizer massflux into the port, Eq. 11.27 may be written:

$$\dot{E}_v = (G_{ox}\pi R_p^2 - \dot{m}_f(O/F)^*)h_v \qquad (11.28)$$

As one might expect, the energy demand for vaporizing the oxidizer is greatest at the entry to the port ($x = 0$). The energy input from combustion increases monotonically with x as we move down the port. We can solve for the critical location where there is just enough combustion energy

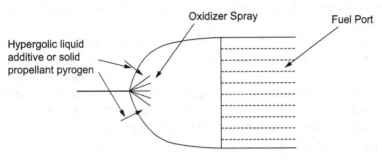

Figure 11.21 Mechanism to prevaporize oxidizer prior to entry into a multi-port fuel section of a large HRE.

to vaporize all of the oxidizer in a given port by equating Eqs. 11.28 and 11.28. Performing this operation, and substituting in Eq. 11.25 for fuel flow, we get:

$$x_{crit} = \frac{h_v G_{ox}^{\,1-n} R_p}{2a(h_c + (O/F)^*(h_c - h_v))} \tag{11.29}$$

Equation 11.29 shows that for a desired massflux level, the larger the port radius, the further down the port liquid oxidizer will penetrate. Oxidizers that have high heat of vaporization tend to exacerbate this issue as well as fuels that have low regression rates (i.e. low a values). Physically, it becomes difficult to hold the flame at the entry of the fuel port with all this cold oxidizer removing heat from the flame and this issue can lead to instability of the motor as the flame is periodically quenched and reignited.

One way this problem might be mitigated is to use alternate combustion sources in a pre-vaporization chamber upstream of the fuel section as highlighted in Figure 11.21. Alternately, one might inject a portion of the oxidizer flow a portion of the way down the fuel port (presumably using a series of smaller tubes injected partway along each fuel port).

FURTHER READING

Heister, D. and Wernimont, E. J. (2007) "Hybrid Rocket Combustion with Storable Oxidizers," in Chiaverini, M. J. and Kuo, K. K. (eds.) *Fundamentals of Hybrid Rocket Combustion*. Progress in Astronautics and Aeronautics, Vol. 218. AIAA, pp. 457–511.

Altman, D., and Holzman, D. (2007) "Overview and History of Hybrid Rocket Propulsion," in *Fundamentals of Hybrid Rocket Combustion and Propulsion*. AIAA, pp. 1–33.

Caravella, J. R., Heister, S. D., and Wernimont, E.J. (1998) "Characterization of Fuel Regression in a Radial Flow Hybrid Rocket," *Journal of Propulsion and Power*, 14(1): 51–56.

Corpening, J. H., Palmer, R. K., Heister, S. D., and Rusek, J. J. (2007) "Combustion of Advanced Non-Toxic Hybrid Propellants," *International Journal of Alternative Propulsion*, 1(2/3): 154–173.

Estey, P., Altman, D., and Mcfarlane, J. (1991) "An Evaluation of Scaling Effects for Hybrid Rocket Motors," in *27th Joint Propulsion Conference*. AIAA.

Humble, R. W., Henry, G. N., and Larson, W. J. (1995) *Space Propulsion Analysis and Design*. McGraw Hill.

Karabeyoglu, M. A. and Cantwell, B. J. (2002) "Combustion of Liquefying Hybrid Propellants: Part 2, Stability of Liquid Films," *Journal of Propulsion and Power*, 18(3): 621–630.

Karabeyoglu, M. A., Altman, D., and Cantwell, B. J. (2002) "Combustion of Liquefying Hybrid Propellants: Part 1, General Theory," *Journal of Propulsion and Power* 18(3): 610–620.

Lee, J., Weismiller, M., Connell, T., Risha, G., Yetter, R., Gilbert, P., and Son, S. (2008) "Ammonia Borane-based Propellants," in *44th AIAA/ASME/SAE/ASEE Joint Propulsion Conference and Exhibit*, Hartford, CT. AIAA.

Marxman, G. and Gilbert, M. (1962) "Turbulent Boundary Layer Combustion and the Hybrid Rocket," in *9th Symposium on Combustion*. Academic Press, p. 371.

Muzzy, R. J. (1972) "Applied Hybrid Combustion Theory," in *8th Joint Propulsion Specialist Conference*. AIAA.

Osmon, F. V. (1966) "An Experimental Investigation of a Lithium Aluminum Hydride-Hydrogen Peroxide Hybrid Rocket," *Journal of Aerospace Engineering*, 62(61): 92–102.

Shark, S. C., Sippel, T. R., Son, S. F., Heister, S. D., and Pourpoint, T. L. (2011) "Theoretical Performance Analysis of Metal Hydride Fuel Additives for Rocket Propellant Applications," in *47th AIAA/ASME/SAE/ASEE Joint Propulsion Conference & Exhibit*. AIAA.

Shark, S. C., Pourpoint, T. L., Son, S. F., and Heister, S. D. (2013) "Performance of Dicyclopentadiene/-Based Hybrid Rocket Motors with Metal Hydride Additives," *Journal of Propulsion and Power*, 29(5): 1122–1129.

Ventura, M. and Heister, S. D. (1995) "Hydrogen Peroxide as an Alternative Oxidizer for a Hybrid Rocket Strap-on Booster," *Journal of Propulsion and Power*, 11(3): 562–565.

Vonderwell, D. J., Murray, I. F., and Heister, S. D. (1995) "Optimization of Hybrid Rocket Engine Fuel Grain Design," *Journal of Spacecraft and Rockets*, 32(6): 964–969.

Wernimont, E.J. and Heister, S.D. (2000) "Combustion Experiments in a Hydrogen Peroxide/ Polyethylene Hybrid Rocket with Catalytic Ignition," *Journal of Propulsion and Power*, 16(2): 318–326.

HOMEWORK PROBLEMS

11.1 A hybrid test engine, using a single circular port as shown in Figure 11.22, is to be throttled during the course of the firing such that the mixture ratio remains constant at a value of 2.5. We want to design a grain that will provide an initial thrust (F_i) of 1000 N and a final thrust (F_f) of 1500 N with a total impulse of 10,000 N-s. The port pressure drop will be assumed to be negligible since the initial oxidizer massflux (G_{oi}) is only 0.03 kg/cm^2 s. For this reason (and the fact that O/F is constant), the I_{sp} will be constant at a value of 250 s. With a regression rate of r_b the fuel is assumed to be proportional to $G_o^{0.8}$, but the constant is not known. Finally, the fuel density (ρ_f) is 0.9 g/cm^3. Referring to Figure 11.22, determine:

   (i) the initial ($\dot{m}_{oi}$, $\dot{m}_{fi}$) and final ($\dot{m}_{of}$, $\dot{m}_{ff}$) oxidizer and fuel flow rates;

   (ii) the initial and final port diameters (D_i, D_f); and

   (iii) the length of the fuel grain, L.

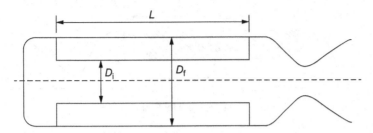

Figure 11.22 Diagram for Problem 11.1.

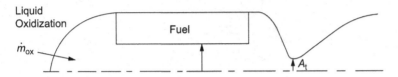

Figure 11.23 Diagram for Problem 11.2.

11.2 Consider the hybrid rocket shown in Figure 11.23. In most cases, the fuel regression rate, $r = a\, G_{\mathrm{ox}}^n$, is correlated with the total massflux of oxidizer, $G_{\mathrm{ox}} = \dot{m}_{\mathrm{ox}}/A_\mathrm{p}$, where A_p is the post cross-sectional area and a, n are correlation constants. We also assume that the fuel, with density ρ_f, has a burning surface area exposed to the flame (A_b). The nozzle throat area (A_t) and characteristic velocity ($c*$) are also known.

   (i) Given this information, obtain a relationship for the chamber pressure in terms of oxidizer flow rate for this engine, assuming steady-state operation.

  (ii) How does your result differ from that of a solid rocket, where we had

$$p_\mathrm{c} = \left[\frac{a\rho_\mathrm{p} A_\mathrm{b} c*}{g A_\mathrm{t}}\right]^{\frac{1}{1-n}}?$$

11.3 Consider the hybrid rocket test motor shown in Figure 11.24. This motor utilizes LOX/HTPB propellants. Data from the NASA thermochemistry code were curvefit for characteristic velocity:

$$c* = -2520 + 6800(\mathrm{O/F}) - 1320(\mathrm{O/F})^2 \quad 2 < \mathrm{O/F} < 3$$

with $c*$ in ft/s. In addition, the fuel density is known to be 0.0325 lb/in^3 and the regression rate (in inches/s) obeys $r = 0.16 G_{\mathrm{ox}}^{0.7}$, where G_{ox} is the oxidizer massflux in lb/(in^2 s). We desire to operate the engine at fixed oxidizer mass flow so we expect mixture ratio variations during the burn. For this reason, we wish to hit the optimum mixture ratio (max. $c*$) at the mid-web location. You may neglect the burning of the end faces of the fuel grain in your analysis. Under these assumptions, determine:

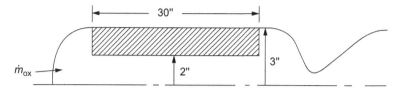

Figure 11.24 Diagram for Problem 11.3.

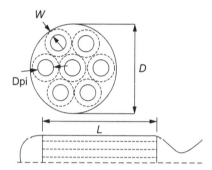

Figure 11.25 Diagram for Problem 11.4.

(i) the optimal O/F for this propellant combination;

(ii) the oxidizer flow rate which maximizes performance at mid-web; and

(iii) the overall O/F shift (max O/F – min O/F) for the firing, assuming the fuel is completely consumed.

11.4 Consider the hybrid rocket using a seven-port cluster as shown in Figure 11.25. We intend to use LOX/HTPB propellants that provide an average I_{sp} of 350 seconds at an O/F ratio of 2.5 for this booster. The average thrust required is 500,000 lbf for a burning duration of 100 seconds. The HTPB fuel density is 0.036 lb/in^3, and the propellant combination produces an average c^* (at optimal O/F) of 5000 ft/s. We wish to design the ports to an initial oxidizer massflux of 1.0 lb/(in^2 s). You may assume a regression behavior of $r = 0.2G_{ox}^{0.68}$ (in inches/s), where G_{ox} is the port massflux based solely on the oxidizer flow.

(i) Compute the initial port diameter, D_{pi}, required to provide this flux level.

(ii) You may assume that the regression rate varies linearly with time in order to compute an average regression rate for the process. In addition, you may neglect any impulse obtained from the fuel slivers resulting from the seven-port grain design and neglect any burning on the end faces of the fuel grain. Under these assumption, determine the total web distance burned, w, as shown in Figure 11.25.

iii) Compute the fuel grain length, L, and overall motor diameter, D, as shown in the figure.

11.5. Consider hybrid rocket combustion inside a simple tubular fuel grain with an initial internal port diameter D_{pi}. Presume that the fuel regression rate follows that for diffusion dominated combustion theory, $r = \alpha(\dot{m}_{tot}/A_p)^\beta$, where $\dot{m}_{tot}$ is the total flow rate, A_p is the current port cross-sectional area, and α, β are constants. Furthermore, presume we are operating at conditions such that the total mass flow remains constant.

(i) Show that the port diameter history can be expressed as $D_p/D_{pi} = 1 + (w/D_{pi})(t/t_b)^\gamma$, where t_b is the total burn time, w is the total web distance consumed, and $\gamma = (1 + 2\beta)^{-1}$.

(ii) Using this result, derive an expression for the total massflux, G/G_i (where G_i is initial massflux) as a function of time. Determine the average massflux ($\overline{G}$) one obtains by integrating this result over the burning duration. You may assume that $w/D_{pi} \ll 1$ to simplify your expression for G/G_i such that an analytic integration can be developed.

(iii) Compare your result in part (ii) with that obtained by simply using the average port diameter in an estimate for $\overline{G}$. What sort of errors are introduced by using this "end-point" method to approximate the nonlinear function G? Here, you may note that combustion theory suggests that $\beta = 0.8$; this may aid you in your assessment of errors introduced in the end-point approach.

11.6 You are required to develop a preliminary design for a hybrid rocket booster using storable oxidizers. Current oxidizer alternatives of greatest interest include hydrogen peroxide (HP) at both 90 and 98% concentrations (the other part is water), hydroxyl ammonium nitrate (HAN) at 95% concentration, and nitrous oxide (N_2O) with the various fuels as shown in Table 11.2. The fuels of interest include polyethylene (PE), hydroxyl-terminated polybutadiene (HTPB), and dicyclopentadiene (DCPD). You should choose a propellant combination from the table shown below.

The booster you are to design will provide an ideal velocity gain (neglecting gravity and drag losses) of 7728 ft/s to a vehicle with an equivalent payload of 219,000 lb. You may assume a mass fraction for your booster of 0.9 in your study. You may assume a fuel grain configuration that uses a seven-port cluster comprised of six ports uniformly spaced at 60° intervals around a port on the centerline and that the overall diameter of the fuel grain is 122 inches. Finally, you may assume a regression law of $r = 0.1G_{ox}^{0.68}$, where r is the

Table 11.2   Performance comparison for several storable propellant combinations, with LOX/HTPB, $p = 1000$, sea-level expansion

Propellant combination	Optimal O/F	ρ_f, g/cc	ρ_{ox}, g/cc	ρ_b, g/cc	I_{sp}, seconds	$\rho_b I_{sp}$, g-s/cc
LOX/HTPB						
90% HP/PE						
98% HP/PE						
98%HP/DCPD						
95%HAN/HTPB						
N_2O/HTPB						

regression rate in inches/s and G_{ox} is the port massflux as computed from oxidizer flow in lb/s/in^2.

   (i) Look up input data on fuel/oxidizer properties and run NASA thermochemical code for various O/F values to fill in the entries in the table above for the propellant selection provided to you in class.

   (ii) Using the optimal I_{sp} value obtained in part (i), compute the propellant weight required for the propellant combination you are studying. Determine both fuel and oxidizer masses required for this combination at the desired O/F.

   (iii) Compute the fuel section geometry including grain length and internal/external diameters of the seven-cylinder cluster. Be sure to include the fuel slivers that are formed between the cylinders and the outer motorcase wall. The initial internal diameter of each cylinder can be sized to accommodate the desired oxidizer flow rate. Assume we desire a constant oxidizer flow rate for a 100 second burning duration at an initial flux level of 0.8 lb/s/in^2. Assuming the oxidizer tank has the same diameter as the fuel section and is comprised of a cylinder with hemispherical domes, determine the overall length of the booster. Make a scale drawing/sketch of your fuel grain and oxidizer tank configuration assuming a ½ diameter space between the two.

   (iv) Derive an expression for the burning surface area of your configuration as a function of web distance including the burning of the slivers.

   (v) Curvefit your performance data near the optimal O/F to define c^* as a function of this parameter.

   (vi) Run a lumped parameter calculation of the ballistics for your design. Size your throat to provide an initial chamber pressure of 1000 psi and plot chamber pressure and sea level thrust histories for the design. Also plot the sea level I_{sp} history and compute the average I_{sp} for the configuration studied. Why does your average I_{sp} differ from the value used in sizing the motor? Why does your burning time differ from the 100 second number assumed in part (iii)?

   (vii) We desire a burning time of 100 seconds and an average thrust roughly twice the booster liftoff mass. What design changes could you recommend to have your booster better match these requirements?

11.7 Consider a hybrid rocket with a simple tubular fuel grain as shown in Figure 11.26. Assume the fuel regression rate is uniform along the length of the grain and obeys

$$r = aG_{ox}^n \tag{1}$$

where G_{ox} is the oxidizer massflux in the fuel port. Assuming the oxidizer mass flow, $\dot{m}_{ox}$, is constant, one can actually solve for the port radius, R, as a function of time using Eq. 1 and the fact that $r = dR/dt$.

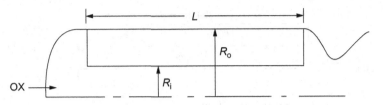

Figure 11.26 Diagram for Problem 11.7.

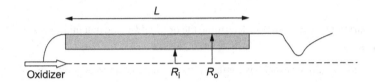

Figure 11.27 Diagram for Problem 11.8.

(i) Show that:

$$R(t) = \left[a(2n+1) \left(\frac{\dot{m}_{ox}}{\pi} \right)^n t + R_i^{2n+1} \right]^{\frac{1}{2n+1}}$$

(ii) Using this result, derive expressions for the fuel flow, $\dot{m}_f$, and mixture ratio, O/F, as functions of time. Is there a special value of n which provides for constant fuel flow and no mixture ratio shifts?

(iii) Suppose $L = 50''$, $R_i = 2''$, $R_o = 5''$, $\rho_f = 1$ g/cc, and $r = 0.1\, G_{ox}^{0.8}$ in inches/s with $G_{ox} \sim$ lb/in^{2}s. Plot $R(t)$, $\dot{m}_f(t)$, and O/F(t) assuming an initial G_{ox} of 1.0 lb/in^2 s.

11.8 Consider the ballistic performance of a simple tubular-fuel grain hybrid with a constant oxidizer flow rate (see Figure 11.27). Since the fuel regression rate decreases more rapidly than the burn surface increases in this design, in general we tend to get an O/F shift toward higher values at the end of the burn. One way to mitigate this fact is to include a variable oxidizer flow capability. Let's assume a fuel regression law of the form $r = a\, G_{ox}^n$ but account for a variable oxidizer flow rate of the form $\dot{m}_{ox} = \dot{m}_{oi} \left(1 - k(t/t_b) \right)$, where $\dot{m}_{oi}$ is the initial oxidizer flow, t_b is the total burn time, and k is a positive constant that addresses the regressivity of the oxidizer flow, i.e. if $k = 0.5$, the oxidizer flow drops by 50% during the burning duration.

(i) Using this formulation, derive expressions for the port radius, fuel flow, and O/F ratio as functions of time. As with the analysis done in class, you can neglect fuel regression on the end faces of the grain.

(ii) Suppose that $L = 50''$, $R_i = 2''$, $R_o = 5''$, $\rho_f = 1$g/cc, $\dot{m}_{oi} = 5$lb/s, and $r = 0.1 G_{ox}^{0.8}$ in inches per second with G_{ox} in lb/in^2-s. What is the burning time of this hybrid rocket? Plot $R(t)$, $\dot{m}_f(t)$, and O/F(t) as functions of time for $k = 0$, 0.2, and 0.4.

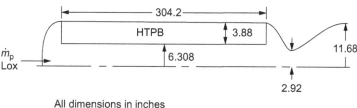

All dimensions in inches
You may assume that end faces don't burn

Figure 11.28 Diagram for Problem 11.9.

(iii) By varying k in your model, determine the optimal k value that minimizes the O/F shift for this rocket.

11.9 Conduct a 1-D ballistics analysis for the simple tubular port motor shown in Figure 11.28 that employs a LOX/HTPB propellant combination. Assume a fuel regression rate of $r_b = 0.104 G_{ox}^{0.681}$, where r_b is in inches/s and G_{ox} is the oxidizer massflux in lb/in^2 s. As the fuel regresses, G_{ox} will drop causing a decrease in regression rate. Offsetting this trend is the tendency of the burning surface area to increase with time.

(i) Let $\dot{m}_p = 35$ lb/s. Plot Mach number and $p_\infty - p$ as a function of distance along the grain at $t = 0$. Here p_∞ is the head-end static pressure entering the port and p is the pressure at some arbitrary location. Repeat the calculation for $\dot{m}_p = 70$ lb/s and plot the results on the same curve.

(ii) Investigate the vacuum I_{sp} shift during the firing for $\dot{m}_p = 35$, 60, and 95 lb/s. Plot I_{sp} vs. t/t_b for these three cases all on one curve. Determine the average $\bar{I}_{sp} = \int F dt / \int \dot{m}_n dt$, where $\dot{m}_n$ is the total mass flow through the nozzle.

(iii) Find the $\dot{m}_p$ value that maximizes $\bar{I}_{sp}$. Explain how to estimate this value a priori.

(iv) Repeat steps (ii) and (iii) using a regression rate of $r_b = 0.08 G_{ox}^{0.4}$. Explain the differences in your results. What tradeoffs are involved in selecting optimal port geometry (G_{ox})? Do we desire high or low exponents on G_{ox} in the regression law?

(v) Attach a listing of your code.

11.10 Design a hybrid rocket capable of generating 100,000 lb of thrust at sea level. Assume a 90% H_2O_2/polyethelyne propellant combination with the grain geometry shown in Figure 11.29. Determine the optimal O/F ratio for this propellant combination assuming a chamber pressure of 60 atm using the NASA Lewis code. Assume a 8% stagnation pressure drop down the grain for thrust calculations and that the grain $L/D = 5$. Determine all dimensions shown below, the mass flow of fuel and oxidizer required, and the ODE specific impulse assuming an average regression rate along the grain of $\bar{r} = 0.06 G^{0.8} L^{-0.2}$ (where $\bar{r}$ is in inches/s, G in lb/in^{2}s, L in inches).

Finally, assume the nozzle is overexpanded at this sea-level condition and that an exit pressure of 4 psi is desired to avert separation.

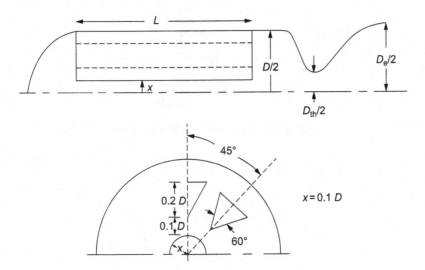

Figure 11.29 Diagram for Problem 11.10.

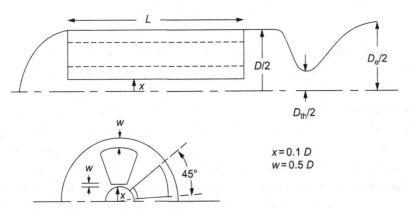

Figure 11.30 Diagram for Problem 11.11.

11.11 Design a hybrid rocket capable of generating 100,000 lb of thrust at sea level. Assume a LOX/HTPB propellant combination with the six-port wagon-wheel grain geometry shown in Figure 11.30. Determine the optimal O/F ratio for this propellant combination assuming a chamber pressure of 40 atm using the NASA Lewis code. Assume a 10% stagnation pressure drop down the grain for thrust calculations and that the grain L/D = 5. Determine all dimensions shown below, the mass flow of fuel and oxidizer required, and the ODE specific impulse assuming an average regression rate along the grain of $\bar{r} = 0.18G^{0.8}L^{-0.2}$ (where $\bar{r}$ is in inches/s, G in $\mathrm{lb/in^2 s}$, L in inches).

Finally, assume the nozzle is overexpanded at this sea-level condition and that an exit pressure of 3.9 psi is desired to avert separation.

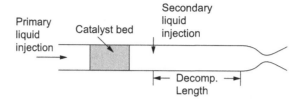

Figure 11.31 Model sketch for Problem 11.12.

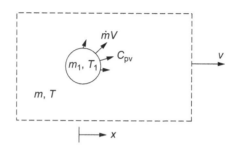

Figure 11.32 Diagram of control volume.

11.12 Our propulsion group does a lot of work with hydrogen peroxide propulsion devices. At present, we are working a project involving thermal decomposition of hydrogen peroxide. This problem looks at minimizing the mass/complexity of systems using hydrogen peroxide as the oxidizer. One of the nice features of HP is that it is a monopropellant, i.e. it gives off energy when decomposed to steam and oxygen. Catalyst beds that employ silver screen meshes are typically used to decompose the fluid, but these devices are heavy and difficult to scale up to high flow rates/thrust levels. As a result, the idea is to use secondary injection of the fluid into the hot decomposition products in order to augment the flow rate/ thrust. A sketch of the concept is shown in Figure 11.31. The hitch is that the fluid needs a certain distance to decompose – this is where you come in.

Model Development

Presume a steady, one-dimensional adiabatic flow for the purposes of the study. Wall friction is neglected as the Reynolds numbers are very large. The aft-injected liquid spray is assumed to be monodisperse with a droplet size corresponding to its Sauter mean diameter. The gas-phase composition is permitted to evolve in time and changes are reflected by vaporized and thermally decomposed hydrogen peroxide. The specific heats of the decomposition products and of the hydrogen peroxide vapor are presumed to be constant in the analysis. You also may neglect the slip between the gas and liquid and assume that both are traveling with constant velocity, v. Under these assumptions, you can model the overall behavior by considering a control volume as shown in Figure 11.32 below in which we consider a single drop and the appropriate amount of gas surrounding this drop. The volume of gas surrounding the drop is set by the secondary/primary mass flow ratio, $\dot{m}_1/\dot{m}_{cat}$ if one assumes that the overall process is steady, i.e. over some interval of time the mass of gas and mass of liquid injected must have the same proportions as the flow rate ratio.

As the droplets evaporate, they will change the composition and mass of gas within the control volume. Hydrogen peroxide is almost always present in an aqueous media (i.e. with water in it). Assuming that 90% hydrogen peroxide (per mass) is decomposed in the catalyst bed and that the same fluid is used for secondary injection, the decomposition reaction for this concentration can be written:

$$H_2O_2 + 0.210 \; H_2O \rightarrow 1.210 \; H_2O + \frac{1}{2} \; O_2$$

so that the initial mass of water and oxygen present in the gas within the control volume may be expressed:

$$m_{wo} = 0.424 \; m_{cat}; \quad m_{oo} = 0.526 \; m_{cat} \tag{1}$$

Water is created via thermal decomposition of hydrogen peroxide vapor and from the portion of water evaporated from the 90% fluid assumed to be injected into the chamber. The resultant water flow can be expressed:

$$\dot{m}_w = 0.1\dot{m}_v - \frac{18}{34}\dot{m}_r \tag{2}$$

where $\dot{m}_r$ is the rate of thermal decomposition of the evaporated gases (a negative number). Similarly, the rate at which oxygen is created from the thermal decomposition reaction can be expressed:

$$\dot{m}_o = \frac{-16}{34}\dot{m}_r \tag{3}$$

Finally, the net rate of peroxide creation (assuming 90% HP in the drops) is:

$$\dot{m}_{HP} = 0.9\dot{m}_v + \dot{m}_r \tag{4}$$

where vaporization increases the peroxide content and the decomposition reaction destroys HP in the control volume. Here we have assumed that the droplet evaporates uniformly in the sense that water is driven off at the equilibrium composition (10% of the droplet mass). In reality, water is evaporated preferentially at least to some extent during the vaporization process. Since the vaporization process happens quickly in the cases of interest, we presume that this effect has only minor impact on the results. Equations 2–4 can be integrated in time subject to the initial conditions given in Eq. 1 and the fact that $m_{HP0} = 0$ (i.e. drops begin to evaporate at $t = 0$).

As the drop evaporates, it will begin to cool the gas and deposit HP vapor into the control volume. The vapor will begin to react (thermally decompose) and add energy back into the system via the heat of reaction, h_{rxn}. You are asked to develop a model of this process for arbitrary initial drop size and secondary/primary flow rate ratios. Take the following steps in your study:

Phase 1 Assignment

In this phase of the process, you are to develop a *methodology* to solve this problem. At the end of Phase 1 you can embark on actually solving the problem at hand. Take the following steps in developing the methodology:

   (i) Develop a list of all known and unknown parameters, i.e. identify the independent and dependent variables in the problem.
  (ii) Derive a relationship for $\dot{m}_r$ based on the kinetics equations presented in the section "Kinetics Model" below, and mass balances, etc. Any additional parameters required for this relationship should be defined and equations developed for their solution using the variables specified above. For example, the mass fractions or molecular weight of the mix must be computed in terms of other independent/dependent variables.
 (iii) Derive an energy equation for the control volume shown in Figure 11.32, including energy gained from thermal decomposition and energy lost from evaporation. Let "m" represent the current mass of gas in the control volume, i.e. at $t = 0$, $m = m_{cat}$. You may neglect kinetic energy contributions in your analysis – you should end up with an equation for dT/dt.
  (iv) Describe a step-by-step procedure to "solve" the problem and determine the temperature, drop diameter, gas-phase composition histories and decomposition lengths for a given set of inputs.

Kinetics Model

The concentration of HP, C_{HP}, can be related to the mass fraction of HP present in the gas within the control volume [turns]:

$$c_{HP} = y_{HP} \frac{P W_{mix}}{R_u T W_{HP}}$$

The unimolecular kinetics reaction can be written:

$$\frac{dC_{HP}}{dt} = -K C_{HP}$$

where $K = A_o e^{-E_a/R_u T} = K(x)$ in the present problem. Data from the literature provide the following for the frequency factor and activation energy:

$$A_o = 10^{13}/s, E_a = 48 \text{ kcal/mol}.$$

Droplet Evaporation Model

Virtually any combustion book should discuss fundamentals of droplet combustion in the famous D^2 law. In this case, the total liquid evaporation rate can be written:

$$\dot{m}_v = 2\pi(k_1 D/C_{pv})\ln(1 + B_q) \qquad (a)$$

where k_1 is the liquid thermal conductivity and B_q is called the *transfer number* or *Spalding number*, defined as:

$$B_q = c_{jw}(T - T_{sat})/(h_v + C_{pl}(T_{sat} - T_1))$$

The Spalding number measures the ratio of the enthalpy driving the evaporation to the enthalpy required to evaporate the fluid in the drop. If T = constant then B_q is also constant and Eq. a can be integrated in time (remember that $D = D(t)$) to show that the drop evaporates at a rate such that D^2 decreases linearly with time. This result is often referred to as the "D^2 law." Note that in our case T is not constant so, Eq. a has to be integrated numerically.

Phase 2

Please use English/Imperial units throughout with the exception of droplet size, which will be expressed in microns. Assume a secondary/primary flow rate ratio of 0.1, an initial drop size of 100 microns, and a gas velocity of 165 ft/s as your primary inputs.

Nomenclature

- C_p – Spec. heat of mixture $C_{pv}Y_v + C_{pp}Y_p$
- C_{pl} – spec. heat of liquid = 0.663 BTU/lb-deg R
- C_{pp} – spec. heat of decomp. Products = 0.4294 BTU/lb-deg R
- C_{pv} – spec. heat of peroxide vapour = 0.438 BTU/lb-deg R
- T_{sat} – peroxide saturation temp = 746 R
- T_{cat} = temp. out of catalyst bed (adiab. Decomp temp) = 1852 R
- ρ_l = liquid density (90%HP) = 85.39 lb/f^3
- h_{rxn} = heat of reaction of 90%HP = 1807 BTU/lb
- C_{HP} – Peroxide vapor concentration
- D – Drop diamter, microns
- h – enthalphy
- h_v – heat of vaporization, 700 BTU/lbm
- K – unimolecular rate = $A_0 e^{-Ea/RuT}$
- k_1 – liquid thermal conductivity = 0.35 Btu/hr*ft*R
- E_a – Activation energy A_0 – frequency factor
- $\dot{m}, m$ – Total gas flow, gas mass in control volume
- $\dot{m}_v$ – Vapor flow rate from evaporation
- $\dot{m}_r$ – Rate of HP vapor reaction(a negative number)
- T – temperature of mix(initial temp of mix is T_{cat})

- v – velocity of gases
- $\dot{m}_l$ - local liquid flow rate
- $\dot{m}_{cat}$ - flowrate out of catalyst bed
- x - axial distance
- W_{HP}, W_{mix} - molecular weight of HP, mixture

(i) Develop and program the algorithm described in the solution and vary the time-step until your solutions become independent of this parameter. Note that these processes will typically occur on a millisecond timescale for the case to be considered, so you will need time-steps of the order of tens to hundreds of microseconds to keep the code stable. [Hint: You will need to include logic in your code to permit it to run after the droplet evaporates completely. Include a lower threshold on the drop mass or size, below which you assume that the drop is essentially gone.]

(ii) Plot gas temperature, peroxide, water, oxygen, and total mass in the control volume, mass flow rates, reaction rate, K, and droplet size as a function of time for the input conditions given.

(iii) Assume that the decomposition length corresponds to the locale where 95% of the peroxide initially in the control volume (i.e. within the droplet) is reacted – determine the decomposition length for the baseline conditions.

(iv) Repeat the analysis with droplets of initial diameters of 200, 300, and 500 microns and plot temperature histories for all cases (including the 100 micron case) on the same plot. Discuss the trends you observe with regard to kinetics and vaporization effects.

(v) Attach a listing of your code including comments to facilitate grading.

11.13 Several companies are currently evaluating the advantages of hybrid propulsion for tactical missiles. This application has traditionally been exclusive domain of the SRM, but the energy management (throttling) capabilities and increased I_{sp} of HREs make them a potential replacement technology. Let's consider a tactical missile with a total impulse requirement of 30,000 lb-s at sea level. Size the throat to provide an initial chamber pressure of 1500 psi. We desire a burn time of 20 seconds for this application.

The large temperature range required for tactical missiles (–160 to +140 °F) severely limits the oxidizers that one might utilize. Currently, researchers are looking at hydroxyl ammonium nitrate (HAN) in an aqueous solution as a potential oxidizer for the extreme temperature range. Let's assume 95% HAN in this case, with HTPB fuel loaded with 10% aluminum for enhanced performance. There is essentially no unclassified regression rate data on HAN-based hybrids, so we will have to use a correlation for LOX instead: $r = 0.104 G_{ox}^{0.681}$, where G_{ox} is in lb/in^2 s, r in inches/s.

The Idea

Figure 11.33 outlines the concept for a hybrid rocket geometry that may be attractive for tactical applications. The oxidizer tank is submerged inside the combustion chamber to

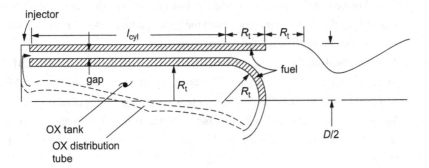

Figure 11.33 Diagram for Problem 11.13.

improve packaging efficiency and to provide a larger burn surface for the fuel. On the end of the tank, a conventional spherical fuel grain (of thickness x) can be assumed. The entire fuel section design is parameterized in terms of the total web thickness, x, the gap which forms the oxidizer port, and the outer radius of the oxidizer tank, R_t. Tactical motors generally need to be long and skinny; the optimal design should have an L/D of 10–15 for the propulsion system.

The challenge in designing a system like this is the strong coupling between the fuel and oxidizer designs. For a given R_t, the oxidizer volume sets the length of the oxidizer tank, yet the fuel volume and burn surface area also depend strongly on this length. For this reason, you may need to make some compromises in the mixture ratio you choose. Obviously, we would like the mixture ratio to be as near the optimal value as possible. By varying "x" and the gap, you can arrive at a suitable design to meet all the requirements except burning time.

The Assignment

You are required to design a system that can provide the required total impulse, burn time, and chamber pressure using the specified propellants. Take the following steps in your study:

(i) Run the TEP code to determine the optimal mixture ratio for the propellant combination. Curvefit c^* and I_{sp} vs. O/F in the range near the optimal O/F value.

(ii) Determine the oxidizer and fuel masses required to deliver the total impulse.

(iii) Solve for the perimeter and port areas in the corrugated fuel section for arbitrary R_t, gap, and web distance, w.

(iv) Guess a tank radius, R_t, and compute the oxidizer tank length to house the necessary oxidizer volume.

(v) Select x and gap values and compute the oxidizer flow rate for the geometry selected.

(vi) Run a steady-state, lumped parameter ballistics calculation assuming that all grain surfaces regress at the same rate based on the regression law provided, i.e. the regression rate is set by the flow properties in the corrugated cylindrical section of the fuel grain. You can assume a constant oxidizer flow rate for the ballistics calculation. Iterate on x and gap dimensions until you match the desired impulse to within 1–2%. Note, that you may be unable to match the desired burn time.

(vii) Repeat the analysis for $R_t = 3$, and 4 inches and sketch the designs you obtain. If you do not obtain any suitable designs, modify your program to handle a conventional cylindrical-port fuel geometry with the oxidizer tank lying ahead of the combustion chamber. Analyze designs near Rt = 3 inches assuming the combustion chamber is the same diameter as the oxidizer tank. Find the overall tank and fuel grain lengths for these cases.

(viii) Document your results – describe what you did and cite the effects of changing R_t on engine design. Include a sample ballistics output (thrust, p_c, A_b, O/F histories) for both annular and conventional fuel grain geometries in your report. Include the equations used in solving for the grain burnback burn surface and port area.

Chapter 12   COMBUSTION INSTABILITY

Combustion instability refers to the growth and presence of high-amplitude pressure oscillations in a combustor. We limit our attention to pressure oscillations that result from unsteady combustion, and occur at discrete frequencies corresponding to resonant modes of the chamber. Although combustion instability is a major issue for all chemical propulsion systems, it is particularly problematic for rocket engine combustors because of their exceedingly high energy release rates, low acoustic damping, and thin thermo-structural design margins.

This chapter provides an introduction to the topic. Section 12.1 provides a brief overview of combustion instability. Section 12.2 provides more detail on the classifications of combustion instability and its phenomenology. The effects of design and operation for the major types of injectors are generalized, and types of damping devices are quickly reviewed. Section 12.3 covers analysis of combustion instability with an emphasis on classical models. The dual time-lag model of Wenzel and Szuch (1965) is introduced for low-frequency instability. Some detail is provided for the linearized acoustics-based models used in the analysis of high-frequency instability, including models for boundary conditions and heat release. Section 12.4 covers testing, and describers accepted practices for pressure measurement and stability rating. Subscale and component level development test practices are also covered. Given the current activity in high-fidelity simulation of combustion instability, a subsection on validation processes is provided at the end of the chapter. More detailed references are provided throughout the text.

12.1 Introduction: Overview and History

Combustion instability is widely observed in a number of combustion devices, but it is particularly problematic in rocket combustors. A useful definition of combustion instability is the condition where the amplitude of pressure oscillations is sufficiently high to cause damage to the combustor or limit its life. Rocket combustors are light-weight and produce a high specific impulse for maximum system-level performance, often resulting in designs with slim margins from thermo-structural failure. Because of acoustic oscillations induced by high-frequency combustion instabilities, heat transfer rates to the chamber wall and face can be significantly increased, and failures can occur in tens of ms (Gross *et al.*, 2003). Furthermore, the onset and growth of combustion instability can be very complex.

Whereas methods for predicting and controlling heat transfer, as well as promoting high performance, are relatively well-established in rocket combustors, the prediction of unsteady combustion and how it couples with other fluctuating flow properties in the chamber remain elusive. It is a result of interactions between gas dynamic modes and the processes of injection, hydrodynamics, mixing, and reaction. These interactions can give rise to high-amplitude, pressure oscillations at high frequencies corresponding to the acoustic modes of the chamber or low frequencies corresponding to the flow residence time. More than 100 years ago, Lord Rayleigh (1896) provided an explanation for the existence of high-frequency oscillations that is still quoted in many scientific papers today: "If heat be given to the air at the moment of greatest condensation, or be taken from it at the moment of greatest rarefaction, the vibration is encouraged. On the other hand, if heat be given at the moment of greatest rarefaction, or abstracted at the moment of greatest condensation, the vibration is discouraged."

Rocket combustors have exceedingly high rates of energy addition. The power generated in a high-pressure main combustor can be more than 10 GW/m^3, and sudden ignition of partly mixed propellants provide a ready source of stochastic pulsations that can trigger instability. Since ignition and extinction occur much faster than convective or even acoustic speeds, localized pressure spikes can have amplitudes that are multiples of the mean chamber pressure; such events are often encountered during development of new designs. Even smaller pressure pulses can force significant gradients and dynamic mixing, resulting in more nonlinear behavior. The interior of a rocket combustor is highly reflective and the loss of acoustic energy through the throat is relatively small. These factors along with the more regular complexities of combustion dynamics combine to produce a very dynamic environment that is difficult to predict or control.

Occurrence of combustion instabilities during rocket combustor development presents high technical and programmatic risks. Instabilities be very difficult to fix and it can take months and even years to eliminate them. Combustion instability jeopardized the success of the Apollo program that landed Neil Armstrong on the Moon – one of the most important feats of the twentieth century, to whom this text is dedicated. The combustion instabilities encountered during the F-1 development is part of rocket lore, and many current practices have their source in that program. Even though some of the most preeminent minds of rocketry were brought together to solve the problem, it was fixed by an arduous and costly process of trial and error, taking over 2000 tests and more than 100 different injector configurations to finally come up with a stable design (Oefelein and Yang, 1993).

The F-1 instability problems were eventually fixed by a combination of manifold and injector design changes, and the incorporation of acoustic baffles on the injector face. To fix the stability problem, the combustion profile down the chamber was changed to be more distributed. The flow areas for LOX and RP-1 injection were increased, and their injection velocities were reduced, resulting in less momentum available to mix the propellants. A consequence of this fix was a reduction in efficiency. Figure 12.1 shows the trend in 1960s era rocket combustor performance in terms of c^* efficiency. To achieve stability, the efficiency was reduced by over 4% compared to the

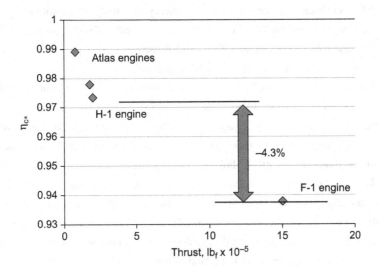

Figure 12.1 Performance trends of mid-twentieth century kerosene-fueled gas generator cycle engines. Source of c* efficiency is NASA Special Publication 8089.

H-1 design that was the basis for the F-1 injector. This reduction in performance corresponds to an additional 100,000 kg of propellant needed for the first stage, approximately three times the mass of the Apollo Command/Service Module that transported the astronauts to the Moon!

Research and development in rocket engine combustion stability reached a peak around the time of the Apollo program due to instabilities encountered during the F-1 development. Crocco (1965) gave an overview of the theoretical studies at the time. The problem complexity and idealized analysis mostly led to frequency domain solutions of linearized problems for design analysis. For the energy source, models for unsteady combustion came directly from test data or attempts to represent a single component process (e.g. chemical kinetics, injection, or drop vaporization). It is noteworthy that the time lag analysis approach developed then provides the basis for much of the analysis used today. Most of the work during this period was summarized in the landmark publication edited by Harrje and Reardon (1972) and published as NASA Special Publication 194. This work stands today as an excellent reference document.

After the end of the Apollo program, research on combustion stability was greatly diminished in the US. In Europe, after combustion instability caused a failure of a Viking engine (Souchier *et al.*, 1982), a comprehensive study was started at ONERA (Habiballah *et al.*, 1991). Yang and Anderson edited an AIAA Progress in Aeronautics and Astronautics series volume (Yang and Anderson, 1995) to summarize work in related topics after SP-194; this volume was also significant because it was the first to include works from the former Soviet Union. Overviews of combustion instability in solid rocket motors were provided by Price (1992), Blomshield (2006), Fabignon *et al.* (2003), and a publication edited by DeLuca *et al.* (1992). In 2006, AGARD published Culick's substantial overview of history, theory, and analysis of combustion instabilities in both rocket engines and motors. More recently, Dranovsky (2007) provided a volume on instability test practices in the former USSR, and Natanzon (2008) provided a volume on the theoretical developments on combustion instability in Russia.

In practice, rocket combustors with pressure oscillations greater than +/– 5% of mean pressure are considered unstable. It can take the form of "rough combustion," manifested by aperiodic pressure oscillations, or pressure oscillations that show a high degree of temporal organization. High-frequency instabilities present the most difficulties. In this case the frequency of the pressure oscillations are typically linked to the acoustic modes of the chamber, and can range from less than 1 kHz for large combustors to over 10 kHz for small thrusters. Combustion instabilities can also take the form of non-acoustic, or "bulk mode" pressure oscillations. The unsteady pressure field in the chamber is spatially uniform but has a distinct frequency content that is related to the gas residence time in the combustor. These latter instabilities are often referred to as "chug" in liquid engines and "chuf" in solid motors.

By making good use of particle damping, and well-formulated grains, the design of composite propellants has advanced to the point where the risk of combustion instability is comparatively low in solid rocket motors. It is still a major technical risk for liquid propellant engines, however. Enabled mostly by computational advances, accurate predictions of growth rates and limit cycles in experimental combustors have recently shown some promise (e.g. Selle *et al.*, 2014), but predictive design analysis models based on physics have not been used in practice. Instead heuristic models dependent on similarity to heritage designs are typically used.

Decades of rocket combustor development have demonstrated that the stability characteristics of a combustor is largely determined by the injector design and the propellant combination. However, because the onset of instability can be very nonlinear due to factors mentioned above, small differences can determine whether the combustor will be unstable. Instabilities also often arise during transient operation, where conditions are difficult to define. Given a nominal operating condition, the design engineer may be satisfied to know whether a given change simply makes the combustor more or less stable, so they can confidently predict the trend-wise effects of design and operational parameters they can control, such as number and type of injector elements; injection pressure drop; and injector element details like orifice diameters and lengths, impingement angles, recess depth etc.

Significant advances in high performance computing over the past decade have allowed detailed studies of the complex behavior of self-excited combustion dynamics using large eddy simulations (Garby *et al.*, 2013; Harvazinski *et al.*, 2015). Recently, Urbano *et al.* (2016) published results from a large eddy simulation of a 42-element rocket combustor using coaxial injector elements and LOX-H_2 propellants. The computation required 70,000,000 cells and 100,000 CPU hours per millisecond of physical time on a BlueGene Q. Although a coarse mesh was used and this type of computation is far out of reach of all but a few research groups, the study demonstrated that CFD predictions of full-scale rocket combustors are possible, at least in principle. Despite these possibilities, model combustors comprising a single element or a set of multiple elements remain more accessible to simulations, and provide valuable insight to the physics of combustion dynamics. Current research is focused on using results of high-fidelity simulations of single- or few-element combustors to produce reduced-order models of unsteady combustion that can be incorporated into a multi-fidelity framework for full-scale prediction.

12.2 BACKGROUND

12.2.1 Classifications and Phenomenology

Combustion instability shows itself by the presence of high-amplitude or damaging pressure oscillations in the combustion chamber. Various criteria for classifying instability have been used to describe instability, most commonly based on the frequency of the pressure oscillations and how they develop. Classification by frequency defines instabilities as either low-frequency or high-frequency instabilities. Low-frequency instabilities are present at frequencies lower than the acoustic modes of the chamber. High-frequency instabilities are usually associated with the acoustic modes of the combustion chamber, although they may also have their source in the hydrodynamics of the propellant flows.

Combustion instabilities can also be classified according to whether they are linear or nonlinear. Linear instabilities grow exponentially in time from low-level noise and have waveforms that are mostly sinusoidal. Nonlinear instabilities include those that appear very suddenly and reach a limit cycle in a few periods of oscillation. They typically appear as steep-fronted waves that are indicative of higher harmonics. Linear instabilities that lose their exponential growth behavior as they reach limit cycle amplitudes are also termed nonlinear. More fundamentally, linear instabilities comprise those coupling events where the response is linearly proportional to the forcing function, with no change in phase. This would generally preclude any case where additional coupling mechanisms come into play as amplitude grows. Strictly speaking, rocket combustion instabilities become nonlinear at relatively low-pressure amplitudes of just a few percent, well within the range that is generally regarded as "stable."

Figure 12.2 shows oscillographs of chamber pressure recorded during instabilities in a two-dimensional combustor. The instability in Figure 12.2(a) starts off as a "spontaneous" or self-excited instability, growing exponentially in time from an initially low amplitude with an approximately sinusoidal wave form, and ends at a limit cycle. The limit cycle represents a state of equilibrium between energy addition and dissipation, and where unsteady flows are stationary and periodic. Since they do represent an equilibrium condition, operation of a rocket combustor at a limit cycle condition may be quite acceptable as long as thermal and structural margins are sufficient and the combustor can meet its lifetime requirements. Figure 12.2(b) indicates a dynamically unstable behavior that is induced by an external pulse; it has a steep-fronted wave form indicating the presence of higher harmonics and reaches a limit cycle amplitude in just five

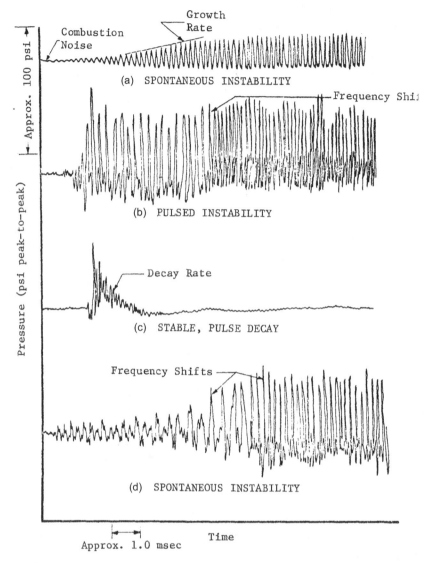

Figure 12.2 Spectrograms of unsteady pressure measurements in unstable and stable combustors (Aerojet, 1970).

cycles. Dynamic instabilities can be the most problematic type because they can be triggered by stochastic events that can cause high-amplitude pulses in the highly energetic flowfield. Figure 12.2(c) shows wave decay from an external pulse, indicating a dynamically stable configuration; dynamic stability is often used as a requirement to demonstrate stability. Figure 12.2(d) shows a self-excited nonlinear instability with steep-fronted waves that grow from a small pressure pulse originating somewhere in the combustor. Similar to the condition in Figure 12.2(b), these types of instability typically concern the design engineer the most because it is difficult to eliminate random pulses.

There are also organized pressure oscillations in liquid rocket engines that occur at very low frequencies, usually less than 100 Hz. These are the so-called "pogo" instabilities that result from coupling between the liquid feed system and the longitudinal vibration mode of the rocket. They can be avoided by exploiting natural damping processes in the fluid systems, or by incorporating damping devices in propellant feedlines, like pogo suppression devices which are spring-loaded pistons or accumulators partly filled with propellant in the gas phase. These are not combustion instabilities, *per se*, and are not covered in this chapter.

The most common types of low-frequency instabilities are called "chug" in liquid rocket engines, and L^* instabilities or "chuffing" in solid motors. Chug and chuff are bulk mode oscillations that are dependent on flow residence time and a representative time scale of combustion, with typical frequencies in the few hundred Hz range. With a bulk mode oscillation, there is no fluctuation in pressure gradient and hence there are no fluctuations in velocity. The instability is associated with a time lag between a fluctuation in the burning rate and massflux through the nozzles; time lag is proportional to the residence time, and hence L^*, for flow in the chamber. They are relatively easy to diagnose and amenable to analysis by the total time lag approach introduced by von Karman and further developed by Summerfield (Summerfield, 1951). Experienced rocket engineers can eliminate them by increasing the flow resistance of propellants entering the chamber (higher injection pressure drop), increasing the burning rate, or by reducing the chamber volume, if permitted by other requirements. They most usually occur at lower power levels with liquid propellants since injection pressure drops scale with the square of the flow rate. Casiano (2010) provides a comprehensive and recent review of chug instabilities.

High-frequency instabilities occur at frequencies close to the acoustic modes of the chamber, and are typically classified according to the associated modes. Longitudinal instabilities occur along the length of the combustion chamber, and transverse instabilities propagate along a direction perpendicular to the center axis of the chamber or parallel to the injector face. The unsteady pressure field can be closely approximated by the solution to the wave equation corresponding to longitudinal, tangential, and radial modes. Mixed modes are also possible. Transverse instabilities can stand or spin. Large combustors like the F-1 thrust chamber can present transverse instabilities at frequencies as low as ~500 Hz, whereas small combustors, used for attitude control for instance, can present instabilities at frequencies greater than 10 kHz. At high amplitudes, shock waves can appear with either longitudinal or transverse instabilities.

Axial modes greater than the first longitudinal frequency (1L) and transverse modes greater than the third tangential (3T) or first radial (1R) modes are rarely seen. Longitudinal modes are often associated with "injection-coupling," whereby the bulk flow of the propellant into the chamber is modulated by periodic variations in injection pressure drop. Transverse modes are more often associated with the unsteadiness of the intrinsic processes of combustion (e.g. atomization, mixing, vaporization, reaction), and their coupling with acoustic modes. They are the most difficult to understand, analyze, and cure.

Non-periodic, high-amplitude pressure pulses can also occur in rocket combustors. In liquid-fueled combustors, these are often associated with an accumulation of unreacted propellants, often a liquid hydrocarbon. Fuel film cooling or poorly mixed propellants are often a root cause, where a local disturbance in the flow could cause a sudden increase in mixing and subsequent ignition leading to the large pressure pulse.

12.2.2 Design Aspects: Effects of Propellant and Injector Types

Before discussing various design guidelines that have been used to obtain stable designs, it is useful to revisit Rayleigh's criterion. Simply stated, for an instability to occur, the cross-section between pressure oscillations and heat addition oscillations, $p'q'$, must be positive and the energy input (or extraction in the case where both p' and q' values are negative) must be sufficient to overcome any losses so that the pressure oscillations can be amplified and sustained. So it should be clear that the spatio-temporal modes of heat addition and acoustics are key. Under most circumstances heat addition modes have a secondary effect on acoustic modes, whereas the acoustic modes can have a major effect on the temporal component of heat release modes through their coupling with hydrodynamics and mixing. The acoustic modes also have an indirect effect on heat release through their effects on unsteady mixing, that in turn determines local properties like temperature and equivalence ratio that determine ignition delay and kinetic rates. From a design point of view, heat addition modes are primarily controlled by injector design and operation, and acoustic modes are determined by chamber geometry and sound speed of the gas.

Combustion stability is achieved by proper injector design and the occasional use of damping devices (baffles and acoustic cavities). Decades of rocket engine developments have yielded a number of design guidelines for stability that are specific to injector types and propellant types. Most rocket combustors use either impinging jets or a pintle injector for pressure-fed or gas generator cycles where propellants are injected into the combustor as a liquid or at a subcritical temperature, and coaxial injectors for staged-combustion cycles where one or both propellants are injected at supercritical temperatures. Each propellant–injector combination may present different types of stability behavior and design guidelines for each type of injector exist.

Pintle injectors are considered "inherently stable," as long as they are designed within certain guidelines. The stability of the pintle may be due to the limited spatio-temporal cross-section between heat release and the acoustic mode. Figure 12.3 shows results from an axisymmetric simulation of a reacting LOX/LCH$_4$ flowfield in a combustor using a pintle injector. It can be seen that the active combustion zones that give rise to high local temperatures exist away from the walls and face of the combustor that naturally provide pressure antinodes of the lower modes of the chamber. Hence the cross-section between unsteady heat and pressure indicated by the Rayleigh criterion, $q'p'$, will be low. Moving the active combustion zone away from reflecting surfaces may also aid stability. The remainder of this section summarizes empirically based rules on the effects of design and operation on stability for combustors that use impinging jet and coaxial injectors.

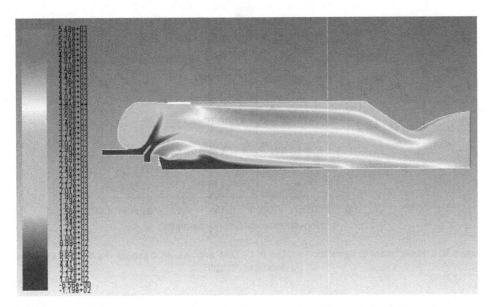

Figure 12.3 Temperature contours in a LOX/methane combustor. High-temperature regions indicated by red show active combustion zones exist away from pressure antinodes in the corners of the combustor.

Impinging jet injectors have been used extensively in pressure-fed and gas-generator cycle engines that use hypergolic propellants and LOX-kerosene. To ensure margin from injection-coupled instabilities with these injectors, injection pressure drops on both circuits of at least 20% of mean chamber pressure are recommended. More problematic are the so-called "intrinsic" instabilities that cannot be linked to injection modulation. A simple view of combustion instability leads to an interest in combustion processes that occur on time scales similar to the acoustic time scale. The ratio of length/velocity gives time, and since the injector is known to control the dynamic behavior a natural scaling parameter is the ratio of injector diameter to injection velocity. The ratio d/v is dominant in correlations for vortex shedding frequency and appears in predictions for reactive stream separation in unlike doublet injectors that use hypergolic propellants. The so-called d/v correlation, also referred to the Hewitt correlation (Anderson et $al.$, 1995), has shown a remarkable ability to match stability characteristics of a number of full-scale combustors that use the most common type of impinging jet injector, like-on-like doublet injectors. It has also been applied to LOX/LCH$_4$ combustors (Hulka and Jones, 2010).

The ratio d/v can presumably be used to predict the highest mode of instability that can be driven in a combustion chamber of a given diameter by an injector of a specific design. In the correlation, the stability characteristics of the injector are related through a stability correlating parameter, the ratio of the orifice diameter, d, to the injection velocity, v. Both d and v relate to the least volatile propellant. The parameter itself does not predict the specific frequency that is driven, per se, rather it predicts a range of frequencies that can be driven. It is presumed that the lowest

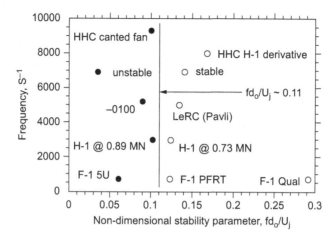

Figure 12.4 Hewitt stability correlation applied to LOX/RP engines using like-on-like doublet injectors (Anderson *et al.*, 1995).

frequency mode is likely to be preferred because it is the least damped mode unless baffles or acoustic cavities are used. It should be noted that this predicted behavior of a "high-frequency cutoff" is fundamentally different than a "preferred frequency" that would be predicted by a sensitive time lag theory.

In Figure 12.4, the stability correlation is reduced to an empirical expression for the maximum instability frequency that can presumably be driven in a combustor that uses an impinging jet injector, i.e. stability occurs for $fd/v > 0.11$. The performance of the correlation for a representative range of injector/combustor combinations using LOX/RP-1 propellants and like-on-like impinging injector elements is indicated: the 1.0 m F-1 (5U, PFRT, and Qual); the 0.53 m diameter H-1; a 0.2 m diameter injector used in an Air Force combustion instability technology program (–0100; Muss and Pieper, 1988); a 0.14 m diameter combustor used in a Lewis Research technology program (LeRC Pavli; Pavli, 1979); and 90 mm and 0.14 m combustors used in a MSFC technology program (HHC H-1 Derivative and HHC canted fan, respectively; Arbit *et al.*, 1991).

The high-frequency cutoff reflected in the d/v correlation is not consistent with the notion of a sensitive time lag. However, its simplicity and applicability over a wide frequency range, along with its functional dependency suggest the presence of some unifying mechanism. The dependency on the least volatile propellant suggests a vaporization-related phenomenon, although it should be noted that most of the combustors shown in Figure 12.4 operate in the supercritical pressure regime.

Combustion instability has not been reported in staged-combustion engines that use warm hydrogen. This may be due to properties of hydrogen gas, like its high diffusivity, wide flammability limits, and rapid chemical kinetics. Combustors that use liquid hydrogen, however, can be unstable, and many studies of combustors used in gas generator cycles have revealed the importance of hydrogen temperature, injector recess, and fuel-to-oxidizer velocity ratio (Hulka and Hutt, 1995). The main general results are that a moderate injector recess of about 5 mm aids stability, and

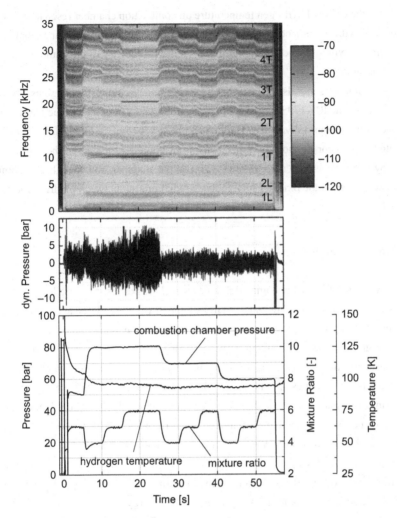

Figure 12.5 Results of self-excited instability tests in BKD combustor tested at DLR Lampoldshausen as a function of chamber pressure, mixture ratio, and hydrogen temperature (Gröning *et al.*, 2016).

that instabilities rarely occur at hydrogen temperatures in excess of 50 K or at velocity ratios in excess of 6. All of these results are consistent with the importance of reliable flameholding in the recess region towards ensuring stability. These revelations also led to the use of hydrogen temperature ramping to test stability margin (Klem and Fry, 1997).

The coupling between the chamber acoustics and the acoustics of the oxidizer flow in its injector tubes has been identified as an important mechanism for combustion instability (Hutt and Rocker, 1995). The research group at the DLR in Lampoldshausen measured self-excited instabilities in a rocket combustor that used LOX and H_2 (Gröning *et al.*, 2016). Contrary to the earlier results on hydrogen temperature ramping, the combustor was stable at low hydrogen temperatures and unstable at temperatures of around 100 K and above, as seen in Figure 12.5. Analysis of the test results showed the effect of hydrogen temperature on combustion chamber resonance frequencies. By varying the hydrogen injection temperature, the frequency of the first tangential mode was shifted to coincide with the second longitudinal resonance frequency of the liquid oxygen injector to cause the instability.

Coaxial injectors are also used in oxidizer-rich staged-combustion engines. Results from component tests of the RD-170 showed the primary importance of oxidizer tube length, as well as tube inlet geometry (Dranovsky, 2007). Harvazinski *et al.* (2015) used high-fidelity simulations of a self-excited combustor to show two mechanisms of coupling, one case where heat release is stimulated by interactions with compression waves in the chamber, and one where a compression traveling downstream in the oxidizer tube induces radial mixing between oxidizer and fuel at the injector exit.

12.2.3 Damping Devices

Damping devices used in rocket engine combustors include baffles, acoustic liners, and quarter-wave resonators. These devices are conceptually simple. Baffles are placed at the injector face and are intended to damp transverse modes. They have the disadvantage of being exposed to the combustion region and need to be cooled. Acoustic liners on the combustion chamber wall have broadband acoustic dissipation characteristics, but can also create serious problems with wall cooling. Quarter-wave cavities are known to efficiently increase the stability margin when placed close to the injection plate. However, these devices have a relatively narrow dissipation bandwidth, and each cavity needs to be tuned to a specific chamber resonance. NASA Special Publication SP-8113, *Liquid Rocket Engine Combustion Stabilization Devices* (NASA, 1974), provides history and design guideline for the devices.

Baffles are partitions that extend from the injector face into the chamber. They are used to damp transverse acoustic instabilities by dissipating the energy of fluctuating transverse oscillations by baffle tip vortices, and interrupting the oscillatory flow. They also shift the acoustic modes inside the baffle compartment to higher frequencies that are uncoupled from combustion processes. Figure 12.6 shows the baffles used in the F-1 and the RD-170. Baffles were also used in the SSME, although it was never shown that it needed baffles. Spokes damp tangential modes and a hub damps the radial modes. An odd number of spokes should be used, and the number of spokes should be at least one greater than the number of nodes in the mode being damped. Spoke and hub-and-spoke configurations are typically used, where the hub will also damp radial modes. The baffle length should be 20–30% of the chamber diameter. Since they extend into the combustion chamber, the baffle must be cooled; film cooling and/or regenerative cooling are used. Ablative baffles have also been used, and the RD-0110 engine used an expendable baffle designed to eliminate the transverse instabilities that occurred during startup (Rubinsky, 1995).

Helmholtz resonators provide broadband damping by two different mechanisms that are additive. The first type of damping is due to viscous drag and heat transfer at the cavity walls, with

(a)

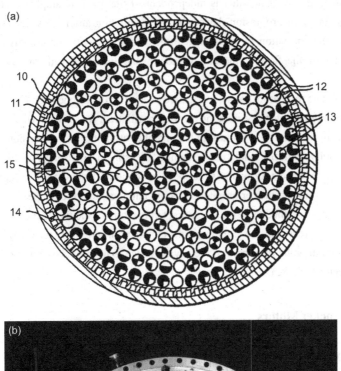

(b)

F-1 Engine / Injector - 1965

Figure 12.6 Hub and blade baffles on RD-170 (a) and F-1 (b). Baffle elements on RD-170 are indicated by hollow circles.

an associated acoustic resistance that is independent of the acoustic amplitude. It is important for the linear stability of propulsion devices at low oscillation amplitudes. The second type of damping is due the formation of vortices at the exit of the cavity. Its effectiveness increases with the velocity of the flow at the cavity mouth, and increases with the amplitude of the oscillation.

Quarter-wave resonators are most effective over a narrow frequency range. They are tuned to provide destructive interference with the acoustic oscillation, for instance where a compression wave will travel up the cavity and return to the chamber at the time an expansion exists near the entrance to the cavity. Multiple modes can be damped by using bi-tuned and even tri-tuned cavities. A key design input to the cavity sizing is the speed of sound in the cavity. To control the speed of sound, fuel can be injected into or around the cavity entrance, and the temperature of the gas or vapor in the cavity is assumed to be the saturation temperature of the fuel at the chamber pressure. Acoustic cavities are relatively easy to tune in development programs by the use of tuning blocks. Another design input is the open area of the cavity, which is usually chosen to be between 5 and 20% of the injector face area.

12.2.4 Solid Rocket Motors

High-frequency combustion instability also occurs in solid rocket motors. The burning rate of solid propellants are characterized in terms of their dependency on pressure, so it may be expected that a main mechanism of combustion instability is coupling with local pressure oscillations of an acoustic field. Blomshield (2006) provides a useful and concise overview of solid propellant instability. Solid propellants with an energetic pressure dependent diffusion flame like AP propellants and high burn rate pressure exponents n are more prone to instability. An increase in mean chamber pressure can lead to reduced stability margin. In addition to the pressure-coupled response, velocity coupling and distributed combustion are included as important mechanisms. Velocity coupling is due to enhanced heat transfer at the propellant surface due to acoustic motions. Distributed combustion is due to the interaction between the acoustic field and burning metal particles away from the propellant surface. It is uncommon and only a factor when metallized propellants are used. Periodic vortex shedding that is induced by acoustic motion or by flow stepping over changes in wall contour can also be a source for amplifying acoustic modes. Fabignon *et al.* (2003) describe its study in segmented motors.

Blomshield (2006) also comments on the effects of particle loading and grain geometry. Avoiding sudden expansions that cause vortex shedding and limiting the amount of burning in the aft end are recommended. The combination of high gas velocity toward the end of a combustor and the throat provide acoustic excitation. Inert particles like Al_2O_3 are used for their significant and nonlinear damping effect on acoustic motions. Smaller particles have more damping at higher frequency oscillations, and larger particles are more successful at damping lower frequency oscillations. The particle size distribution can be tuned for most efficient damping. Related to distributed combustion mentioned above, burning particles are less effective in damping acoustic motions than are inert particles.

12.3 ANALYSIS

In this section, classical models of analyzing combustion instability are reviewed. Low-frequency combustion instability is covered first, with an emphasis on the dual time-lag model of Wenzel and Szuch (1965). High-frequency instability is covered next. After a brief treatment of the calculation of acoustic modes determined by the homogeneous wave equation, a more complete, but still linear, treatment of chamber acoustics is provided along with analytical models of boundary conditions and unsteady heat release that are compatible with either the wave equation approach or Euler computations.

12.3.1 Models for Low-Frequency Instability

Low-frequency combustion instability occurs when the chamber pressure oscillates at a relatively low frequency compared to its acoustic modes. Low-frequency oscillations are called chug in the rocket engine community, and L^* instabilities or chuffing by the solid propellant community. In either case it relates to a cyclic process involving propellant preparation, combustion, and exhaust of products through a choked throat, with pressure oscillations that can reach high amplitudes that can damage the combustor.

In solid motors, chuffing is more prone to occur during the start of the burn when L^* (and gas residence times) is relatively low. In liquid rocket engines, chug tends to occur at lower thrust levels, either during a start transient or during low-power operation for throttled engines. At low flow rates of liquid propellants, the pressure drop across the injector decreases, and the flow is less resistant to modulation. As a corollary, design fixes to chug instability include increasing chamber volume or increasing the pressure drop across the injector.

Models for chug stability have their origin in the time lag model by Summerfield (1951). The time lag accounts for the time between when the propellant was injected and the time it combusts and raises pressure. This obviously simplified approach lumps all the processes of mixing and heat release into a single characteristic time, and presumes the acoustic time scales are much smaller than the period of chug oscillation. A relatively recent review of time lag models, with an emphasis on chug, can be found in Casiano's PhD thesis (Casiano, 2010).

The Wenzel and Szuch (1965) model is essentially a characteristic equation in the frequency domain that accounts for mixing and vaporization of oxidizer and fuel as four separate time lags. As with any stability analysis, if any poles of the system exist in the right half-plane (i.e. have non-negative real parts) then the system is unstable. The poles are the complex solutions to the characteristic equation, which cannot be solved analytically. The characteristic equation is given in Eq. 12.1, while Eq. 12.2 defines residence time:

$$-1 = \frac{e^{-\sigma_m s}}{\theta_g s + 1} \left[e^{-\sigma_{v,o} s} \frac{a}{\frac{\Delta p_{1,o}}{p_c}} \left(\frac{R}{R+1} - R\frac{\frac{dc*}{dR}}{c*} \right) + e^{-\sigma_{v,f} s} \frac{b}{\frac{\Delta p_{1,f}}{p_c}} \left(\frac{1}{R+1} - R\frac{\frac{dc*}{dR}}{c*} \right) \right] \qquad (12.1)$$

$$\theta_g = \frac{c * L * M}{RT_{cg}} \qquad (12.2)$$

The injector exponents, a and b, define how mass flow rate is related to pressure drop across the injector. Mass flow is directly proportional to the pressure drop raised to the exponent power; for injection of liquids the exponents are 0.5. Chamber pressures and pressure drops are defined by the combustor design, and $c*$ can be computed using chemical equilibrium analysis. Mixing time and vaporization time are usually determined from correlations based on injector design and combustor operating conditions, or are sometimes combined into a single time lag.

12.3.2 Linearized Equations and the Wave Equation

High-frequency combustion instability occurs at frequencies that correspond closely to the acoustic modes of the chamber. We have seen from Rayleigh's criterion how heat addition or extraction can amplify or attenuate acoustic modes. Periodic fluctuations in pressure and velocity that are inherent in chamber acoustics also provide a means to synchronize unsteady processes of combustion like bulk flow, atomization, and vortex shedding. Periodic variations in pressure can also strongly affect local chemical equilibria and associated heat input and extraction. It should be noted that the amplitude of the pressure oscillations that occur in a rocket combustor stretch the definition of acoustics even during "stable" operation, and at very high amplitudes frequencies can take a step increase, shock waves appear, and combustion approaches a constant volume process.

To a first approximation, acoustic motions in a combustion chamber can be described by the homogeneous wave equation. The Mach number of the average flow in the chamber is usually no greater than Ma = 0.3, so mean flow effects are small, and the exhaust nozzle is choked and incident waves are efficiently reflected. The homogeneous wave equation is obtained by linearizing the conservation equations by assuming that the instantaneous variables can be expressed in terms of the mean and fluctuating quantities. The fluctuating quantities are assumed to change exponentially with time:

$$p = \bar{p} + \text{Re}(p' e^{st}) \qquad (12.3)$$

where p' is the complex pressure fluctuation and s is a complex eigenvalue, given by:

$$s = \lambda + i\omega \qquad (12.4)$$

where λ is the amplification factor and ω is the angular frequency. Both λ and ω are non-dimensionalized by the reciprocal of the normalization factor of the time. Expressing all the other flow variables in the conservation equations in the form of Eq. 12.3, followed by linearization and rearranging gives the homogeneous wave equation (in terms of eigenvalue s):

$$s^2 p' + \nabla^2 p' = 0 \tag{12.5}$$

Expanding the pressure term as:

$$p' = p_0 + p_1 + p_2 + \cdots \tag{12.6}$$

and

$$s = s_0 + s_1 + s_2 + \cdots \tag{12.7}$$

and neglecting higher-order terms such as p_1/p_0, p_2/p_0, s_1/s_0, s_2/s_0 we get the zeroth-order wave equation:

$$s_0^2 p_0 + \nabla^2 p_0 = 0 \tag{12.8}$$

The zeroth-order equations representing the pure longitudinal motion are given by:

$$\frac{s_0 p_0}{\gamma} + \frac{du_0}{dx} = 0 \tag{12.9}$$

$$s_0 u_0 + \frac{d}{dx}\left(\frac{p_0}{\gamma}\right) = 0 \tag{12.10}$$

The eigenvalue s_0 is always imaginary. For the case where the velocity $u_0 = 0$ at both ends of the duct, i.e. at the injector face and nozzle entrance in case of a combustion chamber, it is given by:

$$s_0 = i\omega_0 = ij(\pi/L) \tag{12.11}$$

where j is the integer that represents number of pressure node locations in the longitudinal mode of interest. Using Eq. 12.10 the frequency of oscillation can be calculated as:

$$f = \frac{j\bar{a}_c}{2L} \tag{12.12}$$

where $\bar{a}_c$ is the speed of sound and L is the distance between the inlet and exit of the system.

The zeroth-order equations representing the pure transverse motions in a cylinder are given by:

$$s_0 u_0 + \frac{d}{dx}\left(\frac{p_0}{\gamma}\right) = s_0 v_0 + \frac{1}{\gamma}\frac{\partial p_0}{\partial r} = s_0 w_0 + \frac{1}{\gamma r}\frac{\partial p_0}{\partial \theta} = 0 \tag{12.13}$$

$$s_0 u_0 + \frac{d}{dx}\left(\frac{p_0}{\gamma}\right) + \frac{1}{r}\frac{\partial(r v_0)}{\partial r} + \frac{1}{r}\frac{\partial(w_0)}{\partial \theta} = 0 \tag{12.14}$$

where, u, v, and w are the velocities in the x, r, and θ directions respectively. The equations are expressed in terms of eigenvalue s_0. The eigenvalue s_0 is given by:

$$s_0 = i\omega_0 = is_{v\eta} \qquad (12.15)$$

where, $s_{v\eta}$ is any root of the equation:

$$\frac{\mathrm{d}J_\eta}{\mathrm{d}Z}(Z) = 0 \qquad (12.16)$$

and v is number of nodal diameters and η is the number of nodal diameters. The roots of Eq. 12.15 are given in Table 12.1.

Using the roots of Eq. 12.15 given in Table 12.1, the frequencies of various transverse modes can be calculated as:

$$f = \frac{s_{v\eta}\overline{a_c}}{2\pi r_c} \qquad (12.17)$$

12.3.3 Acoustic Modes

For simplicity, the acoustic or spatial modes are obtained by solving the homogeneous wave equation for the perturbation pressure; that is $h = 0$ in Eq. 12.5. The solution procedure follows that of classical acoustics for waves traveling through interfaces and area changes (Finch, 2005). In the absence of mean flow or entropy waves, a one-dimensional traveling pressure wave will have the form:

$$p'_n = [Ae^{ik_n z} + Be^{-ik_n z}]e^{i\omega_n t} \qquad (12.18)$$

where A and B are complex amplitude coefficients. Equation 12.18 represents waves traveling in opposite directions and the two wave numbers are defined by $k_z = \omega/a$. Integrating the linearized

Table 12.1   Roots of Eq. 12.15

v	η	$s_{v\eta}$	Transverse character of the mode
1	1	1.8413	First tangential
2	1	3.0543	Second tangential
0	2	3.317	First radial
3	1	4.2012	Third tangential
0	3	7.0156	Second radial
1	2	5.3313	Combined first tangential and first radial
1	3	8.5263	Combined first tangential and second radial
2	2	6.7060	Combined second tangential and first radial

inviscid momentum equation, neglecting the effects of mean flow and using Eq. 12.18, an expression for the velocity can be obtained:

$$u'_n = [Ae^{ik_n z} - Be^{-ik_n z}]e^{i\omega_n t} \tag{12.19}$$

The acoustic mode shapes of pressure and velocity depend on the boundary conditions. As an example, Figure 12.7 shows the pressure and velocity mode shapes corresponding to the first longitudinal mode of a straight duct of length 0.3 m with both ends acoustically closed. An example of the pressure mode shape of first and second transverse modes observed in a thin cylindrical duct (length of duct is much smaller than radius) are shown in Figure 12.8. An example of the pressure mode shape of the radial mode observed in a thin cylindrical duct is shown in Figure 12.9. More details on boundary conditions will be discussed in the following section.

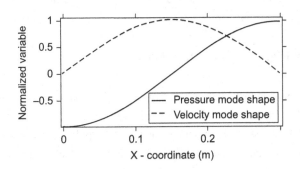

Figure 12.7 Acoustic mode shapes corresponding to first longitudinal mode of straight duct with both ends acoustically closed.

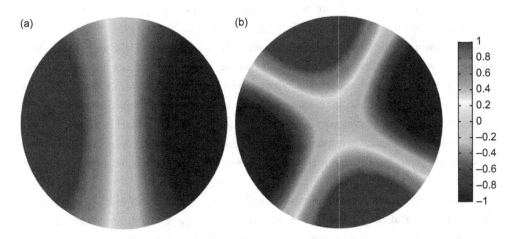

Figure 12.8 Pressure mode shape corresponding to first (a) and second (b) transverse modes of a cylinder.

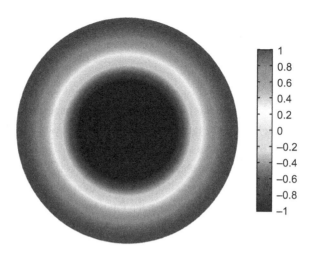

Figure 12.9 Pressure mode shape corresponding to a radial mode of a cylinder.

12.3.4 Models for High-Frequency Instability

High-frequency combustion instability results from a coupling between the resonant gas dynamic modes of the chamber, hydrodynamics that control large-scale mixing between propellants, and unsteady heat release. As such, its modeling in liquid propulsion systems is a complex and challenging task that involves the coupling of diverse physical phenomena such as chemical kinetics, multiphase flows, turbulence and acoustics. In most instances, the precise nature of this coupling is not well understood and the detailed simulations of such effects by fundamental physical models are limited to experimental combustors. In practice, empirically based models are used in analysis.

At the most basic level are the so-called wave equation approaches, pioneered by Crocco (e.g. Crocco, 1951, 1965), Culick (e.g. Culick and Yang, 1995), Flandro (1995), Dowling (Dowling, 1995; Dowling and Morgans, 2005), and others. These methods involve solving the wave equation with a variety of source terms that account for injector and combustion response mechanisms. The response mechanisms range from the simplest and purely empirical correlations to mechanistic models of a single component process that presumably capture the controlling physics. At the most sophisticated level is the full simulation of the unsteady dynamics response, which remains beyond current computing capabilities for general problems but that can presently be applied to simple experimental geometries and configurations.

Historically, wave equation methods have been the cornerstone of much of combustion stability analysis (Culick and Yang, 1995; Dowling and Morgans, 2005). Specifically, several analytical frameworks such as Green functions, weighted residuals, acoustic wave analysis, and separation of variables have been proposed. This paper follows the Galerkin or weighted residual method with acoustic modes as the weighting functions (Culick and Yang, 1995). This method splits the perturbation variables into temporal and spatial modes, and reduces the problem into solving a homogeneous Helmholtz equation for the spatial modes and a second-order inhomogeneous differential equation (ODE) for the temporal modes. The spatial modes are dependent on gas properties and chamber geometry, and the temporal modes are dependent on momentum and energy balances. Source and sink effects, including chamber boundary conditions, momentum transport, and energy addition, appears in the inhomogeneous terms of the ODE. The principal advantages of this method over other analytic models are its term-by-term modeling of various effects, the ability to treat spatial variations of flow properties continuously, and the ability to model nonlinear terms. Time domain models retain a more physical meaning of various terms, whereas frequency domain models are faster.

The Galerkin approach has been extensively applied to combustion instability analysis, including solid rocket engines (Culick and Yang, 1992) and liquid propulsion systems (Culick and Yang, 1995). In many of the studies, several additional assumptions on mean flow effects, acoustic modes, boundary conditions and response functions are traditionally made even though they may not be strictly necessary. Source terms for processes and effects such as gas injection, vaporization, heat release, combustion, vortex shedding, mean flow, and nozzle admittances can be evaluated individually or in combination, thus enabling ease of sensitivity and parametric analysis within a relatively simple, engineering-level code.

First, the formulation of a comprehensive Galerkin method is presented followed by a description of the treatment for the boundary conditions. Portillo *et al.* (2006) provides more detailed derivations of the governing equations, source terms, and boundary conditions, and apply the approach to a variety of flows including an experimental rocket combustor.

12.3.5 Model Formulation

The combustion analysis reduces the two-phase mixed flow conservation equations to a single partial differential equation (PDE) for the gas phase while the effects of the condensed phase are treated as source terms. A detailed derivation of these equations is given in Wicker (1999). The resulting PDE is an inhomogeneous wave equation for a pressure perturbation, p', of the gas phase inside a combustor chamber:

$$\nabla^2 p' - \frac{1}{\bar{a}^2} \frac{\partial^2 p'}{\partial t^2} = h \tag{12.20}$$

with wall boundary conditions:

$$\hat{n} \cdot \nabla p' = -f \tag{12.21}$$

where $\bar{a}$ is the mean sonic velocity and $\hat{n}$ the unit normal vector. The inhomogeneous terms, h and f, include the effects of two-phase flow (injection, atomization, and vaporization), mixing and combustion, mean flow variations, as well as nonlinear effects (acoustics, mean flow/acoustic interactions). In the wave equation approach, realistic boundaries like acoustic cavities, inlets, and nozzles are accounted for by using admittances in the source term f.

The inhomogeneous term from the wave equation, Eq. 2.3, is given by:

$$
\begin{aligned}
h = & -\nabla \cdot \left[\bar{\rho}(\bar{u} \cdot \nabla u' + u' \cdot \nabla \bar{u}) \right] + \frac{1}{\bar{a}^2} \left[\bar{u} \cdot \nabla \frac{\partial p'}{\partial t} + \bar{\gamma} \frac{\partial p'}{\partial t} \nabla \cdot \bar{u} \right] \\
& -\nabla \cdot \left[\bar{\rho}(u' \cdot \nabla u') + \rho' \frac{\partial u'}{\partial t} \right] + \frac{1}{\bar{a}^2} \left[\frac{\partial}{\partial t}(u' \cdot \nabla p') + \bar{\gamma} \frac{\partial}{\partial t}(p' \nabla \cdot u') \right] \\
& +\nabla \cdot \mathbf{M'} - \frac{1}{\bar{a}^2} \frac{\partial \mathbf{E'}}{\partial t}
\end{aligned}
\tag{12.22}
$$

The inhomogeneous term from the wall boundary condition, Eq. 12.21 is:

$$
f = -\mathbf{n} \cdot \left[\left(\bar{\rho} \frac{\partial u'}{\partial t} + \bar{\rho}(\bar{u} \cdot \nabla \, u' + u' \cdot \nabla \, \bar{u}) \right) + \left(\rho' \frac{\partial u'}{\partial t} + \bar{\rho}(u' \cdot \nabla \, u') \right) + \mathbf{M'} \right]
\tag{12.23}
$$

In the above equations, M' and E' represent the perturbation momentum and energy terms from the governing equations (Wicker, 1999). For the solution of Eq. 12.20, Culick uses a Galerkin method, which is basically the method of weighted residuals where the weighting functions are similar to the basis functions. This approach results in the expansion of the perturbation pressure as a combination of temporal, η, and spatial, ψ, modes:

$$
p' = \bar{p} \sum_n \eta_n(t) \psi_n(r)
\tag{12.24}
$$

The error measuring the extent to which Eq. 12.24 satisfies Eq. 12.20 is given by:

$$
\varepsilon_V = \bar{p}[\nabla^2 \psi_n \eta_n - \frac{1}{\bar{a}^2} \psi_n \ddot{\eta}_n] - h
\tag{12.25}
$$

where the double dot corresponds to the second temporal derivative. The weighting function method is used to minimize the weighted average of the error, ε_V, that is:

$$
\int_V [W] \varepsilon_V \mathrm{d}V = 0
\tag{12.26}
$$

where W represents the weighting functions. The Galerkin method then suggests choosing weighting functions similar to the basis functions; the spatial modes, ψ_n, for this case. With this in mind, Eq. 12.26 can be expressed as:

$$
\left[\int_V \psi_m \nabla^2 \psi_n \mathrm{d}V \right] \eta_n - \left[\frac{1}{\bar{a}^2} \int_V \psi_m \psi_n \mathrm{d}V \right] \ddot{\eta}_n = \int_V \psi_m h \mathrm{d}V
\tag{12.27}
$$

Furthermore, if the spatial modes are constructed so that the modes are orthogonal to each other, that is $\int \psi_n \psi_m dV = 0$ for $n \neq m$, and, if it is assumed that they satisfy a homogeneous Helmholtz equation with homogeneous boundary conditions obtained by eliminating all perturbations from Eqs. 12.20 and 12.21, then by expanding the pressure as in Eq. 12.24 the spatial modes are obtained from:

$$\nabla^2 \psi_n + k_n^2 \psi_n = 0 \tag{12.28}$$

with k as the wavenumber. Then, substituting Eq. 12.28 into Eq. 12.27 and using the definition:

$$E_n^2 = \int_V \psi_n^2 dV \tag{12.29}$$

a set of coupled ordinary differential equation for the temporal modes are obtained:

$$\ddot{\eta}_n + \omega_n^2 \eta_n = F_n \tag{12.30}$$

where the angular frequency is related to the wavenumber by the sonic velocity via $\omega = k/\bar{a}$.

The sources, represented by F_n, are then (Wicker, 1999):

$$
\begin{aligned}
F_n = -\frac{\bar{a}^2}{\bar{p} \cdot E_n^2} \Bigg[&\underbrace{\rho k_n^2 \int_V \psi_n (\bar{u} \cdot u') dV}_{[mf\_a]} - \underbrace{\rho \int_V [u' \times (\nabla \times) \bar{u}] \cdot \nabla \psi_n dV}_{[mf\_b]} - \underbrace{\frac{1}{\bar{a}^2} \int_V \frac{\partial p'}{\partial t} \bar{u} \cdot \nabla \psi_n dV}_{[mf\_c]} \\
&+ \underbrace{\frac{\gamma - 1}{\bar{a}^2} \int_V \psi_n \frac{\partial p'}{\partial t} \nabla \cdot \bar{u} dV}_{[mf\_d]} + \underbrace{\int_V \psi_n \nabla \cdot [(\bar{u} \cdot \nabla) \bar{u} \rho'] dV}_{[mf\_e]} + \underbrace{\rho \int_V (u' \cdot \nabla) u' \cdot \nabla \psi_n dV}_{[nl\_a]} \\
&- \underbrace{\frac{1}{\bar{a}^2} \int_V \frac{\partial p'}{\partial t} u' \cdot \nabla \psi_n dV}_{[nl\_b]} + \underbrace{\frac{\gamma - 1}{\bar{a}^2} \int_V \psi_n \frac{\partial p'}{\partial t} \nabla \cdot u' dV}_{[nl\_c]} + \underbrace{\frac{\gamma - 1}{\bar{a}^2} \int_V \psi_n p' \nabla \cdot \frac{\partial u'}{\partial t} dV}_{[nl\_d]} \\
&+ \underbrace{s \psi_n \frac{\partial}{\partial t} \left(\bar{u} \rho' + \rho u' + \rho' u' \right) \cdot dS}_{[bdry\_a]} + \underbrace{s \left[\rho (\bar{u} \cdot u') \right] \nabla \psi_n \cdot dS}_{[bdry\_b]} + \underbrace{s p' \nabla \psi_n \cdot dS}_{[bdry\_c]} \\
&+ \underbrace{s \psi_n [(\bar{u} \cdot \nabla) \bar{u}] \rho' \cdot dS}_{[bdry\_d]} - \underbrace{\int_V M' \cdot \nabla \psi_n dV}_{[mom]} + \underbrace{\frac{1}{\bar{a}^2} \int_V \psi_n \frac{\partial E'}{\partial t} dV}_{[en]} \Bigg]
\end{aligned}
$$

Linear mean flow effects are represented by terms [mf_a] through [mf_d]; the terms [nl_a] thru [nl_d] represent nonlinear sources; momentum and energy sources are represented by [mom] and [*en*] terms; boundaries, which include walls, injector and nozzle, are accounted for in the surface integral terms labeled [*bdry*], note that this term contains the nonlinear interaction between perturbation density and velocity. The last term is the main source of energy and thus perhaps the most important. It also relates Rayleigh's criterion regarding the spatio-temporal cross-section of heat release and pressure, $q'p'$. This product normalized by the product of mean pressure and heat release is also often referred to as the Rayleigh index and used to describe regions of amplification and attenuation in a combustor.

The analysis then reduces to solving both the spatial and temporal equations described by Eq. 12.29 and 12.30 subject to the wall boundary conditions. Similar to the pressure fluctuations, the perturbation velocities are also represented by the temporal and spatial modes resulting in:

$$u' = \sum_n \left(\frac{1}{\bar{\gamma} k_n^2} \dot{\eta}_n \nabla \psi_n - \bar{u} \bar{\gamma} \eta_n \psi_n \right) \quad (12.32)$$

Even though the Galerkin method does not restrict obtaining the acoustic modes from the homogeneous Helmholtz equation, care must be taken in their calculation since the equations used to model the temporal evolution of the disturbances, as well as the velocity expansion into temporal and spatial modes, are based on this homogeneity assumption. In principle though, the acoustic modes can be obtained by different means, such as computational fluid dynamics (CFD) or even experimental measurements, since they represent the somewhat arbitrary weighting functions introduced in the above methodology. However, if the acoustic modes were computed from an inhomogeneous Helmholtz equation, the governing equations should be modified accordingly. This approach allows for treatment of boundary conditions, flow properties, area changes and heat release effects as source terms in the wave equation. The sources can be modeled as functions of pressure and/or velocity fluctuations, and hence can be represented in terms of the normal (ψ) and temporal (η) modes. A detailed description of such treatments follows.

12.3.6 Boundary Conditions

In wave equation solutions, boundary conditions are typically expressed as the ratio of velocity and pressure oscillations, u'/p'. These ratios, commonly known as admittance functions, represent amplitude and phase variations between pressure and velocity at the boundaries. It is important that the admittance formulations incorporate the proper physical boundary conditions corresponding to inflow or outflow in the domain. Values of various admittance functions can be expressed in a generalized equation of the form:

$$u' = \alpha \cdot p'(t - \tau) \quad (12.33)$$

and are presented in Table 12.2. Details of the derivation of some of the boundary types are given by Portillo *et al.* (2006).

Table 12.2   Boundary conditions

Label	Application	Description	α	Reference
BC-1	Inflow/outflow	Closed wall	0	Finch (2005)
BC-2	Inflow/outflow	Open face	$\infty, (p^{'} = 0)$	Finch (2005)
BC-3	Inflow	Choked isentropic flow	$\frac{-\bar{u}}{\bar{\gamma}\bar{p}}$	Portillo *et al.* (2006)
BC-4	Inflow	Choked flow, constant stagnation enthalpy	$\frac{-\bar{u}}{[1+(\bar{\gamma}-1)\cdot\ \overline{M}^{2}]\bar{p}}$	Portillo *et al.* (2006)
BC-5	Inflow	Constant stagnation enthalpy and pressure	$\frac{-\bar{u}}{\bar{\gamma}\overline{M}^{2}\bar{p}}$	Portillo *et al.* (2006)
BC-6	Outflow	Short nozzle	$\frac{\bar{a}(\bar{\gamma}-1)\overline{M}_{nz}}{2\bar{\gamma}\bar{p}}$	NASA SP-194 (Harrje and Reardon 1972)

The proper boundary conditions must be implemented in both the temporal and spatial analyses. Implementation of boundary conditions in the spatial analysis has been explained above. Boundary implementation in the temporal (linear) analysis is made through the surface integral term [$bdry$] in Eq. 12.31. For example, for the short nozzle admittance function (BC-6 in Table 12.2 Table 2.2) and assuming isentropic flow thru the nozzle ($\rho^{'}/\bar{\rho} = p^{'}/\bar{\gamma}\bar{p}$), the corresponding surface integral becomes:

$$F_{\mathrm{nz}} = -\left[\frac{\bar{\gamma}+1}{2}\frac{\bar{a}}{E_n^2}\overline{M}_{\mathrm{nz}}\cdot S_{\mathrm{nz}}\cdot\psi_n^2(z_{\mathrm{nz}})\right]\dot{\eta}_n \qquad (12.34)$$

where $\overline{M}_{\mathrm{nz}}$ and S_{nz} correspond to the Mach number and cross-sectional area at the nozzle's entrance (z_{nz}). As expected, the nozzle has a damping effect when implemented into the temporal ODE, Eq. 12.30. Injector terms, as well as other boundaries, are treated in a similar manner as the above nozzle. Also note that choked flow boundaries correspond to $F_{\mathrm{nz}} = 0$. Finally, we note that the choked, constant stagnation enthalpy boundary condition (BC-4), which is a standard choice when the injector is fed by a choked orifice, is not entirely consistent with the derivation of the governing equations. This is because the derivation implicitly assumes isentropic flow. For this reason, it is not readily apparent how this particular boundary combination can be applied in the present formulation.

12.3.7 Models for Unsteady Heat Release (HR)

At the core of the instability problem is the unsteady addition of heat, q'. It is modeled in the energy [en] source term in Eq. 12.31, giving:

$$F_n = -\frac{R_{\mathrm{u}}/c_{\mathrm{v}}\bar{p}}{E_n^2}\int_V \psi_n \frac{\partial q'}{\partial t}\mathrm{d}V \qquad (12.35)$$

where R_{u} and c_{v} are the mass average gas constant and specific heat release at constant volume.

Typically, heat release models are categorized as either time lag (TL) or transfer function (TF) models. Time lag models, developed first by Crocco and co-workers at Princeton (e.g. Crocco, 1965) for liquid systems, include the phase angle between heat release and pressure or velocity fluctuations as the main parameter, while TF models relate the phase and amplitude between the heat release and the flow properties in the frequency domain. Table 12.3 summarizes a few standard heat response functions found in the literature. Usually the model for heat release is obtained from an experiment that uses a forcing function (like a loudspeaker) to drive p' or u' at frequencies of interest, and measures the flame emission with a photomultiplier tube or high-speed camera. Heat release models obtained this way are used commonly in network models of gas turbine engines where interactions with compressors and turbines are important.

The simplest heat response model is the *pressure time lag* model (HR-01). The reason being that p' does not include any temporal or spatial derivatives since it is expanded as in Eq. 12.24. For this model, α, represents a constant of proportionality, $\bar{\gamma}$ the specific heat ratio, z the axial location of the pressure, and τ the time lag between pressure and heat release fluctuations. The *velocity time lag* model (HR-02) involves temporal and spatial derivatives of the temporal and acoustic modes respectively, for the velocity is expanded as in Eq. 12.32. For this case, $\bar{\rho}$ and $\bar{a}$ represent the mean density and sonic velocity and β is the proportionality constant, the rest of the parameters being similar to the pressure lag model. Pressure time lag models are used most often in rocket engine combustors, namely the n-τ model.

As can be seen in Table 12.3, a common form of algebraically expressing transfer functions consists of using the ratio of two polynomials. Such polynomials are typically first- (HR-03) or second- (HR-04) order in ω and include coefficients that relate to the problem physics or geometry. Of course, transfer functions need not be expressed as polynomial ratios; they can be

Table 12.3   Heat release models

Label	HR model description	$q'/\bar{q}$	Sample reference
HR-1	Pressure time lag	$\frac{2\alpha}{\bar{\gamma}-1}p'(z,t-\tau)/\bar{p}$	Dowling and Stow (2003); Crocco (1965)
HR-2	Velocity time lag	$\beta\frac{\bar{\rho}\bar{a}^2}{\bar{\gamma}-1}u'(z,t-\tau)/\bar{u}$	Dowling and Stow (2003)
HR-3	First-order transfer function	$\left[\frac{1}{1+i\omega\tau_1}\right]u'\bar{u}$	Bloxsidge *et al.* (1988)
HR-4	Second-order transfer function	$\left[\frac{1}{1+i\omega\ \tau_2+(i\omega)^2\tau_3}\right]u'\bar{u}$	Dowling (1999)
HR-5	Analytic transfer function	$\left[\frac{2\omega_0}{(1+R)}\left(\frac{1-Re^{-s/\omega_0}}{s}+\frac{(R-1)\omega_0}{s^2}\right.\right.$ $\left.\left.(1-e^{-s/\omega_0})\right)\right]u'\bar{u}$	Dowling (1999)
HR-6	Mass flow rate	$k\dot{m}'\bar{\dot{m}}e^{-i\omega\tau}$	Dowling and Stow (2003)
HR-7	Vorticity	$C_Q\cdot\sum_j\Gamma_j\delta(L-z_j)\delta(t-t_j)$	Matveev (2004)

obtained analytically, numerically or experimentally (Paschereit *et al.*, 2002). Since commercially available computer software typically treats transfer functions as a polynomial ratio, algebraic manipulation in both the frequency and time domains must be applied to TF models that are not expressed as such (e.g. HR-5).

The *mass flow* model (HR-06) depends on mass flow fluctuations, $\dot{m}'$, frequency, ω and a constant of proportionality, k. Even though it is a convenient way of modeling the heat release, it can be decomposed into pressure and velocity perturbations. Matveev (2004) proposes a *vorticity*-based response function (HR-07) in which the heat release occurs due to a train of vortices being shed from a backward facing step. In other words, it occurs at pulsed time intervals at a fixed spatial location, and is modeled by the Dirac delta function. The magnitude of the heat release is set to be proportional to the vortex strength or circulation, represented by Γ_j, of each individual vortex. Similar to the mass flow model, the vortex strength can also be modeled by the difference between upstream and downstream perturbation velocities.

Most heat release models treat the combustion zone as a single plane; obviously this is not physically possible, so heat response models with spatial dependence should also be considered in the model's development. Matveev (2004), and Bloxsidge *et al.* (1988) provide heat response models with spatial dependence.

12.3.8 Mean Flow Effects, Area Changes, and Other Mechanisms

The above formulation neglects the influence of mean flow on both the spatial and temporal modes. As mentioned previously, the sources that account for mean flow effects in the temporal calculations are given in terms [*mf_a*] thru [*mf_d*] from Eq. 12.31. The first term includes the effects of combustor shape and mean velocity profile, the second the effects of cross flow, the third also includes effects of combustor shape and velocity profile, and the fourth term models the effect of mean flow acceleration as well as the geometry. It is interesting to note that for a one-dimensional analysis with orthogonal mode shapes, if the perturbation velocity is expanded in its temporal and spatial modes without including mean flow effects (the second term of Eq. 12.32 is neglected), then the first and third terms will cancel out. Likewise, mean flow effects can also be accounted for in the calculation of acoustic mode shapes. Other mechanisms, such as atomization, vaporization, heat transfer, etc. can also be included in the momentum [*mom*] or energy [*en*] source terms in Eq. 12.31. Implementation of these mechanisms into the analysis should follow the same approach as for the heat response model, but is not considered here.

To illustrate the use the use of the approach described above, we check Rayleigh's criterion, which states that there is growth when the heat release is in phase with the pressure and decay when they are not in phase. For this purpose, a heat response model based on pressure-lag (HR-01 in Table 12.3) is used. Growth rates for this case are calculated and plotted against time lag in Figure 12.10. Unstable regions are those with a positive growth rate while stable regions correspond to negative values. It is evident that Rayleigh's criterion is satisfied since maximum growth rate occurs when the pressure is in phase with the heat release ($\tau/T = 0$, 1, 2) and minimum growth rate occurs when they are out of phase ($\tau/T = 0.5$, 1.5).

Physical mechanisms are often related to velocity fluctuations in a combustor chamber, e.g. the strength of a vortex shed as flow from an injector enters combustor depends on the local velocity field. Accordingly, we employ the heat release model given by HR-02 in Table 12.3 for this situation. Figure 12.11 shows the corresponding growth rate variation for a range of time-lag magnitudes.

The linearized and simple acoustic models that served as the main means of analysis over the past 50 years were necessary largely because of limited computational capability. Commercial Helmholtz solvers that can be used for complicated, three-dimensional geometries

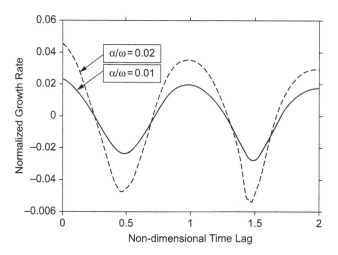

Figure 12.10 Growth rate versus time lag, for pressure time lag model. Key: (–) $\alpha/\omega = 0.01$; (- -) $\alpha/\omega = 0.02$.

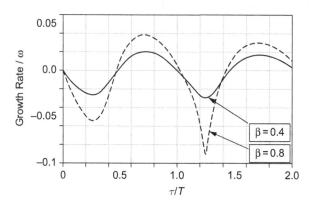

Figure 12.11 Growth rate versus time lag, from velocity time lag model. Key: (–) $\beta = 0.4$; (- -) $\beta = 0.8$.

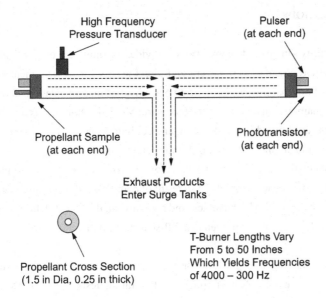

High Frequency
Pressure Transducer

Pulser
(at each end)

Propellant Sample
(at each end)

Phototransistor
(at each end)

Exhaust Products
Enter Surge Tanks

Propellant Cross Section
(1.5 in Dia, 0.25 in thick)

T-Burner Lengths Vary
From 5 to 50 Inches
Which Yields Frequencies
of 4000 – 300 Hz

Figure 12.12 T-burner diagram (Blomshield, 2006).

are widely available and are making their way into combustion stability analysis. Another approach to solve unsteady flows that is easily within modern analysis capabilities is the numerical solution of the Euler equations. There are several specific advantages of an Euler solver. First, mean flow and compressibility effects are completely included in the analysis. A related merit is that it naturally includes entropy waves in addition to the acoustics, and the coupling between these modes that arises at boundaries and interfaces. Pure wave equation methods can include some of these effects, but since they generally do not include the mean flow and entropy waves in the solution, the effects of local sound speed variations cannot be fully included. Second, realistic boundary conditions can be used with Euler models, including boundary conditions that are consistent with more advanced CFD. In contrast, wave equation models use admittance functions to approximate the boundary condition. Effects due to entropy generation at boundaries are thereby generally bypassed in wave equation models. Third, the Euler equations can be solved in multiple, coupled domains, which allows an arbitrary number of different mean flow conditions in the combustor.

Future prospects for combustion instability analysis include multi-fidelity modeling of the combustion chamber along with the reduced-order models (ROM) of unsteady heat release (Huang *et al.*, 2016). The ROM could come from an experiment, or preferably from high-fidelity simulations that have been validated at conditions of interest. In the near term, the use of modal decomposition techniques along with forced, reduced-domain simulations to efficiently map out combustion response appears to be particularly promising (Huang *et al.*, 2017).

12.4 Test Methods

Testing for stability in rocket combustors provides a number of particular challenges. The combustor environment is very severe and direct access for measurement can be difficult. Small thermostructural margins can complicate placement of instrumentation, and penetrations in a highly stressed chamber can add risk to the hardware. Much development is done using heat-sink chambers but these are limited in test duration, and cooling channels can also make measurement difficult. Many nonlinear instabilities result from stochastic events that can't be purposefully reproduced, so a statistical approach requiring many tests or the use of external pulses to may be required. Since high-frequency instability is a result of complex and often unpredictable couplings between chamber acoustics, hydrodynamics, and combustion, the actual stability characteristics of a full-scale combustor can only be proven in a full-scale test, so that the chamber acoustic field and engine transients can be reproduced.

Testing for stability may fall into three general categories: stability rating, component development, and validation. Results from stability rating tests are used to certify that the combustor is stable throughout its intended operating range, or that the presence of instability does not damage the combustor or other parts of the engine or vehicle. The details of stability rating tests may be a part of contractual requirements. Development tests are done with both full-scale and subscale components. For large rocket engines, much development testing in the US is done at the "40K" level, or 40,000 lbf or thrust. Low-frequency stability data from subscale tests are useful, but the smaller diameter chambers do not reproduce the acoustic field of the full scale. Stability characteristics of solid rocket motors and propellants are tested in "T-burners" that can reproduce full-scale acoustic fields. The increased use of CFD to understand instability and to produce results that can be used in design calls for validation tests of the model at representative conditions. These three types of tests are covered in more detail later in this section. Since high-frequency measurement of pressure is a cornerstone of stability testing, it is covered first. More advanced measurement analysis is included in later parts of this section.

12.4.1 Pressure Measurement and Analysis

Piezo-resistive and piezoelectric transducers are commonly used for high-frequency measurements in rocket chamber experiments. They are typically water or helium cooled. The piezo-resistive transducer uses a Wheatstone bridge connected to a diaphragm that flexes under an applied pressure load. The flexing of the diaphragm introduces a resistance change across the previously balanced bridge and the corresponding voltage potential is the proportional to the applied pressure. For a piezoelectric transducer, when the piezoelectric crystal (e.g. quartz) is stressed by applied pressure in the chamber, the displaced electric charge accumulates on the opposite faces of the crystal. This high impedance output of the crystal is amplified and converted to a low-impedance voltage signal that is proportional to the applied pressure signal. Measurement of pressure requires direct access to the fluids inside the chamber, which can be difficult in a water-cooled combustor. In that case externally mounted accelerometers have been used but are not recommended.

The sampling rate of the pressure measurement is of utmost importance. Acoustics-coupled combustion instabilities present at frequencies inversely proportional to chamber size, that can range from a few hundred Hz to over 10 kHz for small thrusters. The Nyquist criterion states that the highest frequency one can accurately resolve is one half of the sampling rate, for example a sampling rate of 20 kHz is necessary to resolve a 10 kHz signal. Under-sampled signals can lead to aliasing, where frequencies higher than the sample rate appear as frequencies lower than one half of the sampling rate. A sampling rate of at least 2.5 times the frequency of interest should be used to accurately resolve the phase of the signal. To accurately measure the shape of a signal the sampling rate should be at least ten times higher than the frequency of interest.

Although modern data acquisition systems can sample at frequencies in the MHz range and high-frequency transducers rated for a few hundred kHz range are readily available, measurement can be limited by the resonant frequency of the port design. For accurate measurements of amplitude, frequency, and phase response, the transducer would ideally be installed flush with the chamber wall, but this is limited by the high heat fluxes in the chamber. The reliability of flush-mounted transducers over even moderate test durations is limited and alternative installation methods where the transducer is protected or cooled are commonly used. Dual diaphragms with water circulation between them, and ablative coatings are ways to isolate the transducer from the high heat fluxes in the chamber. Recessed-cavity installation or semi-infinite tube installation are two common installations. It is critical to ensure that any recess volume between the chamber and sensing element of the transducer doesn't introduce a resonance that can contaminate the measured pressure response.

To evaluate stability characteristics, the analyst should first study the raw signal. If electronic noise is present, the signal may be high-pass filtered at around 150 Hz. After identifying how the signal changes over the test, intervals of similar behavior can be defined. Then the signals from these intervals can be further reduced and analyzed using standard techniques like fast Fourier transforms and filtering. It is important to be aware of any artifacts that can be introduced by the signal analysis technique.

Organized periodic behavior embedded in a signal manifests as an increase in power at the associated frequency. The power spectral density (PSD) of a time-dependent signal is a standard method used to describe the distribution of power (Pa^2-s) across frequency. The frequency of the peaks and their amplitude and shape (in terms of full-width half-maximum, FWHM) are quantifiable and descriptive metrics that can be used to validate simulation results. The FWHM relates to how fixed the signal is to a given frequency and can be used to estimate the amplitude of the signal, but that should be done carefully. The power of a given mode can be plotted against time to yield a spectrogram that shows the temporal evolution of the instability. A more typical method is to filter the data around the central frequency identified from the PSD using a bandpass filter. It is important to use zero-phase digital filtering that doesn't alter the amplitude or the phase of the raw data.

A Butterworth filter with a maximally flat profile over the bandwidth is commonly used. As the amplitude of the pressure signal exceeds a few percent of mean pressure, the sinusoidal signal becomes skewed and higher harmonics begin to form. At that point, reconstructed signals from bandpass filtering will start to deviate from the measurement.

12.4.2 Stability Rating

Stability rating tests are done at full-scale component and/or engines using high-frequency pressure instrumentation. In the case where chamber penetrations are not possible, accelerometers are also used. The high-frequency pressure signal that can be used to determine the presented modes (frequency, type, and amplitude) and their growth rates in the combustor and propellant feed-systems. A common definition of instability is the presence of organized pressure oscillations with a peak-to-peak amplitude greater than 10% of mean value (CPIA 655).

Guidelines for rocket engine combustor stability are provided in CPIA 655, *Guidelines for Combustion Stability Specifications and Verifications Procedures* (Klem and Fry, 1997). Many of these guidelines were based on experiences during the F-1 development program. The document states that any stability test program should include common features that minimize the probability that damage from combustion instability will occur to the propulsion system or test stand; demonstrate that the engine is stable over the entire operating envelope of the engine and expected duty cycle; determines engine stability during the early phases of the development of a combustion chamber; and verifies that the engine maintains stability within the expected range of variation due to manufacturing.

Guidelines for using pressure measurements to diagnose combustion instability are well-documented (e.g. Harrje and Reardon, 1972; Klem and Fry, 1997). Depending upon the mode(s) of instability presented in the chamber – longitudinal, transverse, or combined – a sufficient number of appropriately placed transducers can be used to identify the modes of interest. The selection of the correct pressure transducer and its installation is also critical, especially in the combustion chamber, where the transducer is subjected to extremely high heat fluxes and acceleration forces. CPIA 655 recommends measurements in the combustion chamber and the oxidizer and fuel manifolds. High-frequency measurements of pressure should allow definition of acoustic modes up to the third tangential mode, which can be accomplished with transducers at clocking positions of 0, 135, and 270 degrees.

A major problem that can occur during an engine development program is the unpredict-ability and nonlinearity of combustion instability. For example, instabilities may occur during start transients, or arise from small perturbations during engine operation. Occurrences are not always repeatable, or may occur near the limits of normal engine or thrust chamber operation. An instability that does not present itself until late into the development program, for instance when a thrust chamber is integrated into the engine system, can be disastrous. For these reasons, dynamic stability tests are highly recommended at the component-level testing in the development phase.

In dynamic stability tests, a perturbation from an external source is introduced into the chamber. Typical sources of the perturbation source are bombs and pulse guns. These devices and their use are covered in detail in CPIA 655 and SP-194. Recommended perturbation levels are at least 50% of the mean pressure. According to CPIA 655, the device is dynamically stable if the perturbation is damped within a time $t < (1250/f^{1/2})$, where f is the frequency of the oscillation with the highest amplitude and t has units of ms. If forced perturbation methods are not possible, it is recommended that the combustor be tested along the boundaries of its p_c-MR operating box. For hydrogen-fueled combustors, the stability margin has also been determined by ramping the temperature of the fuel during tests (Hulka and Hutt, 1995). Tests on H_2-fueled combustors using well-designed shear coaxial injectors have shown that increasing the fuel-to-oxidizer velocity ratio leads to stability, and that instability seldom occurs at hydrogen temperatures above about 60 K.

Whereas CPIA 655 and SP-194 cover test practices in the US, AIAA Progress in Aeronautics and Astronautics, Vol. 221 (Dranovsky, 2007) covers testing and development practices in the former Soviet Union. In addition to using some the perturbation methods described above, the text reveals a heavy reliance on theory and statistical approaches. The "oscillation decrement" of a dynamic signal is a passive method of assessing the stability margin in a stable combustor. It was used as a stability rating tool in development combustors and assist in the selection of a viable, stable injector designs, such as in the RD-170 engine (Dranovsky, 2007). The decrement relates stationarity of the pressure oscillation and is essentially the autocorrelation of the signal, comparing the signal to itself as it is shifted through the time domain. The rate of decay of the autocorrelation represents the level of self-similarity and can be used to define a damping rate and hence stability margin. This approach may be particularly useful to discriminate the stability behavior of combustors that present similar levels of pressure oscillation amplitudes.

If a pressure signal is represented by $\eta_i(t)$, then the autocorrelation of $\eta_i(t)$, $C_i(\tau)$, over the period T is defined as:

$$C_i(\tau) = \frac{\int_0^T \eta_i(t)\eta_i(t+\tau)\mathrm{d}t}{\int_0^T \eta_i^2(t)\mathrm{d}t} \tag{12.36}$$

An exponential fit is then calculated for the envelope of the exponential decay, where the argument of the exponential function defines the damping factor. Under stable operating conditions, combustion noise is unorganized, such that the pressure signal is not self-similar. The autocorrelation then decays quickly, representing a high rate of damping. For unstable operation, when amplitudes are high enough to reach the nonlinear limit cycle, organized periodic behavior is present. The autocorrelation results in a much slower decay and lower damping rate in the oscillation decrement.

The center-vented burner, or "T-burner," is the main test device used to rate the combustion stability characteristics of solid propellants. Samples of solid propellant are placed at opposite ends of a combustor to measure the burning response to pressure oscillations. The frequency of the pressure oscillations can be varied by changing the length between the two ends. The stability is characterized by generating a pulse that drives transient acoustic oscillations. The frequency-dependent pressure-coupled response is calculated from the rate of decay of the oscillations (Anon, 1969).

12.4.3 Development testing

Much of the critical development tests use well-instrumented, full-scale, thrust chamber assemblies to test stability characteristics across a wide range of operating conditions. Due to the expense of the full-scale test and the severe consequences of a failed design at that level, stability character-ization at the subscale level is still desirable. Although these tests cannot completely simulate the full-scale environment including all the acoustics, transient operation, and upstream boundary conditions, they can provide valuable insight to the combustion dynamics of the injector. Subscale levels include multiple- or single-element combustors.

Single and multi-element research combustors that are scaled from their full size according to geometric and flow parameters have been used to gain understanding of important physics, acquire validation data for predictive tools, and help develop designs that meet requirements (Hulka, 2008; Dexter *et al.*, 2004). Whereas time-averaged flow characteristics can generally be used to predict performance, heat transfer, and life; combustion dynamics are controlled by complex and nonlinear spatio-temporal couplings between acoustics, hydrodynamics, and heat release that limit the scaling possibilities.

Fourteen rules for scaling experiments for combustion dynamics studies at low pressure were provided by Dexter *et al.* (2004). In their approach, a single injector element, preferably full scale, is tested in an open can combustor that has a natural transverse mode frequency that is close to that of the full-scale combustor. Substituting gas propellants for liquids was allowed to match volumetric flow rates. Since the method was used decades ago for relative comparisons of injector designs, a number of simplifications were used in the approach. Several of the guidelines are still applicable in modern high-pressure tests using more advanced diagnostics. These include the use of injector element scale, which emphasizes the key presumed physics and testing in chambers with acoustic modes in the range of interest. Mixing was also identified as a key physical process, hence matching flow characteristics is important; the momentum flux ratio being a key parameter.

Dexter *et al.* (2004) also describe another methodology for empirical screening of injector designs where full-scale injector elements are used in a representative injector pattern in the subscale tests. The subscale test chamber is designed to so the frequency of its first tangential mode is equal to the frequency of the third tangential mode in the full-scale chamber. If the injector design in the subscale chamber is free from instabilities at this frequency, then it is assumed (on the basis that a high-frequency cutoff for combustion instability exists) the full-scale analog will be free from transverse instabilities of the third acoustic mode and above. It does not presume any instability behavior at the first or second transverse modes in the full-scale chamber. This method was experimentally tested in a LOX/RP-1 combustor experimentally and also used to test the ROCCID predictive model developed for NASA (Muss and Pieper, 1988).

Dexter *et al.* (2004) concluded the following regarding the 1T = 3T methodology:

> One that has been shown to have historical merit is a circular subscale sized so that the third tangential acoustic frequency of the full-scale chamber is made to match the first tangential acoustic frequency of the subscale chamber. An injector pattern that is stable in an undamped chamber of this size demonstrates a minimum level of stability that may be damped successfully in full scale. The previous discussion clearly shows that an injector pattern unstable in this size chamber (or smaller) will present inordinate stability problems at the full scale.

Two-dimensional combustors can approximate the acoustic field of the full-scale combustor in a subscale device (e.g. Muss and Pieper, 1988).

12.4.4 Validation

The rapid advancements in computational capabilities that have occurred over the past two decades, and that will continue to occur, enable detailed simulations of combustion dynamics in both rocket and gas turbine combustors. Confident use of these models call for validation at the benchmark and subsystem levels. Figure 12.13 shows the validation hierarchy defined by Oberkampf and Trucano (2002) to test the accuracy of CFD models.

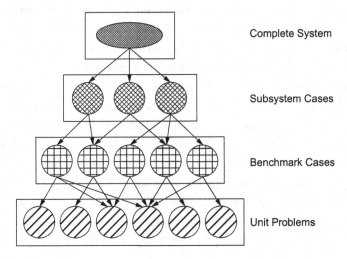

Figure 12.13 Validation hierarchy (Oberkampf and Trucano, 2002).

Simulation at the subsystem and benchmark levels are within the capability of modern computations. Subsystem cases could include subscale combustors containing a multiple set of injector elements, whereas benchmark cases could include single-element combustors. Validation requires that the model be tested at very nearly the same conditions as the simulations. Since combustion instability stems from coupling between flow dynamics, acoustics and the reaction kinetics, the boundary conditions of the chamber play a very important role. Properly devised or measured boundary conditions can help in the goal of isolating and detecting the physics responsible. Poorly matched boundary conditions and/or assumptions can lead to erroneous results far off from the experimental measurements. For these reasons boundary conditions must be controlled and measurable.

Practically it may be necessary to limit and/or simplify the computational domain. Some experiments employ sonically choked inlets to approximate a constant mass flow boundary condition. The choked inlet is a highly reflective acoustic boundary that can be directly simulated with CFD. Such an exit boundary can also be devised to be consistent with a short nozzle approximation, where the nozzle's geometric length is small compared to the acoustic wavelength. Whereas the choked nozzle is fairly simple, the upstream flow dynamics need to be considered for segregating their effects on combustion dynamics. The inlet can present strong hydrodynamic instabilities and other unsteady features along with acoustics as dynamic inputs to the downstream mixing and reaction. Although the energy content of such inputs is negligible compared to the perturbations of the heat release-coupled acoustics, they can be significant in determining the operative driving mechanisms and whether low-level fluctuations decay or grow into a high-amplitude instability.

Thermal conditions at the wall can be very important. Whereas most computations employ an adiabatic wall condition, heat loss to combustor walls affects the acoustic modes that depend on the sound speed and gas temperature, and that couple with the unsteady flow hydrodynamics and heat addition. Wall temperature can also significantly affect the near-wall physics of ignition and flame extinction that may important parts of an instability. For better comparison between simulation and measurement, the use of thermal barrier coatings should be considered in experiments. Whether the chamber is cylindrical or rectangular, and the use of inert flows for cooling windows can have profound effects on the stability behavior that is exhibited.

As a minimum, models should be able to predict the frequency of the instability. Commonly used design analysis models use the acoustics-based models described in Section 12.3, that predict linear growth or decay, hence measured growth rates leading up to the asymptotic or limit cycle condition are valuable validation data, if they exist. Trend-wise data from test configurations that present discriminating stability behavior can be a meaningful test of the prediction, and very valuable to the injector designer. Differences between designs may not be so obvious, for instance in the case where there are no clear instabilities and oscillation amplitudes are similar. More rigorous validation can be obtained with statistics-based analyses such as the oscillation decrement discussed above, or cross-correlations between high-frequency pressure signals.

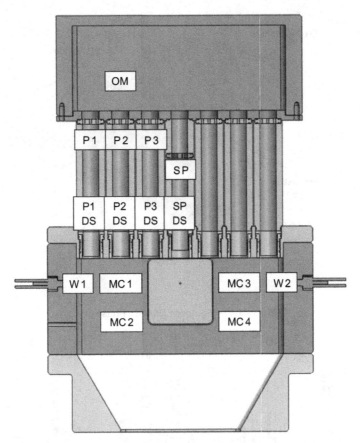

Figure 12.14 Schematic diagram of seven-element combustor and normalized pressure signal and cross-correlation amplitudes for representative test showing self-excited instability.

Wierman *et al.* (2015) used the two-dimensional, multi-element combustor shown in Figure 12.14 to calculate the cross-correlation during a self-excited instability. Results from the analysis of the high-frequency pressure signals are also shown. As the stability behavior transitions between regimes, differences in phase angle between the pressure signal measured at different locations can be seen. Also seen are changes in the magnitude of the cross-correlation between these signals. The normalized cross-correlation shows how temporal coherency between signals change as the instability grows and decays. A more careful examination of the cross-correlation plot shows that the coherency between signals increases prior to growth in amplitude, indicating that the dynamics of the flow become organized in time before the instability grows. Similar to the oscillation decrement described above, the cross-correlation may be used as a measure of stability margin, as well as a possibly predictor of the onset of growth of a combustion instability.

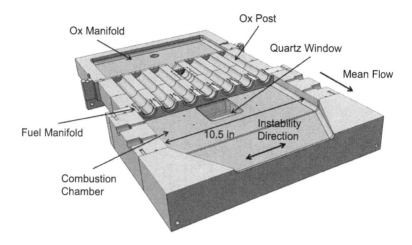

Figure 12.14 (cont.)

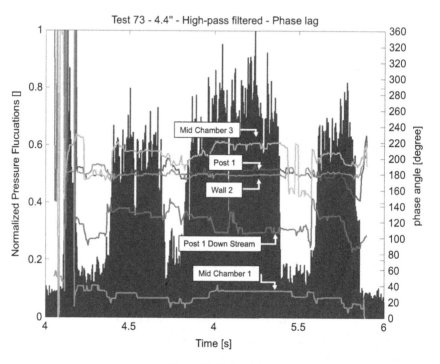

Figure 12.14 (cont.)

   More advanced methods of signal analysis include wavelets, dynamical systems analysis, and complex system theory. A recently developed technique within the last half century, wavelet analysis (Clifton *et al.*, 2007) is being adopted in the study of unstable combustion, predominantly

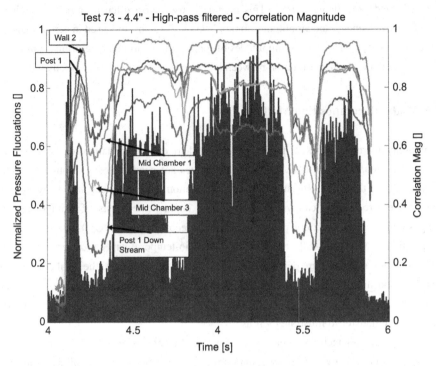

Figure 12.14 (cont.)

in the field of gas turbines. This analysis is used on both pressure and light emission signals from combustion as a detection tool for unsteady combustion. Wavelet analysis provides a means of probing a signal with a small function to study responses at various scales. Signals are decomposed into both time and frequency space, where the result of the wavelet transform is the representation of the wavelet within the probed signal to determine levels of variability in either domain. Unlike Fourier transforms which are localized in frequency, wavelet transforms are localized in both time and frequency. However, there is a tradeoff between time and frequency resolution dependent on the type of wavelet used. Other methods used to characterize the stability behavior of research combustors include intermittency, recurrence, and enhanced correlation between signals to determine impending onset of instabilities (Kabiraj *et al.*, 2016; Nair and Sujith, 2015; Unni and Sujith, 2017), and multi-fractal analysis (Nair and Sujith, 2014).

Validation experiments can be generally split into those that are either self-excited or externally forced. The more general approach used in model combustors has been to artificially force excitation of acoustic resonance modes in the chamber, by means such as pulse guns, explosives, acoustic drivers (speakers) and chopper wheels. These methods are attractive as they allow the measurement of a response to a known forcing frequency. External forcing of model rocket combustors operating at high pressure has been used at DLR and ONERA (Candel *et al.*, 2006; Hardi *et al.*, 2011). A flame transfer function may be obtained if unsteady heat addition can be estimated from flame spectra. This method is used most often in atmospheric flames at relatively low frequencies and amplitudes, for application to combustors used in aeropropulsion and power generating systems. Varying amplitudes of forcing can yield flame describing functions that take into account the nonlinear relation between heat release and pressure or velocity.

12.4.5 Modal Decomposition Methods

Tests of highly instrumented experiments at high sample rates and simulations of the same produce extremely large quantities of data. Analysis techniques are needed to sift through the data and efficiently extract the important and relevant information. Measurement and analysis of high-frequency pressure signals can yield rigorous and sophisticated tests of simulations. Modal decomposition methods, namely proper orthogonal decomposition (POD) and dynamic mode decomposition (DMD), can fully account for the spatio-temporal aspects of pressure, flow, and heat release fields and hence are powerful complements to the analysis of the local pressure signal. Modal decomposition of the flame emission field can yield another level of data for validation. An overview of modal decomposition techniques is provided next, using results from simulations of a research combustor to illustrate their use.

Data processing techniques based on POD and DMD are powerful and elegant methods that can be used to obtain a low-dimensional approximate description of a high-dimensional physical process. POD has been extensively employed in the study of non-reacting flows and has been recently applied to combusting flows to gain insight into flame dynamics. DMD is a newer technique with fewer applications to reacting flow analysis (Motheau *et al.*, 2014).

A primary advantage of the decomposition techniques over filtering is that they work with the entire data set with minimal information loss, unlike traditional spectral methods that analyze data over a limited bandwidth. Moreover, POD and DMD do not require prior knowledge or pre-analysis of the data to obtain the dominant frequencies as do filtering techniques. Another advantage is that decomposition techniques can extract dynamically significant structures from the flowfield of interest. Each decomposed mode can be represented in terms of a spatial response and a temporal response, which provides detailed insight into the dynamics of acoustics and combustion.

For many applications, raw data are available as a series of snapshots. These snapshots can be plots of flow variables from simulations, or can be frames from high-speed imagery or other data collection from experiments. The purpose of the decomposition is to collect data snapshots and decompose them into a reduced form. In the case of POD, this decomposition is done using a singular value decomposition (SVD), while for DMD, the data are reduced using the Arnoldi algorithm. POD decomposes data based on optimality to obtain a set of best representations of the original dataset, whereas DMD decomposes the data based on frequencies. Both POD and DMD result in decomposed modes. POD generates modes which are orthogonal in both space and time, while DMD generates decomposed modes based on discrete frequencies. Huang *et al.* (2016) compared the use of POD and DMD and made following conclusions:

1. POD provides an optimized representation of the dataset and captures the dominant responses efficiently; DMD provides modal decomposition based on frequency content.
2. Multiple frequencies are found in each POD mode. Frequencies of the dominant responses are present in lower modes while weaker responses appear in higher modes. This complicates separating the effects of different physical phenomena which occur at distinct frequencies. With DMD, a direct interpretation of dynamics occurring at discrete frequencies is produced.
3. POD modes of different quantities (e.g. pressure and heat release) are not directly correlated. DMD modes of different quantities can be easily correlated by comparing modes occurring at similar frequencies. The dynamics can be reconstructed accordingly to investigate possible physical coupling.
4. POD requires a complete set of spatial information to produce the correct modes. DMD can be used with partial sets of spatial information.

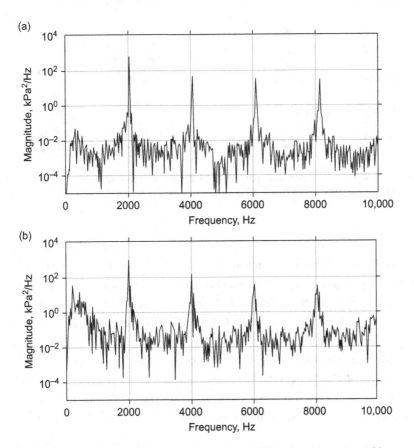

Figure 12.15 Pressure PSD comparison between CFD (a) and experiment (b).

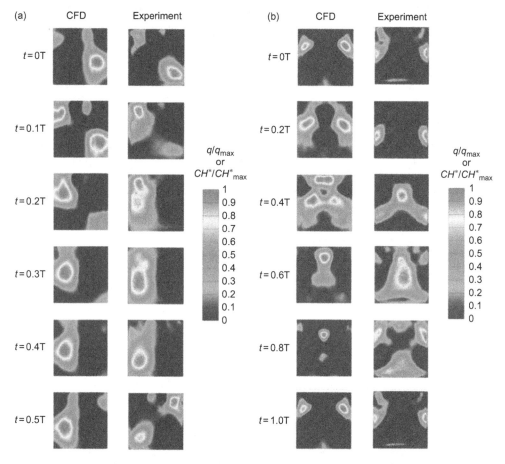

Figure 12.16 Comparison between computed heat release rate from CFD and measured CH* chemiluminescence from experiment using reconstructed DMD responses at (a) 1 W and (b) 2 W frequencies.

For validation, direct and quantitative comparison of predicted values with data is necessary. With regard to predictions of combustion instability, this is presently limited to comparisons of high-frequency pressure. More ideal would be the comparisons between simultaneous measurements of pressure and heat release. Whereas heat release cannot be measured directly, the emission field from short-lived species involved in the combustion process can be used as a marker for reactions actively involved in combustion. Chemiluminescence of OH* and/or CH* is often used.

Morgan *et al.* (2015) applied DMD to compare prediction and measurement of combustion instability in a two-dimensional chamber, similar to the one shown in Figure 12.14. The combustion chamber generates strong self-excited transverse instabilities with the fundamental 1 W frequency at

2000 Hz and harmonics as shown in Figure 12.15. A companion simulation was run, where the only the central three elements were modeled. The flow from the outboard elements was introduced into the chamber as combustion products, and the side walls were moved periodically to produce a close representation of the unsteady transverse flowfield. Reasonable agreement can be seen from the power spectral plots calculated from prediction and experiment.

To compare the unsteady reacting flowfield of the central injector element, frame-by-frame sequences of computed heat release rate and measured CH* data at the frequencies of 1 W and 2 W modes are reconstructed using DMD modes as shown in Figure 12.16. The sequence spans half a cycle of the 1 W frequency and approximately one full cycle of the 2 W frequency. Both the heat release rate and CH* intensity data are normalized by their respective maximum values. It can be readily seen that the reconstructed 1 W and 2 W responses for CFD and experiment show not only similar spatial combustion distribution but also consistent temporal couplings.

Validating the ability of high-fidelity simulations to predict both the unsteady gas dynamic field, and the unsteady heat release field could lead to the use of simulations to provide reliable models for heat release that can be used in multi-fidelity predictions of full-scale combustors.

FURTHER READING

Aerojet (1970) "Design Guide for Stable H_2/O_2 Combustors," Contract NAS 8–20672.

Anon (1969) T-Burner Manual. CPIA Publication No. 191, CPIA, Silver Spring, MD.

Anderson, R. and Santoro, R. (1995) "Impinging Jet Atomization," Chapter 8 in *Liquid Rocket Engine Combustion Instability*, Yang, V. and Anderson, W. E. (eds.). Progress in Aeronautics and Astronautics, Vol. 169. AIAA.

Anderson, R., Santoro, R., and Hewitt, R. (1995) "Combustion Instability Mechanisms in Liquid Rocket Engines Using Impinging Jet Injectors," AIAA 95–2357.

Arbit, H., Tuegel, L., and Dodd, F. (1991) "Heavy Hydrocarbon Main Injector Technology Program Final Report," NASA-CR-184161.

Bilstein, R. (1980) "Stages to Saturn – A Technological History of the Apollo/Saturn Launch Vehicles," NASA Special Publication 4206.

Blomshield, F.S. (2006) "Lessons Learned in Solid Rocket Combustion Instability," in *Proceedings of the 2006 AIAA Missile Sciences Conference*, Monterey, CA, 14–16 November.

Bloxsidge, G. J., Dowling, A. P., and Langhorne, P. J. (1988) "Reheat Buzz: An Acoustically Coupled Combustion Instability. Part 2. Theory," *Journal of Fluid Mechanics*, 193: 445–473.

Candel, S., Juniper, M., Singla, G., Scouflaire, P., and Rolon, C. (2006) "Structure and Dynamics of Cryogenic Flames at Supercritical Pressure," *Combustion Science and Technology*, 178(1–3: 161–192.

Casiano, M. J. (2010) "Extensions to the Time Lag Models for Practical Application to Rocket Engine Stability Design," PhD dissertation, The Pennsylvania State University, College Park, PA.

Clifton, L. A., Yin, H., Clifton, D. A., and Zhang, Y. (2007) "Combined Support Vector Novelty Detection for Multi-channel Combustion Data," in *2007 IEEE International Conference on Networking, Sensing and Control*. IEEE, pp. 495–500.

Crocco, L. (1951) "Aspects of Combustion Stability in Liquid Propellant Rocket Motors: 1. Fundamentals – Low Frequency Instability with Monopropellants," *Journal of the American Rocket Society*, 21(6): 163–178.

Crocco, L. (1965) "Theoretical Studies on Liquid Propellant Rocket Instability," in *Tenth Symposium (International) on Combustion*, pp. 1101–1128.

Culick, F. E. C. and Yang V. (1992) "Prediction of the Stability of Unsteady Motions in Solid-Propellant Rocket Motors," in *Nonsteady Burning and Combustion Stability of Solid Propellants*, Luca, L. D., Price, E. W., and Summerfield, M. (eds.). Progress in Aeronautics and Astronautics, Vol. 143. AIAA, pp. 719–779.

Culick, F. E. C. and Yang, V. (1995) "Overview of Combustion Instabilities in Liquid-Propellant Rocket Engines," in *Liquid Rocket Engine Combustion Instability*, Yang, V. and Anderson, W. E. (eds.). Progress in Aeronautics and Astronautics, Vol. 169. AIAA, pp 3–37.

DeLuca, L., Price, E., and Summerfield, M. (eds.) (1992) *Nonsteady Burning and Combustion Stability of Solid Propellants.* AIAA Progress in Aeronautics and Astronautics Series, Vol. 143. AIAA.

Dexter, C. E., Fisher, M. F., Hulka, J., Denisov, K. P., Shibanov, A. A., and Agarkov, A. F. (2004) "Scaling Techniques for Design, Development, and Test," in *Liquid Rocket Thrust Chambers*. AIAA, pp. 553–600.

Dowling, A. P. (1995) "The Calculation of Thermoacoustic Oscillations," *Journal of Sound and Vibration*, 180(4: 557–581.

Dowling, A. P. (1999) "A Kinematic Model of a Ducted Flame," *Journal of Fluid Mechanics*, 394: 51–72.

Dowling, A. P. and Stow, R. S. (2003) "Acoustic Analysis of Gas Turbine Combustors," *Journal of Propulsion and Power*, 19(5): 751–763.

Dowling, A. P. and Morgans, A. S. (2005) "The Feedback Control of Combustion Oscillations," *Annual Review of Fluid Mechanics*, 37: 151–182.

Dranovsky, M. (2007) "Testing and Development Practices in Russia," in *Combustion Instabilities in Liquid Rocket Engines*, Yang, V., Culick, F., and Talley, D. (eds.). AIAA Progress in Aeronautics and Astronautics Series, Vol. 221. AIAA.

Fabignon, Y., Dupays, J., Avalon, G., Vuillot, F., Lupoglazoff, N., Casalis, G., and Prévost, M. (2003) "Instabilities and Pressure Oscillations in Solid Rocket Motors," *Aerospace Science and Technology*, 7: 191–200.

Finch, R. D. (2005) *Introduction to Acoustics*. Prentice-Hall, Chapter 4.

Flandro, G. A. (1995) "Effects of Vorticity on Rocket Combustion Stability," *Journal of Propulsion and Power*, 11(4): 607–625.

Fleifil, M., Annaswamy, A. M., Ghoniem, Z., and Ghoniem, A. F. (1996) "Response of a Laminar Premixed Flame to Flow Oscillations: A Kinematic Model and Thermoacoustic Instability Results," *Combustion and Flame*, 106(4): 560–573.

Garby, R., Selle, L., and Poinsot, T. (2013) "Large-Eddy Simulation of Combustion Instabilities in a Variable-length Combustor," *Comptes Rendus Mécanique*, 341(1–2): 220–229.

Gröning, S., Hardi, J. Suslov, D., and Oschwald, M. (2016) "Injector-Driven Combustion Instabilities in a Hydrogen/Oxygen Rocket Combustor," *Journal of Propulsion and Power*, 32(3): 59–76.

Gross, A., Osherov, A., and Gany, A. (2003) "Heat Transfer Amplification due to Transverse Mode Oscillations," *Journal of Thermophysics and Heat Transfer*, 17(4): 521–525.

Habiballah, M., Lourme, D. and Pit, F. (1991) "PHEDRE – Numerical Model for Combustion Stability Studies Applied to the Ariane Viking Engine," *Journal of Propulsion*, 7(3): 322–329.

Hardi, J., Oschwald, M., and Dally, B. (2011) "Flame Response to Acoustic Excitation in a Rectangular Rocket Combustor with LOX/H$_2$ Propellants," *CEAS Space Journal*, 2(1): 41–49.

Harrje, D. J. and Reardon, F. H. (1972) *Liquid Propellant Rocket Instability*, NASA SP-194.

Harvazinski, M., Huang, C., Sankaran, V., Feldman, T., Anderson, W., Merkle, C., and Talley, D. (2015) "Coupling Between Hydrodynamics, Acoustics, and Heat Release in a Self-excited Unstable Combustor," *Physics of Fluids*, 27, 041502,.

Huang, C., Anderson, W., and Merkle, C. (2016) "Multi-Fidelity Framework Demonstration for Modeling Combustion Instability," in *AIAA Propulsion and Energy Conference*, July, Salt Lake City, UT. AIAA.

Huang, C., Anderson, W., and Merkle, C. (2017) "Exploration of POD-Galerkin Techniques for Developing Reduced Order Models of Nonlinear Euler Equations," in *53rd AIAA/SAE/ASEE Joint Propulsion Conference*, Atlanta, GA. AIAA 2017–4776.

Hulka, J. and Jones, G. (2010) "Performance and Stability Analyses of Rocket Thrust Chambers with Oxygen/Methane Propellants," in *46th AIAA/ASME/SAE/ASEE Joint Propulsion Conference & Exhibit*, July, Nashville, TN. AIAA 2010–6799.

Hulka, J. (2008) "Scaling of Performance in Liquid Propellant Rocket Engine Combustion Devices," in *44th AIAA/ASME/SAE/ASEE Joint Propulsion Conference & Exhibit, Joint Propulsion Conferences*, Hartford, CT. AIAA 2008–5113.

Hulka, J. and Hutt, J. (1995) "Instability Phenomena in Liquid Oxygen/Hydrogen Rocket Engines," Chapter 2 in *Liquid Rocket Engine Combustion Instability*, Yang, V. and Anderson, W. E. (eds.). Progress in Aeronautics and Astronautics, Vol. 169. AIAA.

Hutt, J. and Rocker, M. (1995) "High-frequency Injection-coupled Combustion Instability," in Liquid Rocket Engine Combustion Instability, Yang V and Anderson W (eds.). Progress in Astronautics and Aeronautics, Vol. 169. AIAA, pp. 345–555.

Kabiraj, L., Saurabh, A., Nawroth, H., Paschereit, C. O., Sujith, R. I., and Karimi, N. (2016) "Recurrence Plots for the Analysis of Combustion Dynamics," in *Recurrence Plots and Their Quantifications: Expanding Horizons: Proceedings of the 6th International Symposium on Recurrence Plots*, Grenoble, France, 17–19 June 2015. Springer, pp. 321–339.

Klem, M.D. and Fry, R.S. (eds.) (1997) *Guidelines for Combustion Stability Specifications and Verification Procedures for Liquid Propellant Rocket Engines*. CPIA Publication 655. The Johns Hopkins University, Chemical Propulsion Information Agency.

Matveev, K. I. (2004) "Vortex-Acoustic Instability in Chambers with Mean Flow and Heat Release," *Technical Acoustics* [online database], available at http://webcenter.ru/~eeaa/ejta.

Morgan, C. J., Shipley, K. J., and Anderson, W. E. (2015) "Comparative Evaluation Between Experiment and Simulation for a Transverse Instability," *Journal of Propulsion and Power*, 31 (6): 1696–1706.

Motheau, E., Nicoud, F., and Poinsot, T. (2014) "Mixed Acoustic–Entropy Combustion Instabilities in Gas Turbines," *Journal of Fluid Mechanics*, 749: 542–576.

Muss, J. (1995) "Instability Phenomena in Liquid Oxygen/Hydrocarbon Rocket Engines," Chapter 3 in *Liquid Rocket Engine Combustion Instability*, Yang, V. and Anderson, W. E. (eds.). Progress in Aeronautics and Astronautics, Vol. 169. AIAA.

Muss, J. and Pieper, J. (1988) "Performance and Stability Characterization of LOX/Hydrocarbon Injectors", *24th Joint Propulsion Conference*, 24th Joint Propulsion Conference, Boston, MA.

Nair, V. and Sujith, R. (2014) "Multifractality in Combustion Noise: Predicting an Impending Combustion Instability," *Journal of Fluid Mechanics*, 747: 635–655.

Nair, V. and Sujith, R. I. (2015) "A Reduced-Order Model for the Onset of Combustion Instability: Physical Mechanisms for Intermittency and Precursors," *Proceedings of the Combustion Institute*, 35(3): 3193–3200.

NASA (1974) *Liquid Rocket Engine Combustion Stabilization Devices*. Special Publication SP-8113. NASA.

Natanzon, M. (2008) *Combustion Instability*. AIAA Progress in Aeronautics and Astronautics Series, Vol. 222, 2008. AIAA.

Oberkampf, W. and Trucano, T. (2007) "Verification and Validation Benchmarks," SAND2007-0853.

Oefelein, J. C. and Yang, V. (1993) "Comprehensive Review of Liquid-Propellant Combustion Instabilities in F-1 Engines," *Journal of Propulsion and Power*, 9(5): 657–677.

Paschereit, C. O., Schuermans, B., Polifke, W., and Mattson, O. (2002) "Measurement of Transfer Matrices and Source Terms of Premixed Flames," *Journal of Engineering for Gas Turbines and Power*, 124: 239–247.

Pavli, A. (1979) "Design and Evaluation of High Performance Rocket Engine Injectors for Use with Hydrocarbon Fuels," NASA TM 79319.

Portillo, J., Sisco, J., Corless, M., Sankaran, V., and Anderson, W. (2006) "Generalized Combustion Instability Model," in *42nd AIAA/ASME/SAE/ASEE Joint Propulsion Conference & Exhibit*, Sacramento, CA, July. AIAA.

Price, E. (1992) "Solid Rocket Combustion Instability – An American Historical Account," Chapter 1 in *Nonsteady Burning and Combustion Stability of Solid Propellants*, Summerfield, M., Price, E., and De Luca, L. (eds.). Progress in Astronautics and Aeronautics, Vol. 143. AIAA, pp. 1–16

Rayleigh, L. (1896) *The Theory of Sound*. London.

Rubinsky, V. (1995) "Combustion Instability in the RD-0110 Engine," in *Liquid Rocket Engine Combustion Instability*, Yang, V. and Anderson, W. E. (eds.). Progress in Aeronautics and Astronautics, Vol. 169. AIAA.

Selle, L., Blouquin, R., Theron, M., Dorey, L.-H., Schmid, M., and Anderson, W. (2014) "Prediction and Analysis of Combustion Instabilities in a Model Rocket Engine," *Journal of Propulsion and Power*, 30(4): 978–990.

Sisco, J., Yu, Y., Sankaran, V., and Anderson, W. (2011) "Examination of Mode Shapes in an Unstable Model Rocket Combustor," *Journal of Sound and Vibration*, 330(1): 61–74.

Souchier, A., Lemoine, J.C. and Dorville, G. (1982) "Resolution du probleme des instabilites sur le moteur Viking," IAF Paper No. 82-363, 33rd IAF Congress, Paris.

Summerfield, M. (1951) "A Theory of Unstable Combustion in Liquid Propellant Rocket Systems," *Journal of the American Rocket Society*, 21(5): 108–114.

Unni, V. R. and Sujith, R. (2017) "Flame Dynamics During Intermittency in a Turbulent Combustor," *Proceedings of the Combustion Institute*, 36(3): 3791–3798.

Urbano, A., Selle, L., Staffelbach, G., Cuenot, B., Schmitt, T., Ducruix, S., and Candel, S. (2016) "Exploration of Combustion Instability Triggering Using Large Eddy Simulation of a Multiple Injector Liquid Rocket Engine," *Combustion and Flame*, 169: 129–140.

Wenzel, L. M. and Szuch, J. R. (1965) "Analysis of Chugging in Liquid-Bipropellant Rocket Engines using Propellants with Different Vaporization Rates," NASA TN D-3080, Cleveland, OH, October.

Wicker, J. M. (1999) "Triggered Instabilities in Rocket Motors and Active Combustion Control for an Incinerator Afterburner Flow," Ph.D. Dissertation, Pennsylvania State University, College Park, PA.

Wierman, M. K., Hallum, W. Z., Anderson, W. E., and Austin, B. L. (2015) "Oxidizer Post Resonance Response of a High Pressure Combustor," in *51st AIAA/SAE/ASEE Joint Propulsion Conference*, Orlando, FL. AIAA.

Yang, V. and Anderson, W. (1995) "Combustion Instability in the RD-0110 Engine," in *Liquid Rocket Engine Combustion Instability*, Yang, V. and Anderson, W. E. (eds.). Progress in Aeronautics and Astronautics, Vol. 169. AIAA.

CHAPTER 13   ELECTRIC PROPULSION FUNDAMENTALS

Electric propulsion (EP) for spacecraft is defined as:

> the use of energy from a source not stored in the chemical bonds of the propellants to augment or provide the exhaust velocity of the propulsion device.

Examples of EP include arcjets, Hall thrusters, and ion engines. The beauty of EP is that in space there is a free and plentiful source of energy – namely the Sun. EP has a great impact on the ability to execute space missions (Rayman, 2018). It is not surprising then to learn that, as early as 1964, EP devices were being flown as experiments and in 1974 the Zond-2 Russian Mars probe was launched. With its launch, Zond-2 became the first spacecraft to rely on EP in an operational role with six Teflon pulsed plasma thrusters (PPTs) used for attitude control (Zak, 2016). Perhaps more surprising is that using EP was not a new idea in the 1960s. Early rocketry pioneers such as Konstantin Tsiolkovsky and Hermann Oberth both discussed the possibility of its use (Mel'kumov, 1965; Oberth, 1929). This caused even the great Werner von Braun to be intrigued by the possibilities and in 1947 he assigned one of his key engineers, Dr. Ernst Stuhlinger, to assess the potential for using EP on missions including exploration of Mars. When Stuhlinger seemed a bit reluctant to take on the assignment, von Braun urged him on by stating, "Professor Oberth has been right with so many of his early proposals, I would not be at all surprised if we flew to Mars electrically" (Edgar, 2005).

In this chapter, we will introduce the reader to the fundamentals of electric propulsion. An introduction to the concept and its advantages is included in Section 13.1. Section 13.2 provides a discussion of the background and historical development. Section 13.3 provides a top level overview of the fundamentals of electric propulsion, including the three principal types of acceleration: electrothermal, electrostatic, and electromagnetic. Section 13.4 builds on this discussion by describing the typical thrusters used in each of these categories, while Section 13.5 delves into the applications of EP. Section 13.6 addresses the issues of spacecraft interaction that must be considered when using EP systems.

13.1 INTRODUCTION

We can think of electric propulsion as an "end-run" around the limitations placed on our ability to increase the performance of rocket engines. In Chapter 3, we learned of the dominant role I_{sp} plays as it appears in the exponential of the rocket equation and this factor is perhaps never more important than in space where all mass is precious due to the high delivery cost. In the ever important quest to reduce the mass of spacecraft systems, the propulsion engineer often looks to reduce propellant mass, because it costs just as much to launch as does useful mass into orbit. For example, consider the options shown in Table 13.1 for a satellite control thruster, assuming a 5-year north–south stationkeeping (NSSK) maneuver in geosynchronous orbit for a 2000 kg beginning of life (BOL) satellite mass.

The advantages of EP over chemical propulsion – nearly a factor of 10 reduction in propellant required to perform the same function – can be seen from the example in Table 13.1. Figure 13.1 graphically illustrates the overall system comparison of a chemical propulsion system and an EP system. In the chemical system, much of the launch mass is propellant and the associated

Table 13.1 Comparison of attitude control propulsion options

Propellant	I_{sp}, s	Total propellant mass for 5-year NSSK, kg
Monopropellant hydrazine	220	219
Bipropellant (MMH/NTO)	300	163
Bipropellant (N2H4/LOX)	340	144
Hydrazine arcjet	600	83
Xenon hall thruster	1800	28

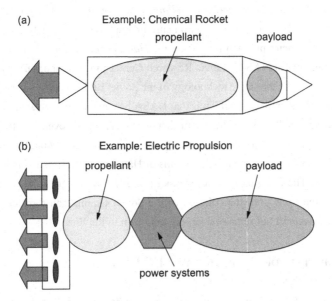

Figure 13.1 EP systems greatly reduce the proportion of a spacecraft dedicated to propellant. (a) Diagram of a chemical rocket. (b) An EP system.

structure and systems to manage the propellant and house the thrusters. By contrast, the EP system has a much lower percentage of the total launch mass devoted to propellant and associated systems. It does, however, have to add in the mass of the power processing electronics, which are not required for the chemical system.

The advantage, in a mass-constrained system, is that more mass is devoted to useful payload and less is devoted to propellant. This is true in general, even though the EP system must accommodate the added mass of the power system to provide the power to the thrusters. The savings in propellant mass and associated systems is more than enough to accommodate the extra power system mass.

13.2 Background: Historical Developments

As mentioned above, early experimental satellites were flying EP in the 1960s in both the West (US) and the East (Soviet Union). Such experimental flights as the Space Electric Rocket Test (SERT I) launched on July 20, 1964 (interestingly exactly 5 years to the day before Neil Armstrong's first steps on the Moon) and the Meteor series of test flights in the Soviet Union established a baseline upon which the foundation of flight EP systems could be built. SERT I flew two *gridded ion engine* designs (Cybulski *et al.*, 1965; Gold *et al.*, 1965). The first was a cesium-fueled contact-type ion engine. It suffered a failure due to a high-voltage short circuit. The second device operated for over 31 minutes of the sounding rocket flight, making it the first successful test of an EP device on a Western spacecraft. It was based on the mercury-fueled electron bombardment design ion engine pioneered by NASA's Lewis Research Center in Cleveland, OH. Readers should take note of the fuel (or more accurately *propellant*) types – cesium and mercury.

Meanwhile on the other side of the iron curtain, the Soviets were actively developing their own EP technology. Their approach hinged on the *Hall effect thruster* developed and originally patented by an engineer at NASA's Lewis Research Center in Cleveland, OH. The story of how a US invention came to be the staple technology of the Soviet EP industry is a very interesting (and little known) tale. The summary of which is that NASA became more interested in the gridded ion engine and never pursued the Hall thruster further. The Soviets, however, saw the US Patent and had several of their prominent physicists working on it throughout the 1960s. The Soviet physicists were able to solve some of the daunting problems of Hall thruster design and make a system that worked adequately. The first Russian Hall thruster (an SPT-60) flew in 1971 on the Meteor 1–10 spacecraft. From that time until the collapse of the Soviet Union, Hall thrusters were regularly flown on Soviet spacecraft but remained largely unknown in the West.

13.3 Fundamentals of Operation for EP Devices

There are two basic types of electric propulsion: devices where electrical energy is used to add heat to a propellant in a conventional thermal expansion – *electrothermal thrusters*; and devices where electrical energy is used to ionize and accelerate an ionized propellant – *electromagnetic* and *electrostatic thrusters*. Before we delve further into the types of EP and how it is used, we should delve a bit deeper into the fundamentals. Things such as:

- How does an EP device generate thrust?
- What are the various discharge types?
- What sort of losses occur in EP devices and how can they be minimized?
- What are cathodes and neutralizers?
- What are the lifetime considerations/limitations for various EP devices?
- How important is power processing and how is it usually implemented?
- How important is propellant selection and flow control?

13.3.1 Electrothermal Acceleration

Up until now, we have focused on propulsion systems that react chemical propellants in some fashion to generate enthalpy and develop thrust via thermal expansion through a nozzle. EP devices can also utilize this type of thermodynamic mechanism to generate thrust, via either heaters or direct arc heating of propellant to add enthalpy. Such devices can either augment an already heated flow or provide all of the heat to the propellant prior to expansion in a nozzle. This general category is classified as electrothermal propulsion. As noted in Chapter 8, the increase in chamber temperature (ΔT_c) can be simply estimated by equating electrical energy (P_e) added to the mixture to the change in thermal energy:

$$\Delta T_c = P_e/(\dot{m}c_p) \tag{13.1}$$

The associated increase in I_{sp} can then be estimated recognizing that this parameter scales with $\sqrt{T_c/\mathfrak{M}}$. In the case of a hydrazine resistojet, the base T_c is determined by the decomposition of hydrazine propellant as it flows over a catalyst. This T_c is raised via the addition of heat as the decomposition gases flow over a surface heated by a resistive coil to a temperature above the base T_c. The I_{sp} increases from 220 s to 300 s through the addition of heat to raise the T_c. Such a device is shown later in the chapter in Figure 13.8.

A second general category of EP devices use ionization of the propellant and acceleration of the ions by means of an electric field. This category is known as electrostatic. Finally, there are EP devices which ionize the propellant and accelerate it through the action of crossed electric and magnetic fields. This category is known as the electromagnetic thrusters.

13.3.2 Fundamentals of Plasmas and Plasma Acceleration

The first thing we need to establish before we discuss how electric propulsion devices (other than electrothermal) develop thrust is some basic facts about plasmas. Plasmas are defined as a quasi-neutral gas of charged particles (+ and –) and neutral particles that exhibit collective behavior (Krall and Trivelpiece, 1986). Two terms from that definition require their own definitions:

Quasi-neutral: zero net charge (+ and –) within a volume larger than a characteristic dimension known as the Debye length (l_d), which is a measure of the distance the electron thermal energy can maintain between the ions and the electrons.

Collective behavior: gas properties resulting from the interactions between charged particles and the self- and externally generated electromagnetic fields (note: plasmas do not require collisions to cause transport or motion!).

The charged particles present in the plasma make it electrically conductive. If there is electrical conductivity that implies that there are moving charge carriers and thus a current. So let us analyze a bit further. We can express a generalized Ohm's law:

$$V = IR \tag{13.2}$$

This is the way we express Ohm's law for a circuit with an applied voltage V, a current I, and a known resistance R. In a plasma, we have n_q charged particles q, which yield a conductivity, σ. We can then define the current density $\vec{j}$ in terms of the imposed electric field $\vec{E}$ and the conductivity of the plasma:

$$\vec{j} = \sigma\vec{E} \tag{13.3}$$

The force felt by a current driven within the gas is given by:

$$\vec{F} = \vec{j} \times \vec{B} \tag{13.4}$$

This definition is key to understanding the acceleration of bulk plasma in electromagnetic devices. If we introduce motion of the plasma, we add a term and achieve the classic equation describing the force acting on charges in a flowing plasma – the *Lorentz force* (Lorrain and Corson, 1970):

$$\vec{F_p} = \sigma\left(\vec{E} + \vec{\upsilon} \times \vec{B}\right) \tag{13.5}$$

where $\vec{\upsilon}$ is the velocity of the flowing plasma and $\vec{B}$ is the magnetic field

This provides the means to accelerate the charged particles simply through the action of the electric and the magnetic fields. No collisional or thermal mechanisms need to be involved. The vector product is illustrated simply in Figure 13.2.

Electrostatic acceleration is accomplished in gridded ion engines by imposing a strong negative bias to a physical grid, which attracts the ions from the chamber. To control the discharge, a second electrode is used to provide beam shaping. This configuration is shown in Figure 13.3.

It can be a daunting task to try and define the plasma conditions and electric field interactions over this large array of extraction elements. However, it has been shown that one can define an individual aperture and a "beamlet" that flows through that aperture and analyze the situation as a one-dimensional beam. When this approach is taken, we get as a result the Child–Langmuir space charge limited current equation. The situation that we are analyzing is illustrated in Figure 13.4.

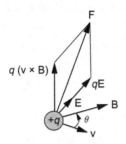

Figure 13.2 Coordinates describing the Lorentz force action on a charged particle (arbitrary E and B fields).

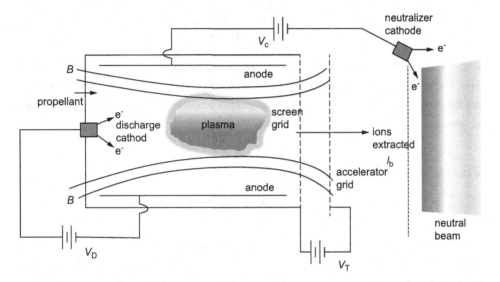

Figure 13.3 Schematic diagram of gridded ion engine showing magnetic fields and electric circuits.

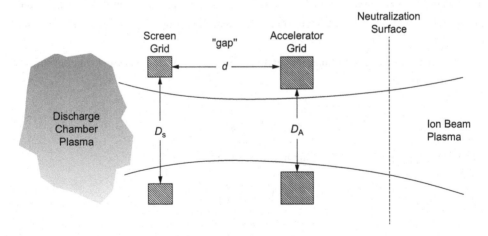

Figure 13.4 Simplified model of ion acceleration using a single "beamlet."

To do this we can start with the following expressions for a plasma (Chen, 1984):

Poisson's equation in the gap:

$$\frac{d^2\varphi}{dx^2} = \frac{-en_i}{\varepsilon_0}$$ (13.6)

Ion continuity:

$$j = en_i v_i \equiv \text{constant}$$ (13.7)

Electrostatic ion exhaust velocity:

$$v_i = \sqrt{\frac{2e(-\varphi)}{m_i}}$$ (13.8)

where e is the unit charge, φ is the accelerating potential, and ε_0 is a constant known as the *permittivity of free space* and has units of farads per meter (Lorrain and Corson, 1970). Using these equations, we can obtain a second-order, nonlinear differential equation for $\varphi(x)$ (Jahn, 1968). The boundary conditions are:

$$\varphi(0) = 0 \text{ and } \varphi(x = d) = -V_T$$

We can also assert the condition that the field must be zero at screen grid:

$$\frac{d\varphi}{dx_{x=0}} = 0$$ (13.9)

This assumption arises from the operation of the ion engine at space charge limited conditions (Spitzer, 1962). Assuming that we have an ion source that keeps producing ions at a sufficient rate, applying a negative screen field extracts more ions, thus increasing positive space charge in the gap. This tends to reduce the negative screen field, and the process stops only when this field is driven to near zero. If the field were driven positive, ion flow would cease. At that point, the grids are by definition extracting the highest current density possible, and are said to be "space charge limited."

Using these three conditions, integration of Eqs. 13.5–13.7 will yield the voltage profile and also the current density j. The result is:

$$j = \frac{4}{9}\varepsilon_0 \sqrt{2\frac{e}{m_i}} \frac{V_T^{3/2}}{d^2}$$ (13.10)

which is recognized as the Child–Langmuir law, and also:

$$\varphi(x) = -V_T \left(\frac{x}{d}\right)^{4/3}$$ (13.11)

$$E(x) = \frac{4}{3}\frac{V_T}{d} \left(\frac{x}{d}\right)^{1/3}$$ (13.12)

Equation 13.11 shows that the field is zero (as imposed) at $x = 0$, and is $\frac{4}{3}(V_T/d)$ at $x = d$ (the accelerator grid). This now allows us to calculate the net electrical force per unit area on the ions in the gap:

$$\frac{F}{A} = \frac{1}{2}\varepsilon_0 \left(\frac{4}{3}\frac{V_T}{d}\right)^2 = \frac{8}{9}\varepsilon_0 \left(\frac{V_T}{d}\right)^2 \tag{13.13}$$

which is of course the same as the thrust (assuming no other force acts on the ions once they are past the accelerator).

We should then be able to obtain the same result from the classical rocket thrust equation:

$$F = \dot{m}u_e \tag{13.14}$$

The mass flow rate is:

$$\dot{m} = jA\frac{m_i}{e} \tag{13.15}$$

and the ion exit velocity is:

$$u_e = \sqrt{\frac{2eV_T}{m_i}} \tag{13.16}$$

giving:

$$\frac{F}{A} = \frac{\dot{m}}{A}u_e = j\frac{m_i}{e}\sqrt{\frac{2eV_T}{m_i}} \tag{13.17}$$

Substituting for j using the Child–Langmuir law, Eq. 13.10, this becomes:

$$\frac{F}{A} = \frac{4}{9}\varepsilon_0\sqrt{2\frac{e}{m_i}}\frac{V_T^{3/2}}{d^2}\frac{m_i}{e}\sqrt{\frac{2eV_T}{m_i}} \tag{13.18}$$

which reduces to the identical result:

$$\frac{F}{A} = \frac{8}{9}\varepsilon_0 \left(\frac{V_T}{d}\right)^2 \tag{13.19}$$

Therefore, the force per unit area of a gridded ion engine can be expressed in terms of the voltage V_T and the gap dimension d. Other important parameters in calculating electrostatic engine performance include the charge-to-mass ratio e/m_i and the ionization production cost.

For this reason, propellant selection is driven to high molecular weight elements such as mercury, cesium, and xenon which have a high charge-to-mass ratio. Note that this is exactly the opposite of what we desire for a chemical or electrothermal thruster, where a low molecular weight species is used to raise the sound speed and hence the exit velocity. We will cover this subject in more depth later in Section 13.3.5.

13.3.3 Loss Mechanisms in Electric Thrusters

Losses in electrothermal thrusters are similar to those in chemical rocket engines. A large contributor is the *frozen flow loss*. The short residence time in the constrictor and nozzle does not allow for recovery of energy from modes such as recombination. This is evident from the bright plume of a high-power ammonia arcjet seen in Figure 13.5. The pink color is attributable to the hydrogen Balmer series emission occurring downstream of the nozzle exit. Also, there is a significant viscous flow loss component due to the low Reynolds number flow (Re $\gg$ 1000) in these devices. Other losses occur owing to electrode sheaths at the cathode and anode, with the losses at the anode being particularly of concern in arcjet thrusters.

Anode losses result from sheath phenomena, radiative and convective heat transfer from the gas to the anode, and viscous losses from the flow. The result is power deposited into the anode that is no longer available to increase the enthalpy of the gas. As seen in Figure 13.5, the anode in an arcjet thruster gets quite hot to the point that it glows incandescently. Anode losses can be characterized with the following expression:

$$P_a = P_{rad} + kA\left(\eta_{ps} - \eta_a\right) + I\left(\frac{5kT_e}{2e} + V_f + \phi\right) \tag{13.20}$$

where P_a is the power deposited in the anode, P_{rad} is the power transferred by radiation from the gas to the anode, and the next two terms represent power deposited by conduction and electron current heating, respectively.

For electrostatic and electromagnetic thrusters, losses are more related to the ionization and acceleration processes. These can vary from one thruster type to another but are generally in three classes. The electrical energy must be partitioned into amounts required to produce ions and electrons, to provide and sustain electric and magnetic fields in the discharges, and to neutralize the ion beams produced.

Figure 13.5 High-power ammonia arcjet firing shows hydrogen Balmer series recombination occurring in plume far downstream of nozzle exit (Image courtesy of Aeroject Rocketdyne).

In electro static thrusters, this is typically expressed in terms of two parameters – *mass utilization efficiency* and *discharge loss*. Mass utilization efficiency is the measure of the fraction of the propellant mass that is converted into ions and accelerated by the thruster. It can be defined in the following way:

$$\eta_m = \frac{\dot{m}_i}{\dot{m}_p} = \frac{I_b}{e} \frac{M}{\dot{m}_p} \qquad (13.21)$$

where $\dot{m}_i$ is the mass flow rate of ions, $\dot{m}_p$ is the total mass flow rate of propellant, I_b is the beam current, e is the charge (for a singly ionized case), and M is the ion mass.

The electrical efficiency of the thruster is defined as the ratio of the power in the beam to the total input power, which can be simply expressed as:

$$\frac{P_b}{P_T} = \frac{I_b V_b}{I_b V_b + P_{other}} \qquad (13.22)$$

where P_{other} represents the other combined power input to the thruster to include the electrical cost of producing the ions, the power to the cathode heater and keeper, and the grid currents in ion thrusters.

The cost of producing ions is described by a measure of ion production efficiency, referred to in the literature as the discharge loss:

$$\eta_d = \frac{\text{power to produce ions}}{\text{current of ions produced}} = \frac{P_d}{I_b} \qquad (13.23)$$

The units of η_d are therefore watts per ampere, which can also be expressed as electron-volts per ion. This term is different from most efficiency terms in that it is desirable to have its value be as small as possible. This is true because it represents a power loss.

Example: If an ion engine takes a 15 A, 30 V discharge to produce 3 A of ions in the beam, the discharge loss would be:

$$\eta_d = \frac{15\ A \times 30\ V}{3\ A} = 150\frac{W}{A} \text{ or } 150\frac{eV}{ion}$$

13.3.4   Thrust Efficiency

The overall thrust efficiency of an electric thruster is defined as the ratio of the power in the beam effecting a thrust upon the spacecraft to the input electrical power. This can be written as follows:

$$\eta_T = \frac{P_{beam}}{P_i} \qquad (13.24)$$

For electric thrusters P_{beam} is defined in terms of measurable quantities thrust T and mass flow $\dot{m}$:

$$P_{beam} = \frac{T^2}{2\dot{m}} \qquad (13.25)$$

Therefore the total efficiency can be expressed as:

$$\eta_T = \frac{T^2}{2\dot{m}P_i} \tag{13.26}$$

Recall that I_{sp} is defined as:

$$I_{sp} = \frac{T}{\dot{m}g} \tag{13.27}$$

We now have:

$$\eta_T = \frac{g}{2}I_{sp}\frac{T}{P_i} \tag{13.28}$$

This is a very useful expression for calculating the efficiency of any electric thruster from commonly measured parameters. It is also evident from Eq. 13.28 that for a given thruster efficiency, I_{sp} and T/P are inversely related. If I_{sp} increases, T/P must decrease and vice versa. This is an important rule of thumb to know when comparing electric thrusters for different kinds of mission applications.

13.3.5 Neutralization (Cathodes and Cathode Phenomena)

Cathodes play an important role in almost all EP devices with plasma plumes. They are the source of the electrons that create the ions and that also neutralize the plasma beam as it leaves the spacecraft. There are many types of electron-emitting devices that can be used. However, because of its favorable lifetime and reliable operational characteristics, the *hollow cathode* has become the standard for almost all flight systems.

The basics of hollow cathode operation are shown in Figure 13.6. A small amount of propellant flow is used to provide the discharge medium. Initially, to get the cathode up to temperatures where thermionic emission can occur, an external heater is used. An anode (called a keeper) located just downstream of the orifice is also required. Electrons are emitted from the insert surface via a field-enhanced thermionic emission process. These electrons are accelerated through the cathode sheath at the electrode surface and ionize neutral atoms. Practically all of the electrons produced via emission and through the ionization process leave the cathode through the orifice. This is true because the cathode sheath blocks their flow to the insert and other internal surfaces that are at cathode potential. The electrons see the positive potential of the anode (keeper) and are drawn toward it. Ions inside the cathode bombard the insert and cathode interior surfaces, thereby heating them and sustaining the thermionic emission process. For neutralization, the electrons produced form a plasma bridge to the main discharge and equalize the net charge on the spacecraft. To do this it is necessary that the beam current and the electron emission current be equal.

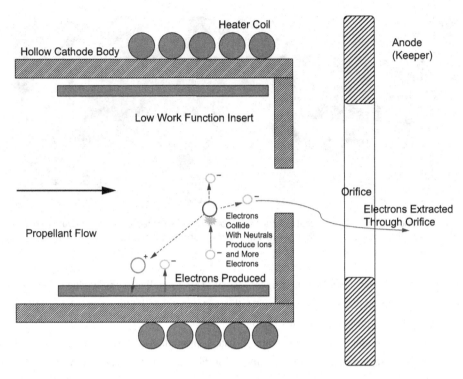

Figure 13.6 Key features of a hollow cathode showing electron production and extraction.

From a system standpoint this creates an interesting complication. The flow must be split off from the main discharge and regulated separately to feed the cathode at the appropriate flow rate. To avoid this complication, there have been cathodes proposed which require no propellant to operate. Examples include field emission cathodes made of carbon nanotubes. These have thus far been designed to operate up to only a few mA of current, which limits their application to low-power EP systems.

Cathode erosion can be the one of the main lifetime limitations in EP thrusters. The main wearout mechanism is erosion of the keeper electrode by ions which are accelerated by the cathode fall voltage. An example of this phenomena is shown in the wear test results of the neutralizer cathode for the Dawn mission as evident from the pre-test and post-test photos in Figure 13.7.

13.3.6 Propellant Selection for EP Thrusters

For electrothermal thrusters, the same rules of thumb for propellants apply as they do for chemical thrusters. The lowest molecular weight gas possible will be the best performer. In this case, hydrogen would be the propellant of choice. However, since few satellites or spacecraft carry hydrogen because it is difficult to store for long times in space, a compromise is made to a space

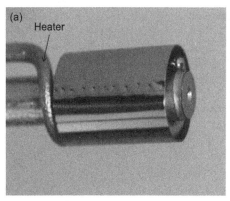

Figure 13.7 Evidence from life test that discharge cathode must be protected from ion sputtering. (a) Pre-test discharge cathode. (b) Post-test discharge cathode. (Image courtesy of NASA.)

storable propellant that has a high hydrogen content. This propellant is hydrazine (N_2H_4) and is commonly used on satellites and spacecraft. Both resistojet and arcjet thrusters were developed to work with hydrazine propellant for this reason.

For electromagnetic and electrostatic thrusters, propellant selection becomes more about *ion production cost* and *charge-to-mass ratio*. This facilitates high thruster efficiency. A high propellant molecular mass facilitates operation at high propellant utilization efficiency and if specific impulse is held at its optimal value, it assures operation at a high net acceleration voltage, V_N, where thruster electrical efficiency is high. Achievement of this criterion is also aided by low first ionization and high first excitation potentials. Such potentials facilitate minimal unwanted radiation losses and a low ion production cost.

Another important operating feature is a high *thrust-to-power ratio*. High molecular mass propellants also help to achieve this. High thrust-to-power is important because most EP systems are power constrained and maximizing thrust keeps trip times (or maneuver durations) shorter. Additionally, monoatomic propellant species are preferred because they ionize predictably and their expulsion is easily measured onboard the spacecraft to allow for operational controllability.

Propellants with a low first ionization potential and a high second ionization potential are favored because doubly charged ions and charge exchange ions cause excessive erosion of the thruster surfaces. This is one of the main contributors to thruster wearout mechanisms that limit lifetime.

Other criteria include compatibility with spacecraft surfaces, materials, and functions. Propellants which may react, condense, or degrade surfaces of solar panels or thermal radiators are not desirable. Also storability over many years of operation in space is important. Finally, cost is always a consideration.

This results in the shortlist of interesting propellants shown in Table 13.2.

Table 13.2 Propellants most favorable for electrostatic and electromagnetic thrusters

Propellant	Molecular weight	Storability	Compatibility	Comments
Cesium	133	Good	Poor	Condensing on S/C
Mercury	201	Very good	Good	Toxicity limits application
Xenon	131	Very good	Excellent	Common use today
Krypton	84	Good	Excellent	Can be used in blend with Xe
Iodine	254	Excellent	Poor	Condensing on S/C

13.4 TYPES OF ELECTRIC PROPULSION DEVICES

13.4.1 Electrothermal

Electrothermal thrusters utilize electrical energy to impart heating to a gas. This propellant gas can be of any type, including decomposition products of monopropellant hydrazine thrusters. Propellant selection depends on many factors, but in general lower molecular weight propellants provide higher performance because the thrust is derived from expansion through a nozzle. There are two types of electrothermal thrusters in use today.

Resistojets
Resistojets employ a heater coil to provide the additional heating of the propellant gas. The name derives from the fact that the coil is a resistor in the electrical circuit and heats up when current is passed through it, much like the filament of an incandescent light bulb. The coil can either be enclosed in a sealed cavity and radiatively transfer heat to the walls of the flow passage or it can be immersed in the flowing gas. A typical immersed heater resistojet design is shown in Figure 13.9. It is more compact and simpler in concept than the radiative heater design, but early attempts to fly immersed heaters found problems with the hot coil material reacting with the gas or being eroded by the flowing gas. This lead to the selection of the radiatively heated resistojet as the primary choice for flight designs. A detailed cross-section of the Aerojet Rocketdyne MR-501, highlighting the gas path is shown in Figure 13.8. The figure shows the coil which heats the inner walls of the annular flow passage prior to the gas expanding out the nozzle. The MR-501 may be the most flown EP device in terms of cumulative numbers, since its use goes back into the 1980s on geosynchronous communications satellites and since it was also used on the first generation Iridium satellite constellation. In all, more than 280 MR-501 and MR-502 (an improved version) resistojets have flown.

Another unique feature of the resistojet thruster is that it can accommodate use of a wide variety of propellants. The MR-501 described above operates on hydrazine decomposition products (N_2, NH_3, and H_2); however, other tests have been run on diverse propellants such as CO_2 and even water. At one time, NASA studied the use of a resistojet design for the space station that would

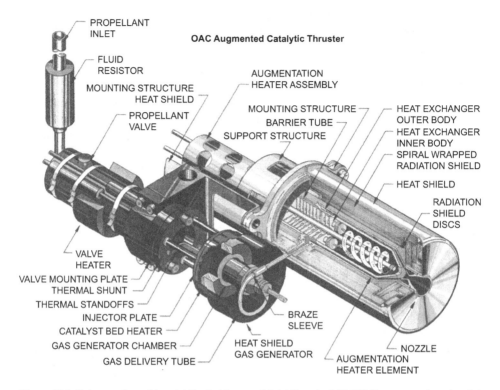

Figure 13.8 Cutaway view of Aerojet Rocketdyne resistojet thruster MR-501 (Image courtesy Aerojet Rocketdyne).

use waste gases from the environmental system while providing some of the reboost propulsion to keep the station in its desired orbit (Morren *et al.*, 1989). These resistojets were also flown on the first generation of the Iridium[TM] satellites, where they provide repositioning and deorbit propulsion for the 70+ satellite constellation.

Arcjets

Arcjet thrusters take the electrical augmentation of a propellant flow beyond the limitations of the resistojet thruster by replacing the coil and heat exchanger with a direct arc heating mechanism. This process is illustrated in Figure 13.9. The propellant flows in an annular path around a center-body cathode, then through a narrow constrictor passage. Vortex flow generated by the gas injection ports is used to help maintain arc stability. The arc is struck between the cathode and the anode, which is also the nozzle of the device. (A note on convention here: for EP devices cathodes are defined as the negative electrode in the circuit and anodes the positive. Electrons always flow to the anode.) As the gas flows through the constrictor, the arc ionizes a small column of gas and directly heats the remainder. The heated gas is then expanded in the nozzle to produce thrust via conversion of enthalpy to kinetic energy. Loss mechanisms are primarily dominated by anode heating and

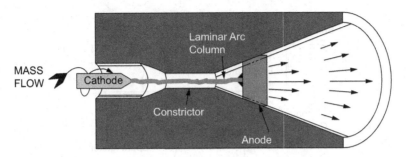

Figure 13.9 Diagram of arcjet thruster showing arc discharge heating propellant flow.

frozen flow losses. First used on the Telstar-401 satellite in 1994, there are now more than 200 arcjet systems flying on various satellites.

13.4.2   Electromagnetic

Electromagnetic thrusters generate thrust through an interaction of crossed electric and magnetic fields. The first thing that must happen is to ionize the propellant gas to produce a plasma. There are many different ways to approach this but two of the more common are: using a high voltage to create an arc discharge; and using a radio frequency (RF) field to add energy to free electrons which ionize the gas via collisions. Once the plasma is established, current can flow and magnetic fields (either self-generated or applied) can react with the electric field present in the discharge to produce a force on the charged particles which is perpendicular to both the electric and magnetic fields.

Pulsed Plasma Thrusters
Pulsed plasma thrusters (PPTs) have been designed with many configurations, but the most common is the Teflon pulsed plasma thruster shown in Figure 13.10. The PPT is very simple in design, with the fuel bar as the only moving part. A capacitor is charged from the spacecraft power bus, then discharged by means of a spark igniter across the face of the Teflon fuel bar. As the arc discharge passes from the cathode to the anode, it ablates material off the face of the fuel bar and ionizes it. The flowing current sets up a magnetic field and the interaction of the current and the magnetic fields ($j \times B$) produces a body force on the plasma which accelerates it out at high velocity. This discharge is shown in Figure 13.11. The capacitor is then recharged and the process is repeated. This cycle can be run at multi-hertz (1–10 hZ) frequency if power is available. The individual pulses produce a very small impulse-bit, which makes the PPT attractive for attitude control or fine pointing applications. Recent demonstrations on the Earth Observing-1 mission proved that a PPT could replace the function of a momentum wheel with no degradation of image quality from the advanced Landsat imager (Meckel *et al.*, 1997). In three-axis stabilized satellites, momentum wheels are commonly used to compensate for small disturbances in each of the three spacecraft axes, providing a very stable platform for imagers and other instruments.

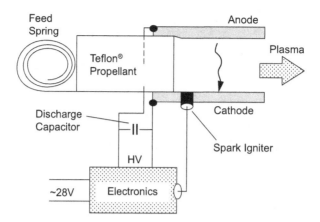

Figure 13.10 Simplified schematic of a solid propellant (Teflon) pulsed plasma thruster. A coiled spring at the left forces the Teflon propellant into position as portions are ablated away during thruster operation.

Figure 13.11 PPT discharge in vacuum chamber test (Image courtesy Aerojet Rocketdyne).

Magnetoplasmadynamic (MPD) Thrusters

Magnetoplasmadynamic (MPD) thrusters have been of great interest as a high-power EP device since the 1960s. They are simple, cylindrically symmetric channels with a center-body cathode and a ring shaped anode, as shown in Figure 13.12. Gas is injected through an insulated backplate and is ionized and accelerated via a $j \times B$ interaction. The magnetic field can be either self-generated or applied through external magnets. It is attractive because it has the potential to process very large amounts of power in a compact size. These features belie the difficulties that have limited the application of the MPD thruster.

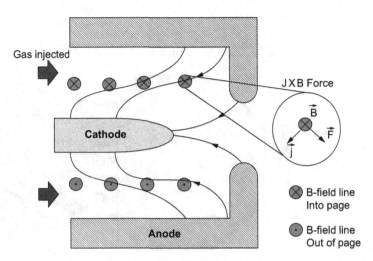

Figure 13.12 Simplified diagram of MPD thruster operation showing current streamlines and self-induced magnetic field.

The first major obstacle is the high power required to efficiently operate an MPD thruster. Self-field MPD thrusters operate best at above 1 MW. Applied field MPD thrusters have been run at powers as low as hundreds of kilowatts, but even that is much higher than what is available from spacecraft today. Until such time as there is a nuclear power source for space, we are unlikely to see applications of MPD thrusters.

13.4.3 Electrostatic

Electrostatic thrusters or engines use an electric field imposed by a physical set of grids to accelerate ions from plasma formed inside the discharge chamber to high exhaust velocity. The ions thus extracted are neutralized by electrons from an external cathode. The processes taking place in an electrostatic thruster are summarized in Figure 13.13. Neutral propellant is injected into an ionization chamber, which may operate on a variety of principles: electron bombardment (as shown in the figure), contact ionization, radio-frequency ionization, or any other means of ionizing a neutral gas. Even if the plasma contained in the discharge chamber is weakly ionized, we may assume that only ions and electrons leave this chamber because only ions are extracted by the imposed field.

Gridded Ion Engines
Ion engines are possibly the most famous EP devices. Gene Roddenberry knew about them and included them in his *Star Trek* screenplays as the "impulse drive" on the USS *Enterprise* (Coon, 1968).

Referring to Figure 13.13, neutral gas is introduced into the discharge chamber and a stream of electrons is produced by the discharge hollow cathode. Ions are produced by collisions with the

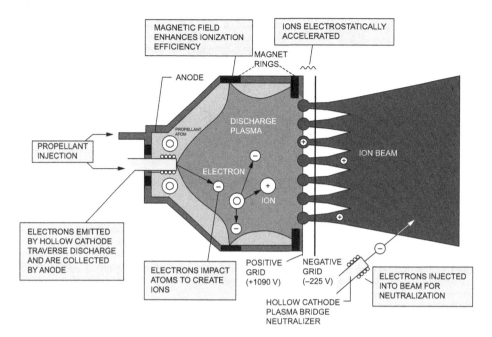

Figure 13.13 Operating principles of a gridded ion engine.

electrons – the so-called "bombardment" of the neutrals by the electrons. The ions are then accelerated out of the discharge chamber by the electric field potential applied at the grids. This same potential restrains the movement of the electrons and keeps them within the discharge chamber. They are eventually collected and returned to the electric circuit at the anode of the thruster. Downstream of the grids, a neutralizer cathode supplies more electrons to prevent negative charging of the spacecraft, which would occur if only ions were accelerated. The resultant beam is a stream of ions and electrons that is overall neutral in charge. The reaction to the momentum flux of this plasma beam produces the thrust of the ion engine. The high I_{sp} of the ion engine made it an attractive option for satellite buses with bipropellant propulsion systems, because the addition of a small tank for the xenon propellant was not difficult to implement. To date, some 212 ion engines have flown, mostly on geosynchronous communications satellites but also on the Deep Space-1, Hayabusa, and DAWN spacecraft.

Hall Effect Thrusters

As discussed in the introduction, Hall thrusters have enjoyed a resurgence in the West after the opening up of the former Soviet Union exposed American and European researchers to its unique attributes. The Hall thruster operates on a similar principle as the gridded ion engine, but does not require physical grids to accelerate the plasma. Instead, it utilizes the Hall effect to cause electrons to be delayed in reaching the anode thus creating an electric field gradient that accelerates the ions in the plasma, as illustrated in Figure 13.14.

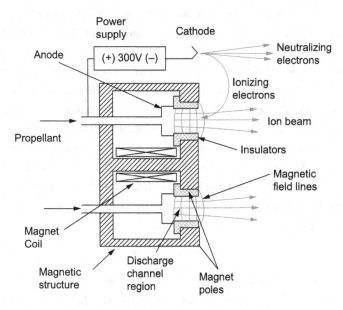

Figure 13.14 Hall thruster operating principles.

Hall thrusters have an advantage over ion engines in that the acceleration can be accomplished without the limitations imposed by physical grids, such as minimum gap distances and sputtering erosion. They can also be scaled to higher powers without the same Child–Langmuir law limitations on space charge which cause ion engine diameters to grow very rapidly. Also, the I_{sp} range of operation of a Hall thruster is in the 1800–2200 s range, which tends to be the sweet spot for large orbit maneuvers in a gravity well. This has made them popular in recent years among satellite manufacturers.

13.5 ELECTRIC PROPULSION APPLICATIONS

13.5.1 Launch – Railgun/EM Launchers

Electric propulsion is primarily used for in-space maneuvers, due to its inherently low thrust. However, there are situations where it might be possible to use a form of $j \times B$ acceleration to launch payloads or to provide a first-stage acceleration to high velocity. There have been numerous proposals to examine this concept and smaller scale accelerators have been built. The principal limitation of EP for launch is the extreme acceleration and the amount of air drag present at the exit of the railgun.

Extreme acceleration occurs because the payload must achieve orbital velocity before it reaches the end of the rail, when acceleration stops. Given an orbital velocity of 17,000 mph and a rail length of 10 miles, it is necessary to achieve an acceleration of over 180g. Doubling the rail length to 20 miles makes the construction challenges much more difficult, but drops the required acceleration to only 90g. Still many payloads, and certainly human beings, would have a hard time withstanding such high loads for even the short 4–10 seconds they would be imparted over.

Air drag is a problem because, unlike traditional rocket launches where the acceleration occurs as the vehicle is climbing higher into the atmosphere, the rail launcher remains attached firmly to the ground. For that reason, it is often proposed that such a launcher be placed on a high mountain so that the payload encounters less air resistance when it emerges from the exit. Even so, imagine a railgun that has its exit at 28,000 ft near the top of Mt. Everest. The air density there is 490 g/m^3 and the projectile is travelling at 17,000 mph or 7600 m/s. In order to have a useful payload, we might imagine a projectile that has a diameter of 1 m, therefore a cross-sectional area of 0.785 m^2. Assuming a streamlined body with a drag coefficient of 0.04, we can calculate the air drag on the departing object:

$$F_D = \tfrac{1}{2}\rho u^2 C_D A \tag{13.29}$$

with ρ = 490 g/m^3, u = 7,600 m/s, C_D = 0.04, and A = 0.785 m^2.

The resulting air drag of approximately 444,000 N would slow the payload down from 17,000 mph to nearly zero in only a little more than 30 seconds. It is left as an exercise for the reader to calculate the heat generated which would have to be dissipated from the moving body, which obviously poses a severe issue as well. For these reasons, electromagnetic launch has never been seriously considered except possibly for raw materials from the surface of the Moon. Obviously, less velocity is required to place objects into lunar orbit since the Moon has only one-sixth the gravity of Earth. The fact that the Moon has no sensible atmosphere eliminates the concern over air drag at the exit of the rail launcher.

13.5.2 Satellite Stationkeeping and Position Maintenance

This is the predominant use for EP today and will likely continue to be for the near future. Satellites experience disturbance forces after they are placed in their desired orbits. These forces include such things as: solar pressure, Earth oblateness effects, atmospheric drag, and outgassing (Burns, 1976). Small correction maneuvers are required to keep the satellite in its proper location. Some of these perturbations are secular and some are periodic. Secular disturbances cause drift in the same direction, while periodic disturbances drift first one direction, then the other. Depending on the satellite's mission, either can require some form of counteracting force to be provided by onboard thrusters.

Stationkeeping

For GEO satellites, the predominant *stationkeeping maneuver* is to maintain satellite position within the assigned box to avoid interference with other neighboring satellites. Usually this box is something like 0.1 degrees of arc and the satellite must be maintained on a frequent basis to stay

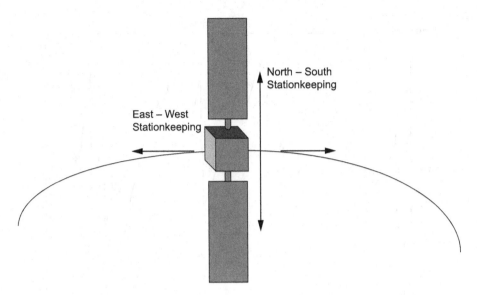

North – South
Stationkeeping

East – West
Stationkeeping

Figure 13.15 Satellite stationkeeping definitions for a typical GEO satellite.

within it. Typical operator strategies allow for drift toward one side of the box until it reaches about 0.05 degrees off the nominal position, then a propulsive maneuver is used to position it near the other edge of the box. Over the course of a year these maneuvers total about 50 m/s. The example shown previously in Table 13.1 showed the impact of going from a thruster with a 220 s specific impulse to a 300 s resistojet or a 600 s arcjet was significant in terms of the propellant mass saved. In fact, this savings has been used to extend the lifetimes of GEO communications satellites to nearly 15 years on average, an increase of 300% over what was possible with chemical thrusters. *North–south stationkeeping* is performed in the vertical axis relative to the orbit plane, while *east–west stationkeeping* is performed tangentially to the orbit. This is illustrated in Figure 13.15.

Drag Make-up

Drag make-up is an issue for low Earth orbit (LEO) satellites. The atmosphere continues to cause drag on satellites out to about 400 km altitude. An estimate of the drag force can be determined from Eq. 13.19 if the atmospheric density, the satellite frontal area and orbital velocity are known. A good example of a LEO satellite which undergoes drag is the International Space Station (ISS). Figure 13.16 shows a plot of the orbital altitude versus time of the ISS with periodic thruster firings to provide reboost. For satellites that need to maintain more precise orbital altitudes, this problem is even more pronounced. Some satellites even want to be maintained continuously in the same orbital altitude – the so-called "drag-free" satellite. This establishes a continuum of propulsion require-ments that puts a premium on efficient use of propellant. In the 1970s, RCA Astro Space pioneered the drag-free satellite with the use of PPTs on the NOVA spacecraft (Ebert *et al.*, 1989). NOVA was

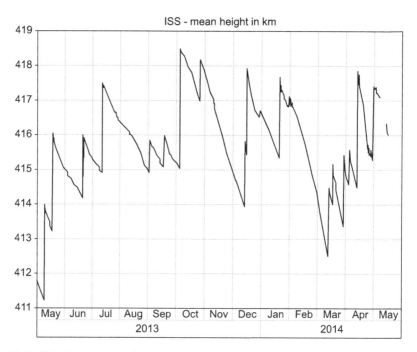

Figure 13.16 Altitude versus time for international space station showing effect of atmospheric drag.

part of a satellite constellation called TRANSIT used by the Navy prior to GPS to aid in navigation of ships on the oceans. More recently, the European Space Agency (ESA) used small ion engines on its Gravity Field and Steady-State Ocean Circulation Explorer (GOCE) spacecraft to allow it to fly much lower (229 km mean altitude) than is typically possible (Drinkwater, 2007).

13.5.3 Satellite Orbital Maneuvers

Satellite operators have become increasingly comfortable with the idea of performing larger orbit maneuvers with EP. This is largely because of the money that can be saved through lower-cost launches. Occasionally it is driven by necessity due to the very large size of a spacecraft and the limitations of available launchers. As we discussed previously, using EP to perform some or all of the on-orbit maneuvers such as the transfer from where the launcher drops the satellite off to its final destination (usually GTO to GEO for communications satellites) can save a very large amount of propellant. Current chemical bipropellant systems make up about 50% of the mass of a GEO satellite when it is delivered to geosynchronous transfer orbit (GTO). The downside of using EP for the on-orbit transfer is that it takes significantly more time. With chemical apogee engines, such transfers are accomplished in hours or at most a few days. For EP systems the time is dependent on the amount of power which can be devoted to the thrusters, but typical transfer times are 4–8 months. When comparing EP devices for *orbit raising maneuvers*, a thruster with a higher T/P is more desirable than one with higher I_{sp}. In fact, many manufacturers are designing thrusters that can operate in two modes – one which maximizes T/P during orbit raising and then operates in a higher I_{sp} mode for stationkeeping.

13.5.4 Deep Space Propulsion

As mankind ventures ever further into the solar system, the need is growing to provide additional impulse to spacecraft after escaping Earth's sphere of influence (SOI) and achieving a heliocentric trajectory. A great example is the Dawn mission, which travelled to two separate dwarf planets, Ceres and Vesta, in the main asteroid belt and orbited both. Dawn used gridded ion engines to accomplish this feat. It was a first in space exploration to send a spacecraft to one body, orbit it, and then depart from it to visit another body. In the process of performing these maneuvers, Dawn's ion engines delivered more than 9 km/s of velocity increase to the spacecraft (Garner et al., 2015). That is more than the Delta II launch vehicle which was originally used to send Dawn into space.

It is easy to see from a quick calculation that such a large amount of velocity change would not be possible with chemical propulsion. We start with the rocket equation:

$$\frac{m_f}{m_o} = e^{-\left(\frac{\Delta v}{g I_{sp}}\right)} \tag{13.30}$$

Let us assume an initial satellite mass of $m_o = 1500$ kg. Solving for m_f will provide the final mass after all propellant is consumed. We set $\Delta v = 9000$ m/s and g is the acceleration due to gravity 9.81 m/s^2.

Therefore, if $I_{sp} = 300$ s, $m_f = 70.5$ kg while if $I_{sp} = 2800$ s, $m_f = 1081$ kg. The mass of propellant required is $m_o - m_f$.

For a 300 s chemical bipropellant system, $m_o - m_f = 1429.5$ kg of chemical propellant.

For a 2800 s ion engine, $m_o - m_f = 419$ kg of xenon propellant.

If the satellite bus and instruments have a mass of 800 kg, then clearly the propellant exceeds the mass for the chemical system while there is mass margin with the ion engines. In this hypothetical example, the satellite needed to be 1500 kg or less to be launched into the correct trajectory by the Delta II launcher. With a chemical propulsion system, 1430 kg of that 1500 kg would be propellant, leaving only 70 kg for everything else. But the bus and payload need 800 kg. So you either have to increase the mass of the satellite and use a bigger launcher or you don't do all those maneuvers that allow Dawn to orbit two different asteroids.

The ion engine propulsion system requires only 419 kg of the 1500 kg for propellant. Assuming there are some additional items required in the power system, it may be reasonable to assess the ion propulsion system as requiring 500 kg, but this still leaves 1000 kg for everything else. The mission designer can add more science instruments, reduce the mass of the satellite, and get better launch insertion, or some combination of the three.

13.6 SYSTEM DESIGN AND SPACECRAFT INTERACTIONS

EP is more integrated into spacecraft bus design than other forms of propulsion because it requires power from the spacecraft, has to dissipate heat from electronics as well as thruster mounting brackets, and has a conductive plume which may interact with other spacecraft surfaces. Because of this, when considering use of EP for use on a spacecraft, designers must take into account such things as where the PPU will be located relative to the thrusters so that a good path to the thermal bus is achieved. Trades must be considered between a shorter power cable (to reduce electrical losses) and the relative positions of the thruster and PPU. Thrusters may need to be positioned outboard on mechanisms to point them and keep surfaces such as radiators and solar panels out of the plume impingement zone. In addition, for most types of EP thrusters, a fluid connection must be made between the tanks and the thrusters. Figure 13.17 shows the block diagram of a typical EP system.

   As is evident from the figure, there are two resources that must be provided to any EP device: power and propellant. The third interface is the command and control of the EP system by the spacecraft. This requires primarily digital signals being passed back and forth between the spacecraft computer and the EP system. Sometimes this is accomplished inside the PPU and sometimes there is a separate digital control box to perform the function.

   EP systems engineers must consider several constraints when thinking of applications. The first and most important is how much power is available to the EP device. Spacecraft solar arrays and power systems are usually sized to provide power for a payload. Because such systems add to the cost and mass of the spacecraft, their design is usually limited to this required power level

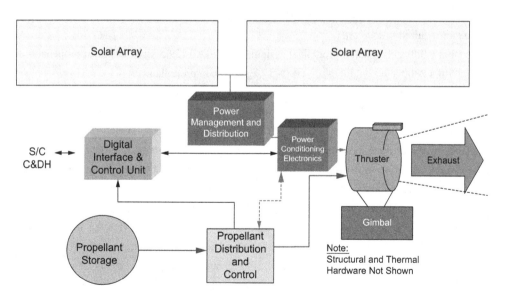

Figure 13.17 Simplified block diagram of typical EP system.

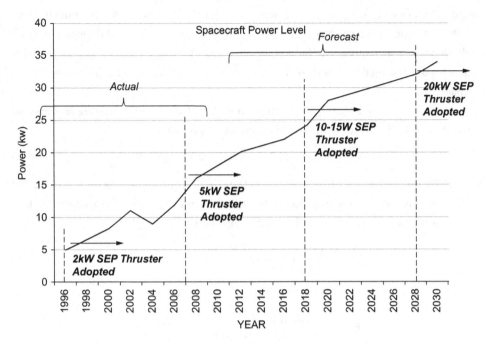

Figure 13.18 Trends in spacecraft power set the stage for adoption of SEP.

and no more. Therefore, the EP system needs to be matched to the power level. Figure 13.18 shows the trend in spacecraft power level over a recent 15-year period and the corresponding adoption of higher power EP systems.

The second important constraint is the propellant availability aboard the spacecraft. This is especially true in the case of electrothermal thrusters because they are often used as part of a broader system of small chemical thrusters which perform other functions such as momentum wheel unloading and reaction control. It was this constraint that lead to the selection of hydrazine as the propellant of choice for resistojets and arcjet thrusters. Both of these devices had been more widely tested in the laboratory with other propellants such as hydrogen and ammonia, but implementing a completely separate propellant storage and feed system in addition to the hydrazine system already used for the small chemical thrusters was impractical. Therefore, development work was performed to ensure that these thrusters could be made to work on decomposition products of hydrazine. For ion engines and Hall thrusters, both of which now operate primarily on xenon, this is not an issue because the I_{sp} of the devices reduces the amount of propellant required and xenon stores easily, so the required tank sizes make it practical to implement a separate propellant system.

Thermal considerations play an important role in EP systems integration, especially as power levels increase. Overall efficiency of the EP system is typically >60%. This breaks down into component efficiencies such as PPU efficiency and thruster efficiency. PPUs are usually greater than 90% efficient; however, they also reject heat at a low temperature and through a relatively small footprint. This requires careful consideration and discussion with the spacecraft systems engineer and the thermal designers to ensure that heat can be effectively removed from the PPU baseplate.

For example, in a 5 kW EP system, with a 92% efficient PPU operating at a baseplate temperature of 45 °C, the required thermal dissipation is given by:

$$P_T = (1 - 0.92)P_i = (0.08)(5 \text{ kW}) = 400 \text{ W} \tag{13.31}$$

Assume all heat is transferred by conduction to the spacecraft thermal radiator and the PPU baseplate area is 0.12 m². If the spacecraft bus temperature local to the PPU baseplate is 40 °C and the equivalent thickness d is 0.035 m, then using Fourier's law we have:

$$\frac{Q}{t} = 400 \text{ W} = \frac{kA(T_{PPU} - T_{bus})}{d} \tag{13.32}$$

We can solve for the required thermal conductivity k at the interface:

$$k = \frac{(400 \text{ W})(0.035\text{m})}{(0.12 \text{ m}^2)(45 \text{ °C} - 40 \text{ °C})} = 23.33 \frac{\text{W}}{\text{m°C}} \tag{13.33}$$

The final spacecraft interaction that we will consider for EP systems involves the interaction of the plumes with the spacecraft surfaces. EP devices, as we have seen, involve charged particles moving at very high velocities. The electrical conductivity of the plume can lead to interactions such as arcing between two sections of the spacecraft separated by a dielectric element. The impact of high-velocity particles (regardless of whether they are charged or neutral) can result in sputtering of surfaces. Other spacecraft surfaces may be cold relative to the plume and thereby attract slow moving neutral particles that can coat out and degrade surfaces such as radiators. Designers considering use of EP systems need to understand the plume characteristics and consider this when selecting a location to mount the hardware on the spacecraft. Mechanisms such as gimbals are often used (as shown in Figure 13.17) to accommodate the need to point the thrusters as a means of mitigating such issues.

This chapter provides an overview and an introduction to electric propulsion and some of the fundamentals and applications that are relevant to the field. A student interested in pursuing more detailed understanding would do well to read further in the references cited. For plasma physics fundamentals and electrode phenomena, Chen (1984) is especially enlightening. For more detailed explanations of the processes as they are applied to electric propulsion devices, Jahn (1968) remains the standard almost 50 years since it was first published. The Electric Rocket Propulsion Society also maintains a website (erps.spacegrant.org) where the collected papers of the International Electric Propulsion Conference (IEPC) are archived and available for free download. These provide a wealth of information from researchers around the world working in the EP field.

FURTHER READING

Burns, J. A. (1976) "Elementary Derivation of the Perturbation Equations of Celestial Mechanics," *Am. J. Physics*, 44, 10, 944–949.

Chen, F. (1984) *Introduction to plasma physics and controlled fusion, Volume 1, Plasma Physics*. Plenum Press.

Choueiri, E. (2005) *A Critical History of EP – The First Fifty Years (1906–1956)*. Princeton University.

Coon, G. (1968) "Spock's Brain," Episode 1 of Season 3, *Star Trek*, air date September 20.

Cybulski, R. J. *et al*. (1965) "Results from SERT I Ion Rocket Flight Test," NASA Technical Note D-2718.

Drinkwater, M. R., Haagmans, D., Muzi, D., Popescu, A., Floberghagen, R., Kern, M., and Fehringer, M. (2007) "The GOCE Gravity Mission: ESA's First Core Earth Explorer," in *Proceedings of the 3rd International GOCE User Workshop*, 6–8 November, Frascati, Italy. ESA Sepcial Publication SP-627. European Space Agency, pp. 1–8. https://esamultimedia.esa .int/docs/GOCE_3rd_Workshop_Drinkwater_et_al.pdf

Ebert, W. L., Kowal, S. J., and Sloan, R. F. (1989) *"Operational NOVA Spacecraft Teflon Pulsed Plasma Thruster System,"* AIAA, 89–2487, July.

Garner, C., Rayman, M., and Brophy, J. (2015) "In-Flight Operation of the Dawn Ion Propulsion System – Arrival at Ceres," in *IEPC 2015–88, 34th IEPC*, Kobe, Japan.

Gold, H. *et al*. (1965) "Description and Operation of Spacecraft in SERT I Ion Thruster Flight Test," NASA Technical Memo. X-1077.

Jahn, R. G. (1968) *The Physics of Electric Propulsion*. McGraw Hill.

Krall, N. and Trivelpiece, A. (1986) *Principals of Plasma Physics*. San Francisco Press, Inc.

Lorrain, P. and Corson, D. R. (1970) *Electromagnetic Fields and Waves*. Freeman.

Meckel, N. J., Cassady, R. J., Osborne, R. D., Hoskins, W. A., and Myers, R. M. (1997) "Investigation of Pulsed Plasma Thrusters for Spacecraft Attitude Control," IEPC-97–128.

Mel'kumov, T. M. (ed.) (1965) *Pioneers of Rocket Technology, Selected Works*. Academy of Sciences of the USSR, Institute for the History of Natural Science and Technology, Moscow; translated from the 1964 Russian text by NASA as NASA TT F-9285.

Morren, W. E., Hay, S. S., Haag, T. W., and Sovey, J. S. (1989) "Performance Characterization of an Engineering Model Multipropellant Resistojet," *Journal of Propulsion and Power*, 5(2): 197–203.

Oberth, H. (1929) "Wege zur Raumschiffahrt." R. Oldenbourg.

Rayman, M. (2018) Dawn Mission Journal, June 30, https://dawn.jpl.nasa.gov/mission/journal.asp

Spitzer, L. (1962) *Physics of Fully Ionized Gases*. Interscience.

Zak, A. (2016) "Russia in Space: The Past Explained and the Future Explored," *Apogee Prime*.

HOMEWORK PROBLEMS

13.1 For a Hall thruster operating at 2.75 kW input power, the thrust efficiency is 58%. In testing, the mass flow rate is varied between 30 mg/s and 33 mg/s and the corresponding thrust measurements are 163 mN and 180 mN. Assuming the efficiency does not change over this range of operating conditions, what is the I_{sp} range of the thruster at these operating points?

13.2 If the I_{sp} is increased to 2200 s, what would the corresponding thrust be?

13.3 You want to design an ion engine that will produce 50 mN of thrust per meter of grid area. If the applied voltage V_T is 500 V, what gap distance d will be required?

13.4 You are planning a new constellation of low Earth orbiting satellites. As the systems engineer, you must decide on the form of propulsion that will be used to keep the satellites in their proper orbit. Assuming that the desired orbit is 600 km altitude and the perturbing force is principally atmospheric drag, calculate the required thrust of the ion engine to perform continuous drag cancellation. Based on that thrust level, and assuming a thruster diameter of 10 cm and a gap size of 20 mm, determine the required operating voltage V_T.

13.5 An arcjet thruster has thermal losses of 30%, frozen flow losses of 25%, and electrode losses of 10%. Calculate the overall efficiency of the thruster. Given this efficiency, determine the thrust produced if the I_{sp} is 580 s, and the power input is 2500 W.

13.6. Calculate the size of the thermal baseplate of a Hall thruster PPU if the input power is 6000 W and the PPU efficiency is 88%. How much smaller could the baseplate be if the PPU efficiency is increased to 92%?

APPENDIX

This section provides readers with more detailed information on several topics. In Appendix A.1 we discuss numerical solutions to nonlinear algebraic equations and numerical integration of parabolic (time-dependent) ordinary differential equations as these topics have arisen in several chapters of the book. In Appendix A.2 we provide a series of links to flow tables, fluid properties, heats of formation of propellant ingredients, and general resources that are commonly used by propulsion students.

A.1 NUMERICAL METHODS

Modern computers are powerful tools that provide capabilities to solve problems that don't have analytic solutions. While there are a large number of texts devoted to this subject, a brief overview is contained in this appendix to serve as a supplement for students who have not previously studied this area. Modern tools such as Matlab or Mathematica provide pre-programmed routines for many numerical techniques, but the user needs to be aware of the inner workings of these routines, as well as their limitations, in order to be fully productive in this area. Several of the exercises contained in this text require expertise in numerical solution of nonlinear algebraic equations and in the integration of parabolic (time dependent) ordinary differential equations. Appendix A.1.1 discusses the former topic while Appendix A.1.2 provides background on the latter. Students are encouraged to take formal coursework in this topic as the brief review contained herein is hardly sufficient to provide a comprehensive view of the subject.

A.1.1 Numerical Solution of Nonlinear Algebraic Equations

In algebra class, you learned to manipulate algebraic equations to solve for a dependent variable. For example, you learned that the equation: $4x + 7 = 0$ can be manipulated to directly solve for x ($x = -7/4$, right?). Such an equation is called a *linear algebraic equation*. In engineering and science, we frequently encounter more complex relationships that cannot be readily rearranged as in our simple example. For instance, suppose we modify our example equation: $4x + 7 \tan(x) = 0$. This simple alteration leads to a situation where we can't rearrange the equation to solve for x directly, due to the presence of the transcendental function $\tan(x)$. Equations of this type are called *nonlinear algebraic equations*. An excellent practical example of this type of equation is the Mach number/area ratio relationship derived in Chapter 2 (Eq. 2.9):

$$A/A_{\mathrm{t}} = \frac{1}{M}\left\{\frac{2 + (\gamma - 1)M^2}{(\gamma + 1)}\right\}^{\frac{\gamma+1}{2(\gamma-1)}}$$

If the nozzle design is known (i.e. the area ratio is known), the equation is nonlinear as it cannot be rearranged to solve for the corresponding Mach number.

In many cases like this, solutions to the equation do exist and are physically meaningful; however, iterative techniques are required for their solution. Your first inclination is likely to think of iterating on x until you find a value that satisfies the equation – a technique like this is called an *exhaustive search*. In many cases, an exhaustive search can be truly exhaustive (for the computer at least), and approaches that speed up this process are highly desirable. In this appendix, we will introduce you to two very powerful techniques called *Newton's method* and the *Bisector method* as a couple alternatives to solve nonlinear algebraic equations.

Newton's Method

Consider the nonlinear algebraic equation:

$$F(x) = 0 \tag{A.1}$$

where we desire to find the x value(s) that satisfy the equation, i.e. Eq. A.1 will be satisfied when we find the x value (let's call it x^*) that gives $F(x^*) = 0$. All iterative techniques require an initial guess, and unless we are incredibly lucky, our initial x guess (let's call it x_i) will not be equal to x^*. For this reason, the situation might look as that illustrated in Figure A.1 where our initial guess in this case is giving $F(x_i) > 0$. As F is presumed to be a smooth, single-valued algebraic function, the derivative of the function $dF/dx = F'$ exists. The derivative is simply the local slope of the curve as illustrated by the dashed line in Figure A.1. Newton's method is basically a linear extrapolation toward the desired solution, i.e. we project a straight line to the axis of the curve as shown graphically in Figure A.1. The resulting iteration is then:

$$x_{i+1} = x_i - F(x_i)/F'(x_i) \tag{A.2}$$

If the function $F(x)$ is a straight line, we will get to the solution in a single step, a more complex curve such as that in Figure A.1 will require additional iterations. The iteration is typically terminated when the value of $F(x)$ is sufficiently close to zero or when the change in x, i.e. $|x_{i-1} - x_i| < $ tol, where "tol" is a user-defined tolerance level. For example, in the Mach number/area ratio relation, let's define $F(M)$ as the function:

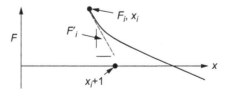

Figure A.1 Local behavior of nonlinear function F near its zero point.

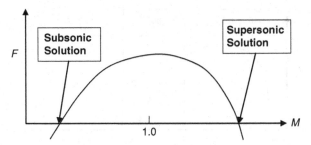

Figure A.2 Graphical depiction of Mach number/area ratio relation in Eq. A.3.

$$F = A/A_t - \frac{1}{M} \left\{ \frac{2 + (\gamma - 1)M^2}{(\gamma + 1)} \right\}^{\frac{\gamma+1}{2(\gamma-1)}} \qquad (A.3)$$

so that when $F(M) = 0$ we have a proper solution to Eq. 2.9. In this case we might terminate the process if the change in Mach number between successive iterations is less than say, 0.001, i.e. the tolerance value would be set to 0.001.

In most cases, Newton's method converges in only a few iterations, thereby making it vastly superior to an exhaustive search. As with all iterative techniques, the method is sensitive to the initial guess and users should always endeavor to use the best initial guess possible. In the Mach Number/area ratio equation, the function $F(M)$ generally has two roots as shown in Figure A.2.

Here, the two solutions are the subsonic root that occurs upstream of the nozzle throat and the supersonic root that occurs downstream of the throat. In this case, we would want to choose a very small Mach number as initial guess if we were searching for the subsonic root, and a Mach number substantially greater than 1.0 to find the supersonic root. If you are stepping along the length of the nozzle where the area is known as a function of x, i.e. $A = A(x)$, then the decision on the initial guess is as simple as knowing the x location where the throat lies.

Like all numerical schemes, Newton's method is not infallible. Figure A.3 provides a couple examples of functions that will challenge the scheme. Because we are dividing by the slope (in Eq. A.2), the iteration becomes very sensitive near local minima and maxima such as those shown in Figure A.3a. While we may have chosen an initial guess in Region I in this figure, if the iteration nears the local minima it may cause a large correction and jump to Region II. In this case, the iteration may never enter Region III where the solution lies.

Another situation that occurs frequently is when the function $F(x)$ contains periodic functions. In this case, there are an infinite number of solutions (i.e. $\sin(x) = 0$ has solutions of $x = 0 + 2n\pi$ for $n =$ any integer). In this case, the initial guess must be chosen to obtain convergence to the root of interest.

In some cases, the function $F(x)$ can be so complex that it is even challenging to compute its derivative. In this case, one might approximate the local derivative using a finite difference approximation, i.e. approaches using this methodology are referred to as the *Secant method*.

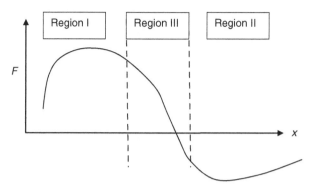

Figure A.3 Graphical depiction of potentially problematic cases for Newton's method.

Bisector Method

While not as efficient as Newton's method, the bisector method provides a practical alternative, particularly in situations where a large number of roots exist but the user can bracket the problem to the range of x values over which the particular solution of interest lies. For example, if we are solving for an angle and we know it lies in the range 0–90°, this provides a logical range over which to limit the search. In the Mach number/area ratio relation, we would logically limit the range:

$0.001 \leq M* \leq 1$ for the region upstream of a choked throat

$1 \leq M* \leq 10$ for the region downstream of a choked throat

where we have arbitrarily assumed that chamber Mach numbers would be no lower than 0.001 and nozzle exit Mach numbers no higher than 10. Given these or other user-defined ranges, the bisection begins by evaluating the function so as to eliminate half the interval in question. Let's say we were searching in the nozzle and that our solution was at $M* = 2$. The initial guess we might apply would be in the middle of the interval between $M = 1$ and $M = 10$, i.e. at $M = 5.5$. Evaluating $F(M)$ at this location will allow us to determine if it is positive or negative and thereby we can eliminate half the interval, i.e. all M values above $M = 5.5$ in this case. The next step is to move the upper bound of the interval accordingly so that the upper bound is now $M = 5.5$ instead of $M = 10$. Proceeding with another bisection allows another assessment of the bounds of the interval and further refinement of the region where the solution must lie. A pseudo-language algorithm of the process is as follows:

```
Mhigh = 10 (initial upper bound of interval)
Mlow = 1   (initial lower bound of interval)
Msav = 0.0 (saved value of Mach from prior solution)
Begin loop/iteration
While err > tol (looping until the error is less than the desired
tolerance)
M = 0.5(Mhigh + Mlow) (bisection process)
```

$F = A/A_\mathrm{t} - \frac{1}{M}\left\{\frac{2+(\gamma-1)M^2}{(\gamma+1)}\right\}^{\frac{\gamma+1}{2(\gamma-1)}}$ (function evaluation)

If F < 0 Mhigh = M (refining bounds of bisection interval)

If F > 0 Mlow = M (refining bounds of bisection interval)

err = |M − Msav| (evaluating error/change in Mach)

Msav = M (saving current M for evaluating subsequent errors)

End Loop

The only real difficulty in implementing the bisection is in deciding how to limit the interval based on the sign of $F(M)$ (refining bounds steps in above algorithm). There are only two alternatives here and we advise students to simply try one – if it doesn't work, then the other alternative must be correct. If you plot the function $F(M)$ you can then decide a priori which way to limit the interval.

A.1.2 Numerical Solution of Parabolic Ordinary Differential Equations (ODEs)

Ordinary differential equations (ODE) that can be solved via time-stepping procedures belong to a class of problems referred to as being "parabolic" in nature. The parabola has a single characteristic that permits a sequential process of integration. While many time-dependent problems fall into this category, we can also think of 1-D internal flow problems in which the distance along the duct/combustor/nozzle is the parameter advanced sequentially. If we let "t" represent the independent variable, a general ODE that represents this class of problems can be written:

$$\frac{du}{dt} = u' \quad = f(u, t); u(t_0) = u_0 \tag{A.4}$$

where the initial condition u_0 must be specified for a well-posed problem. An obvious example of this type of problem is the integration of the acceleration history (du/dt) of a particle or fluid based on time and its current velocity. Many numerical schemes are derived from a Taylor expansion. For example, we may approximate the value of a discrete value of u (say u_i) in terms of neighboring values and its derivatives:

$$u_{i+1} = u_i + \Delta t \left(\frac{du}{dt}\right)_i + \frac{\Delta t^2}{2}\left(\frac{d^2u}{dt^2}\right)_i + \frac{\Delta t^3}{6}\left(\frac{d^3u}{dt^3}\right)_i + \cdots + \frac{\Delta t^n}{n!}\left(\frac{d^nu}{dt^n}\right)_i \tag{A.5}$$

Equation A.5 is said to be a Taylor series approximation to u_{i+1} via an expansion about u_i. Solving this relation for du/dt gives:

$$\left(\frac{du}{dt}\right)_i = \frac{u_{i+1} - u_i}{\Delta t} - \frac{\Delta t}{2}\left(\frac{d^2u}{dt^2}\right)_i + \text{HOT} \tag{A.6}$$

where "HOT" means higher-order terms (proportional to Δt^2 and higher). We can see from Eq. A.6 that if we retain just the first term on the right-hand side we have an approximation for the derivative of u. The approximation becomes more exact as the *time-step*, Δt, vanishes.

The time-step is the lone numerical parameter in parabolic ODEs – it is something we choose in order to write approximations of the form inferred above. If we retain just the first term on the right-hand side of Eq. A.6 our error in doing so is proportional to the size of Δt. Schemes of this type are said to be *first order*, i.e. the error in the approximation is proportional to Δt^1. In this case, the term $\Delta t/2 \left(d^2 u/dt^2 \right)_i$ is called the *leading truncation error*. In this case, we can approximate the slope of a line with just the first term of the right-hand side of Eq. A.6 because for a simple line $d^2 u/dt^2 = 0$. More complex (and hence interesting) functions will have nonzero higher derivatives, the magnitude of these derivatives determines the overall truncation error of the scheme.

Euler Methods

The Euler schemes are first-order methods based on the suggestion of the above paragraph. A *forward Euler scheme* results from this process and the approximation for u at the new time level becomes:

$$u_{i+1} = u_i + \Delta t \left(\frac{du}{dt} \right)_i = u_i + \Delta t f(u_i, t_i) \tag{A.7}$$

This scheme tends to be quite unstable in that the new value of u is determined solely by the slope determined at the prior time level. We can remedy this problem if we expand our Taylor series in Eq. A.4 about u_{i-1} instead of u_i. In this case, you should be able to show (by shifting indexing) that:

$$u_{i+1} = u_i + \Delta t \left(\frac{du}{dt} \right)_{i+1} = u_i + \Delta t f(u_{i+1}, t_{i+1}) \tag{A.8}$$

The scheme shown in Eq. A.8 is referred to as a backward Euler method. The forward Euler scheme is called an *explicit method* because we can immediately evaluate the right-hand side as only time level "i" information is required. In contrast, the backward Euler method is an *implicit scheme* as we need to compute the derivative at the new time level $(i + 1)$, which isn't known until we compute u at that time level. Implicit schemes are inherently more stable, but in general require iteration for solution. In the case of Eq. A.8, we might employ a Newton iteration to solve for the derivative at the new time level:

$$F = u_{i+1} - u_i - \Delta t f(u_{i+1}, t_{i+1}) = 0 \tag{A.9}$$

so

$$u_{i+1}^{m+1} = u_{i+1}^m - F(u_{i+1}^m)/F'(u_{i+1}^m) \tag{A.10}$$

where the superscript "m" denotes the Newton iteration number and:

$$F'(u_{i+1}^m) = 1 - \Delta t \frac{\partial}{\partial u_{i+1}} f(u_{i+1}, t_{i+1}) \tag{A.11}$$

Higher-Order Schemes – Trapezoidal Integration

Being only first order, the Euler schemes are not used very frequently to solve parabolic ODEs. A good second-order scheme (leading truncation error proportional to Δt^2) is derived via trapezoidal integration.[1] Integrating Eq. A.5 between time levels i and $i + 1$ gives:

$$u_{i+1} - u_i = \int_{t_i}^{t_{i+1}} f(u, t)dt \tag{A.12}$$

which can be integrated using trapezoidal quadrature[1] to give:

$$u_{i+1} - u_i = \frac{\Delta t}{2}(f(u_i, t_i) + f(u_{i+1}, t_{i+1})) \tag{A.13}$$

The trapezoidal quadrature scheme is implicit and can formally be shown to be second-order accurate. We can obtain an explicit variant on this scheme by approximating $f(u_{i+1}, t_{i+1})$ to just first order as it is already multiplied by Δt in Eq. A.13. Using a forward Euler scheme for this approximation we arrive at Huen's method:

$$u_{i+1} - u_i = \frac{\Delta t}{2}(f(u_i, t_i) + f(u*, t_{i+1})) \tag{A.14}$$

where

$$u* = u_i + \Delta t f(u_i, t_i) \tag{A.15}$$

For many problems in this text Huen's method will suffice to obtain adequate accuracy. It can be proven that the scheme is formally second-order accurate. To achieve second-order accuracy, two evaluations of the derivative $(f(u_i, t_i), f(u*, t_{i+1}))$ are required each time level.

Runge–Kutta Method

Probably the most popular time integration scheme is the fourth-order accurate Runge–Kutta method. This scheme is imbedded in the popular Matlab routine ODE45. To achieve fully fourth-order accuracy, four function evaluations are required at each time level. The scheme may be written as follows:

$$u_{i+1} = u_i + \frac{\Delta t}{6}(k_1 + 2k_2 + 2k_3 + k_4) \tag{A.16}$$

where

$$\begin{aligned}
k_1 &= f(u_i, t_i) \\
k_2 &= f\left(u_i + \frac{\Delta t}{2}k_1, t_i + \frac{\Delta t}{2}\right) \\
k_3 &= f\left(u_i + \frac{\Delta t}{2}k_2, t_i + \frac{\Delta t}{2}\right) \\
k_4 &= f(u_i + \Delta t k_3, t_{i+1})
\end{aligned} \tag{A.17}$$

---

[1]   To get the area of a trapezoid (area under integral of the slope history), we simply average the height of the trapezoid (slope) and multiply by the length of the interval. See a numerical methods text for more discussion.

If very long time integrations are required than higher accuracy is preferred if not required. An extreme example of this are requirements for integrations of trajectories of space probes. Integration of trajectories over what can be many years under very small gravitational forces demands sixth-order or even higher accuracy.

Applying Newton's Method to Lumped Parameter Systems Modeling

In Section 7.8, we discussed systems modeling to obtain transient local pressures and flow rates using the lumped parameter assumption. In this context, a system of first-order ODEs results from the problem formulation. The form of the equations is analogous to Eq. A.2, but in this case the unknown x is a vector of the pressures or flow rates at various nodes that comprise the engine system model. In addition, the slope $F'(x)$ becomes a 2-D array we call the Jacobian, J. Therefore, we write the Newton iteration as:

$$x_n^{i+1} = x_n^i - F(x_n^i)/J(x_n^i) \tag{A.18}$$

Here, the time level is shown as a superscript and the subscript n indicates the node within the network that comprises the system to be modeled, i.e. for N total elements in the network $1 \le n \le N$. The x_n values represent unknown pressures or flow rates at various nodes in the system. The Jacobian matrix represents the sensitivity of changes in one element relative to another:

$$J(x_n^i) = \left\{ \begin{matrix} \partial F_1/\partial x_1 & \dots & \partial F_1/\partial x_N \\ \vdots & \ddots & \vdots \\ \partial F_N/\partial x_1 & \dots & \partial F_N/\partial x_N \end{matrix} \right\}^i \tag{A.19}$$

Numerical issues with the technique are similar to those discussed with the single parameter Newton's method; in regions where the elements of the Jacobian become small, the $F(x_n^i)/J(x_n^i)$ term becomes large and numerical instabilities can result. Problems of this sort are said to be *numerically stiff*, implying derivatives that are very large. Shocks and combustion fronts are also examples of stiff regions where fluid properties change drastically over short distances. Here, the stiffness occurs in time – for example, the chamber pressure can climb hundreds or thousands of atmospheres per second during ignition. As with all numerical schemes, an obvious correction here is to reduce the time-step so that the amount of the correction to the x_n values is also reduced.

A.1.3 Grid Convergence Tests

We save the most important topic for this appendix for last. We introduced the time-step as a numerical parameter to approximate derivatives of an ODE for the purpose of time integration. However, the solution of the ODE should be independent of this arbitrary parameter, i.e. the problem stated in Eq. A.4 does not *require* a time-step value to be well posed. To ensure that the solution you obtain from your numerical integration is the solution, you *must* conduct a *grid convergence study*. By simply plotting your solution for different assumed time-steps you can

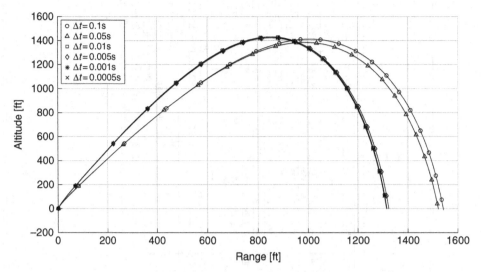

Figure A.4 Grid function convergence test for a rocket trajectory calculation.

readily confirm that the solution is in fact independent of the numerical parameter. Figure A.4 depicts a typical convergence process for a trajectory integration per the discussion in Chapter 3. This figure shows that a time-step of 0.001 seconds or less gives essentially the same results while substantial/noticeable errors result from a larger time-step of 0.05 seconds.

It is in the grid convergence study that the higher-order schemes tend to shine. Halving the time-step in a second-order scheme reduces errors by a factor of four, while in a fourth-order scheme this change reduces errors by $2^4 = 16$. For this reason, the higher the accuracy of the method, the larger the time-step one can reasonably employ.

In general, grid function convergence studies should be repeated when attributes of the problem change significantly. A time-step that is adequate to resolve acceleration of a launch vehicle over an 8 minute ascent to orbit is clearly much different than that required to resolve a maneuver of a kinetic kill vehicle that occurs on a millisecond timescale. When in doubt, perform the test to confirm that your solutions contain negligible numerical errors.

A.2   Fluid Properties and Other Resources

A.2.1 Properties of Liquid Propellants

Chapter 9 provides an extensive discussion of liquid propellants, but we include a summary table here for ease of access and applicability to hybrid and electric rockets as well. Table A.2.1 below provides a wealth of basic information such as freezing and boiling points, physical properties, and toxicity/flammability limits for fluids currently used or in development for rocket propulsion applications.

Table A.2.1 Properties of Liquid Rocket Propellants

Name	Class	Symbol	Molecular Weight (g/mol)	Freezing Point (°C)	Boiling Point (°C)	Density (g/cm3) at 1 atm and:		Vapor Pressure (atm) at:		Viscosity (cP) at:		Specific Energy (MJ/kg)	Heat of Form. (kJ/kg)	Isp (s), vac with [O/F, complement]	Autoign. Temp. (°C)	Flammability Limits (vol. %)		MAC (ppm)	PEL	LC_{50}, rat, 4 hr, inhalation	Effect of Contact
						20 C	70 C	20 C	70 C	25 C	70 C					Lower	Upper				
Fuels																					
Hydrogen		H_2	2.0	-259.2	-252.8	8.99E-05		13		0.009	0.01	142	0	456 [5.1/O_2]	500	4	76			15000 ppm	Frostbite
RP-1	Hydrocarbon	C_nH_m [a]	170.3	-40.0	149.9	0.806	0.767	0.002	0.03	2.2	0.88	42.8		367 [2.6/O_2]	210	0.7	5	500	500 ppm	5.28 mg/L	-
JP-8	Hydrocarbon	C_nH_m	132.4	-50.0	140.0	0.8	0.77	0.002	0.03	2.2	0.88	42.8		367 [2.6/O_2]	210	0.7	5	500	500 ppm	5.28 mg/L	-
Methane		CH_4	16.0	-182.6	-161.5	7.17E-04		613		1.09E-02	1.33E-02	55.5	-4680	379 [3.3/O_2]	537	5.3	14			500000 ppm	Frostbite
Methanol	Alcohol	CH_4O	32.0	-97.8	64.7	0.787		0.12		0.544		19.7	-7470	192 [0.8/O_2]	464	6	36.5			64000 ppm	-
Hydrazine	Monoprop	N_2H_4	32.1	2.0	113.5	1	0.964	0.021	0.2	0.97	0.51	19.4	-1580	234 [mono]	270	2.9	98	1	1 ppm	252 ppm	Severe Irritation
UDMH	Amine	$C_2H_8N_2$	60.1	-58.0	63.9	0.793	0.751	0.135	1.31	0.492		32.9	-804	342 [2.8/NTO]	249	2	95	0.5	0.5 ppm	172 ppm	Severe Irritation
MMH	Amine	CH_6N_2	46.1	-52.4	87.5	0.877	0.832	0.05	0.69	0.775	0.47	28.3	-1170	344 [2.3/NTO]	195	2.5	92	<0.5	0.2 ppm	0.68 mg/L	Severe Irritation
Ammonia		NH_3	17.0	-77.7	-33.3	0.604	0.524	9.87	35	0.276		22.5	-2700	343 [1.5/O_2]	630	15	28	100	50 ppm	4.92 mg/L	Severe Irritation
AF-M315E	Monoprop.	HAN Based	96.04[b]	-22.0	N.A.	1.465	N.A.	0.016[c]	0.117[d]	> Hydrazine	N.A.	N.A.	N.A.	250 [mono][e]	1800[f]	N.A.	N.A.	N.A.	N.A.	550 mg/kg	Slight irritant
LMP-103s	Monoprop.	ADN Based[g]	46.96[g]	-90.0	120.0	1.24	N.A.	0.128[h]	N.A.	N.A.	N.A.	N.A.	N.A.	256 [mono]	1600[f]	N.A.	N.A.	N.A.	N.A.	823 mg/kg	None reported
FLP-106	Monoprop.	ADN Based[i]	N.A.	N.A.	N.A.	1.357	N.A.	N.A.	N.A.	3.7	N.A.	N.A.	N.A.	259 [mono]	1880[f]	N.A.	N.A.	N.A.	N.A.	N.A.	N.A.
Oxidizers																					
Oxygen		O_2	32.0	-218.4	-183.0	1.43E-03		50.1		0.021				456 [5.1/H_2]					40 pph		Frostbite
Nitric Acid		HNO_3	63.0	-41.6	83.0	1.51		0.083		0.746			-2130	327 [2.6/MMH]				10	2 ppm	65 ppm	Severe Irritation
H_2O_2, 90%	Monoprop	H_2O_2	32.4	-11.0	141.0	1.44		0.002		1.24			-4940	360 [12/H_2]					1		Severe Irritation
H_2O_2, 98%	Monoprop	H_2O_2	33.7	-0.4	152.1	1.45	1.39	0.001	0.04	1.1	0.64		-4190	372 [12/H_2]					1		Severe Irritation
NTO		N_2O_4	92.0	-9.3	21.2	1.45	1.31	1.19	7.6	0.396	0.22		-99	344 [2.3/MMH]				5	5 ppm (NO_2)	315 ppm (NO_2)	Severe Irritation
MON-3	MON	N_2O_m	90.2	-15.7		1.44		1.3					-186	345 [2.3/MMH]				5	5 ppm (NO_2)	315 ppm (NO_2)	Severe Irritation
MON-10	MON	N_2O_m	85.8	-24.5		1.41		2.2					-390					5	5 ppm (NO_2)	315 ppm (NO_2)	Severe Irritation
MON-25	MON	N_2O_m	76.5	-55.6		1.37		5.4					-827	350 [2.4/MMH]				5	5 ppm (NO_2)	315 ppm (NO_2)	Severe Irritation

Notes:

a typical formulation for RP-1 is $CH_{1.95}$
b for HAN only
c for 60% aqueous HAN, at 25°C
d for 60% aqueous HAN, at 65°C
e for 22 N Aerojet Rocketdyne Thruster
f Adiabatic Combustion Temperature
g 63% ADN/18.4% CH3OH/14% H2O/4.6% NH3
h for methanol
i 64.6% ADN/23.9% H2O/11.5% hydrocarbon

A.2.2 Web Resources

Recognizing the ease of access to information on the internet, we include a number of links here to useful tables and data that have, in the past, been appended to many text books. Not only do we save on printing all of this information, but it is arguably more useful and convenient to provide in this form. At Purdue, we developed a Propulsion Web Page (https://engineering.purdue.edu/~propulsi/propulsion/), which has a wealth of information that is useful for students. Information that is directly applicable to the materials discussed in Chapters 2 and 3 of this text can be found at this link; a screen shot of the region of interest is shown in Figure A.5.

The isentropic flow tables provide tabular values of flow properties (p, T, ρ, A) as a function of Mach number from Eqs. 2.5–2.8. Tables are included for $\gamma = 1.2$, 1.3, and 1.4. While not used directly in this text, other 1-D compressible flow results are included for jump conditions across normal shocks, for 1-D flow with friction in a constant area pipe (Fanno flow) and for 1-D flow with heat addition/subtraction in a constant area pipe (Rayleigh flow). Thrust coefficient data (Eqs. 2.33 and 2.34) are provided as a function of nozzle expansion area ratio ε, for $\gamma = 1.2$, 1.3, and 1.4. A US

Useful Propulsion Data

Fundamental Properties of Internal Fluid Flows

- 1-D Isentropic Flow
  - Gamma=1.4
  - Gamma=1.3
  - Gamma=1.2
- Normal Shocks
- Fanno Flow
- Rayleigh Flow
- Rocket Thrust Coefficient Tables
  - Gamma=1.4
  - Gamma=1.3
  - Gamma=1.2
- U.S. Standard Atmosphere

Combustion and Thermochemical Data

- Periodic Table of the Elements
- Specific Heat Data for Common Gases
- NASA Thermochemistry Code - Sample Input Files
- Heats of Formation and Chemical Composition
- Constant Pressure Equilibrium Constants (Kp Data)

Figure A.5 Screen shot of a portion of the of Purdue Propulsion Web Page (https://engineering.purdue .edu/~propulsi/propulsion/).

Standard Atmosphere table is included so that students can correct thrust coefficient (and hence I_{sp}) values based on local atmospheric pressure.

   A significant amount of thermochemical data are also available on the website, including c_p data for common gases, sample input files for CEA code, a large database of heats of formation and chemical composition for numerous reactants and some curvefits of k_p vs. T for some of the more common dissociation/recombination reactions. One issue to note on the heat of formation database is that values are given in cal/g whereas the CEA input file demands cal/mol. For this reason, one must multiply the values in the table by the molecular weight of the material in question.

Fluid Properties Links

There are two websites that are quite useful and popular for providing fluid properties with just a few mouse clicks. One is maintained by the US National Institute of Standards and Technology and is called the NIST Web book. The Web book contains a monstrous amount of information on a given atom or molecule, but the most useful part for propulsion types is the thermophysical properties database located at: http://webbook.nist.gov/chemistry/fluid/. On this site, information for the following parameters can be generated along constant pressure (isobar), constant temperature (isotherm), or constant volume (isochor) lines:

- density
- c_p
- enthalpy
- internal energy
- viscosity
- Joule–Thomson coefficient

- specific volume
- c_v
- entropy
- speed of sound
- thermal conductivity
- surface tension (saturation curve only)

Data can be generated in English or SI units and can be displayed graphically or in HTML table form. A screenshot of an isotherm of liquid/gaseous oxygen is provided in Figure A.6 as an example of the output from a given query.

   Another very valuable resource is the Engineering Toolbox (www.engineeringtoolbox.com/), which includes a wealth of information including, but not limited to:

- characteristics (Young's, shear and bulk modulus) of common structural materials
- moments of inertia of common structural material cross-sections
- standard dimensions, weights and volumes of ANSI pipes
- thermal conductivity of gases and common construction and insulating materials
- viscosity of numerous fluids
- flanges and bolt dimensions ASME/ANSI B16.5 – Class 150 to 2500
- material properties for gases, fluids and solids – densities, specific heats, viscosities and more
- online P&ID drawing tool
- unit conversion chart.

Fluid Data

Isothermal Data for T = 150.00 K

Temperature (K)	Pressure (MPa)	Density (mol/l)	Volume (l/mol)	Internal Energy (kJ/mol)	Enthalpy (kJ/mol)	Entropy (J/mol*K)	Cv (J/mol*K)	Cp (J/mol*K)	Sound Spd. (m/s)	Joule-Thomson (K/MPa)	Viscosity (uPa*s)	Therm. Cond. (W/m*K)	Phase
150.00	1.0000	0.86814	1.1519	2.9212	4.0731	164.70	21.661	32.956	224.36	9.2643	11.965	0.014659	vapor
150.00	2.0000	1.9189	0.52113	2.6957	3.7380	157.36	22.791	39.355	213.92	9.4246	12.866	0.015995	vapor
150.00	3.0000	3.3048	0.30259	2.4025	3.3103	151.85	24.577	53.033	201.35	9.4979	14.195	0.018398	vapor
150.00	4.0000	5.6598	0.17668	1.9122	2.6189	145.66	29.731	118.09	181.72	9.0645	16.918	0.024439	vapor
150.00	4.2186	6.7170	0.14888	1.6938	2.3219	143.45	33.574	212.01	172.82	8.6358	18.361	0.028695	vapor
150.00	4.2186	21.110	0.047372	-0.41330	-0.21346	126.54	28.982	174.84	273.80	1.7389	52.886	0.064843	liquid
150.00	5.0000	22.415	0.044613	-0.58849	-0.36543	125.29	27.320	115.72	330.93	1.1579	57.362	0.066672	liquid
150.00	6.0000	23.380	0.042772	-0.72370	-0.46707	124.32	26.627	94.379	374.30	0.83147	60.894	0.069342	liquid
150.00	7.0000	24.068	0.041548	-0.82272	-0.53188	123.61	26.299	84.159	406.38	0.63813	63.564	0.071850	liquid
150.00	8.0000	24.616	0.040624	-0.90280	-0.57781	123.03	26.113	77.885	432.63	0.50417	65.795	0.074180	liquid
150.00	9.0000	25.076	0.039878	-0.97090	-0.61200	122.53	26.000	73.544	455.16	0.40364	67.755	0.076355	liquid
150.00	10.000	25.476	0.039253	-1.0306	-0.63807	122.10	25.932	70.316	475.05	0.32438	69.532	0.078396	liquid

Figure A.6 Tabular output of NIST thermochemical data for liquid/vapor oxygen at specified temperature of 150 K. Note that density (and other property) variations are also important in the liquid phase for high-pressure rocket conditions.

NTRS and NASA SP Series

The NASA Technical Reports Server (NTRS) provides a very useful link to a large number of NASA documents (https://ntrs.nasa.gov). Of particular interest for students is the compilation of NASA Special Publications (SPs) that were written in the early 1970s as design guides for the community. In many cases, these documents have stood the test of time and still represent valuable resources of introductory material and guidelines on rocket propulsion technologies. The following list is a summary of all the NASA SPs in chemical propulsion:

SP-8087   *Liquid Rocket Engine Fluid-Cooled Combustion Chambers*, April 1972

SP-8113   *Liquid Rocket Engine Combustion Stabilization Devices*, November 1974

SP-8107   *Turbopump Systems for Liquid Rocket Engines*, August 1974

SP-8109   *Liquid Rocket Engine Centrifugal Flow Turbopumps*, December 1973

SP-8052   *Liquid Rocket Engine Turbopump Inducers*, May 1971

SP-8110   *Liquid Rocket Engine Turbines*, January 1974

SP-8081   *Liquid Propellant Gas Generators*, March 1972

SP-8048   *Liquid Rocket Engine Turbopump Bearings*, March 1971

SP-8101   *Liquid Rocket Engine Turbopump Shafts and Couplings*, September 1972

SP-8100   *Liquid Rocket Engine Turbopump Gears*, March 1974

SP-8088   *Liquid Rocket Metal Tanks and Tank Components*, May 1974

SP-8094   *Liquid Rocket Valve Components*, August 1973

SP-8097   *Liquid Rocket Valve Assemblies*, November 1973

SP-8090   *Liquid Rocket Actuators and Operators*, May 1973

SP-8112   *Pressurization Systems for Liquid Rockets*, October 1975

SP-8080   *Liquid Rocket Pressure Regulators, Relief Valves, Check Valves, Burst Disks, and Explosive Valves*, March 1973

SP-8064   *Solid Propellant Selection and Characterization*, June 1971

SP-8075   *Solid Propellant Processing Factors in Rocket Motor Design*, October 1971

SP-8076   *Solid Propellant Grain Design and Internal Ballistics*, March 1972

SP-8073   *Solid Propellant Grain Structural Integrity Analysis*, June 1973

SP-8039   *Solid Rocket Motor Performance Analysis and Prediction*, May 1971

SP-8051   *Solid Rocket Motor Igniters*, March 1971

SP-8025   *Solid Rocket Motor Metal Cases*, April 1970

SP-8115   *Solid Rocket Motor Nozzles*, June 1975

SP-8114   *Solid Rocket Thrust Vector Control*, December 1974

SP-8041   *Captive-Fired Testing of Solid Rocket Motors*, March 1971

Other Useful Links

We asked our rocket propulsion students at the Zucrow lab to populate the table below with other sites that they frequently visit. We assume no responsibility for the operability/availability of these sites, but from our perspective these are all reliable firms that will tend to be vibrant for years and decades to come. Enjoy.

Table A.1   Useful links

http://astronautix.com/
Mark Wade's encyclopedia of astronautics
http://webbook.nist.gov/chemistry/fluid/
NIST webbook of fluid properties
https://engineering.purdue.edu/~propulsi/propulsion/index.html
Purdue Propulsion Web Page (this website also contains a list of the links given here)
https://youtu.be/vs3zNwXhzSA
NASA film on NERVA nuclear thermal rocket program
www.braeunig.us/space/propuls.htm
Rocket fundamentals site
www.dept.aoe.vt.edu/~devenpor/aoe3114/calc.html
Compressible flow function calculator
www.grc.nasa.gov/WWW/CEAWeb/ceaHome.htm
Online CEA
www.grc.nasa.gov/WWW/K-12/airplane/index.html
NASA beginners guide to aeronautics
www.rocketmime.com/rockets/rckt_eqn.html
Rocket trajectory simulator
www.rocketmime.com/rockets/RocketEquations.pdf
Rocket equations in condensed form
www.spacex.com/sites/spacex/files/falcon_9_users_guide_rev_2.0.pdf
Falcon 9 payload planners' guide
www.spg-corp.com/hybrid-rocket-propulsion.html
Hybrid rocket info
www.ulalaunch.com/uploads/docs/atlasvusersguide2010.pdf
Atlas V payload planners' guide
www.ulalaunch.com/uploads/docs/DeltaIIPayloadPlannersGuide2007.pdf
Delta II payload planner's guide
www.ulalaunch.com/uploads/docs/launch_vehicles/delta_iv_users_guide_june_2013.pdf
Delta IV payload planners' guide

INDEX

Printed in the United States
by Baker & Taylor Publisher Services